Lecture Notes in Mathematics

For information about Vols. 1–461, please contact your bookseller or Springer-Verlag.

Vol. 462: P. Gérardin, Construction de Séries Discrètes p-adiques. »Sur les séries discrètes non ramifiées des groupes réductifs déployés p-adiques«. III, 180 pages. 1975.

Vol. 463: H.-H. Kuo, Gaussian Measures in Banach Spaces. VI, 224 pages. 1975.

Vol. 464: C. Rockland, Hypoellipticity and Eigenvalue Asymptotics. III, 171 pages. 1975.

Vol. 465: Séminaire de Probabilités IX. Proceedings 1973/74. Edité par P. A. Meyer. IV, 589 pages. 1975.

Vol. 466: Non-Commutative Harmonic Analysis. Proceedings 1974. Edited by J. Carmona, J. Dixmier and M. Vergne. VI, 231 pages. 1975.

Vol. 467: M. R. Essén, The Cos $\pi\lambda$ Theorem. With a paper by Christer Borell. VII, 112 pages. 1975.

Vol. 468: Dynamical Systems – Warwick 1974. Proceedings 1973/74. Edited by A. Manning. X, 405 pages. 1975.

Vol. 469: E. Binz, Continuous Convergence on C(X). IX, 140 pages. 1975.

Vol. 470: R. Bowen, Equilibrium States and the Ergodic Theory of Anosov Diffeomorphisms. III, 108 pages. 1975.

Vol. 471: R. S. Hamilton, Harmonic Maps of Manifolds with Boundary. III, 168 pages. 1975.

Vol. 472: Probability-Winter School. Proceedings 1975. Edited by Z. Ciesielski, K. Urbanik, and W. A. Woyczyński. VI, 283 pages. 1975.

Vol. 473: D. Burghelea, R. Lashof, and M. Rothenberg, Groups of Automorphisms of Manifolds. (with an appendix by E. Pedersen) VII, 156 pages. 1975.

Vol. 474: Séminaire Pierre Lelong (Analyse) Année 1973/74. Edité par P. Lelong. VI, 182 pages. 1975.

Vol. 475: Répartition Modulo 1. Actes du Colloque de Marseille-Luminy, 4 au 7 Juin 1974. Edité par G. Rauzy. V, 258 pages. 1975. 1975.

Vol. 476: Modular Functions of One Variable IV. Proceedings 1972. Edited by B. J. Birch and W. Kuyk. V, 151 pages. 1975.

Vol. 477: Optimization and Optimal Control. Proceedings 1974. Edited by R. Bulirsch, W. Oettli, and J. Stoer. VII, 294 pages. 1975.

Vol. 478: G. Schober, Univalent Functions – Selected Topics. V, 200 pages. 1975.

Vol. 479: S. D. Fisher and J. W. Jerome, Minimum Norm Extremals in Function Spaces. With Applications to Classical and Modern Analysis. VIII, 209 pages. 1975.

Vol. 480: X. M. Fernique, J. P. Conze et J. Gani, Ecole d'Eté de Probabilités de Saint-Flour IV-1974. Edité par P.-L. Hennequin. XI, 293 pages. 1975.

Vol. 481: M. de Guzmán, Differentiation of Integrals in R^n. XII, 226 pages. 1975.

Vol. 482: Fonctions de Plusieurs Variables Complexes II. Séminaire François Norguet 1974–1975. IX, 367 pages. 1975.

Vol. 483: R. D. M. Accola, Riemann Surfaces, Theta Functions, and Abelian Automorphisms Groups. III, 105 pages. 1975.

Vol. 484: Differential Topology and Geometry. Proceedings 1974. Edited by G. P. Joubert, R. P. Moussu, and R. H. Roussarie. IX, 287 pages. 1975.

Vol. 485: J. Diestel, Geometry of Banach Spaces – Selected Topics. XI, 282 pages. 1975.

Vol. 486: S. Stratila and D. Voiculescu, Representations of AF-Algebras and of the Group U (∞). IX, 169 pages. 1975.

Vol. 487: H. M. Reimann und T. Rychener, Funktionen beschränkter mittlerer Oszillation. VI, 141 Seiten. 1975.

Vol. 488: Representations of Algebras, Ottawa 1974. Proceedings 1974. Edited by V. Dlab and P. Gabriel. XII, 378 pages. 1975.

Vol. 489: J. Bair and R. Fourneau, Etude Géométrique des Espaces Vectoriels. Une Introduction. VII, 185 pages. 1975.

Vol. 490: The Geometry of Metric and Linear Spaces. Proceedings 1974. Edited by L. M. Kelly. X, 244 pages. 1975.

Vol. 491: K. A. Broughan, Invariants for Real-Generated Uniform Topological and Algebraic Categories. X, 197 pages. 1975.

Vol. 492: Infinitary Logic: In Memoriam Carol Karp. Edited by D. W. Kueker. VI, 206 pages. 1975.

Vol. 493: F. W. Kamber and P. Tondeur, Foliated Bundles and Characteristic Classes. XIII, 208 pages. 1975.

Vol. 494: A Cornea and G. Licea. Order and Potential Resolvent Families of Kernels. IV, 154 pages. 1975.

Vol. 495: A. Kerber, Representations of Permutation Groups II. V, 175 pages. 1975.

Vol. 496: L. H. Hodgkin and V. P. Snaith, Topics in K-Theory. Two Independent Contributions. III, 294 pages. 1975.

Vol. 497: Analyse Harmonique sur les Groupes de Lie. Proceedings 1973–75. Edité par P. Eymard et al. VI, 710 pages. 1975.

Vol. 498: Model Theory and Algebra. A Memorial Tribute to Abraham Robinson. Edited by D. H. Saracino and V. B. Weispfenning. X, 463 pages. 1975.

Vol. 499: Logic Conference, Kiel 1974. Proceedings. Edited by G. H. Müller, A. Oberschelp, and K. Potthoff. V, 651 pages 1975.

Vol. 500: Proof Theory Symposion, Kiel 1974. Proceedings. Edited by J. Diller and G. H. Müller. VIII, 383 pages. 1975.

Vol. 501: Spline Functions, Karlsruhe 1975. Proceedings. Edited by K. Böhmer, G. Meinardus, and W. Schempp. VI, 421 pages. 1976.

Vol. 502: János Galambos, Representations of Real Numbers by Infinite Series. VI, 146 pages. 1976.

Vol. 503: Applications of Methods of Functional Analysis to Problems in Mechanics. Proceedings 1975. Edited by P. Germain and B. Nayroles. XIX, 531 pages. 1976.

Vol. 504: S. Lang and H. F. Trotter, Frobenius Distributions in GL_2-Extensions. III, 274 pages. 1976.

Vol. 505: Advances in Complex Function Theory. Proceedings 1973/74. Edited by W. E. Kirwan and L. Zalcman. VIII, 203 pages. 1976.

Vol. 506: Numerical Analysis, Dundee 1975. Proceedings. Edited by G. A. Watson. X, 201 pages. 1976.

Vol. 507: M. C. Reed, Abstract Non-Linear Wave Equations. VI, 128 pages. 1976.

Vol. 508: E. Seneta, Regularly Varying Functions. V, 112 pages. 1976.

Vol. 509: D. E. Blair, Contact Manifolds in Riemannian Geometry. VI, 146 pages. 1976.

Vol. 510: V. Poènaru, Singularités C^∞ en Présence de Symétrie. V, 174 pages. 1976.

Vol. 511: Séminaire de Probabilités X. Proceedings 1974/75. Edité par P. A. Meyer. VI, 593 pages. 1976.

Vol. 512: Spaces of Analytic Functions, Kristiansand, Norway 1975. Proceedings. Edited by O. B. Bekken, B. K. Øksendal, and A. Stray. VIII, 204 pages. 1976.

Vol. 513: R. B. Warfield, Jr. Nilpotent Groups. VIII, 115 pages. 1976.

Vol. 514: Séminaire Bourbaki vol. 1974/75. Exposés 453 – 470. IV, 276 pages. 1976.

Vol. 515: Bäcklund Transformations. Nashville, Tennessee 1974. Proceedings. Edited by R. M. Miura. VIII, 295 pages. 1976.

Vol. 516: M. L. Silverstein, Boundary Theory for Symmetric Markov Processes. XVI, 314 pages. 1976.

Vol. 517: S. Glasner, Proximal Flows. VIII, 153 pages. 1976.

Vol. 518: Séminaire de Théorie du Potentiel, Proceedings Paris 1972–1974. Edité par F. Hirsch et G. Mokobodzki. VI, 275 pages. 1976.

Vol. 519: J. Schmets, Espaces de Fonctions Continues. XII, 150 pages. 1976.

Vol. 520: R. H. Farrell, Techniques of Multivariate Calculation. X, 337 pages. 1976.

continuation on page 285

Lecture Notes in Mathematics

Edited by A. Dold and B. Eckmann

689

Cabal Seminar 76–77

Proceedings, Caltech-UCLA Logic Seminar 1976–77

Edited by
A. S. Kechris and Y. N. Moschovakis

Springer-Verlag Berlin Heidelberg GmbH 1978

Editors
Alexander S. Kechris
Department of Mathematics
California Institute of Technology
Pasadena, CA 91125/USA

Yiannis N. Moschovakis
Department of Mathematics
University of California at Los Angeles
Los Angeles, CA 90024/USA

AMS Subject Classifications (1970): 02-02, 02Fxx, 02Hxx, 02Kxx

ISBN 978-3-540-09086-1 ISBN 978-3-540-35626-4 (eBook)
DOI 10.1007/978-3-540-35626-4

Originally published by Springer-Verlag Berlin Heidelberg New York in 1978.
2141/3140-543210

Keep it simple but make me believe it.

- Alan Arkin in Hearts of the West

INTRODUCTION

This volume contains material presented in the Caltech-UCLA Logic Seminar in 1976-1977, except for Kechris' "On Spector classes" which was presented in the same seminar the preceding year and "Notes on the theory of scales" which dates back to 1971. There is also an Appendix with the Victoria Delfino problems.

Aside from several new results, there is a good deal of exposition in these papers, including a treatment of several basic consequences of determinacy hypotheses whose publication had not been allowed until now.

Los Angeles
February 1978

Alexander S. Kechris
Yiannis N. Moschovakis

TABLE OF CONTENTS

NOTES ON THE THEORY OF SCALES

Alexander S. Kechris
Department of Mathematics
California Institute of Technology
Pasadena, California 91125

Yiannis N. Moschovakis
Department of Mathematics
University of California
Los Angeles, California 90024

These informal notes were written in the Summer of 1971 and were distributed fairly widely. Despite our original intention not to publish them, it seemed like a good idea to include them in these proceedings, particularly since there are many references to them in the literature. It still seems that they are a good source of basic information about scales for logicians.

The notes are reproduced in their original form, except for minor corrections and the addition of a few footnotes indicating the progress achieved since 1971 on some of the open questions discussed in the manuscript.

NOTES ON THE THEORY OF SCALES

Alexander S. Kechris and Yiannis N. Moschovakis[1]

It is the purpose of these notes to give an informal exposition of several recent results in Descriptive Set Theory, all centering around the notion of a scale. This was first isolated explicitly in the generalization of the Uniformization Theorem on the hypothesis of projective determinacy [14], but is surely implicit in some of the classical proofs. It has turned out that scales have many applications beyond the Uniformization Theorem, both in producing new results and in providing more elegant proofs of known results. Among the new results the Kunen-Martin theorem on the length of wellfounded relations is perhaps the most important. As for new proofs, one can now establish the beautiful results of Solovay on the regularity of $\underset{\sim}{\Sigma}^1_2$ sets (Lebesgue measurability, property of Baire, etc.) without any use of Cohen's forcing - in fact the new arguments are very simple and much in the spirit of classical Descriptive Set Theory. (Some of these new arguments are due to Solovay again.)

This paper is not meant for publication; Moschovakis wants to keep his results for his forthcoming book [15], Kechris is holding his for his Ph.D. thesis and the theorems which belong to neither of us will be presumably written up by their authors. In many ways this can be considered a first draft of parts of [15]. The point of making it available now is that sometimes books remain "forthcoming" for a long time, despite the best of authors' intentions. We believe that these results are interesting enough to deserve early - if incomplete - dissemination.

This draft should be comprehensible to one with some knowledge of classical Descriptive Set Theory, recursive functions with real arguments and at least the basic definitions of games and determinacy. Some parts depend on a knowledge of the theory of indiscernibles for L, but they are independent of the main results. Except for a few inessential changes (explained in §1) we shall follow the notation and terminology of §1, §2 of [13]; it will be convenient and space-saving to assume that the reader is familiar with this material, although the rest of [13] is not relevant to this work, except for one result which will be identified when used.

1. Preliminaries. As usual, $\omega = \{0,1,2,\ldots\}$ and $\mathcal{R} = {}^{\omega}\omega$ = the set of reals. We study subsets of the product spaces

[1]Moschovakis is a Sloan Foundation Fellow. During the preparation of this manuscript both authors were partially supported by NSF Grant #GP-27964.

$$\mathcal{X} = X_1 \times \cdots \times X_k \qquad (X_i = \omega \text{ or } X_i = \mathcal{R})$$

which we call pointsets. Sometimes we think of these as relations and write interchangeably,

$$x \in A \Leftrightarrow A(x).$$

A pointclass is a class of pointsets, not necessarily all in the same product space. Thus Σ^1_1 consists of all relations expressible in the form $(\exists\alpha)(\forall n)R(\bar{\alpha}(n),x)$ (R recursive) and similarly for Π^1_1, etc.

If $A \subseteq \mathcal{X} \times \omega$, put

$$\exists^\omega A = \{x : \exists\, nA(x,n)\},$$

$$\forall^\omega A = \{x : \forall\, nA(x,n)\}$$

and if $A \subseteq \mathcal{X} \times \mathcal{R}$, put

$$\exists^{\mathcal{R}} A = \{x : \exists\, \alpha A(x,\alpha)\},$$

$$\forall^{\mathcal{R}} A = \{x : \forall\, \alpha A(x,\alpha)\}.$$

If Γ is a pointclass, let

$$\check{\Gamma} = \{\mathcal{X} - A : A \subseteq \mathcal{X},\ A \in \Gamma\} = \text{The dual class of } \Gamma,$$

$$\underset{\sim}{\Gamma} = \{A : \text{for some } B \subseteq \mathcal{R} \times \mathcal{X} \text{ and some } \alpha_0 \in \mathcal{R},\ x \in A \Leftrightarrow (\alpha_0,x) \in B\}$$

and for any operation Φ on pointsets,

$$\Phi\Gamma = \{\Phi A : A \in \Gamma\}.$$

A pointclass Γ is adequate if it contains all recursive sets and is closed under disjunction, conjunction, bounded number quantification of both kinds and substitution of recursive functions. All the usual arithmetical, analytical and projective classes are adequate.

If κ is an ordinal, then κ^+ is the least cardinal greater than κ. By "κ^λ" we always mean ordinal exponentiation.

We work here entirely within ZF + DC, Zermelo-Fraenkel set theory with Dependent Choices,

(DC) $$(\forall u \in x)(\exists\, v)(u,v) \in r \Rightarrow (\exists\, f)(\forall n)(f(n),f(n+1)) \in r.$$

We always state all additional hypotheses, including the full axiom of choice, AC, when we need them.

2. Norms and the Prewellordering property. The Prewellordering property on a pointclass Γ was formulated in order to extend elegantly to Σ^1_2 some of the basic results about Π^1_1. It was later shown that if Projective Determinacy (PD) holds, then the Prewellordering property can be established for all Π^1_n (odd n) and Σ^1_k (even k), so that the same results could be extended to all analytical classes of the right kind and index. Our main purpose in this section is to establish the elementary facts about the prewellordering property and prove this theorem.

2A. Definition and elementary properties. A norm on a pointset A is any function $\varphi : A \twoheadrightarrow \varkappa$ from A onto some ordinal $\varkappa$, the length of φ. Each norm φ determines uniquely a prewellordering (reflexive, transitive, connected, wellfounded relation) $\leq^{\varphi}$ on A given by

$$x \leq^{\varphi} y \Leftrightarrow \varphi(x) \leq \varphi(y);$$

conversely each prewellordering $\lesssim$ on A determines a unique norm φ such that $\lesssim \, = \, \leq^{\varphi}$.

There are, of course, many trivial norms on a pointset, for example the constant 0 function. The concept becomes nontrivial when we place definability conditions on a norm in the following way:

Let Γ be a pointclass, $\varphi : A \twoheadrightarrow \varkappa$ a norm on some pointset. We all φ a Γ-norm if there exist relations $\leq^{\varphi}_{\Gamma}$, $\leq^{\varphi}_{\check{\Gamma}}$ in Γ and $\check{\Gamma}$ respectively such that for every y,

(1) $$y \in A \Rightarrow \forall x\{[x \in A \ \& \ \varphi(x) \leq \varphi(y)] \Leftrightarrow x \leq^{\varphi}_{\Gamma} y \Leftrightarrow x \leq^{\varphi}_{\check{\Gamma}} y\}.$$

Notice that if Γ is adequate and φ is a Γ-norm on A we can also define relations $<^{\varphi}_{\Gamma}$, $<^{\varphi}_{\check{\Gamma}}$ in Γ, $\check{\Gamma}$ respectively such that for every y,

(2) $$y \in A \Rightarrow \forall x\{[x \in A \ \& \ \varphi(x) < \varphi(y)] \Leftrightarrow x <^{\varphi}_{\Gamma} y \Leftrightarrow x <^{\varphi}_{\check{\Gamma}} y\}.$$

In fact we put

$$x <^{\varphi}_{\Gamma} y \Leftrightarrow x \leq^{\varphi}_{\Gamma} y \ \& \ \neg(x \leq^{\varphi}_{\check{\Gamma}} y)$$

$$x <^{\varphi}_{\check{\Gamma}} y \Leftrightarrow x \leq^{\varphi}_{\check{\Gamma}} y \ \& \ \neg(x \leq^{\varphi}_{\Gamma} y).$$

It is quite important for the applications that the definition of a Γ-norm be precisely that given by (1). Notice that for Γ adequate and $A \in \Gamma$, this is stronger than simply requiring that $\leq^{\varphi} \in \Gamma$, but weaker than insisting that $\leq^{\varphi} \in \Gamma \cap \check{\Gamma}$ (which implies $A \in \Gamma \cap \check{\Gamma}$).

Finally put:

Prewellordering(Γ) $\Leftrightarrow$ Every pointset A in Γ admits a Γ-norm.

This notion is not interesting unless Γ is at least adequate. For adequate, parametrized Γ it is equivalent to that defined in §2 of [13], where there is also a discussion of some properties that it implies (see also [2]). We only give here a short proof of the Reduction property from this form of the Prewellordering property.

(2A-1) Theorem: Assume Γ is adequate and Prewellordering(Γ); then Reduction(Γ), i.e. if $A,B \in \Gamma$, $A \subseteq \mathfrak{X}$, $B \subseteq \mathfrak{X}$, then there exist $A_1 \subseteq A$, $B_1 \subseteq B$, $A_1 \in \Gamma$, $B_1 \in \Gamma$ such that $A_1 \cap B_1 = \emptyset$, $A_1 \cup B_1 = A \cup B$.

Proof: Let $A,B \subseteq \mathfrak{X}$, $A,B \in \Gamma$ be given. Define $C = (A \times \{0\}) \cup (B \times \{1\})$ (notice that $C \subseteq \mathfrak{X} \times \omega$). Since Γ is adequate $C \in \Gamma$. Let φ be a Γ-norm on C, and put

$$x \in A_1 \Leftrightarrow x \in A \ \& \neg((x,1) \leq_{\Gamma}^{\varphi} (x,0))$$

$$x \in B_1 \Leftrightarrow x \in B \ \& \neg((x,0) <_{\Gamma}^{\varphi} (x,1)).$$

An easy checking shows that A_1, B_1 have all the required properties. ⊣

The following trivial observation will save us having to deal separately, with the "lightface" and "boldface" cases:

(2A-2) Proposition: If Γ is adequate and Prewellordering(Γ), then Prewellordering($\utilde{\Gamma}$).

Proof: Let $A \in \utilde{\Gamma}$, $A \subseteq \mathfrak{X}$. Then for some α_0 and some $B \subseteq \mathcal{R} \times \mathfrak{X}$, $B \in \Gamma$ we have $A = \{\alpha : (\alpha_0,\alpha) \in B\}$. Find a Γ-norm $\overline{\varphi}$ for B and define the following ordinal map on A

$$\psi(\alpha) = \overline{\varphi}(\alpha_0,\alpha).$$

Then let $\alpha \lesssim \beta \Leftrightarrow \psi(\alpha) \leq \psi(\beta)$ and let φ be the norm on A such that $\lesssim \, = \, \leq^{\varphi}$. (We define the norm in this roundabout way because ψ need not be onto an ordinal). Clearly φ is a $\utilde{\Gamma}$-norm on A. ⊣

2B. Establishing the Prewellordering property.

We first prove

(2B-1) Theorem: Prewellordering(Π_1^1).

Proof: Let $A \in \Pi_1^1$, $A \subseteq \mathfrak{X}$. Put

$$LOR = \{\alpha : \{(m,n) : \alpha(\langle m,n \rangle) = 0\} = \leq_\alpha \ \text{is an ordering}\}$$

and

$$WO = \{\alpha : \leq_\alpha \ \text{is a wellordering}\}.$$

Then for some recursive function $f : \mathcal{X} \to \mathcal{R}$, $f(x) \in \mathrm{LOR}$, for all $x \in \mathcal{X}$ and $x \in A \Leftrightarrow f(x) \in \mathrm{WO}$. Define on A,

$$\psi(x) = |f(x)|$$

where for $\alpha \in \mathrm{WO}$, $|\alpha|$ = length of $\leq_\alpha$. Then let $x \lesssim y \Leftrightarrow \psi(x) \leq \psi(y)$ and let φ be the norm of A such that $\lesssim \, = \, \leq^\varphi$.

To see that φ is actually a Π^1_1-norm, we use the well-known fact that there exist relations $Q_{\Pi^1_1}$, $Q_{\Sigma^1_1}$ in Π^1_1, Σ^1_1 respectively, such that for $\beta \in \mathrm{WO}$,

$$[\alpha \in \mathrm{WO} \;\&\; |\alpha| \leq |\beta|] \Leftrightarrow Q_{\Pi^1_1}(\alpha,\beta) \Leftrightarrow Q_{\Sigma^1_1}(\alpha,\beta). \qquad \dashv$$

We now give two theorems which establish the prewellordering property for some pointclasses closed under $\exists^{\mathcal{R}}$.

(2B-2) <u>Theorem</u>: Assume Γ is adequate, $A \in \Gamma$ and A admits a Γ-norm. Then $\exists^{\mathcal{R}} A$ admits an $\exists^{\mathcal{R}}\forall^{\mathcal{R}}\Gamma$-norm. (Moschovakis, see [13].)

<u>Corollary</u>: Γ adequate, Prewellordering(Γ), $\forall^{\mathcal{R}}\Gamma \subseteq \Gamma \Rightarrow$ Prewellordering($\exists^{\mathcal{R}}\Gamma$).

<u>Corollary</u>: Prewellordering(Σ^1_2).

<u>Proof of the theorem</u>: Let $A \subseteq \mathcal{X} \times \mathcal{R}$, $A \in \Gamma$, $B = \exists^{\mathcal{R}} A = \{x : \exists\alpha(x,\alpha) \in A\}$. Let φ be a Γ-norm on A, define on B

$$x \lesssim y \Leftrightarrow \min\{\varphi(x,\alpha) : (x,\alpha) \in A\} \leq \min\{\varphi(y,\beta) : (y,\beta) \in A\}$$

and let

$$\psi : B \twoheadrightarrow \lambda$$

be the unique norm such that $\leq^\psi \, = \, \lesssim$.

To check that this is actually a $\exists^{\mathcal{R}}\forall^{\mathcal{R}}\Gamma$-norm, let $\leq^\varphi_\Gamma$, $\leq^\varphi_{\check\Gamma}$, $<^\varphi_\Gamma$, $<^\varphi_{\check\Gamma}$ be the relations associated with φ and notice that for $y \in B$,

$$[x \in B \;\&\; \psi(x) \leq \psi(y)] \Leftrightarrow (\exists\alpha)((x,\alpha) \in A \;\&\; \forall\beta[\neg\,(y,\beta) <^\varphi_{\check\Gamma} (x,\alpha)])$$

$$\Leftrightarrow (\forall\beta)((y,\beta) \in A \Rightarrow \exists\alpha[(x,\alpha) \leq^\varphi_{\check\Gamma} (y,\beta)]). \qquad \dashv$$

Let $\leq$ be a wellordering of $\mathcal{R}$ of order-type $\aleph_1$; put

$$\text{In.Segment}^{\leq}(\gamma,\alpha) \Leftrightarrow \{\beta : \beta \leq \alpha\} = \{(\gamma)_n : n = 0,1,2,\ldots\}.$$

We call $\leq$ Γ-<u>good</u> if both the relation $\leq$ and the relation In.Segment$^{\leq}$ are in $\Gamma \cap \check\Gamma$.

(2B-3) <u>Theorem</u>: Assume Γ is adequate, closed under both $\exists^\omega$, $\forall^\omega$ and $\check{\Gamma} \subseteq \exists^{\mathcal{R}}\Gamma$ and there is some wellordering $\leq$ of $\mathcal{R}$ which is Γ-good; then Prewellordering($\exists^{\mathcal{R}}\Gamma$). (Essentially Addison [1].)

<u>Proof</u>: If $A = \exists^{\mathcal{R}}B$ with $B \in \Gamma$, define on A

$$x \precsim y \Leftrightarrow \min\{\alpha : (x,\alpha) \in B\} \leq \min\{\beta : (y,\beta) \in B\},$$

where the minima are taken in the good wellordering $\leq$ and let ψ be the associated norm. The computation is easy. ⊣

<u>Corollary</u>: $V = L \Rightarrow$ Prewellordering(Σ^1_k), $k \geq 2$.

<u>Proof</u>: $V = L \Rightarrow$ There exists a wellordering of $\mathcal{R}$ which is Δ^1_2-good. ⊣

<u>Corollary</u>: $V = L[\mu] \Rightarrow$ Prewellordering(Σ^1_k), $k \geq 2$. (Silver, [17].) (Here μ is a normal κ-additive measure on some cardinal.)

<u>Proof</u>: $V = L[\mu] \Rightarrow$ There exists a wellordering of $\mathcal{R}$ which is Δ^1_3-good. ⊣

2C. <u>The First Periodicity Theorem</u>.

(2C-1) <u>Theorem</u>: Assume Γ is adequate and Determinacy($\underset{\sim}{\Gamma} \cap \underset{\sim}{\check{\Gamma}}$). Then if $A \in \Gamma$ and A admits a Γ-norm, $\forall^{\mathcal{R}}A$ admits a $\forall^{\mathcal{R}}\exists^{\mathcal{R}}\Gamma$-norm (Martin, Moschovakis [9], [2].)

<u>Corollary</u>: Γ adequate, Prewellordering(Γ), Determinacy($\underset{\sim}{\Gamma} \cap \underset{\sim}{\check{\Gamma}}$), $\exists^{\mathcal{R}}\Gamma \subseteq \Gamma \Rightarrow$ Prewellordering($\forall^{\mathcal{R}}\Gamma$).

<u>Proof of the theorem</u>: Assume the hypotheses hold and let ψ be a Γ-norm on $A \subseteq \mathfrak{X} \times \mathcal{R}$, $A \in \Gamma$. Let $B = \forall^{\mathcal{R}}A = \{x : \forall\alpha(x,\alpha) \in A\}$. Instead of giving directly a norm φ on A we shall define the associated prewellordering $\leq^\varphi$, for simplicity $\precsim$.

Given $x,y \in \mathfrak{X}$ consider the game $G(x,y)$

I	II
$\alpha(0)$	$\beta(0)$
$\alpha(1)$	$\beta(1)$
$\alpha(2)$	$\beta(2)$
$\vdots$	$\vdots$
α	β ;

I plays α, II plays β and II wins if

$$(y,\beta) \notin A \ \underline{\text{or}} \ [(y,\beta) \in A \ \& \ (x,\alpha) \in A \ \& \ \psi(x,\alpha) \leq \psi(y,\beta)].$$

Now for $x,y \in B$ put

$$x \precsim y \Leftrightarrow \text{II} \ \underline{\text{has a winning strategy in}} \ G(x,y).$$

The idea is that $x \precsim y$ if there is some procedure (strategy) τ which to each α assigns bit-by-bit some $[\alpha] * \tau$ so that $\psi(x,\alpha) \leq \psi(y,[\alpha] * \tau)$; in some sense, $\sup_\alpha\{\psi(x,\alpha)\} \leq \sup_\beta\{\psi(y,\beta)\}$ effectively.

1. $\forall x \in B \ (x \precsim x)$.

Proof: In the game $G(x,x)$, where $x \in B$, II has always a trivial winning strategy, namely copying down the moves of I.

2. $\forall x,y,z \in B \ (x \precsim y \ \& \ y \precsim z \Rightarrow x \precsim z)$.

Proof: Let $x,y,z \in B \ \& \ x \precsim y \ \& \ y \precsim z$. Then II has winning strategies in both $G(x,y)$ and $G(y,z)$. Fix one in each game and consider the diagram:

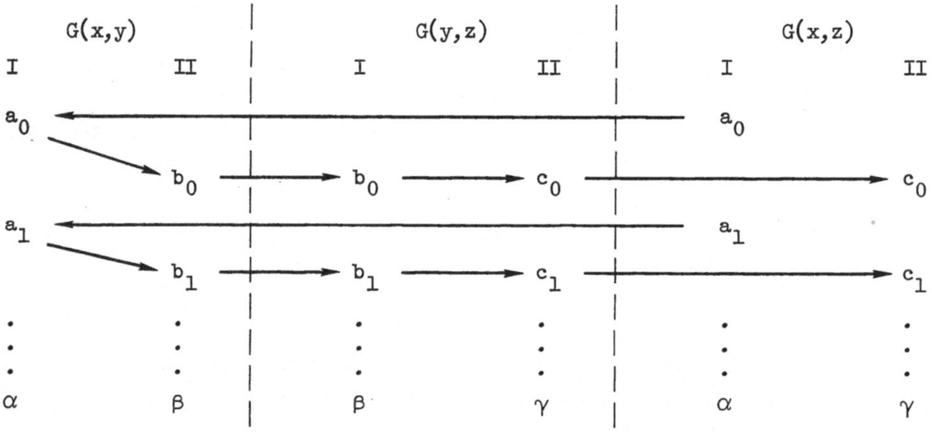

We describe a strategy for II in $G(x,z)$ as follows: I plays in $G(x,z)$ a_0. Then I copies in $G(x,y)$ a_0 and II answers in $G(x,y)$ by his winning strategy to give b_0. I plays in $G(y,z)$ this b_0 and II answers in $G(y,z)$ by his winning strategy to give c_0. II's answer in $G(x,z)$ is this c_0, and so on as in the diagram. After the game is over, reals α, β, γ result and since II played with his winning strategy in $G(x,y)$ and $G(y,z)$ it follows that $\psi(x,\alpha) \leq \psi(y,\beta) \leq \psi(z,\gamma)$ (recall that $x,y,z \in B$). Thus $\psi(x,\alpha) \leq \psi(z,\gamma)$ and II wins $G(x,z)$.

We described a winning strategy for II in $G(x,z)$, so $x \precsim z$.

3. For any x, y, $G(x,y)$ is determined. If $x,y \in B$, then $x \prec y \Leftrightarrow$ I has

a winning strategy in $G(y,x)$, where $x < y \Leftrightarrow x \precsim y \;\&\; \neg(y \precsim x)$. Thus, $\forall x,y \in B\ (x \precsim y$ or $y \precsim x)$.

Proof: Let $x,y \in \mathfrak{X}$ be given. If $y \notin B$, pick β_0 such that $(y,\beta_0) \notin A$; II then wins by playing β_0. If $y \in B$ then we have

$$\text{II wins } G(x,y) \Leftrightarrow [(x,\alpha) \in A \;\&\; \psi(x,\alpha) \leq \psi(y,\beta)]$$

$$\Leftrightarrow (x,\alpha) \leq^{\psi}_{T} (y,\beta) \Leftrightarrow (x,\alpha) \underset{\sim}{\leq}^{\psi}_{T} (y,\beta).$$

Thus the game is in $\underset{\sim}{\Gamma} \cap \check{\underset{\sim}{\Gamma}}$ and is determined.

Now assume $x,y \in B$. If $x < y$ then $\neg(y \precsim x)$, so II does not have a winning strategy in $G(y,x)$ and I has a winning strategy in $G(y,x)$. Conversely assume that I has a winning strategy in $G(y,x)$. We shall show that $x < y$ i.e. II has a winning strategy in $G(x,y)$, but II has no winning strategy in $G(y,x)$. The last statement is obvious so we proceed to invent a winning strategy for II in $G(x,y)$.

Fix a strategy for I in $G(y,x)$. Consider the diagram

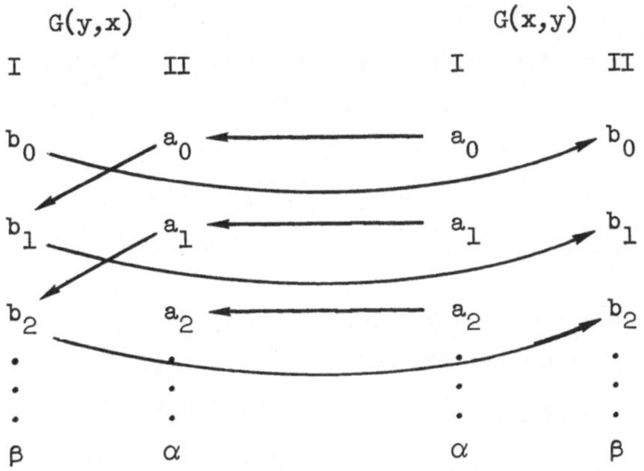

We describe a strategy for II in $G(x,y)$ as follows: I plays in $G(x,y)$ a_0. Then I plays b_0 by his strategy in $G(y,x)$ and II plays this b_0 in $G(x,y)$ answering a_0. Then I plays a_1 in $G(x,y)$. II copies a_0 in $G(y,x)$ and I answers by his strategy in $G(y,x)$ to give b_1. II plays this b_1 in $G(x,y)$ answering to a_1. Then I gives a_2 in $G(x,y)$, II copies a_1 in $G(y,x)$ and I answers in $G(y,x)$ by his strategy to give b_2. II then copies b_2 in $G(x,y)$ etc. After the end of the game reals α, β result as in the picture. Since I played by his winning strategy in $G(y,x)$ we have $\psi(y,\beta) > \psi(x,\alpha)$; thus II wins $G(x,y)$. Thus the above described strategy is a winning one for II in $G(x,y)$ and we are done.

4. The relation $\precsim$ is wellfounded.

Proof: We have to show that there is no infinite descending chain

$x_0 > x_1 > x_2 > \cdots$. Assume not, towards a contradiction. If $x_0 > x_1 > x_2 > \cdots$ then I wins $G(x_i, x_{i+1})$ for each $i \geq 0$. Fix winning strategies for I in each one of these games.

Consider the diagram:

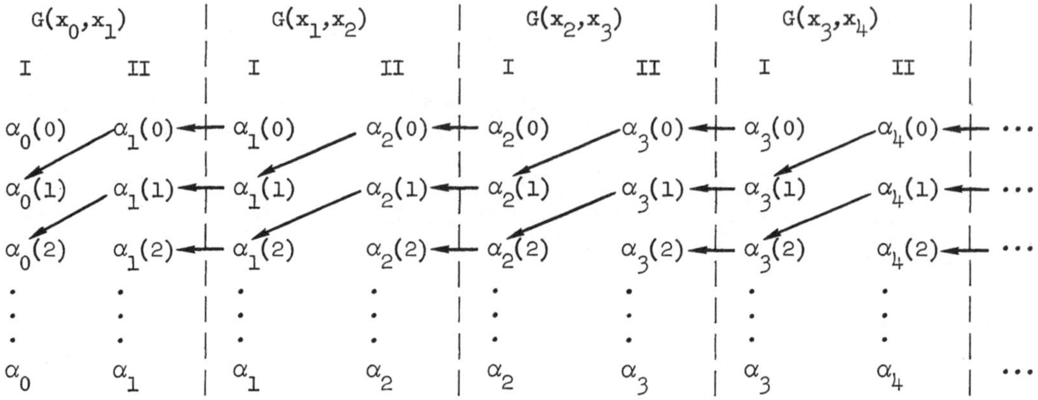

Let first I play $\alpha_0(0), \alpha_1(0), \alpha_2(0), \ldots$ in $G(x_0,x_1), G(x_1,x_2), G(x_2,x_3), \ldots$ resp. by following his winning strategies. Then let II play $\alpha_1(0), \alpha_2(0), \alpha_3(0), \ldots$ in $G(x_0,x_1), G(x_1,x_2), G(x_2,x_3), \ldots$ resp. I answers by his winning strategies to give $\alpha_0(1), \alpha_1(1), \alpha_2(1), \ldots$ in $G(x_0,x_1), G(x_1,x_2), G(x_2,x_3), \ldots$ resp. Then II plays $\alpha_1(1), \alpha_2(1), \alpha_3(1), \ldots$ in $G(x_0,x_1), G(x_1,x_2), G(x_2,x_3), \ldots$ resp. I answers by his strategies to give $\alpha_0(2), \alpha_1(2), \alpha_2(2), \ldots$ in $G(x_0,x_1), G(x_1,x_2), G(x_2,x_3), \ldots$ resp., etc.

After all these moves have been played reals $\alpha_0, \alpha_1, \alpha_2, \ldots$ are created as in the picture and since I wins all the games $G(x_i, x_{i+1})$, $(i \geq 0)$ we have $\psi(x_0,\alpha_0) > \psi(x_1,\alpha_1) > \psi(x_2,\alpha_2) > \cdots$ which is a contradiction.

5. The norm associated with $\lesssim$ is a $\forall^{\mathbb{R}}\exists^{\mathbb{R}}\Gamma$-norm on B.

Proof: Notice first that for $y \in B$, $x \in B$ & $x \lesssim y \Longleftrightarrow$ II has a winning strategy in $G(x,y) \Leftrightarrow \exists\tau\forall\alpha((x,\alpha) \leq^{\psi}_{\Gamma} (y,[\alpha] * \tau))$. But since $G(x,y)$ is determined, II has a winning strategy in $G(x,y) \Leftrightarrow$ I has no winning strategy in $G(x,y) \Leftrightarrow \forall\sigma\exists\beta((x,\sigma * [\beta]) \leq^{\psi}_{\Gamma} (y,\beta))$. Thus for $y \in B$,

$$x \in B \;\&\; x \lesssim y \Leftrightarrow \exists\tau\,\forall\alpha((x,\alpha) \leq^{\psi}_{\Gamma} (y,[\alpha] * \tau))$$

$$\Leftrightarrow \forall\sigma\exists\beta((x,\sigma * [\beta]) \leq^{\psi}_{\Gamma} (y,\beta))$$

and we are done. ⊣

2D. The zig-zag picture. It is not hard to verify that the Reduction property cannot hold both for Σ^1_n and Π^1_n for any n, hence the same is true for the Prewellordering property. If we make a diagram of the classes Σ^1_n, Π^1_n and circle those which have the Prewellordering property, we get the following two pictures, under the hypotheses $V = L$ (or $V = L[\mu]$) and PD:

$V = L$ or $V = L[\mu]$

PD

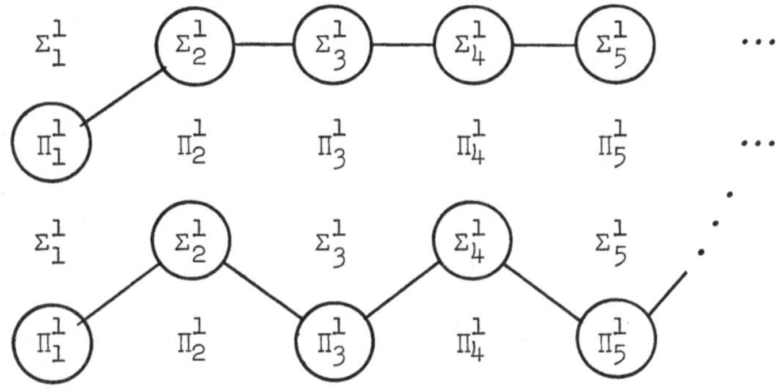

This second picture is the motivation for the name "Periodicity Theorem." We will see later that (assuming PD) we can construct models in which this picture has any finite predetermined number of "teeth."

3. Scales.

In this section we define scales and the property Scale(Γ), we prove that Scale(Γ) $\Rightarrow$ Uniformization(Γ) for suitable Γ (in particular $\Pi^1_1, \Pi^1_3, \Pi^1_5, \ldots$) and we establish Scale(Γ) for $\Gamma = \Pi^1_1, \Sigma^1_2, \Pi^1_3, \Sigma^1_4, \ldots$ under the hypothesis PD. The elementary theory of scales is quite similar to that of norms, so the structure of this section is parallel to that of §2.

3A. Definitions and basic properties. A *scale* on a pointset A is a sequence of norms $\{\varphi_n\}_{n\in\omega}$ on A with the following *limit property*:

> If $x_0, x_1, x_2, \ldots \in A$, if $\text{limit}_{i\to\infty} x_i = x$, if for each n and all large i, $\varphi_n(x_i) = \lambda_n$, then $x \in A$ and for each n, $\varphi_n(x) \leq \lambda_n$.

It is easy to see that assuming AC every pointset admits a scale (take every φ_n to be equal to a fixed $1-1$ mapping from the pointset onto an ordinal). In order to get something interesting we place definability conditions on a scale as follows:

Let Γ be a pointclass, $\{\varphi_n\}_{n\in\omega}$ a scale on a pointset A. We call $\{\varphi_n\}_{n\in\omega}$ a *Γ-scale* if there exist relations S_Γ, $S_{\check{\Gamma}}$ in Γ and $\check{\Gamma}$ respectively such that, for every y,

$$(3) \qquad y \in A \Rightarrow \forall x\{[x \in A \;\&\; \varphi_n(x) \leq \varphi_n(y)] \Leftrightarrow S_\Gamma(n,x,y) \Leftrightarrow S_{\check{\Gamma}}(n,x,y)\}.$$

Finally we put:

$$\text{Scale}(\Gamma) \Leftrightarrow \text{Every pointset } A \text{ in } \Gamma \text{ admits a } \Gamma\text{-scale}.$$

The notion of a scale and the associated scale property were introduced in connection with the uniformization problem in [14]. If Γ, Γ^* are pointclasses, put

$$\underline{\text{Uniformization}}(\Gamma,\Gamma^*) \Leftrightarrow \underline{\text{For every}}\ P \in \Gamma,\ P \subseteq \mathfrak{X} \times \mathfrak{Y}\ \underline{\text{we can find}}$$

$P^* \in \Gamma^*$, $P^* \subseteq \mathfrak{X} \times \mathfrak{Y}$ <u>such that</u>

$$P^* \subseteq P$$

<u>and</u>

$$\forall x[\exists y\, P(x,y) \Leftrightarrow \exists\, !y\, P^*(x,y)].$$

(In this case we say that P^* <u>uniformizes</u> P.) In this definition we allow $P \subseteq \mathfrak{Y}$ in which case $P^* \subseteq P$ and $\exists\, y\, P(y) \Leftrightarrow \exists\, !y\, P^*(y)$ i.e. P^* is a singleton contained in P, if $P \neq \emptyset$.

We abbreviate

$$\underline{\text{Uniformization}}(\Gamma) \Leftrightarrow \underline{\text{Uniformization}}(\Gamma,\Gamma).$$

(3A-1) <u>Theorem</u>: Assume Γ is adequate and closed under $\forall^{\mathcal{R}}$. Then,

$$\underline{\text{Scale}}(\Gamma) \Rightarrow \underline{\text{Uniformization}}(\Gamma).$$

<u>Proof</u>: Let $P \in \Gamma$, $P \subseteq \mathfrak{X} \times \mathfrak{Y}$ where for simplicity and w.l.o.g. we can assume $\mathfrak{Y} = \mathcal{R}$. Let $\{\varphi_n\}_{n\in\omega}$ be a Γ-scale for P with associated relations S_Γ, $S_{\check{\Gamma}}$. Fix $x \in \mathfrak{X}$ and define inductively the following (where we agree that $\min(\emptyset) = 0$):

$$\begin{cases} P_0^x = \{\alpha : P(x,\alpha)\} \\ k_0^x = \min\{\alpha(0) : \alpha \in P_0^x\} \\ \lambda_0^x = \min\{\varphi_0(x,\alpha) : \alpha \in P_0^x \ \&\ \alpha(0) = k_0^x\}, \end{cases}$$

$$\begin{cases} P_{n+1}^x = \{\alpha : \alpha \in P_n^x \ \&\ \alpha(n) = k_n^x \ \&\ \varphi_n(x,\alpha) = \lambda_n^x\} \\ k_{n+1}^x = \min\{\alpha(n+1) : \alpha \in P_{n+1}^x\} \\ \lambda_{n+1}^x = \min\{\varphi_{n+1}(x,\alpha) : \alpha \in P_{n+1}^x \ \&\ \alpha(n+1) = k_{n+1}^x\}. \end{cases}$$

Finally put

$$P_\infty^x = \bigcap_{n\in\omega} P_n^x.$$

We have now the following:

1. $P_0^x \supseteq P_1^x \supseteq P_2^x \supseteq \cdots$.

2. $\exists\alpha\, P(x,\alpha) \Rightarrow \forall n\,(P_n^x \neq \emptyset)$.

3. Assume $\exists\alpha\, P(x,\alpha)$. Let $\alpha^x = (k_0^x, k_1^x, k_2^x, \ldots)$. Then $P_\infty^x = \{\alpha^x\}$.

Proof: If $\alpha \in P_\infty^x$ then for each n, $\alpha(n) = k_n^x$, i.e. $\alpha = \alpha^x$. Conversely, if $\exists\alpha\, P(x,\alpha)$ pick reals $\alpha_i \in P_i^x$, $i = 0,1,\ldots$. Then for $i > n$, $\alpha_i(n) = k_n^x$, thus $\alpha_i \to \alpha^x$ and $(x,\alpha_i) \to (x,\alpha^x)$. Also for $i > n$, $\varphi_n(x,\alpha_i) = \lambda_n^x$, hence by the limit property of scales, $(x,\alpha^x) \in P$ and $\varphi_n(x,\alpha^x) \leq \lambda_n^x$. Then certainly $\alpha^x \in P_0^x$. But also $\alpha^x(0) = k_0^x$ and $\varphi_0(x,\alpha^x) \leq \lambda_0^x$, i.e. $\varphi_0(x,\alpha^x) = \lambda_0^x$. Thus $\alpha^x \in P_1^x$. A similar argument shows inductively that for all n, $\alpha^x \in P_n^x$.

Put now

$$P^*(x,\alpha) \Leftrightarrow \exists\alpha\, P(x,\alpha) \;\&\; \alpha = \alpha^x.$$

Clearly $P^* \subseteq P$ and

$$\exists\alpha\, P(x,\alpha) \Rightarrow \exists\,!\,\alpha P^*(x,\alpha).$$

To complete the proof it will be enough to show that $P^* \in \Gamma$. It is easier to show that the complement of P^* is in $\check{\Gamma}$. And this follows from the computation

$$\neg P^*(x,\alpha) \Leftrightarrow \neg P(x,\alpha) \;\underline{\text{or}}$$

$$\{P(x,\alpha) \;\&\; \exists n \exists\beta\{P(x,\beta) \;\&\; [(\forall i < n)[\alpha(i) = \beta(i) \;\&\; \varphi_i(x,\alpha) = \varphi_i(x,\beta)]$$

$$\&\; [\beta(n) < \alpha(n) \;\underline{\text{or}}$$

$$(\beta(n) = \alpha(n) \;\&\; \varphi_n(x,\beta) < \varphi_n(x,\alpha))]]\}$$

$$\Leftrightarrow \neg P(x,\alpha) \;\underline{\text{or}}$$

$$\{\exists n \exists\beta\{[(\forall i < n)[S_{\check{\Gamma}}(i,x,\beta,x,\alpha) \;\&\; S_{\check{\Gamma}}(i,x,\alpha,x,\beta) \;\&\; \alpha(i) = \beta(i)]$$

$$\&\; [\beta(n) < \alpha(n) \;\underline{\text{or}}$$

$$(\beta(n) = \alpha(n) \;\&\; S_{\check{\Gamma}}(n,x,\beta,x,\alpha) \;\&\; \neg S_{\Gamma}(n,x,\beta,x,\alpha))]]\}. \quad \dashv$$

Again we should mention the trivial observation that

$$\Gamma \text{ adequate, } \underline{\text{Scale}}(\Gamma) \Rightarrow \underline{\text{Scale}}(\underset{\sim}{\Gamma}).$$

3B. Establishing the Scale property.

(3B-1) Theorem: $\underline{\text{Scale}}(\Pi_1^1)$.

Corollary: Uniformization(Π^1_1). (The classical Novikoff-Kondo-Addison Theorem.)

Proof: Let $A \subseteq \mathfrak{X}$, $A \in \Pi^1_1$. Then for some recursive function $f : \mathfrak{X} \to \mathcal{R}$

$$x \in A \Leftrightarrow f(x) \in WO.$$

For $\alpha \in WO$ put $\alpha \restriction n = \{m : m <_\alpha n\}$ (where $m <_\alpha n \Leftrightarrow m \leq_\alpha n \;\&\; m \neq n$) and $|\alpha \restriction n|$ = the length of $\alpha \restriction n$ ($\alpha \restriction n$ is an initial segment of $\leq_\alpha$). If $n \notin \mathrm{Field}(\leq_\alpha)$, i.e. if $\alpha(\langle n,n\rangle) \neq 0$, then of course $|\alpha \restriction n| = 0$. Define now the following prewellorderings on A

$$x \leq_n y \Leftrightarrow |f(x)| < |f(y)| \;\underline{\text{or}}$$
$$[\,|f(x)| = |f(y)| \;\&\; |f(x) \restriction n| \leq |f(y) \restriction n|\,].$$

Let φ_n be the norm on A such that $\leq^{\varphi_n} = \leq_n$. We will show that $\{\varphi_n\}_{n\in\omega}$ is a Π^1_1-scale on A.

1. $\{\varphi_n\}_{n\in\omega}$ is a scale.

Proof: Assume $x_i \in A$ for every i, $x_i \to x$ and for some $\{\lambda_n\}_{n\in\omega}$, $\varphi_n(x_i) = \lambda_n$ for all sufficiently large i. This implies that $|f(x_i)| = \lambda'$ for some λ' and all sufficiently large i and also that for some $\{\lambda'_n\}_{n\in\omega}$, $|f(x_i) \restriction n| = \lambda'_n$ for all sufficiently large i. We show first that $x \in A$, i.e. that $f(x) \in WO$, by proving that the mapping

$$n \to \lambda'_n$$

is order preserving on the field of $\leq_{f(x)}$. In fact, let $n <_{f(x)} m$ i.e. $f(x)(\langle n,m\rangle) = 0 \;\&\; f(x)(\langle m,n\rangle) \neq 0$. Since f is continuous and $x_i \to x$ clearly for all sufficiently large i, $f(x_i)(\langle n,m\rangle) = 0$ and $f(x_i)(\langle m,n\rangle) \neq 0$, i.e. $n <_{f(x_i)} m$. But then $|f(x_i) \restriction n| < |f(x_i) \restriction m|$. Taking i sufficiently large this shows that $\lambda'_n < \lambda'_m$ and we are done.

Finally we have to show that for each n

$$|f(x)| < \lambda' \;\underline{\text{or}}\; [\,|f(x)| = \lambda' \;\&\; |f(x) \restriction n| \leq \lambda'_n].$$

Since $n \to \lambda'_n$ is order preserving on the field of $\leq_{f(x)}$ it follows that $|f(x) \restriction n| \leq \lambda'_n$ for each n. But $\lambda'_n \leq \lambda'$ for each n (since for large enough i, $\lambda'_n = |f(x_i) \restriction n|$ while $\lambda' = |f(x_i)|$). Thus

$$|f(x)| = \sup\{|f(x) \restriction n| : n \in \omega\} \leq \lambda'.$$

2. $\{\varphi_n\}_{n\in\omega}$ is a Π^1_1-scale.

This is proved by a computation similar to that in the proof of (2B-1). ⊣

It should be pointed out here that the proofs of Theorems (3B-1) and (3A-1) taken together constitute the classical proof of the Novikoff-Kondo-Addison theorem.

We proceed now to prove the analog of Theorem (2B-2).

(3B-2) Theorem: Assume Γ is adequate, $A \in \Gamma$ and A admits a Γ-scale; then $\exists^{\mathcal{R}}A$ admits an $\exists^{\mathcal{R}}\forall^{\mathcal{R}}\Gamma$-scale. (Moschovakis, [14].)

Corollary: Γ adequate, $\forall^{\mathcal{R}}\Gamma \subseteq \Gamma$, $\underline{\text{Scale}}(\Gamma) \Rightarrow \underline{\text{Scale}}(\exists^{\mathcal{R}}\Gamma)$.

Corollary: $\underline{\text{Scale}}(\Sigma^1_2)$.

Proof of the theorem: If λ is any ordinal and $n \in \omega$, consider the lexicographical wellordering of ${}^n\lambda$,

$$(\xi_1,\ldots,\xi_n) \leq (\eta_1,\ldots,\eta_n) \Leftrightarrow \xi_1 < \eta_1 \ \underline{\text{or}}\ (\xi_1 = \eta_1 \ \&\ \xi_2 < \eta_2)\ \underline{\text{or}}$$

$$\cdots\ \underline{\text{or}}\ (\xi_1 = \eta_1 \ \&\ \cdots\ \&\ \xi_{n-1} = \eta_{n-1}\ \&\ \xi_n \leq \eta_n).$$

This wellorders ${}^n\lambda$ with ordinal λ^n and we let $\langle\xi_1,\ldots,\xi_n\rangle$ = ordinal of $(\xi_1,\ldots,\xi_n)$ in the lexicographical wellordering. Each $\vartheta < \lambda^n$ can be written uniquely as $\vartheta = \langle\xi_1,\ldots,\xi_n\rangle$ for $\xi_i < \lambda$.

Now let $A \subseteq \mathfrak{X} \times \mathcal{R}$, $A \in \Gamma$ and assume $\{\varphi_n\}_{n\in\omega}$ is a Γ-scale on A. If $B = \exists^{\mathcal{R}}A = \{x : \exists\alpha(x,\alpha) \in A\}$, define for $x \in B$

$$\psi'_n(x) = \min\{\langle\varphi_0(x,\alpha),\alpha(0),\varphi_1(x,\alpha),\alpha(1),\ldots,\varphi_n(x,\alpha),\alpha(n)\rangle : (x,\alpha) \in A\}.$$

Let $\leq_n$ be the prewellordering (on B)

$$x \leq_n y \Leftrightarrow \psi'_n(x) \leq \psi'_n(y)$$

and let ψ_n the associated norm. We will prove that $\{\psi_n\}_{n\in\omega}$ is a $\exists^{\mathcal{R}}\forall^{\mathcal{R}}\Gamma$-scale on A.

1. $\{\psi_n\}_{n\in\omega}$ is a scale.

Proof: Assume $x_i \in B$, and $x_i \to x$ and for some $\{\lambda_n\}_{n\in\omega}$ we have for each n, $\psi_n(x_i) = \lambda_n$, for all large enough i. In fact we may assume w.l.o.g. that for $i \geq n$, $\psi_n(x_i) = \lambda_n$. This implies that for some $\{\lambda_n\}_{n\in\omega}$ and all $i \geq n$, $\psi'_n(x_i) = \lambda'_n$. Each λ'_n can be written uniquely as

$$\lambda'_n = \langle\lambda^n_0,k^n_0,\ldots,\lambda^n_n,k^n_n\rangle.$$

We claim that for $i \geq n$, $\underline{\lambda^i_n = \lambda^n_n}$ and $\underline{k^i_n = k^n_n}$. Because if $i \geq n$, $\psi'_n(x_i) = \lambda'_n =$

$$\min\{(\varphi_0(x_i,\alpha),\alpha(0),\ldots,\varphi_n(x_i,\alpha),\alpha(n)) : (x_i,\alpha) \in A\}$$

and

$$\psi_i'(x_i) = \lambda_i' = \min\{\langle\varphi_0(x_i,\alpha),\alpha(0),\ldots,\varphi_n(x_i,\alpha),\alpha(n),\ldots,\varphi_i(x_i,\alpha),\alpha(i)\rangle : (x_i,\alpha) \in A\}.$$

Thus if α_i is such that $(x_i,\alpha_i) \in A$ and $\psi_i'(x_i) = \lambda_i' = \langle\varphi_0(x_i,\alpha_i),\alpha_i(0),\ldots,$ $\varphi_i(x_i,\alpha_i),\alpha_i(i)\rangle$, then we must have $\langle\varphi_0(x_i,\alpha_i),\alpha_i(0),\ldots,\varphi_n(x_i,\alpha_i),\alpha_i(n)\rangle = \lambda_n'$ (since to minimize a sequence lexicographically one has to minimize first all its initial segments). Thus $\lambda_n^i = \varphi_n(x_i,\alpha_i) = \lambda_n^n$ and $k_n^i = \alpha_i(n) = k_n^n$, and the claim is proved.

Now find α_i such that $(x_i,\alpha_i) \in A$ and $\psi_i'(x_i) = \lambda_i' = \langle\varphi_0(x_i,\alpha_i),\alpha_i(0),\ldots,$ $\varphi_i(x_i,\alpha_i),\alpha_i(i)\rangle$. As shown above, for $i \geq n$, $\alpha_i(n) = k_n^n$ and therefore $\alpha_i \to \alpha = (k_0^0,k_1^1,k_2^2,\ldots)$. Thus $(x_i,\alpha_i) \to (x,\alpha)$. Moreover, for $i \geq n$, $\varphi_n(x_i,\alpha_i) = \lambda_n^n$ so that by the limit property of scales

$$(x,\alpha) \in A \;\&\; \varphi_n(x,\alpha) \leq \lambda_n^n.$$

But then $x \in B$ and

$$\begin{aligned}\psi_n'(x) &= \min\{\langle\varphi_0(x,\alpha),\alpha(0),\ldots,\varphi_n(x,\alpha),\alpha(n)\rangle : (x,\alpha) \in A\}\\ &\leq \langle\lambda_0^0,\alpha(0),\lambda_1^1,\alpha(1),\ldots,\lambda_n^n,\alpha(n)\rangle\\ &= \langle\lambda_0^n,k_0^n,\lambda_1^n,k_1^n,\ldots,\lambda_n^n,k_n^n\rangle = \lambda_n'.\end{aligned}$$

Thus $x \in B$ and $\psi_n(x) \leq \lambda_n$ and we are done.

<u>2.</u> $\{\psi_n\}_{n\in\omega}$ <u>is a</u> $\exists^{\mathcal{R}}\forall^{\mathcal{R}}\Gamma$-<u>scale</u>.

This is very similar to the computation done in the proof of Theorem (2B-2) and we omit the details. ⊣

The direct analog of (2B-3) is true, but we have no use for it. The useful analog of (2B-3) gives uniformization directly.

(3B-3) <u>Theorem</u>: Assume Γ is adequate, closed under both $\exists^\omega$, $\forall^\omega$, $\check{\Gamma} \subseteq \exists^{\mathcal{R}}\Gamma$, $\exists^{\mathcal{R}}(\Gamma \cap \check{\Gamma}) = \Gamma$ and there is a wellordering $\leq$ of $\mathcal{R}$ which is Γ-good. Then Uniformization($\exists^{\mathcal{R}}\Gamma,\exists^{\mathcal{R}}\Gamma$). (Essentially Addison [1]).

<u>Proof</u>: Since for Γ, Γ' adequate

$$\underline{\text{Uniformization}}(\Gamma,\Gamma') \Rightarrow \underline{\text{Uniformization}}(\exists^{\mathcal{R}}\Gamma,\exists^{\mathcal{R}}\Gamma')$$

it will be enough to prove <u>Uniformization</u>$(\Gamma \cap \check{\Gamma},\Gamma \cap \check{\Gamma})$. But if $P \subseteq \mathfrak{X} \times \mathfrak{Y}$ is in $\Gamma \cap \check{\Gamma}$, where we can assume w.l.o.g. $\mathfrak{Y} = \mathcal{R}$, let

$$P^*(x,\alpha) \Leftrightarrow P(x,\alpha) \ \& \ (\forall \beta < \alpha) \neg P(x,\beta).$$

Then P^* uniformizes P and clearly $P^* \in \Gamma \cap \check{\Gamma}$. ⊣

Corollary: (a). $V = L \Rightarrow$ Uniformization(Σ^1_k,Σ^1_k) $(k \geq 2)$. (Addison, [1]).
(b) $V = L[\mu] \Rightarrow$ Uniformization(Σ^1_k,Σ^1_k) $(k \geq 2)$. (Silver, [18]).

3C. The Second Periodicity Theorem.

(3C-1) Theorem: Assume Γ is adequate and Determinacy$(\underset{\sim}{\Gamma} \cap \check{\underset{\sim}{\Gamma}})$. Then if $A \in \Gamma$ and A admits a Γ-scale, $\forall^{\mathcal{R}}A$ admits a $\forall^{\mathcal{R}}\exists^{\mathcal{R}}\Gamma$-scale. (Moschovakis, [14]).

Corollary: Γ adequate, Scale(Γ), Determinacy$(\underset{\sim}{\Gamma} \cap \check{\underset{\sim}{\Gamma}})$, $\exists^{\mathcal{R}}\Gamma \subseteq \Gamma \Rightarrow$ Scale$(\forall^{\mathcal{R}}\Gamma)$.

Corollary: PD $\Rightarrow$ Uniformization(Π^1_{2n+1}), $n \geq 1$.

Proof of the theorem: Enumerate in a 1 - 1 recursive way all the finite sequences of integers, say $u_0,u_1,u_2,\ldots$ such that $u_0 = (\ )$ (the empty sequence) and if u_i is a proper initial segment of u_j then $i < j$.

Put $B = \forall^{\mathcal{R}}A \subseteq \mathfrak{X}$ and define

$$x \in B_n \Leftrightarrow (\forall \alpha \supseteq u_n) A(x,\alpha)$$

(where $\alpha \supseteq u_n \Leftrightarrow \alpha$ extends u_n). Notice that $B_0 = B$ and for every n, $B \subseteq B_n$. We shall define norms on each B_n by considering games as follows: For each $n \in \omega$ and $x,y \in \mathfrak{X}$ let $G_n(x,y)$ be the following game

u_n	u_n
I	II
α'	β'

I plays α', II plays β' and if we call $\alpha = u_n \frown \alpha'$, $\beta = u_n \frown \beta'$ then

$$\text{II wins iff } \neg A(y,\beta) \text{ or}$$

$$[A(y,\beta) \ \& \ A(x,\alpha) \ \& \ \langle \varphi_0(x,\alpha),\ldots,\varphi_n(x,\alpha)\rangle \leq \langle \varphi_0(y,\beta),\ldots,\varphi_n(y,\beta)\rangle],$$

where $\{\varphi_n\}_{n\in\omega}$ is a Γ-scale on A.

Finally put for $x,y \in B_n$

$$x \leq_n y \Leftrightarrow \text{II has a winning stragegy in } G_n(x,y).$$

Proofs similar to those in (2C-1) show that each $\leq_n$ is a prewellordering on

B_n and that for some S, S' in $\forall^{\mathbb{R}}\exists^{\mathbb{R}}\Gamma$, $\exists^{\mathbb{R}}\forall^{\mathbb{R}}\check{\Gamma}$ respectively we have

$$y \in B_n \Rightarrow \forall x[(x \in B_n \;\&\; x \leq_n y) \Leftrightarrow S(n,x,y) \Leftrightarrow S'(n,x,y)].$$

Call ψ_n the norm associated with $\leq_n$. Suppose we can prove that $\underline{x_i \in B}$, $x_i \to x$ and $\psi_n(x_i) = \lambda_n$ for all large enough $i \Rightarrow x \in B$ and $\psi_n(x) \leq \lambda_n$. Then if we put for $x \in B$

$$\psi_n'(x) = \langle \psi_0(x), \psi_n(x) \rangle$$

and let ψ_n'' be the norm on B such that

$$\psi_n''(x) \leq \psi_n''(y) \Leftrightarrow \psi_n'(x) \leq \psi_n'(y)$$

it is easy to check that $\{\psi_n''\}_{n\in\omega}$ is a $\forall^{\mathbb{R}}\exists^{\mathbb{R}}\Gamma$-scale on B.

Thus, to complete the proof, we assume $x_i \in B$, $x_i \to x$ and $\psi_n(x_i) = \lambda_n$ for all $i \geq n$, and try to show that $x \in B$ and $\psi_n(x) \leq \lambda_n$.

<u>1</u>. $x \in B$.

<u>Proof</u>: We have to show that for every α, $(x,\alpha) \in A$. Fix α. Let n_i be such that $u_{n_i} = \bar{\alpha}(i)$. Notice that $n_0 = 0 \;\&\; n_0 < n_1 < n_2 < \cdots$. Consider the subsequence $(x_{n_0}, x_{n_1}, x_{n_2}, \ldots)$. Then $\psi_{n_i}(x_{n_i}) = \psi_{n_i}(x_{n_{i+1}})$. Thus II has a winning strategy in all the games $G_{n_i}(x_{n_{i+1}}, x_{n_i})$ (since $x_{n_i} \geq_{n_i} x_{n_{i+1}}$).

Fix strategies for II in each one of the games $G_{n_i}(x_{n_{i+1}}, x_{n_i})$ and consider the diagram below:

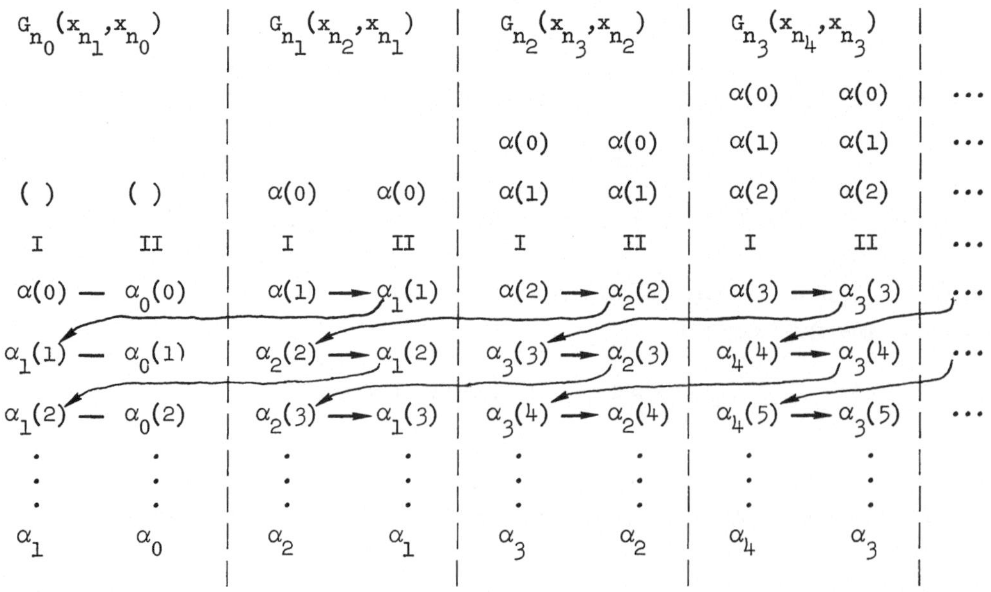

I plays $\alpha(0)$ in $G_{n_0}(x_{n_1},x_{n_0})$ and II answers with his strategy to give $\alpha_0(0)$. Then I plays $\alpha(1)$ in $G_{n_1}(x_{n_2},x_{n_1})$ and II answers with his strategy to give $\alpha_1(1)$ etc. After these moves have been played I plays $\alpha_1(1)$ in $G_{n_0}(x_{n_1},x_{n_0})$ and II answers with his strategy to give $\alpha_0(1)$. Then I plays $\alpha_2(2)$ in $G_{n_1}(x_{n_2},x_{n_1})$ and II answers by his strategy to give $\alpha_1(2)$ etc. Finally reals $\alpha_0,\alpha_1,\alpha_2,\ldots$ are formed (where e.g. $\alpha_4 = (\alpha(0),\alpha(1),\alpha(2),\alpha(3),\alpha_4(4),\ldots))$. Clearly $\alpha_i \to \alpha$. Moreover since II always wins we have

$$\varphi_0(x_{n_1},\alpha_1) \leq \varphi_0(x_{n_0},\alpha_0)$$

$$\varphi_0(x_{n_2},\alpha_2) \leq \varphi_0(x_{n_1},\alpha_1)$$

$$\varphi_0(x_{n_3},\alpha_3) \leq \varphi_0(x_{n_2},\alpha_2)$$

$$\vdots$$

Thus for all large enough i, $\varphi_0(x_{n_i},\alpha_i)$ is constant. Looking at such i's and since again II always wins we have

$$\varphi_1(x_{n_{i+1}},\alpha_{i+1}) \leq \varphi_1(x_{n_i},\alpha_i)$$

$$\varphi_1(x_{n_{i+2}},\alpha_{i+2}) \leq \varphi_1(x_{n_{i+1}},\alpha_{i+1})$$

$$\vdots$$

Thus again $\varphi_1(x_{n_i},\alpha_i)$ becomes eventually constant. Then we look at φ_2 etc. Thus for the sequence $\{(x_{n_i},\alpha_i)\}_{i\in\omega}$ we have $(x_{n_i},\alpha_i) \in A$, $(x_{n_i},\alpha_i) \to (x,\alpha)$ and for each n, $\varphi_n(x_{n_i},\alpha_i)$ becomes eventually constant. So $(x,\alpha) \in A$.

2. For each n, $\psi_n(x) \leq \lambda_n$.

Proof: Since for $i \geq n$, $\psi_n(x_i) = \lambda_n$, it will be enough to prove that for each n, $x \leq_n x_n$, i.e. that II has a winning strategy in $G_n(x,x_n)$.

Since $\psi_k(x_k) = \psi_k(x_m)$, for $k \leq m$ we have $x_m \leq_k x_k$ for $m \geq k$, thus II has a winning strategy in each one of the games $G_k(x_m,x_k)$ for $m \geq k$. Fix strategies for II in each one of these games. Fix n. In order to invent a strategy for II in $G_n(x,x_n)$ consider the diagram

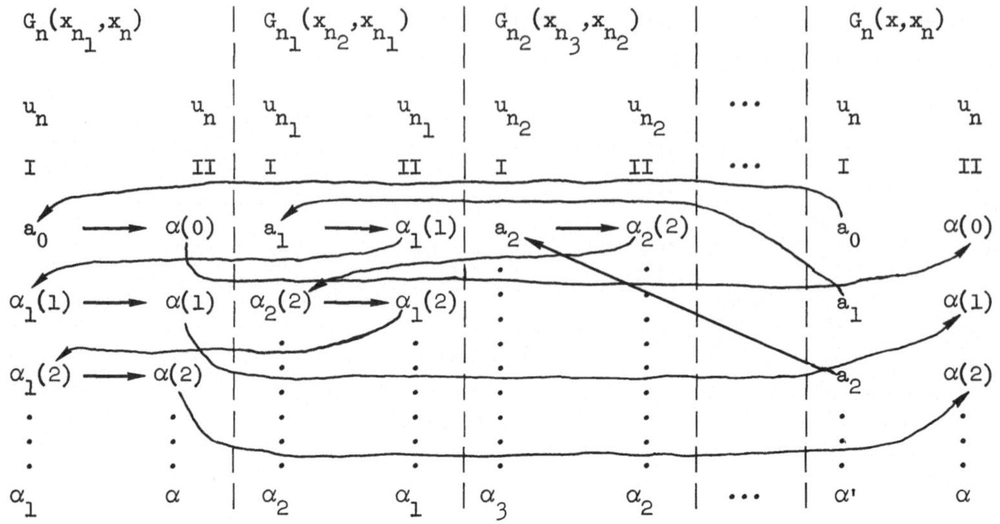

Let I play a_0 in $G_n(x,x_n)$. Let $u_{n_1} = u_n \frown a_0$. Then $n_1 > n$. Consider the game $G_n(x_{n_1},x_n)$ and let I play in $G_n(x_{n_1},x_n)$ a_0 and II answer in this game by his winning strategy to give $\alpha(0)$. This $\alpha(0)$ is II's answer to a_0 in $G_n(x,x_n)$. Then I plays a_1 in $G_n(x,x_n)$. Let $u_{n_2} = u_n \frown a_0 \frown a_1$. Then $n_2 > n_1$. Consider the game $G_{n_1}(x_{n_2},x_{n_1})$ and let I play a_1 in $G_{n_1}(x_{n_2},x_{n_1})$ and II answer in this game by his winning strategy to give $\alpha_1(1)$. Then I plays in $G_n(x_{n_1},x_n)$ this $\alpha_1(1)$ and II answers by his strategy in $G_n(x_{n_1},x_n)$ to give $\alpha(1)$ which is II's next move in $G_n(x,x_n)$ etc. After all these moves have been played $\alpha',\alpha,\alpha_1,\alpha_2,\ldots$ are created (e.g. $\alpha_2 = u_{n_1} \frown (a_1,\alpha_2(2),\alpha_2(3),\ldots) = u_n \frown (a_0,a_1,\alpha_2(2),\alpha_2(3),\ldots)$). Clearly $\alpha_n \to \alpha' = u_n \frown (a_0,a_1,a_2,a_3,\ldots)$ and since II always wins we have $\varphi_0(x_n,\alpha) \geq \varphi_0(x_{n_1},\alpha_1) \geq \varphi_0(x_{n_2},\alpha_2) \geq \cdots$. Thus after a while $\varphi_0(x_{n_i},\alpha_i)$ becomes constant. Then we look at $\varphi_1(x_{n_i},\alpha_i)$ for such i's; it is nonincreasing with i, thus becomes eventually constant etc. Thus for each n, $\varphi_n(x_{n_i},\alpha_i)$ becomes eventually constant. But also $(x_{n_i},\alpha_i) \to (x,\alpha')$, so $(x,\alpha') \in A$ and $\varphi_n(x,\alpha') \leq \lim_i \varphi_n(x_{n_i},\alpha_i)$. Then, as we saw above, $\varphi_0(x_n,\alpha) \geq \lim_i \varphi_0(x_{n_i},\alpha_i) \geq \varphi_0(x,\alpha')$. If $\varphi_0(x_n,\alpha) > \varphi_0(x,\alpha')$ clearly II wins the game $G_n(x,x_n)$ and we are done. If $\varphi_0(x_n,\alpha) = \varphi_0(x,\alpha')$, then for all i, $\varphi_0(x_{n_i},\alpha_i) = \varphi_0(x_n,\alpha) = \varphi_0(x,\alpha')$. But then if $n \geq 1$, $\varphi_1(x_n,\alpha) \geq \varphi_1(x_{n_1},\alpha_1) \geq \varphi_1(x_{n_2},\alpha_2) \geq \cdots$, thus $\varphi_1(x_n,\alpha) \geq \varphi_1(x,\alpha')$. If again $\varphi_1(x_n,\alpha) > \varphi_1(x,\alpha')$ we are done, otherwise $\varphi_1(x_n,\alpha) = \varphi_1(x_{n_i},\alpha_i) = \varphi_1(x,\alpha')$ and then we look (if $n \geq 2$) at φ_2 etc. In any case this shows that

$$\langle \varphi_0(x_n,\alpha),\varphi_1(x_n,\alpha),\ldots,\varphi_n(x_n,\alpha)\rangle \geq$$

$$\langle \varphi_0(x,\alpha'),\ldots,\varphi_n(x,\alpha')\rangle \quad \text{i.e. II wins.}$$

Thus we have described a winning strategy for II in $G_n(x,x_n)$, so $x \leq_n x_n$. ⊣

3D. The zig-zag picture. It follows from the results of this section that the pictures given in 2D for the Prewellordering property hold also for the Scale and Uniformization properties, i.e. under the stated hypotheses these properties hold for the circled pointclasses. That, assuming PD, they hold only for the circled pointclasses we will prove in the next section.

3E. The Martin-Solovay Uniformization Theorem. From the results of this section it is obvious that

$$\text{Determinacy}(\mathbf{\Delta}^1_2) \Rightarrow \text{Uniformization}(\Pi^1_2,\Pi^1_3).$$

However Martin and Solovay had obtained a similar theorem from weaker hypotheses before these results were proved, namely

$$\forall\alpha[\alpha^\# \text{ exists}] \Rightarrow \text{Uniformization}(\Pi^1_2,\Delta^1_4),$$

see [11]; this in turn was strengthened by Mansfield [8] to

$$\forall\alpha[\alpha^\# \text{ exists}] \Rightarrow \text{Uniformization}(\Pi^1_2,\Pi^1_3).$$

These proofs (and in fact the statements of the theorems) involve the theory of indiscernibles with which we are not concerned here. We will state one by-product of this work which will be useful later.

(3E-1) Theorem: Assume that there exists a measurable cardinal or (the weaker hypothesis) that for each α, $\alpha^\#$ exists, let $u_1 = \aleph_1, u_2, u_3, \ldots, u_\omega$ be the first $\omega + 1$ uniform indiscernibles. Then:

(1) $u_n \leq \aleph_n$, $u_\omega \leq \aleph_\omega$ and cofinality(u_{n+1}) = cofinality(u_2).

(2) If AC holds, then $u_\omega < \aleph_3$.

(3) Every Π^1_2 set A admits a Π^1_3 scale on u_ω, i.e. a scale $\{\varphi_n\}_{n\in\omega}$ with each $\varphi_n : A \to u_\omega$.

((1) and (2) are due to Solovay, (3) is implicit in [11], [8]).

4. Bases. One of the most interesting corollaries of uniformization results is the computation of bases for pointclasses. If Γ is a pointclass and C a set of reals, put

$$\underline{\text{Basis}}(\Gamma,C) \Leftrightarrow \text{for each } A \in \Gamma,\ A \subseteq \mathcal{R},\ A \neq \emptyset \Rightarrow A \cap C \neq \emptyset.$$

If Λ is a pointclass, we often abbreviate

$$\underline{\text{Basis}}(\Gamma,\Lambda) \Longleftrightarrow \underline{\text{Basis}}(\Gamma,\{\alpha : \text{the set } \{(n,m) : \alpha(n) = m\} \in \Lambda\}).$$

4A. Computation of bases. It is immediately obvious that

(4A-1) $$\underline{\text{Uniformization}}(\Gamma,\Gamma') \Rightarrow \underline{\text{Basis}}(\Gamma,\{\alpha : \{\alpha\} \in \Gamma'\})$$

and it is easy to see that

(4A-2) Γ adequate, $\underline{\text{Basis}}(\Gamma,C) \Rightarrow$

$$\underline{\text{Basis}}(\exists^{\mathcal{R}}\Gamma,\{\alpha : (\exists\beta)(\beta \in C \text{ and } \alpha \text{ is recursive in } \beta)\}).$$

From this and the results in §3 it is clear that

(4A-3) $$\underline{\text{Determinacy}}(\underset{\sim}{\Delta}^1_{2n}) \Rightarrow \underline{\text{Basis}}(\Sigma^1_{2n+2},\Delta^1_{2n+2})$$

and

$$\text{PD} \rightarrow \underline{\text{Basis}}(\Sigma^1_{2n},\Delta^1_{2n}), \qquad n \geq 2.$$

On the other hand we have

(4A-4) Theorem: $\underline{\text{Determinacy}}(\underset{\sim}{\Delta}^1_{2n}) \Rightarrow \underline{\text{not}}\ \underline{\text{Basis}}(\Sigma^1_{2n+1},\Delta^1_{2n+1})$.

Proof: Since $\underline{\text{Determinacy}}(\underset{\sim}{\Delta}^1_{2n})$, we have $\underline{\text{Prewellordering}}(\Pi^1_{2n+1})$. But this has a consequence that $\{\alpha : \alpha \in \Delta^1_{2n+1}\} \in \Pi^1_{2n+1}$. (This is announced in [2]). From this the result follows immediately. (For another proof see [11]). ⊣

The periodicity phenomenon is again clear in (4A-3) and (4A-4).

(4A-5) Theorem: (a) $\forall\alpha(\alpha^{\#}$ exists) $\Rightarrow$ There exists a fixed Π^1_3 singleton α_0 (i.e. $\{\alpha_0\} \in \Pi^1_3$) such that

$$\underline{\text{Basis}}(\Sigma^1_3,\{\beta : \beta \text{ is recursive in } \alpha_0\}).$$

(Martin-Solovay [11], Mansfield [8].)

(b) $\underline{\text{Determinacy}}(\underset{\sim}{\Delta}^1_{2n}) \Rightarrow$ There exists a fixed Π^1_{2n+1} singleton α_0 such that

$$\underline{\text{Basis}}(\Sigma^1_{2n+1},\{\beta : \beta \text{ is recursive in } \alpha_0\}).$$

(Moschovakis, [14].)

Proof: By (4A-2) it will be enough to find a Π^1_{2n+1} singleton α_0 such that every Π^1_{2n} set contains a real recursive in α_0.

Let $B \subseteq \omega \times \mathcal{R}$ be a universal Π^1_{2n} set. Uniformize B by some $B^* \in \Pi^1_{2n+1}$. Then $B^* \subseteq B$ and $\exists\alpha B(n,\alpha) \Longleftrightarrow \exists ! \alpha B^*(n,\alpha)$. Define $B^{**}(n,\alpha) \Longleftrightarrow B^*(n,\alpha)$ <u>or</u> $\{\forall\beta(\neg B(n,\beta))\ \&\ \alpha = \lambda t 0\}$. Then $B^{**} \in \Pi^1_{2n+1}$ and $\forall n \exists ! \alpha B^{**}(n,\alpha)$. Put

$$\alpha \in C \Longleftrightarrow \forall n B^{**}(n,(\alpha)_n).$$

For this proof choose $(\alpha)_n$ so that α is completely determined by $\{(\alpha)_n : n \in \omega\}$. Thus C is a singleton and $C \in \Pi^1_{2n+1}$. If $C = \{\alpha_0\}$ we show that every Π^1_{2n} set A contains a real recursive in α_0. In fact if $A \in \Pi^1_{2n}$ we have $a \in A \Longleftrightarrow (n_0,\alpha) \in B$, for some n_0. Then $A \neq \emptyset \Rightarrow \exists\alpha(n_0,\alpha) \in B$, so $(n_0,(\alpha)_{n_0}) \in B$ i.e. $(\alpha)_{n_0} \in A$. $\dashv$

<u>Open problem:</u>[1)] A well known basis theorem says that for some fixed Σ^1_1 subset of ω, say A_0, we have

$$\underline{\text{Basis}}(\Sigma^1_1,\{\alpha : \alpha \text{ is recursive in } A_0\}).$$

Does this generalize (under any reasonable hypothesis) to Σ^1_{2n+1}, $n \geq 1$?

4B. <u>Independence results.</u> It is clear that the weakest basis result one can expect for a ("lightface") pointclass Γ is

$$\underline{\text{Basis}}(\Gamma,\{\alpha : \alpha \text{ is ordinal definable}\}).$$

But even such a weak result is not provable in ZFC for Γ beyond Σ^1_2 as the next theorem shows.

(4B-1) <u>Theorem:</u> In ZFC alone we cannot prove

$$\underline{\text{Basis}}(\Pi^1_2,\{\alpha : \alpha \text{ is ordinal definable}\}).$$

(Levy, [5].)

<u>Proof:</u> It is enough to show that if M is a countable model of ZF + V = L and α is a real Cohen generic over M then in $N = M[\alpha]$ there is a Π^1_2 set containing no ordinal definable real. In fact, <u>in N</u> consider the set $A = \{\beta : \beta \notin L\}$. Then $A \in \Pi^1_2$ and $A \neq \emptyset$. But A cannot contain an ordinal definable real, since all such reals belong already to $M = L^N$ (because the notion

---

[1)] Martin and Solovay have shown in 1972 that this generalization is false for $n \geq 1$, granting Determinacy($\underset{\sim}{\Delta}^1_{2n}$). They also show that in (4A-5)(b) α_0 can be <u>any</u> Π^1_{2n+1} singleton which is not Δ^1_{2n+1}. This turns out to be the correct generalization of the Kleene Basis Theorem for Σ^1_1. See their "Basis theorems for Π^1_{2k} sets of reals", to appear.

of forcing is homogeneous). ⊣

Of course we have a basis theorem for Σ^1_3 assuming, for example, that there exists a measurable cardinal (Theorem (4A-5)). Unfortunately we cannot go further even with this stronger hypothesis.

(4B-2) Theorem: In ZFC + There exists a measurable cardinal we cannot prove

$$\underline{\text{Basis}}(\Pi^1_3, \{\alpha : \alpha \text{ is ordinal definable}\}).$$

(Levy's method for (4B-1) using a key result of Silver [18].)

Proof: Repeat the proof of (4B-1), but now start with an M which is a countable model of $ZF + V = L[\mu]$, where μ is a normal measure on a cardinal κ. ⊣

5. Partially playful universes. We outline here a construction which (granting PD) yields for each $n \geq 3$ a model M^n of $ZF + AC$ such that

$$M^n \models \underline{\text{Determinacy}}(\underset{\sim}{\Delta}^1_{n-1}),$$

$$M^n \models \mathbb{R} \text{ } \underline{\text{admits a } \Sigma^1_{n+1}\text{-good wellordering}}.$$

In particular the zig-zag picture of 2D for the scale property in M^n has only finitely many teeth, i.e. the scale property settles on the Σ side for $k \geq n + 1$. The results are due to Moschovakis.

Fix $n \geq 3$, let

$$k = \underline{\text{largest even integer}} < n \qquad (k = n - 1 \text{ or } k = n - 2)$$

and assume $\underline{\text{Determinacy}}(\underset{\sim}{\Delta}^1_k)$. By the Second Periodicity Theorem we have

$$\underline{\text{Uniformization}}(\Pi^1_{n-1}, \Pi^1_n)$$

whether n is odd or even, so let $P_{n-1}(m,\alpha,\beta)$ be the standard Π^1_{n-1} universal relation and let $P^*_{n-1}(m,\alpha,\beta)$ be the Π^1_n relation that comes out of the proof of the Second Periodicity Theorem such that

$$(*) \qquad P^*_{n-1}(m,\alpha,\beta) \Rightarrow P_{n-1}(m,\alpha,\beta),$$

$$(**) \qquad (\exists\beta)P_{n-1}(m,\alpha,\beta) \Rightarrow (\exists!\beta)P^*_{n-1}(m,\alpha,\beta).$$

Finally define

$$F_n^*(m,\alpha) = \begin{cases} \underline{\text{the unique}}\ \beta\ \underline{\text{such that}}\ P_{n-1}^*(m,\alpha,\beta), \\ \qquad\qquad \underline{\text{if}}\ (\exists\beta)P_{n-1}(m,\alpha,\beta), \\ \lambda t0\ \text{ if }\ \forall\beta\ \neg P_{n-1}(m,\alpha,\beta). \end{cases}$$

Clearly F_n^* is a function whose graph is Π_n^1.

Let M be a model of ZF, transitive and containing all ordinals (for brevity, standard model). We call M Σ_n^1-correct if for every Σ_n^1 formula $\theta(\alpha_1,\ldots,\alpha_\ell)$,

$$\alpha_1,\ldots,\alpha_\ell \in M \Rightarrow [\theta(\alpha_1,\ldots,\alpha_\ell) \Leftrightarrow M \models \theta(\alpha_1,\ldots,\alpha_\ell)].$$

(5-1) Lemma: Assume Determinacy($\underset{\sim}{\Delta}_k^1$). A standard model M of ZF + DC is Σ_n^1-correct if and only if M is closed under F_n^*, i.e.

$$\alpha \in M \Rightarrow F_n^*(m,\alpha) \in M.$$

Proof: Assume first that M is Σ_n^1-correct. Notice that if for some $\alpha \in M$ and some m_1, m_2,

$$\forall\beta[P_{n-1}(m_1,\alpha,\beta) \Leftrightarrow \neg P_{n-1}(m_2,\alpha,\beta)],$$

then the same equivalence holds in M (it is expressible by a Π_n^1 formula); thus (m_1,m_2,α) codes a $\underset{\sim}{\Delta}_{n-1}^1$ set in M if and only if it does in the world. This applies to $\underset{\sim}{\Delta}_k^1$ sets, since $k \leq n-1$, and it is now easy to verify that

$$M \models \text{Determinacy}(\underset{\sim}{\Delta}_k^1).$$

Hence the Second Periodicity Theorem holds in M, so that (*) and (**) hold. Now if for some $m,\alpha \in M$, $(\exists\beta)P_{n-1}(m,\alpha,\beta)$, then $M \models (\exists\beta)P_{n-1}(m,\alpha,\beta)$, hence for some $\beta \in M$, $M \models P_{n-1}^*(m,\alpha,\beta)$, hence $P_{n-1}^*(m,\alpha,\beta)$ in the world and $\beta = F_n^*(m,\alpha) \in M$.

To prove the converse, assume that M is closed under F_n^* and then show by induction on $i \leq n$ that M is Σ_i^1-correct. This part of the proof does not need the assumption that $M \models$ DC. We omit the details. ⊣

(Actually neither direction of the equivalence needs the assumption $M \models$ DC, but the proof is a bit more complicated.)

Define by induction on the ordinal ξ,

$$M_0^n = \emptyset ,$$

$$M_{\xi+1}^n = M_\xi^n \cup \{x \subseteq M_\xi^n : x \text{ \underline{is definable in} } \langle M_\xi^n, \in \restriction M_\xi^n \rangle\}$$

$$\cup \{F_n^*(m,\alpha) : \alpha \in M_\xi^n\},$$

$$M_\eta^n = \bigcup_{\xi<\eta} M_\xi^n \quad \text{if} \quad \eta = \bigcup \eta > 0 ,$$

and put

$$M^n = \bigcup_\xi M_\xi^n .$$

(5-2) <u>Theorem</u>: Let $n \geq 3$, assume Determinacy($\underset{\sim}{\Delta}_k^1$), with k = largest even integer $< n$, let M_ξ^n, M^n be defined as above.

1. M^n is a standard model of ZF, it is closed under F_n^* and it is Σ_n^1-correct.
2. The relation "$x \in M_\xi^n$" is definable by a formula which is absolute for all Σ_n^1-correct models of ZF that are closed under F_n^*.
3. $M^n \models \forall x \, \exists \xi (x \in M_\xi^n)$
4. $M^n \models$ AC.
5. M^n is the smallest standard model of ZF + DC which is Σ_n^1-correct.
6. $M^n \models$ Determinacy($\underset{\sim}{\Delta}_k^1$) and if Determinacy($\underset{\sim}{\Delta}_{n-1}^1$) holds in the world it also holds in M^n.
7. $M^n \models$ Scale(Π_i^1) for i odd, $i \leq n$,
   $M^n \models$ Scale(Σ_i^1) for i even, $i \leq n + 1$.
8. $M^n \models$ Generalized Continuum Hypothesis.
9. $M^n \models \mathbb{R}$ admits a Σ_{n+1}^1-good wellordering.
10. $M^n \models$ Scale(Σ_i^1) for $i \geq n + 1$.

<u>Outline of proof</u>: 1-5 are easy by standard methods, 6 follows by the remarks on absoluteness made in the proof of Lemma (5-1) and 7 follows from this. The key to 8 and 9 is a version of the Gödel Condensation Lemma that is appropriate to Σ_n^1-correct models. First notice that there is a finite subset Φ_0 of the axioms in ZF + DC + <u>Determinacy</u>($\underset{\sim}{\Delta}_k^1$) such that the function $\xi \to M_\xi^n$ is absolute for transitive sets which are models of Φ_0 and closed under F_n^* and hence the transitive models of $\Phi_0 + \forall x \, \exists \xi (x \in M_\xi^n)$ which are closed under F_n^* are precisely of the form M_ξ^n. Now get the Condensation Lemma as usually, except that in taking elementary submodels close under F_n^*. We omit the details. ⊣

We draw the zig-zag picture for the scale property for the models M^3, M^4:

$V = M^3$
or
$V = M^4$

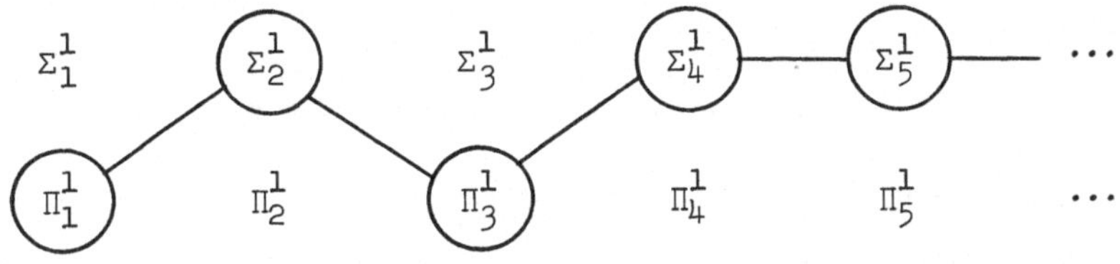

Another interesting model, M^ω, can be obtained by closing under all F_n^*, $n = 3,4,5,\ldots$. This satisfies PD and has the same zig-zag picture as V (assuming PD of course), but in M^ω, $\mathcal{R}$ admits a very simple (hyperanalytic) good wellordering.

Kechris has shown by indiscernibility considerations that

$$M^3 \subsetneq M^4 \subsetneq M^5 \subsetneq \cdots$$

and in fact for each $n \geq 3$, $(\exists\alpha)[\alpha \in M^{n+1} - M^n]$.

6. Trees. We show here that the existence of a scale on a set A yields a representation for A in terms of a tree on ordinals which is very similar to the classical representations for $\undertilde{\Sigma}_1^1$ sets. This is the key to the applications of scales described in the remainder of this paper.

6A. Notation for trees. A tree on some set C is a set T of finite sequences from C such that

$$(c_0,c_1,\ldots,c_k) \in T \ \& \ i \leq k \Rightarrow (c_0,\ldots,c_i) \in T;$$

in particular every non-empty tree contains the empty sequence ( ).

A branch through (or of) a tree T on C is any function $f \in {}^\omega C$ such that

$$\text{for all } n,\ (f(0),\ldots,f(n)) \in T;$$

put

$$[T] = \text{the set of all branches through } T$$

and call T wellfounded if $[T] = \emptyset$, i.e. if T has no infinite branches. The idea here is a bit clearer if we consider the relation $\succ$ of proper extension on finite sequences.

$$(c_0,\ldots,c_k) \succ (d_0,\ldots,d_\ell) \Longleftrightarrow k < \ell \ \& \ c_0 = d_0 \ \& \cdots \& \ c_k = d_k;$$

T is wellfounded if and only if $\prec \restriction T$ has no infinite descending chains, i.e. if and only if $\prec \restriction T$ is wellfounded. We can now assign an ordinal rank to every sequence of a wellfounded tree in the canonical way we do this for any wellfounded relation,

$$|u|_T = \text{supremum}\{|v|_T + 1 : v \in T,\ u > v\}$$

(where $\sup(\emptyset) = 0$) and define the rank of T,

$$|T| = \text{supremum}\{|u|_T : u \in T\} = |(\ )|_T.$$

By convention let also $|u|_T = -1$, if $u \notin T$.

We shall often look at the subtree of T starting from some sequence,

$$T_u = \{v : u \frown v \in T\},$$

where $u \frown v$ is concatenation of sequences.

Most useful for us will be trees of pairs, i.e. trees on sets $C = A \times B$ - usually $C = \omega \times \kappa$ for some ordinal κ. A typical member of a tree T on $A \times B$ is a sequence

$$((a_0,b_0),(a_1,b_1),\dots,(a_n,b_n))$$

and a branch through T is a function $f \in {}^{\omega}(A \times B)$. It will be convenient to represent each brach f by the pair (g,h), $g \in {}^{\omega}A$, $h \in {}^{\omega}B$ which determines it,

$$f(n) = (g(n),h(n)).$$

For each fixed $g \in {}^{\omega}A$ now, we can define a new tree $T(g)$ on B by

$$T(g) = \{(b_0,\dots,b_n) : ((g(0),b_0),\dots,(g(n),b_n)) \in T\}.$$

In the typical case when T is a tree on $\omega \times \kappa$, for each $\alpha \in \mathcal{R}$ we will have a tree on κ

$$T(\alpha) = \{(\xi_0,\dots,\xi_n) : ((\alpha(0),\xi_0),\dots,(\alpha(n),\xi_n)) \in T\};$$

notice that the function

$$\alpha \mapsto T(\alpha)$$

is continuous in a strong sense, i.e.

$$(\xi_0,\dots,\xi_n) \in T(\alpha)\ \&\ \bar{\alpha}(n+1) = \bar{\beta}(n+1) \Rightarrow (\xi_0,\dots,\xi_n) \in T(\beta).$$

6B. κ-scales and their trees. Let $\{\varphi_n\}_{n\in\omega}$ be a scale on A; we call $\{\varphi_n\}_{n\in\omega}$ a κ-scale, if every φ_n is a function on A into κ, i.e. if the length of each prewellordering $\leq^{\varphi_n}$ is $\leq \kappa$. With each κ-scale $\{\varphi_n\}_{n\in\omega}$ on A we define the associated tree T on $\omega \times \kappa$ by

$$T = \{((\alpha(0),\varphi_0(\alpha)),(\alpha(1),\varphi_1(\alpha)),\ldots,(\alpha(n),\varphi_n(\alpha))) : \alpha \in A\}.$$

(6B-1) <u>Theorem</u>: Let A be a pointset, $A \subseteq \mathcal{R}$, $\{\varphi_n\}_{n\in\omega}$ a κ-scale on A, T the associated tree. Then

$$\alpha \in A \Longleftrightarrow T(\alpha) \text{ \underline{is not wellfounded}}$$

$$\Longleftrightarrow (\exists f)(\alpha,f) \in [T].$$

(This is an idea implicit in many of the classical proofs.)

<u>Proof</u>: If $\alpha \in A$, then $(\varphi_0(\alpha),\varphi_1(\alpha),\varphi_2(\alpha),\ldots)$ is a branch through $T(\alpha)$. Conversely, suppose $(\xi_0,\xi_1,\xi_2,\ldots)$ is a branch through $T(\alpha)$, i.e. for each n,

$$((\alpha(0),\xi_0),\ldots,(\alpha(n),\xi_n)) \in T;$$

by the definition of T, there must exist reals $\alpha_0,\alpha_1,\ldots$ in A, so that for each n,

$$((\alpha_n(0),\varphi_0(\alpha_n)),(\alpha_n(1),\varphi_1(\alpha_n)),\ldots,(\alpha_n(n),\varphi_n(\alpha_n)))$$

$$= ((\alpha(0),\xi_0),(\alpha(1),\xi_1),\ldots,(\alpha(n),\xi_n)).$$

This implies immediately that $\text{limit}_n\, \alpha_n = \alpha$ and for $m \leq n$, $\varphi_m(\alpha_n) = \xi_m$, so that by the basic property of scales $\alpha \in A$. ⊣

Kechris has shown that a converse to (6B-1) is true, namely: if $\alpha \in A \Longleftrightarrow T(\alpha)$ <u>is not wellfounded</u>, where T is a tree on $\omega \times \kappa$, then A admits a κ^ω-scale. This shows a connection between the notion of scale and some ideas of Mansfield in [7].

6C. <u>Computing lengths of scales.</u>

(6C-1) <u>Theorem</u>: If $A \subseteq \mathfrak{X} \times \mathcal{R}$ admits a κ-scale, $\kappa \geq \omega$, then $\exists^{\mathcal{R}} A$ admits a κ^ω-scale.

<u>Proof</u>: See the proof of Theorem (3B-2). ⊣

(6C-2) <u>Theorem</u>: Every Σ^1_1 set admits an ω^ω-scale.

<u>Proof</u>: Every closed set admits an ω-scale. ⊣

Now define

$$\delta^1_n = \sup\{\xi : \xi \text{ \underline{is the length of a} } \Delta^1_n$$

$$\text{\underline{prewellordering of} } \mathcal{R}\}.$$

Classically it is known that $\underset{\sim}{\delta}^1_1 = \aleph_1$.

Clearly every $\underset{\sim}{\Pi}^1_{2n+1}$-norm on a set has length $\leq \underset{\sim}{\delta}^1_{2n+1}$.

Thus:

(6C-3) Theorem: Assume Determinacy($\underset{\sim}{\Delta}^1_{2n}$). Then every $\underset{\sim}{\Pi}^1_{2n+1}$ set admits a $\underset{\sim}{\delta}^1_{2n+1}$-scale (by the Periodicity Theorem (3C-1)).

Corollary: (a) Every $\underset{\sim}{\Pi}^1_1$ set admits a $\aleph_1$-scale.
(b) Every $\underset{\sim}{\Sigma}^1_2$ set admits a $\aleph_1^\omega$-scale.

Corollary: Assume Determinacy($\underset{\sim}{\Delta}^1_{2n}$). Then every $\underset{\sim}{\Sigma}^1_{2n+2}$ set admits a $(\underset{\sim}{\delta}^1_{2n+1})^\omega$-scale.

From 3E we also have

(6C-4) Theorem: $\forall\alpha(\alpha^\#$ exists) $\Rightarrow$ Every $\underset{\sim}{\Pi}^1_2$ set admits a u_ω-scale. (Martin-Solovay [11]).

Corollary: $\forall\alpha(\alpha^\#$ exists) $\Rightarrow$ Every $\underset{\sim}{\Sigma}^1_3$ set admits a $(u_\omega)^\omega$-scale. If we also assume AC, then every $\underset{\sim}{\Sigma}^1_3$ set admits a $\varkappa$-scale, with $\varkappa < \aleph_3$. (Martin [10]).

The reader should have noticed that in this section a considerable change in our attitude towards scales has happened. We started worrying not only about definability of a scale but also about its length. Later sections will show why.

7. Computing lengths of wellfounded relations.

7A. The Kunen-Martin theorem. Recall that for a wellfounded relation $<$ we put for $x \in \text{Field}(<)$,

$$|x|_< = \sup\{|y|_< + 1 : y < x\}$$

and

$$|<| = \text{length of } < = \sup\{|x|_< : x \in \text{Field}(<)\}.$$

(7A-1) Theorem: Let $< \subseteq \mathcal{R} \times \mathcal{R}$ be a wellfounded relation and assume $<$ (as a pointset) admits a $\varkappa$-scale. Then $|<| < \varkappa^+$. (Kunen, Martin, independently, unpublished; the proof below is Kunen's).

Proof: To the wellfounded relation $<$ associate a tree $\dot{T}$ of reals as follows:

$$T = \{(\alpha_0,\alpha_1,\ldots,\alpha_n) : \alpha_0,\alpha_1,\ldots,\alpha_n \in \text{Field}(<) \ \&\ \alpha_0 > \alpha_1 > \cdots > \alpha_n\}.$$

Notice that T is also wellfounded and in fact $|<| \leq |T|$. To prove the last statement one can show by $<$-induction that for any $\alpha \in \text{Field}(<)$ and any $\alpha_0,\ldots,\alpha_n$

such that $\alpha_0 > \alpha_1 > \cdots > \alpha_n > \alpha$ we have $|\alpha|_< = |(\alpha_0,\alpha_1,\ldots,\alpha_n,\alpha)|_T$.

We shall define a mapping

$$f : T \twoheadrightarrow S$$

where S is a set of finite sequences from an ordinal $\lambda < \kappa^+$ such that

$$(\alpha_0,\alpha_1,\ldots,\alpha_n) \succ (\beta_0,\ldots,\beta_m) \Rightarrow f((\alpha_0,\ldots,\alpha_n)) \succ f((\beta_0,\ldots,\beta_m))$$

at least when $m \geq 1$. Of course $u \succ v$ means u is a proper initial segment of v. Then we will show that $\prec \restriction S$ is wellfounded, therefore $|T| \leq |\prec \restriction S| + 1$; but $|\prec \restriction S| < \kappa^+$ and the proof will be complete.

Let $\{\varphi_n\}_{n\in\omega}$ be a κ-scale on $>$. To simplify the definition of f (though it is not essential) we put for $\alpha > \beta$

$$\psi_n(\alpha,\beta) = \langle \alpha(0),\beta(0),\varphi_0(\alpha,\beta),\ldots,\alpha(n),\beta(n),\varphi_n(\alpha,\beta)\rangle .$$

Then notice the following limit property of $\{\psi_n\}_{n\in\omega}$:

if $\alpha_i > \beta_i$ for all i, and for each n, $\psi_n(\alpha_i,\beta_i)$ is eventually constant, then $\lim_i(\alpha_i,\beta_i) = (\alpha,\beta)$ exists and $\alpha > \beta$.

Define now f by induction on the length of sequences:

$$f((\ )) = (\ )$$

$$f((\alpha_0)) = (\ )$$

$$f((\alpha_0,\alpha_1)) = (\psi_0(\alpha_0,\alpha_1))$$

$$f((\alpha_0,\alpha_1,\alpha_2)) = (\psi_0(\alpha_0,\alpha_1),\psi_1(\alpha_0,\alpha_1),\psi_1(\alpha_1,\alpha_2),\psi_0(\alpha_1,\alpha_2)),$$

and in general

$$f((\alpha_0,\ldots,\alpha_{n-1},\alpha_n)) = f((\alpha_0,\ldots,\alpha_{n-1}))$$

$$\frown (\psi_{n-1}(\alpha_0,\alpha_1),\psi_{n-1}(\alpha_1,\alpha_2),\ldots,\psi_{n-1}(\alpha_{n-1},\alpha_n),\psi_{n-2}(\alpha_{n-1},\alpha_n),\ldots,\psi_0(\alpha_{n-1},\alpha_n)).$$

The idea is to include in $f((\alpha_0,\ldots,\alpha_n))$ all $\psi_j(\alpha_i,\alpha_{i+1})$ for $i \leq n-1$, $j \leq n-1$. The diagram below explains the way we have done it:

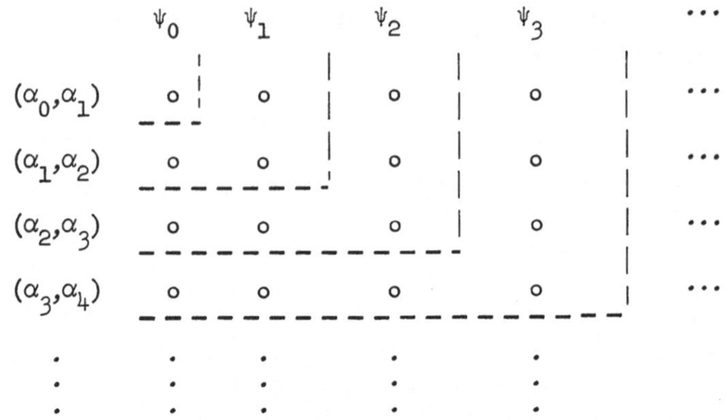

Clearly f is an $\prec$-preserving (on $T - \{(\ )\}$) map from T onto a set of finite sequences S on $\kappa^{\omega} = \lambda$. Thus it will be enough to show that $\prec \upharpoonright S$ is wellfounded. Assume not, towards a contradiction. Then we have

$$f((\alpha_0^0)) \succ f((\alpha_0^1,\alpha_1^1)) \succ f((\alpha_0^2,\alpha_1^2,\alpha_2^2)) \succ \cdots$$

for some $\alpha_0^0,\alpha_0^1,\alpha_1^1,\ldots,$ such that $\alpha_0^1 > \alpha_1^1$, $\alpha_0^2 > \alpha_1^2 > \alpha_2^2$ etc. Then in the diagram

$$\begin{array}{lllll}
\psi_0(\alpha_0^1,\alpha_1^1) & & & & \\
\psi_0(\alpha_0^2,\alpha_1^2) & \psi_1(\alpha_0^2,\alpha_1^2) & \psi_1(\alpha_1^2,\alpha_2^2) & \psi_0(\alpha_1^2,\alpha_2^2) & \\
\psi_0(\alpha_0^3,\alpha_1^3) & \psi_1(\alpha_0^3,\alpha_1^3) & \psi_1(\alpha_1^3,\alpha_2^3) & \psi_0(\alpha_1^3,\alpha_2^3) & \psi_2(\alpha_0^3,\alpha_1^3) \quad \cdots \\
\vdots & \vdots & \vdots & \vdots & \vdots
\end{array}$$

each column consists of identical ordinals, thus for each n and for each j, $\psi_n(\alpha_j^i,\alpha_{j+1}^i)$ becomes constant for large enough i. Thus for each j, $(\alpha_j^i,\alpha_{j+1}^i) \to (\alpha_j,\alpha_{j+1})$ and $\alpha_j > \alpha_{j+1}$ i.e. $\alpha_0 > \alpha_1 > \alpha_2 > \cdots$, a contradiction. $\dashv$

<u>Corollary</u>: Every $\underset{\sim}{\Sigma}_1^1$ wellfounded relation has length $< \aleph_1$ (classical result).

<u>Corollary</u>: Every $\underset{\sim}{\Sigma}_2^1$ wellfounded relation has length $< \aleph_2$. Thus if there exists a $\underset{\sim}{\Sigma}_2^1$ wellordering of $\mathbb{R}$, the Continuum Hypothesis holds (Martin; by an unpublished forcing argument before scales were introduced.)

<u>Corollary</u>: Assume $\forall\alpha(\alpha^{\#}$ exists). Then every $\underset{\sim}{\Sigma}_3^1$ wellfounded relation has length $< (u_\omega)^+$. If we also assume AC, then every $\underset{\sim}{\Sigma}_3^1$ wellfounded relation has length $< \aleph_3$. (That $\underset{\sim}{\delta}_3^1 \leq (u_\omega)^+$ was already shown in Martin [10].)

7B. <u>Projective ordinals</u>. We introduced in §6 the <u>projective ordinals</u> $\underset{\sim}{\delta}_n^1$

and we mentioned that $\underset{\sim}{\delta}_1^1 = \aleph_1$ (this follows also independently from our first corollary in 7A). By the results in 7A it is then clear that

(7B-1) $$\underset{\sim}{\delta}_2^1 \leq \aleph_2 \qquad \text{(Martin)}$$

(7B-2) $$\forall\alpha(\alpha^{\#} \text{ exists}) + AC \Rightarrow \underset{\sim}{\delta}_3^1 \leq \aleph_3 \qquad \text{(Martin)}$$

(7B-3) $$\underline{\text{Determinacy}}(\underset{\sim}{\Delta}_{2n}^1) \Rightarrow \underset{\sim}{\delta}_{2n+2}^1 \leq (\underset{\sim}{\delta}_{2n+1}^1)^+ \qquad \text{(Kunen, Martin)}$$

(To prove (7B-3) recall (6C-3).)

(7B-4) $$\underline{\text{Determinacy}}(\underset{\sim}{\Delta}_2^1) + AC \Rightarrow \underset{\sim}{\delta}_4^1 \leq \aleph_4 \qquad \text{(Kunen, Martin).}$$

Open problem. Is it true that assuming AC (and any other reasonable hypotheses), $\underset{\sim}{\delta}_n^1 \leq \aleph_n$, $n \geq 5$?

We shall mention some other known results about the projective ordinals in the last section.

8. Construction principles.

A construction principle for a pointclass Γ asserts, roughly speaking, that every set in Γ can be expressed, in some canonical way, in terms of sets in a simpler pointclass Γ'. A classical example is the result that every analytic ($\underset{\sim}{\Sigma}_1^1$) set can be expressed both as a union and an intersection of $\aleph_1$ Borel sets.

8A. Inductive analysis of projection of trees. Let T be a tree on $\omega \times \varkappa$. We write

$$A = p[T]$$

iff $\alpha \in A \Longleftrightarrow \exists f(\alpha,f) \in [T] \Longleftrightarrow T(\alpha)$ is not wellfounded.

(8A-1) Theorem: Let T be a tree on $\omega \times \varkappa$ and $A = p[T]$. Put for $1 \leq \xi < \varkappa^+$ and u a finite sequence from $\varkappa$,

$$A_u^\xi = \{\alpha : |T(\alpha)_u| < \xi\},$$

where for any tree J we abbreviate

$$|J| < \xi \Longleftrightarrow J \text{ is wellfounded and } |J| < \xi.$$

Then, if $\text{length}(u) = n$ we have

$$A_u^0 = \{\alpha : ((\alpha(0),u_0),\ldots,(\alpha(n),u_{n-1})) \notin T\}$$

$$A_u^{\xi+1} = A_u^\xi \cup \bigcap_{\eta<\kappa} A_{u^\frown\eta}^\xi$$

$$A_u^\lambda = \bigcup_{\xi<\lambda} A_u^\xi \quad \text{if} \quad \lambda = \bigcup\lambda > 0$$

and

$$\mathcal{R} - A = \bigcup_{\xi<\kappa^+} A_{(\ )}^\xi .$$

(For $\kappa = \omega$ this is apparently due to Sierpinski, see [4] p. 32; Martin [10] first applied these methods to $\kappa > \omega$.)

Proof: Notice that $\alpha \notin A \Longleftrightarrow T(\alpha)$ is wellfounded

$$\Longleftrightarrow (\exists\, \xi < \kappa^+)(|T(\alpha)| < \xi) \Longleftrightarrow (\exists\, \xi < \kappa^+)(|T(\alpha)_{(\ )}| < \xi). \qquad \dashv$$

If λ is an ordinal put

$\mathcal{B}(\lambda)$ = the smallest Boolean algebra containing all closed sets and closed under unions of length $< \lambda$.

Let also

$$\mathcal{B}_n = \mathcal{B}(\underset{\sim}{\delta}_n^1).$$

(8A-2) Theorem: Let T be a tree on $\omega \times \kappa$ and $A = p[T]$. Then A is both the union and the intersection of κ^+ sets in $\mathcal{B}(\kappa^+)$. (Sierpinski for $\kappa = \omega$.)

Proof: In the notation of (8A-1), we clearly have A_u^1 clopen, for all u. Thus $A_u^\xi \in \mathcal{B}(\kappa^+)$, for any ξ and u. So A is the intersection of κ^+ sets in $\mathcal{B}(\kappa^+)$.

Now put

$$B_\xi = \{\alpha : |T(\alpha)| < \xi\} \cup \{\alpha : (\exists\, u)(|T(\alpha)_u| = \xi)\}.$$

Since

$$B_\xi = A_{(\ )}^\xi \cup \bigcup_u (A_u^{\xi+1} - A_u^\xi),$$

clearly each B_ξ is in $\mathcal{B}(\kappa^+)$. It is then easy to check that

$$\mathcal{R} - A = \bigcap_{\xi<\kappa^+} B_\xi;$$

thus A is the union of κ^+ sets in $\mathcal{B}(\kappa^+)$. $\dashv$

Corollary: (a) Each Σ^1_1 set is both the union and the intersection of $\aleph_1$ Borel sets. (Classical)

(b) Each Σ^1_2 set is both the union and the intersection of $\aleph_2$ sets in $\mathcal{B}(\aleph_2)$.

(c) $\forall\alpha(\alpha^\#$ exists) $\Rightarrow$ Each Σ^1_3 set is both the union and the intersection of u^+_ω sets in $\mathcal{B}(u^+_\omega)$. Thus if also AC holds, each Σ^1_3 set is both the union and the intersection of $\aleph_3$ sets in $\mathcal{B}(\aleph_3)$. (Martin [10].)

(d) $\mathrm{Det}(\Delta^1_{2n}) \Rightarrow$ Every Σ^1_{2n+2} set is both the union and the intersection of $(\delta^1_{2n+1})^+$ sets in $\mathcal{B}((\delta^1_{n+1})^+)$. Thus $\mathrm{Det}(\Delta^1_2)$ + AC $\Rightarrow$ Every Σ^1_4 set is both the union and the intersection of $\aleph_4$ sets in $\mathcal{B}(\aleph_4)$.

In 8B and 8C we will see how most of the results of this corollary can be improved.

8B. An extension of Suslin's Theorem.

(8B-1) Theorem: Assume Determinacy(Δ^1_{2n}). Then

$$\Delta^1_{2n+1} \subseteq \mathcal{B}_{2n+1}.$$

(Martin [10] for $n = 1$; Moschovakis [14] in general).

Proof: (In this proof we use essentially Lemmas 9 and 10 of [13].)

Let $A \subseteq \mathcal{R}$, $A \in \Delta^1_{2n+1}$. Find a Π^1_{2n+1}-scale on $\mathcal{R} - A$, say $\{\varphi_m\}_{m\in\omega}$. Since $\mathcal{R} - A \in \Delta^1_{2n+1}$, each prewellordering $\leq^{\varphi_m}$ is actually a Δ^1_{2n+1} prewellordering, thus it has length $\xi_m < \delta^1_{2n+1}$. Since obviously cofinality $(\delta^1_{2n+1}) > \omega$, $\sup_m \xi_m < \delta^1_{2n+1}$ say $\xi_m < \kappa < \delta^1_{2n+1}$, for all m, and $\{\varphi_m\}_{m\in\omega}$ is a κ-scale.

For each α put

$$J(\alpha) = \{(\xi_0,\dots,\xi_k) : \xi_i < \kappa \ \&\ [(\exists \alpha_0 \cdots \alpha_k \in \mathcal{R} - A)(\forall i \leq k(\bar{\alpha}_i(i) = \bar{\alpha}(i))$$

$$\&\ (\forall i \leq k\, \forall j)(i \leq j \leq k \Rightarrow \varphi_i(\alpha_j) = \xi_i))]\}.$$

The mapping $\alpha \mapsto J(\alpha)$, from $\mathcal{R}$ into trees on κ, is continuous i.e. for each $(\xi_0,\dots,\xi_k)$, $\{\alpha : (\xi_0,\dots,\xi_k) \in J(\alpha)\}$ is open, and

$$\alpha \notin A \Leftrightarrow J(\alpha) \text{ is not wellfounded;}$$

thus $\alpha \in A \Leftrightarrow J(\alpha)$ is wellfounded. Put $\lambda = \sup\{J(\alpha) + 1 : \alpha \in A\}$. Suppose we can prove $\lambda < \delta^1_{2n+1}$. Then as in (8A-1) we can define $A^\xi_u = \{\alpha : |J(\alpha)_u| < \xi\}$ for $1 \leq \xi < \lambda$ and u a finite sequence from κ and prove that $A^\xi_u \in \mathcal{B}_{2n+1}$ (since $\kappa, \lambda < \delta^1_{2n+1}$). Since $A = \bigcup_{\xi<\lambda} A^\xi_{(\ )}$ we have $A \in \mathcal{B}_{2n+1}$.

We prove now $\lambda < \underset{\sim}{\delta}^1_{2n+1}$. Define the tree

$$J^*(\alpha) = \{(\alpha_0,\ldots,\alpha_k) : (\alpha_0,\ldots,\alpha_k) \in \mathcal{R} - A \ \& \forall i \leq k(\bar{\alpha}_i(i) = \bar{\alpha}(i))$$

$$\& \forall i \leq k \forall j \forall \ell(i \leq j \leq \ell \leq k \Rightarrow \varphi_i(\alpha_j) = \varphi_i(\alpha_\ell))\}.$$

Then there is an obvious surjective map f from $J^*(\alpha)$ onto $J(\alpha)$, namely

$$f((\alpha_0,\ldots,\alpha_k)) = (\varphi_0(\alpha_0),\ldots,\varphi_k(\alpha_k)),$$

which is clearly $<$-preserving. Thus $J^*(\alpha)$ is wellfounded for any $\alpha \in A$ and it is easy to check that for $\alpha \in A$

$$|(\alpha_0,\ldots,\alpha_k)|_{J^*(\alpha)} = |f((\alpha_0,\ldots,\alpha_k))|_{J(\alpha)}.$$

Thus $|J(\alpha)| = |J^*(\alpha)|$, for every $\alpha \in A$. It will then be enough to show $\sup\{|J^*(\alpha)| + 1 : \alpha \in A\} < \underset{\sim}{\delta}^1_{2n+1}$. To prove this define

$$(\alpha,(\alpha_0,\ldots,\alpha_k)) > (\beta,(\beta_0,\ldots,\alpha_m))$$

$$\Leftrightarrow \alpha = \beta \in A \ \& (\alpha_0,\ldots,\alpha_k),(\beta_0,\ldots,\beta_m) \in J^*(\alpha)$$

$$\& (\alpha_0,\ldots,\alpha_k) > (\beta_0,\ldots,\beta_m).$$

Then $<$ is a $\underset{\sim}{\Delta}^1_{2n+1}$ wellfounded relation, if we code the finite sequences by single reals. But a simple variation of Lemma 10 in [13] shows that every $\underset{\sim}{\Delta}^1_{2n+1}$ wellfounded relation has length $< \underset{\sim}{\delta}^1_{2n+1}$, thus $|<| < \underset{\sim}{\delta}^1_{2n+1}$ and we are done, since $|J^*(\alpha)| \leq |<|$, for each $\alpha \in A$. ⊣

<u>Corollary</u>: Assume Determinacy($\underset{\sim}{\Delta}^1_{2n}$). Then every $\underset{\sim}{\Sigma}^1_{2n+2}$ set is the union of $\underset{\sim}{\delta}^1_{2n+1}$ sets in $\mathcal{B}_{2n+1}$ (Moschovakis [14]).

<u>Proof</u>: It follows from <u>Prewellordering</u>($\underset{\sim}{\Pi}^1_{2n+1}$) that every $\underset{\sim}{\Pi}^1_{2n+1}$ set is the union of $\underset{\sim}{\delta}^1_{2n+1}$ $\underset{\sim}{\Delta}^1_{2n+1}$ sets (see remarks before (6C-3)). It also follows from <u>Prewellordering</u>($\underset{\sim}{\Pi}^1_{2n+1}$) that a 1 - 1 continuous image of a $\underset{\sim}{\Delta}^1_{2n+1}$ set is also a $\underset{\sim}{\Delta}^1_{2n+1}$ set. We can get then the result easily by applying the uniformization theorem and (8B-1). ⊣

<u>Remark</u>: Put

$$\text{AD} \Leftrightarrow \text{Every set is determined.}$$

Martin [10] has shown,

$$AD \Rightarrow \underset{\sim}{\Delta}^1_{2n+1} \supseteq \mathcal{B}_{2n+1} \; ; \quad \text{thus}$$

$$AD \Rightarrow \underset{\sim}{\Delta}^1_{2n+1} = \mathcal{B}_{2n+1} \; .$$

Moschovakis has shown in [13], that

$$AD \Rightarrow \text{Every union of } \underset{\sim}{\delta}^1_{2n+1} \text{ sets in } \underset{\sim}{\Delta}^1_{2n+1} \text{ is in } \underset{\sim}{\Sigma}^1_{2n+2}; \text{ thus}$$

$$AD \Rightarrow (A \in \underset{\sim}{\Sigma}^1_{2n+2} \Leftrightarrow A \text{ is the union of } \underset{\sim}{\delta}^1_{2n+1} \text{ sets in } \mathcal{B}_{2n+1}).$$

8C. Unions of Borel sets.

(8C-1) Theorem: Assume AC. If a pointset A admits a κ-scale with $\kappa < \aleph_{n+1}$, $0 < n \in \omega$, then A is the union of $\aleph_n$ Borel sets. (Martin [10].)

Corollary: (a) Assume AC + $\forall\alpha(\alpha^{\#}$ exists). Then every $\underset{\sim}{\Sigma}^1_3$ set is the union of $\aleph_2$ Borel sets.

(b) Assume AC + Determinacy($\underset{\sim}{\Delta}^1_2$). Then every $\underset{\sim}{\Sigma}^1_4$ set is the union of $\aleph_3$ Borel sets.

Proof of the theorem: Assume $A \subseteq \mathcal{R}$ admits a κ-scale with $\kappa < \aleph_{n+1}$. Then for some tree T, $\alpha \in A \Leftrightarrow T(\alpha)$ is not wellfounded, where we may assume that T is actually a tree on $\omega \times \aleph_n$ (we replace if necessary the tree coming from the scale on A by an isomorphic tree, noticing that $|\kappa| \leq \aleph_n$). Since $n > 0$, $\text{cof}(\aleph_n) > \omega$, thus if $T(\alpha)$ is not wellfounded there is a $\xi < \aleph_n$ such that $T^{\xi}(\alpha)$ is not wellfounded, where

$$T^{\xi} = \{((k_0,\xi_0),\ldots,(k_m,\xi_m)) \in T : \xi_0,\ldots,\xi_m \leq \xi\}.$$

Thus

$$\alpha \in A \Leftrightarrow \exists \xi_n < \aleph_n [T^{\xi_n}(\alpha) \text{ is not wellfounded}].$$

If $\xi_n < \aleph_n$ we can replace T^{ξ_n} by an isomorphic tree on $\omega \times \aleph_{n-1}$, say $T_1^{\xi_n}$. Then

$$\alpha \in A \Leftrightarrow (\exists \xi_n < \aleph_n)[T_1^{\xi_n}(\alpha) \text{ is not wellfounded}].$$

If $n - 1 > 0$ we have again

$$T_1^{\xi_n}(\alpha) \text{ is not wellfounded}$$

$$\Leftrightarrow (\exists \xi_{n-1} < \aleph_{n-1})((T_1^{\xi_n})^{\xi_{n-1}} \text{ is not wellfounded})$$

and we proceed similarly. After at most n steps we get

$$\alpha \in A \iff \exists \xi_n < \aleph_n \, \exists \xi_{n-1} < \aleph_{n-1} \cdots \exists \xi_1 < \aleph_1$$

$$(T^{\xi_n,\dots,\xi_1}(\alpha) \text{ is not wellfounded})$$

where $T^{\xi_n,\dots,\xi_1}$ is a tree on $\omega \times \omega$. But then $\{\alpha : T^{\xi_n,\dots,\xi_1}(\alpha)$ is not wellfounded$\}$ is a Σ^1_1 set, thus it is the union of $\aleph_1$ Borel sets and the proof is complete. ⊣

Open problem. Prove assuming AC (and any other reasonable hypotheses) that every Σ^1_{n+1} set is the union of $\aleph_n$ Borel sets, $n \geq 4$. Notice that a solution to the problem at the end of §7 solves this problem too.

9. Constructibility in the tree associated with a scale. In §6 we associated with each κ-scale $\{\varphi_n\}_{n\in\omega}$ on a set $A \subseteq \mathcal{R}$ a tree T on $\omega \times \kappa$. We introduce and study here the models $L[T]$, where T comes from a complete Π^1_{2n+1} set, granting Determinacy(Δ^1_{2n}) - these are the basic tools for the results in the next two sections. The key theorem in the present section is that the tree that comes from a complete Π^1_1 set is in fact constructible.

9A. The models $L[T^{2n+1}]$.

(9A-1) Theorem: Let T be the tree associated with a κ-scale on some set A, let $Q \subseteq \mathcal{R} \times \mathcal{R}$ and assume that for some recursive $f : \mathcal{R} \times \mathcal{R} \to \mathcal{R}$,

$$Q(\alpha,\beta) \iff f(\alpha,\beta) \in A.$$

Then for some tree $S \in L[T]$ on $\omega \times \kappa$

(4) $\exists\beta\, Q(\alpha,\beta) \iff S(\alpha)$ is not wellfounded.

(5) $\iff (\exists\beta \in L[T,\alpha])Q(\alpha,\beta)$.

(Folk-type result.)

Proof: We have

$$\exists\beta\, Q(\alpha,\beta) \iff \exists\beta\,(f(\alpha,\beta) \in A)$$

$$\iff \exists\beta\,\exists\gamma\,(f(\alpha,\beta) = \gamma \;\&\; \gamma \in A)$$

$$\iff \exists\beta\,\exists\gamma\,(\forall n\, R(\bar\alpha(n),\bar\beta(n),\bar\gamma(n)) \;\&\; T(\gamma) \text{ is not wellfounded})$$

where R is recursive. Put

$$S' = \{((a_0,(b_0,c_0,\xi_0)),\ldots,(a_k,(b_k,c_k,\xi_k))) : ((c_0,\xi_0),\ldots,(c_k,\xi_k)) \in T$$

$$\& \; R(\langle a_0,\ldots,a_k\rangle, \langle b_0,\ldots,b_k\rangle, \langle c_0,\ldots,c_k\rangle)\}.$$

Then S' is a tree on $\omega \times (\omega \times \omega \times \varkappa)$, $S' \in L[T]$ and

$$\exists\beta\, Q(\alpha,\beta) \iff S'(\alpha) \text{ is not wellfounded}$$

$$\iff \exists\beta \in L[T,\alpha] Q(\alpha,\beta)$$

where the last equivalence follows from the usual absoluteness of wellfoundedness. To get instead of S' a tree on $\omega \times \varkappa$ fix a 1 - 1 mapping from $\omega \times \omega \times \varkappa$ onto $\varkappa$ <u>in</u> $L[T]$ and replace S' by an isomorphic tree on $\omega \times \varkappa$, call it S. ⊣

Fix now a <u>complete</u> Π^1_{2n+1} set $\mathcal{P}_{2n+1} \subseteq \mathcal{R}$. ($\mathcal{P}_{2n+1}$ is such that for every $A \subseteq \mathcal{X}$, $A \in \Pi^1_{2n+1}$ we can find $f : \mathcal{X} \to \mathcal{R}$ recursive such that $x \in A \iff f(x) \in \mathcal{P}_{2n+1}$.) Assuming <u>Determinacy</u>($\underset{\sim}{\Delta}^1_{2n}$), let $\{\varphi_n\}_{n\in\omega}$ be a fixed for the discussion Π^1_{2n+1} scale on $\mathcal{P}_{2n+1}$ and let T^{2n+1} be the associated tree. T^{2n+1} is a tree on $\omega \times \underset{\sim}{\delta}^1_{2n+1}$.

(9A-2) <u>Theorem</u>: Assume Determinacy($\underset{\sim}{\Delta}^1_{2n}$). Then for every Σ^1_{2n+2} set A we can find a tree $S \in L[T^{2n+1}]$ such that

$$\alpha \in A \iff S(\alpha) \text{ is not wellfounded.}$$

(9A-3) <u>Theorem</u>: Assume Determinacy($\underset{\sim}{\Delta}^1_{2n}$). Then Σ^1_{2n+2} formulas are absolute for $L[T^{2n+1}]$.

<u>Proof</u>: We show successively that for $2n + 2 \geq k \geq 2$, Σ^1_k formulas are absolute for $L[T^{2n+1}]$. For $k = 2$ this is Shoenfield's theorem, while for $k \geq 3$ we proceed using (4) and (5) of (9A-1). ⊣

Unfortunately, except for the case $n = 0$, which we shall study in the rest of this section, there is practically nothing known about the <u>internal structure</u> of $L[T^{2n+1}]$.[2)]

[2)] It has been recently shown by Harrington and Kechris that for all $n \geq 0$ $\mathcal{R} \cap L[T^{2n+1}]$ is the largest countable Σ^1_{2n+2} set of reals (as conjectured by Moschovakis) and that additionally,

$$L[T^{2n+1}] \models \text{"}\mathcal{R} \text{ has a } \Delta^1_{2n+2}\text{-good wellordering"}.$$

These results suggest that $L[T^{2n+1}]$ is a correct higher level analog of L. Their proof uses determinacy of all hyperprojective sets. See their "Ordinal quantification and the models $L[T^{2n+1}]$", mimeo-grafed note, January 1977.

9B. Absoluteness of closed games. Let $\mathfrak{g}$ be a set of even finite sequences from a set A. We define the game $G_{\mathfrak{g}}$ as follows:

I	II	
		I plays $a_0, a_1, \ldots$ and II plays $b_0, b_1, \ldots$,
a_0	b_0	$a_i, b_i \in A$. Then I wins iff for some n,
a_1	b_1	$(a_0, b_0, \ldots, a_n, b_n) \in \mathfrak{g}$. Clearly the game is open
$\vdots$	$\vdots$	in I.

The following is a folk-type result.

(9B-1) Theorem: Let $M \models ZF + DC$ and $M \supseteq$ Ordinals. Let $A, \mathfrak{g} \in M$, and assume A is wellorderable in M. Then

I has a winning strategy in $G_{\mathfrak{g}} \Leftrightarrow M \models$ I has a winning strategy in $G_{\mathfrak{g}}$

and similarly for II. Moreover the player who has a winning strategy has a winning strategy (for the game in the world) which lies in M.

Proof: For each $(a_0, b_0, \ldots, a_n, b_n)$ consider the subgame $G_{\mathfrak{g}}(a_0, b_0, \ldots, a_n, b_n)$ defined by:

I	II	
		I plays α, II plays β and I wins iff for some m
α	β	$(a_0, b_0, \ldots, a_n, b_n) \frown (\alpha(0), \beta(0), \ldots, \alpha(m), \beta(m)) \in \mathfrak{g}$.

Then define

$$\mathfrak{g}^0 = \mathfrak{g}$$

$$\mathfrak{g}^\xi = \{(a_0, b_0, \ldots, a_n, b_n) : \exists a_{n+1} \in A \, \forall b_{n+1} \in A$$
$$\exists \eta < \xi((a_0, b_0, \ldots, a_{n+1}, b_{n+1}) \in \mathfrak{g}^\eta)\}.$$

Then for each ξ, $(a_0, b_0, \ldots, a_n, b_n) \in \mathfrak{g}^\xi \Rightarrow$ I has a winning strategy in $G_{\mathfrak{g}}(a_0, b_0, \ldots, a_n, b_n)$. Using this we show:

II has a winning strategy in $G \Leftrightarrow \forall \xi[(\ ) \notin \mathfrak{g}^\xi]$.

Proof: If II has a winning strategy in $G_{\mathfrak{g}} = G_{\mathfrak{g}}((\ ))$ then I has no winning strategy in $G_{\mathfrak{g}}((\ ))$, thus for all ξ, $(\ ) \notin \mathfrak{g}^\xi$. Conversely assume that for each ξ, $(\ ) \notin \mathfrak{g}^\xi$. We describe a winning strategy for II in $G_{\mathfrak{g}}$ as follows: If I plays a_0 II plays the least b_0 (in a fixed wellordering of A) such that

$\forall\xi(a_0,b_0) \notin \mathfrak{g}^\xi$. Such a b_0 exists, because otherwise for all b, there exists a ξ such that $(a_0,b) \in \mathfrak{g}^\xi$. Let $g(b)$ = least such ξ and find $\xi_0 >$ all $g(b)$, $b \in A$. Then $\forall b \exists \xi < \xi_0(a_0,b) \in \mathfrak{g}^\xi$, thus $(\ ) \in \mathfrak{g}^{\xi_0}$, a contradiction. Similarly if I plays a_1, II picks the least b_1 such that $\forall\xi(a_0,b_0,a_1,b_1) \notin \mathfrak{g}^\xi$, etc.

Since the above equivalence was proved under the assumption "ZF + DC + A is wellorderable" and since $\xi \mapsto \mathfrak{g}^\xi$ is clearly an absolute map and $M \supseteq$ Ordinals, it is immediate that "II has a winning strategy" is absolute for M, thus the same is true for "I has a winning strategy." Moreover the argument above clearly provides a winning strategy for II which lies in M and wins in the world, thus it will be enough in order to complete the proof to show that when I has a winning strategy we can find one (who wins in the world also) in M. Notice that

$$\text{I has a winning strategy} \Longleftrightarrow \exists\xi[(\ ) \in \mathfrak{g}^\xi]$$

and check that the following is a winning strategy for I which lies in M. Put ξ_0 = least ξ such that $(\ ) \in \mathfrak{g}^\xi$. If $\xi_0 = 0$, I has already won. If $\xi_0 > 0$, let I play the least a_0 such that for every b $\exists\xi < \xi_0(a_0,b) \in \mathfrak{g}^\xi$. If now II plays b_0, let ξ_1 = least $\xi < \xi_0$ such that $(a_0,b_0) \in \mathfrak{g}^\xi$. If $\xi_1 = 0$ I has already won, otherwise let I play the least a_1 such that for all b, $\exists\xi < \xi_1(a_0,b_0,a_1,b) \in \mathfrak{g}^\xi$ etc. (Notice that $\xi_0 > \xi_1 > \cdots$, so this cannot go on.) ⊣

9C. <u>Proof that</u> $T^1 \in L$. Suppose $A \subseteq \aleph_1$. We let

$$\underline{\text{Code}}(A) = \{\alpha \in WO : |\alpha| \in A\}.$$

Similarly if T is a tree on $\omega \times \aleph_1$ we let

$$\underline{\text{Code}}(T) = \{(k_0,\alpha_0,\ldots,k_n,\alpha_n) : ((k_0,|\alpha_0|),\ldots,(k_n,|\alpha_n|)) \in T\}.$$

We say that <u>Code(T)</u> <u>is in</u> Γ iff

$$\{\langle k_0,\alpha_0,\ldots,k_n,\alpha_n\rangle : (k_0,\alpha_0,\ldots,\alpha_n) \in \text{Code}(T)\} \in \Gamma$$

where $\langle k_0,\alpha_0,\ldots,k_n,\alpha_n\rangle = (n,k_0,\ldots,k_n,\alpha_0(0),\ldots,\alpha_n(0),\alpha_0(1),\ldots,\alpha_n(1),\ldots) \in \mathcal{R}$.

(9C-1) <u>Lemma</u>: Let $A \subseteq \mathcal{R}$, $A \in \Pi^1_1$ and assume $\{\varphi_n\}_{n\in\omega}$ is a Π^1_1-scale on A. Then the tree T associated with $\{\varphi_n\}_{n\in\omega}$ is Σ^1_2 in the codes. (Kechris).

<u>Proof</u>: We have

$$(k_0,\alpha_0,\dots,k_n,\alpha_n) \in \underline{\text{Code}}(T) \Longleftrightarrow \alpha_0,\dots,\alpha_n \in WO$$

$$\& (\exists\alpha)(\alpha \in A \ \& \ \varphi_0(\alpha) = |\alpha_0| \ \& \ \alpha(0) = k_0$$

$$\& \cdots \& \ \varphi_n(\alpha) = |\alpha_n| \ \& \ \alpha(n) = k_n).$$

The result follows immediately if we can show that for each n,

$$\alpha \in A \ \& \ \beta \in WO \ \& \ \varphi_n(\alpha) = |\beta|$$

is a Σ^1_2 relation in α, β, n. But each φ_n is a Π^1_1-norm, thus every initial segment of $\leq^{\varphi_n} = \leq_n$ will have countable length (since $\underset{\sim}{\delta}^1_1 = \aleph_1$). From this we have

$$\alpha \in A \ \& \ \beta \in WO \ \& \ \varphi_n(\alpha) = |\beta| \Longleftrightarrow \alpha \in A \ \& \ \beta \in WO$$

$$\& \ \exists\ \delta[(\forall m)(m \leq_\beta m \Rightarrow (\delta)_m \leq_n \alpha)$$

$$\& \ (\forall\gamma)(\gamma \leq_n \alpha \Rightarrow (\exists\ m)(m \leq_\beta m \ \& \ (\delta)_m \leq_n \alpha \ \& \ \alpha \leq_n (\delta)_m)$$

$$\& \forall m, \ell\ (m <_\beta \ell \Longleftrightarrow (\delta)_m <_n (\delta)_\ell)]$$

(where of course $\delta <_n \varepsilon \Longleftrightarrow \varphi_n(\delta) < \varphi_n(\varepsilon)$) and the proof is complete.

(9C-2) Theorem: Suppose $A \subseteq \aleph_1$ and Code(A) is Σ^1_2. Then A is in L. Similarly if T is a tree on $\omega \times \aleph_1$ and Code(A) is Σ^1_2, then $T \in L$. (For $A \subseteq \aleph_1$ with $\text{Code}(A) \in \Pi^1_1$ the result is implicit in methods (using forcing) of Solovay. The game method used below as a substitute of forcing was used by Moschovakis to prove a version of the corollary below and traces to [12]. The present version of the theorem is due to Kechris.)

Corollary: $T^1 \in L$ (Moschovakis).

Proof of the theorem: We give the proof for $A \subseteq \aleph_1$, the case of a tree being similar.

Thus let $A \subseteq \aleph_1$ and $\text{Code}(A) = P \in \Sigma^1_2$. Then

$$\xi \in A \Longleftrightarrow (\exists\ \alpha)(\alpha \in P \ \& \ |\alpha| = \xi).$$

Let $\alpha \in P \Longleftrightarrow \exists\beta\ Q(\alpha,\beta) \Longleftrightarrow \exists\beta(f(\alpha,\beta) \in WO)$, where $Q \in \Pi^1_1$ and $f : \mathcal{R} \times \mathcal{R} \to \mathcal{R}$ is recursive and for all α, β, $f(\alpha,\beta) \in LOR$. Then

$$\xi \in A \Longleftrightarrow \exists\alpha\,\exists\beta(f(\alpha,\beta) \in WO \ \& \ |\alpha| = \xi).$$

Consider the following game G_ξ:

I			II		
ξ_0	a_0	b_0	η_0	θ_0	k_0
ξ_1	a_1	b_1	η_1	θ_1	k_1
$\vdots$	$\vdots$	$\vdots$	$\vdots$	$\vdots$	$\vdots$
	α	β			

I and II play as in the diagram natural numbers and ordinals $< \aleph_1$ and II wins iff <u>for every</u> n,

<u>either</u> for some $i \leq n$, $\xi_i \geq \xi$ <u>or</u> all the following are true:

(a) The mapping $i \to \eta_i$ $(i \leq n)$ is order preserving on the part of $\leq_{f(\alpha,\beta)}$ already determined by $((a_0,\ldots,a_n),(b_0,\ldots,b_n))$ (notice that f is continuous).

(b) The mapping $i \to \theta_i$ $(i \leq n)$ is order preserving on the part of $\leq_\alpha$ already determined by $(a_0,\ldots,a_n)$ and $\theta_i < \xi$, for each $i \leq n$.

If $k_i \leq n$, then $\theta_{k_i} = \xi_i$ and if $\langle k_i,k_i \rangle = j \leq n$, then $a_j = 0$.

Notice now the following: <u>For</u> $\xi < \aleph_1$, $\xi \in A \Longleftrightarrow$ II <u>has a winning strategy in</u> G_ξ.

<u>Proof</u>: Assume $\xi \in A$; let α, β be such that $f(\alpha,\beta) \in WO$ and $|\alpha| = \xi$ and let $i \to \eta_i$ be an order preserving map on $\leq_{f(\alpha,\beta)}$ into $\aleph_1$ and $i \to \theta_i$ a mapping from ω into ξ such that its restriction to Field$(\leq_\alpha)$ is an order preserving bijection onto ξ, with inverse g. Consider the following strategy for II in G_ξ and verify easily that it is winning: If I plays ξ_0, II plays $\alpha(0),\beta(0),\eta_0,\theta_0$, $g(\xi_0)$ (unless $\xi_0 \geq \xi$ in which case II plays anything). If I plays ξ_1, II gives $\alpha(1),\beta(1),\eta_1,\theta_1,g(\xi_1)$ etc.

Conversely assume II has a winning strategy. Let I play $\xi_0,\xi_1,\ldots$ enumerating without repetitions ξ i.e. $\xi = \{\xi_0,\xi_1,\xi_2,\ldots\}$. Then II plays by his winning strategy and produces $\alpha,\beta,(\eta_0,\eta_1,\ldots),(\theta_0,\theta_1,\ldots),(k_0,k_1,\ldots)$, such that $i \to \eta_i$ is order preserving on $\leq_{f(\alpha,\beta)}$, thus $f(\alpha,\beta) \in WO$, $i \to \theta_i$ is order preserving on $\leq_\alpha$ into ξ, thus $\alpha \in WO$, and finally $\xi_i \to k_i$ is an inverse to $i \to \theta_i$ on Field$(\leq_\alpha)$, thus $|\alpha| = \xi$ and the proof is complete.

It is clear now that $G_\xi = G_{\mathfrak{S}_\xi}$ where $\mathfrak{S}_\xi$ is a set of finite sequences and moreover the map $\xi \to \mathfrak{S}_\xi$ is absolute for L. Thus <u>for</u> $\xi < \aleph_1$,

$$\xi \in A \Longleftrightarrow \text{II has a winning strategy in } G_{\mathfrak{S}_\xi}$$

$$\Longleftrightarrow L \models \text{II has a winning strategy in } G_{\mathfrak{S}_\xi}.$$

So A is definable in L, therefore $A \in L$. Notice that the definition of A involves as the only parameter $\aleph_1$, thus $A = \tau^L(\aleph_1)$ for some term τ. $\dashv$

The following converse to (9C-2) was proved by Kechris:

$$\forall\alpha(\alpha^{\#} \text{ exists}) \Rightarrow \text{Every } A \in L,\ A \subseteq \aleph_1 \text{ has } \text{Code}(A) \in \undertilde{\Sigma}^1_2.$$

The proof (as also the proof of (9C-2)) relativizes to any real and thus gives the elegant characterization:

If $\tilde{L} = \bigcup_{\alpha\in\mathbb{R}} L[\alpha]$ and $\forall\alpha[\alpha^{\#}$ exists], then

$$A \subseteq \aleph_1 \Rightarrow [A \in \tilde{L} \Longleftrightarrow \text{Code}(A) \in \undertilde{\Sigma}^1_2].$$

Since T^{2n+1} is a tree definable by some formula of set theory, it makes sense to talk of the tree $(T^{2n+1})^M$ for any model M of ZF + DC + Determinacy($\undertilde{\Delta}^1_{2n}$). In particular, for the models M^n introduced in §5, we have (by (5-2) and (9A-3)) that

$$M = L[(T)^M]$$

if $M = M^{2n+1}$ or $M = M^{2n+2}$ and $T = T^{2n+1}$. Thus the models of §5 are ordinary relative constructibility models, but we do not have an independent characterization of the trees $(T^{2n+1})^M$ for these M's.

Finally, we should memtion that in contrast to $L = L[T^1]$, Kechris has noticed that

$$M^{2n} \subsetneqq L[T^{2n-1}] \qquad n \geq 2;$$

this is because $M^{2n} \subseteq L[\alpha]$ for some α.

10. Lebesgue measurability and the property of Baire. We prove here Solovay's results that $\undertilde{\Sigma}^1_2$ sets are Lebesgue measurable and have the property of Baire, if $\forall\alpha[\aleph_1^{L[\alpha]} < \aleph_1]$. These will appear as corollaries of more general results about approximations of sets which admit definable scales, but we do not know any other applications of these general theorems. The forcing-free proofs given here are an adaptation by Moschovakis of the original Solovay proofs which used the forcing method.

10A. Sets which are ∞-Boolean over a model of ZFC. We define a set of codes $\mathcal{C}(\kappa)$ and for each $a \in \mathcal{C}(\kappa)$ a set B_a in the Boolean algebra $\mathcal{B}(\kappa)$, by the induction:

1. $(1, \langle n_0,\ldots,n_k\rangle) \in \mathcal{C}(\kappa)$ and

$$B_{(1,\langle n_0,\ldots,n_k\rangle)} = \{\alpha : \bar{\alpha}(k+1) = \langle n_0,\ldots,n_k\rangle\},$$

2. If $a \in \mathcal{C}(\varkappa)$, then $(2,a) \in \mathcal{C}(\varkappa)$ and

$$B_{(2,a)} = \mathcal{R} - B_a.$$

3. If $f : \xi \to \mathcal{C}(\varkappa)$ is a function with domain some $\xi < \varkappa$ and values in $\mathcal{C}(\varkappa)$, then $(3,f) \in \mathcal{C}(\varkappa)$ and

$$B_{(3,f)} = \bigcup_{\eta<\xi} B_{f(\eta)}.$$

Clearly, using AC,

$$\mathcal{B}(\varkappa) = \{B_a : a \in \mathcal{C}(\varkappa)\}.$$

Moreover the relation "$a \in \mathcal{C}(\varkappa)$" is definable by a formula which is absolute for models of ZFC.

In particular the set $\mathcal{C}(\omega + 1)$ gives canonical codes to the algebra $\mathcal{B}(\omega + 1)$ of Borel sets.

Suppose $\mathfrak{M}$ is a (transitive, containing all ordinals) model of ZFC. A set of reals A is $\varkappa$-Boolean over $\mathfrak{M}$, if

$$A = B_a \text{ for some } a \in \mathcal{C}(\varkappa) \cap \mathfrak{M}.$$

A is ∞-Boolean over $\mathfrak{M}$ if it is $\varkappa$-Boolean over $\mathfrak{M}$ for some $\varkappa$. A is $\omega + 1$-Boolean over $\mathfrak{M}$ if it is "Borel rational over $\mathfrak{M}$" in Solovay's terminology, [19].

We would expect that sets which are ∞-Boolean over "thin," definable models have some regularity properties. But first let us show how we can get such sets.

(10A-1) Theorem: Let T be a tree on $\omega \times \varkappa$ and $A = p[T] = \{\alpha : T(\alpha)$ is not wellfounded$\}$. Then A is $\varkappa^+ + 1$-Boolean over $L[T]$.

Proof: Immediate from Theorem (8A-1). ⊣

Corollary: If $A \in \Sigma^1_2$, then A is $\aleph_1 + 1$-Boolean over L.

Remark: We know already that a Σ^1_2 set is the union of $\aleph_1$ Borel sets, a seemingly better representation than the above. Its only disadvantage is that we cannot have $A = \bigcup_{\xi<\aleph_1} B_{f(\xi)}$, where $f \in L$, i.e. we cannot control this representation from inside L, unless $\aleph_1^L = \aleph_1$.

10B. Ideals of Borel sets which are suitable over a model of ZFC. We isolate in a definition the key properties of the ideals of sets of measure 0 and sets of the first category that we need.

Let $\mathfrak{M}$ be a fixed model of ZFC, $\mathfrak{J}$ a collection of Borel sets. We call $\mathfrak{J}$ suitable over $\mathfrak{M}$ if the following conditions hold:

<u>1</u>. $\mathfrak{J} \neq \emptyset$ and $\mathfrak{J}$ is closed under subsets and countable unions (i.e. $\mathfrak{J}$ is a σ-<u>ideal</u> in the Boolean algebra of Borel sets).

<u>2</u>. There is a definable functor

$$G : \mathfrak{M} \to \mathfrak{M}$$

(i.e. an operation on $\mathfrak{M}$ definable by a formula, perhaps with parameters from $\mathfrak{M}$) such that if

$$f : \xi \to C(\omega + 1), \qquad f \in \mathfrak{M}$$

is a function in $\mathfrak{M}$ which maps some ordinal ξ in the Borel codes of $\mathfrak{M}$, then

$$G(f) = \{\eta_i\}_{i \in \omega}$$

gives a countable sequence of ordinals less than ξ such that

$$\text{for all } \eta < \xi, \ B_{f(\eta)} - \bigcup_i B_{f(\eta_i)} \in \mathfrak{J}.$$

This last condition says in effect that the Boolean algebra $\mathcal{B}(\omega + 1)/\mathfrak{J}$ has the <u>countable chain condition</u> for wellordered sequences of Borel sets in $\mathfrak{M}$. It is the key property that we need.

If $\mathfrak{J}$ is the ideal of Borel sets of measure 0 (where we can take the measure on $\mathcal{R}$ as coming from the measure on the true reals via the standard topological identification of $\mathcal{R}$ with the irrationals <u>or</u> we can take only the subsets of $^{\omega}2$ with the product measure) and if $\mathfrak{J}$ is the ideal of Borel sets of the first category, then $\mathfrak{J}$ is suitable over every $\mathfrak{M}$ <u>which admits a definable</u> (with parameters from $\mathfrak{M}$) <u>wellordering</u>, in particular any $L[X]$. This is not hard to prove from standard analytic and topological facts about these ideals, e.g. see [19], [4].

Let $\mathfrak{J}$ be suitable over $\mathfrak{M}$ and put

$$\underline{\mathrm{Alg}}(\mathfrak{M}, \mathfrak{J}) = \{\alpha : \underline{\text{for some}} \ a \in C(\omega + 1) \cap \mathfrak{M}, \ B_a \in \mathfrak{J} \ \underline{\text{and}} \ \alpha \in B_a\}.$$

The reals in <u>Alg</u>($\mathfrak{M}, \mathfrak{J}$) are called <u>$\mathfrak{J}$-algebraic over</u> $\mathfrak{M}$. If we think of sets in $\mathfrak{J}$ as <u>small</u> sets, they are the reals which can be <u>approximated</u> in $\mathfrak{M}$, in the sense that they belong to a small set with code in $\mathfrak{M}$. Let

$$\underline{\mathrm{Trans}}(\mathfrak{M}, \mathfrak{J}) = \{\alpha : \alpha \notin \underline{\mathrm{Alg}}(\mathfrak{M}, \mathfrak{J})\}$$

be the set of reals <u>$\mathfrak{J}$-transcendental</u> over $\mathfrak{M}$. In the case of sets of measure 0 or of the first category these are Solovay's <u>random</u> and <u>Cohen generic</u> (over $\mathfrak{M}$) reals, respectively. This is the key notion needed for the statement and proof of the approximation theorem we want.

(10B-1) Theorem: Let $\mathfrak{M}$ be a model of ZFC. Let $\mathfrak{J}$ be a σ-ideal of Borel sets which is suitable over $\mathfrak{M}$. Then for every set A which is ∞-Boolean over $\mathfrak{M}$ there is a Borel set A^*, rational over $\mathfrak{M}$, such that

$$A \triangle A^* = (A - A^*) \cup (A^* - A) \subseteq \underline{\mathrm{Alg}}(\mathfrak{M}, \mathfrak{J}).$$

If in addition $|\mathfrak{M} \cap \mathfrak{R}| < \aleph_1$, then

$$A \triangle A^* \in \mathfrak{J}.$$

Proof: Let A be $\varkappa$-Boolean over $\mathfrak{M}$. We assign by induction in $\mathfrak{M}$ on the codes $C(\varkappa) \cap \mathfrak{M}$, to each $a \in C(\varkappa) \cap \mathfrak{M}$ a code $a^* \in C(\omega + 1) \cap \mathfrak{M}$ such that $B_a \triangle B_{a^*} \subseteq \underline{\mathrm{Alg}}(\mathfrak{M}, \mathfrak{J})$ i.e. for any $\alpha \in \underline{\mathrm{Trans}}(\mathfrak{M}, \mathfrak{J})$

(*) $$\alpha \in B_a \Longleftrightarrow \alpha \in B_{a^*}.$$

Case 1: $a = (1, \langle n_0, \ldots, n_k \rangle)$. Put $a^* = a$.

Case 2: $a = (2,b)$. Put $a^* = (2,b^*)$.

Case 3: $a = (3,f)$. Put $f^*(\xi) = (f(\xi))^*$. Then let $\{\eta_i\}_{i\in\omega} = G(f^*)$ and $f_1(i) = f^*(\eta_i)$. Clearly $f_1 : \omega \to C(\omega + 1) \cap \mathfrak{M}$ and we define

$$a^* = (3,f_1).$$

The verification that (*) holds is trivial, except for the third case. Thus let $a = (3,f) \in \mathfrak{M}$, where $f : \xi \to C(\varkappa) \cap \mathfrak{M}$, and by induction hypothesis assume that for $\alpha \in \underline{\mathrm{Trans}}(\mathfrak{M}, \mathfrak{J})$

$$\alpha \in B_{f(\eta)} \Longleftrightarrow \alpha \in B_{(f(\eta))^*} \Longleftrightarrow \alpha \in B_{f^*(\eta)}.$$

We have to show that for $\alpha \in \underline{\mathrm{Trans}}(\mathfrak{M}, \mathfrak{J})$,

$$\alpha \in B_a \Longleftrightarrow \alpha \in B_{(3,f_1)}.$$

Let $\alpha \in \underline{\mathrm{Trans}}(\mathfrak{M}, \mathfrak{J})$ and $\alpha \in B_a$. Then $\alpha \in B_{f(\eta)}$ for some $\eta > \xi$. Thus $\alpha \in B_{f^*(\eta)}$. Then by 2 above $\alpha \notin B_{f^*(\eta)} - \bigcup_i B_{f^*(\eta_i)}$, so $\alpha \in \bigcup_i B_{f^*(\eta_i)} = B_{(3,f_1)}$.

Conversely assume $\alpha \in \underline{\mathrm{Trans}}(\mathfrak{M}, \mathfrak{J})$ and $\alpha \in B_{(3,f_1)}$. Then $\alpha \in \bigcup_i B_{f^*(\eta_i)}$, so for some i, $\alpha \in B_{f^*(\eta_i)}$, thus $\alpha \in B_{f(\eta_i)}$ and $\alpha \in \bigcup_{\eta<\xi} B_{f(\xi)} = B_a$. ⊣

Corollary: (a) Assume $\forall\alpha(\aleph_1^{L[\alpha]} < \aleph_1)$. Then every $\underset{\sim}{\Sigma}^1_2$ set is Lebesgue Measurable and has the property of Baire. (Solovay, unpublished).

(b) Assume Determinacy($\underset{\sim}{\Delta}^1_{2n}$) and that for each α, $|\mathfrak{R} \cap L[T^{2n+1},\alpha]| = \aleph_0$.

Then every $\underset{\sim}{\Sigma}^1_{2n+2}$ set is Lebesgue Measurable and has the property of Baire. (Solovay, unpublished).

It should be pointed out that the implication

PD $\Rightarrow$ all projective sets are Lebesgue measurable and have the property of Baire

has been known for some time, Mycielski-Swierczkowski [17]. The proof of (b) above seems to use "less determinacy" (none if $n = 0$) and has a different flavor.

<u>11. Perfect subsets of pointsets.</u> The main results here are that if PD holds and for each n, $|L[T^{2n+1}] \cap \mathbb{R}| = \aleph_0$, then every uncountable projective set has a perfect subset and there exist largest countable Σ^1_{2n+2} sets. The first result does not need the hypothesis $|L[T^{2n+1}] \cap \mathbb{R}| = \aleph_0$, see [3], [16], but the proof given here (due to Solovay and Mansfield) uses "less determinacy," none if $n = 0$.

11A. <u>The theorem on perfect sets.</u> Solovay in [20] proved that, if $\forall\alpha[\aleph_1^{L[\alpha]} < \aleph_1]$, then every uncountable $\underset{\sim}{\Sigma}^1_2$ set contains a perfect subset. His method was one of the earliest applications of forcing to the proof of positive results. A few months later Mansfield obtained a similar theorem (see [6]) and in [7] he generalized the result to (11A-1) below. His proof also used forcing. Finally Solovay obtained a new forcing-free proof of Mansfield's result. This is essentially the proof reproduced below with one alteration: Solovay's "inductive analysis" was replaced by the notion of "derivation on a tree"; as a result the proof becomes astonishingly similar to Cantor's proof of the Cantor-Bendixson theorem.

(11A-1) <u>Theorem</u>: Assume T is a tree on $\omega \times \kappa$ and $A = p[T]$. Then if A contains an element not in $L[T]$, A contains a perfect set. (Mansfield, [7].)

<u>Proof</u>: For any tree T on $\omega \times \kappa$ we define the <u>derivative</u> T' of T as follows: $((k_0,\xi_0),\ldots,(k_n,\xi_n)) \in T' \Leftrightarrow$ There are two, incompatible in the first coordinate, extensions of $((k_0,\xi_0),\ldots,(k_n,\xi_n))$, both in T i.e. we can find $((k'_0,\xi'_0),\ldots,(k'_m,\xi'_m)),((k''_0,\xi''_0),\ldots,(k''_\ell,\xi''_\ell)) \in T$ extending $((k_0,\xi_0),\ldots,(k_n,\xi_n))$, such that $(k'_0,\ldots,k'_m)$ is incompatible with $(k''_0,\ldots,k''_\ell)$. Notice that T' is a tree and $T' \subseteq T$.

Then we define a la Cantor the ξth-<u>derivative</u> of T by

$$T^0 = T$$

$$T^{\xi+1} = (T^\xi)'$$

$$T^\lambda = \bigcap_{\xi<\lambda} T^\xi, \quad \text{if } \lambda = \bigcup\lambda > 0.$$

It is then clear that $\xi \to T^\xi$ is a function absolute for any model containing T, in particular for $L[T]$. Moreover $T^0 \supseteq T^1 \supseteq T^2 \supseteq \cdots \supseteq T^\xi \supseteq T^{\xi+1} \supset \cdots$; thus let ξ_T be the least ξ such that $T^\xi = T^{\xi+1}$.

Case 1. $T^{\xi_T} = \emptyset$. Then consider $\alpha \in A$. Since $A = p[T]$, we can find f such that $(\alpha, f) \in [T]$. Since $(\alpha, f) \notin [T^{\xi_T}] = \emptyset$, let $\xi < \xi_T$ be such that $(\alpha, f) \in [T^\xi] - [T^{\xi+1}]$. Let n be the least integer such that $((\alpha(0),f(0)),\ldots,(\alpha(n),f(n))) \notin T^{\xi+1}$. Since $((\alpha(0),f(0)),\ldots,(\alpha(n),f(n))) \in T^\xi$ it is clear that all branches of T^ξ extending $((\alpha(0),f(0)),\ldots,(\alpha(n),f(n)))$ have the same real part, namely α, therefore $p[T^\xi_{((\alpha(0),f(0)),\ldots,(\alpha(n),f(n)))}] = \{\beta\}$ where $\alpha = (\alpha(0),\ldots,\alpha(n)) \frown \beta$. But then clearly $\beta \in L[T]$, since β is definable absolutely from elements of $L[T]$; thus $\alpha \in L[T]$. So in this case $A \subseteq L[T]$.

Case 2. $T^{\xi_T} \neq \emptyset$. Then $T^{\xi_T} = (T^{\xi_T})' \neq \emptyset$ i.e. every sequence in T^{ξ_T} has two extensions in T^{ξ_T} which are incompatible in the first coordinate. Then it is easy to show that $p[T^{\xi_T}]$ ($\subseteq p[T] = A$) contains a perfect set. $\dashv$

Corollary: (a) Every Σ^1_2 set with an element not in L contains a perfect set; thus, $\forall\alpha[\aleph_1^{L[\alpha]} < \aleph_1] \Rightarrow$ Every uncountable $\underset{\sim}{\Sigma}^1_2$ set contains a perfect subset. (Solovay, [20].)

(b) Assume Determinacy($\underset{\sim}{\Delta}^1_{2n}$). Then every Σ^1_{2n+2} set with an element not in $L[T^{2n+1}]$ contains a perfect set. Thus if $|L[T^{2n+1},\alpha] \cap \mathcal{R}| = \aleph_0$, for all $\alpha \in \mathcal{R}$, every uncountable $\underset{\sim}{\Sigma}^1_{2n+2}$ set contains a perfect subset.

11B. Largest countable Σ^1_{2n} sets. The following is also a corollary of Theorem (11A-1).

(11B-1) Theorem: Assume $\aleph_1^L < \aleph_1$. Then there exists a largest countable Σ^1_2 set of reals, namely $\{\alpha : \alpha \in L\}$. (Solovay, [20].)

The next result extends (11B-1) to higher levels.

(11B-2) Theorem: Assume Determinacy($\underset{\sim}{\Delta}^1_{2n}$). If $|\mathcal{R} \cap L[T^{2n+1}]| = \aleph_0$, then there exists a largest countable Σ^1_{2n+2} set. (Kechris-Moschovakis).

Proof: Notice first that Uniformization(Π^1_{2n+1}) implies that for every countable Σ^1_{2n+2} set A we can find a countable Π^1_{2n+1} set B, so that every real in A is recursive in some real in B. Thus it will be enough to find a countable Σ^1_{2n+2} set C which contains all countable Π^1_{2n+1} sets. Then $C^* = \{\alpha : (\exists\beta)(\beta \in C \ \& \ \alpha \text{ is recursive in } \beta)\}$ is the largest countable Σ^1_{2n+2} set.

It will be convenient for this proof to choose a particular Π^1_{2n+1}-complete set $\mathcal{P}_{2n+1}$ and a Π^1_{2n+1}-scale on it as follows: Let $W_{2n+1} \subseteq \omega \times \mathcal{R}$ be universal for

Π^1_{2n+1} subsets of $\mathcal{R}$ and put

$$\alpha \in \mathcal{P}_{2n+1} \Longleftrightarrow (\alpha(0),\alpha') \in W_{2n+1},$$

where $\alpha' = (\alpha(1),\alpha(2),\ldots)$. Let also $\{\varphi_n\}_{n\in\omega}$ be a Π^1_{2n+1}-scale on $\mathcal{P}_{2n+1}$. Let T^{2n+1} be the tree associated with this scale.

We define now C and then we show that it works:

$$\alpha \in C \Longleftrightarrow \exists m \exists \xi [\,|\{m \frown \beta \in \mathcal{P}_{2n+1} : \varphi_0(m \frown \beta) \leq \xi\}| \leq \aleph_0 \;\&\; \varphi_0(m \frown \alpha) \leq \xi].$$

1. $C \in \Sigma^1_{2n+2}$.

Proof: Notice that

$$\alpha \in C \Longleftrightarrow \exists m \exists \beta [\beta \in \mathcal{P}_{2n+1} \;\&\; \varphi_0(m \frown \alpha) \leq \varphi_0(\beta)$$

$$\& \;\exists \gamma \forall \delta [\varphi_0(m \frown \delta) \leq \varphi_0(\beta) \Rightarrow \exists k(\delta = (\gamma)_k)]].$$

2. C <u>contains every countable</u> Π^1_{2n+1} <u>set</u>.

Proof: Let $B \in \Pi^1_{2n+1}$, $B \subseteq \mathcal{R}$, $|B| \leq \aleph_0$. Find m such that $\beta \in B \Longleftrightarrow (m,\beta) \in W_{2n+1} \Longleftrightarrow m \frown \beta \in \mathcal{P}_{2n+1}$. If $B \not\subseteq C$, let $\beta_0 \in B - C$. Put $\xi = \varphi_0(m \frown \beta)$. Then since $\beta_0 \notin C$, $|\{m \frown \beta \in \mathcal{P}_{2n+1} : \varphi_0(m \frown \beta) \leq \xi\}| > \aleph_0$; but $B \supseteq \{m \frown \beta \in \mathcal{P}_{2n+1} : \varphi_0(m \frown \beta) \leq \xi\}$, a contradiction.

3. $C \subseteq L[T^{2n+1}]$; <u>thus</u> $|C| = \aleph_0$.

Proof: It is enough to show that if for some m, ξ $|\{m \frown \beta \in \mathcal{P}_{2n+1} : \varphi_0(m \frown \beta) \leq \xi\}| \leq \aleph_0$, then $\{m \frown \beta \in \mathcal{P}_{2n+1} : \varphi_0(m \frown \beta) \leq \xi\} \subseteq L[T^{2n+1}]$. Put

$$(T^{2n+1})_{m,\xi} = \{((k_0,\xi_0),\ldots,(k_\ell,\xi_\ell)) \in T^{2n+1} : k_0 = m \;\&\; \xi_0 \leq \xi\}.$$

Clearly $(T^{2n+1})_{m,\xi} \in L[T^{2n+1}]$ and the limit property of scales shows that

$$\alpha \in p[(T^{2n+1})_{m,\xi}] \Longleftrightarrow \alpha \in \mathcal{P}_{2n+1} \;\&\; \varphi_0(\alpha) \leq \xi \;\&\; \alpha(0) = m.$$

Thus $\{m \frown \beta \in \mathcal{P}_{2n+1} : \varphi_0(m \frown \beta) \leq \xi\} = p[(T^{2n+1})_{m,\xi}]$, so by (11A-1)

$$|\{m \frown \beta \in \mathcal{P}_{2n+1} : \varphi_0(m \frown \beta) \leq \xi\}| \leq \aleph_0$$

$$\Rightarrow \{m \frown \beta \in \mathcal{P}_{2n+1} : \varphi_0(m \frown \beta) \leq \xi\} \subseteq L[(T^{2n+1})_{m,\xi}] \subseteq L[T]. \qquad \dashv$$

Open problem.[3] It is a well known result, that every countable Σ^1_1 set contains only Δ^1_1 reals. Thus there is no largest countable Σ^1_1 set (since $\{\alpha : \alpha \in \Delta^1_1\} \in \Pi^1_1 - \Sigma^1_1$). Does any of these results generalize to Σ^1_{2n+1} $(n \geq 1)$, under any reasonable hypotheses?

12. A summary of results about projective ordinals. We give here a list of theorems about the projective ordinals. Proofs are omitted but many results follow from what we have already done.

1. $\underset{\sim}{\delta}^1_1 = \aleph_1$ (classical)

2. (a) $\underset{\sim}{\delta}^1_2 \leq \aleph_2$ (Martin, unpublished)

   $\forall \alpha(\alpha^{\#}$ exists$) \Rightarrow \underset{\sim}{\delta}^1_2 \leq u_2$ (Martin, unpublished)

   (b) $\forall \alpha(\alpha^{\#}$ exists$) \Rightarrow \underset{\sim}{\delta}^1_2 \geq u_2$ (Kechris, Martin, unpublished). Thus $\forall \alpha(\alpha^{\#}$ exists$) \Rightarrow \underset{\sim}{\delta}^1_2 = u_2$.

   (c) $\forall \alpha(\alpha^{\#}$ exists$) \Rightarrow \underset{\sim}{\delta}^1_n = u_{\underset{\sim}{\delta}^1_n}$, $n \geq 3$ (Kechris), where $u_1, u_2, \ldots, u_\xi, \ldots$ is the increasing enumeration of the uniform indiscernibles.

3. $\forall \alpha(\alpha^{\#}$ exists$) + AC \Rightarrow \underset{\sim}{\delta}^1_3 \leq \aleph_3$ (Martin, [10]).

4. $PD \Rightarrow \underset{\sim}{\delta}^1_{2n+2} \leq (\underset{\sim}{\delta}^1_{2n+1})^+$ (Kunen, Martin, unpublished).

5. Determinacy$(\underset{\sim}{\Delta}^1_2) + AC \Rightarrow \underset{\sim}{\delta}^1_4 \leq \aleph_4$ (Kunen, Martin, unpublished).

6. (a) $PD \rightarrow \underset{\sim}{\delta}^1_{2n+1} < \underset{\sim}{\delta}^1_{2n+2}$ (Moschovakis, [13]).

   (b) $PD \rightarrow \underset{\sim}{\delta}^1_{2n} < \underset{\sim}{\delta}^1_{2n+1}$ (Kechris).

7. $PD \rightarrow$ Every Π^1_{2n+1}-norm on a universal Π^1_{2n+1}-set has length $\underset{\sim}{\delta}^1_{2n+1}$ (Moschovakis, [13]).

If we now assume full determinacy (AD), the results have a different flavor.

---

[3] Kechris (The theory of countable analytical sets, Trans. Amer. Math. Soc., 202 (1975), 259-297) has proved from PD that for each $n \geq 1$ there is no largest countable Σ^1_{2n+1} set. Martin (Countable Σ^1_{2n+1} sets, circulated note, 1973) then showed from PD that every countable Σ^1_{2n+1} set of reals contains only Δ^1_{2n+1} reals (Moschovakis has earlier shown this result for countable Δ^1_{2n+1} sets).

Assume AD; then:

1'. (a) $\underset{\sim}{\delta}^1_n$ is a cardinal and for n odd regular (Moschovakis, [13]).

(b) $\underset{\sim}{\delta}^1_n$ is regular, for n even (Kunen, unpublished).

2'. $\underset{\sim}{\delta}^1_2 = \aleph_2$ $(= u_2)$ (Martin).

3'. (a) $u_\omega = \aleph_\omega$; cofinality$(\aleph_n) = \aleph_2$ $(n \geq 2)$ (Martin, [10]).

(b) $u_n = \aleph_n$ (Kunen, Solovay, unpublished).

4'. $\underset{\sim}{\delta}^1_3 = \aleph_{\omega+1}$ (Martin, [10]).

5'. (a) $\underset{\sim}{\delta}^1_{2n+2} = (\underset{\sim}{\delta}^1_{2n+1})^+$ (Kunen, Martin, unpublished).

(b) $\underset{\sim}{\delta}^1_4 = \aleph_{\omega+2}$ (Kunen, Martin, unpublished).

6'. $\underset{\sim}{\delta}^1_{2n+1} = (\lambda_n)^+$, where λ_n is a cardinal and cofinality$(\lambda_n) = \omega$; thus $\underset{\sim}{\delta}^1_{2n+1} \geq \aleph_{\omega n+1}$ (Kechris).

7'. (a) $\aleph_1$ $(= \underset{\sim}{\delta}^1_1)$, $\aleph_2$ $(= \underset{\sim}{\delta}^1_2)$ are measurable (Solovay; for $\aleph_1$ see [21], for $\aleph_2$ unpublished).

(b) $\underset{\sim}{\delta}^1_{2n+1}$ is measurable (Martin, unpublished).

(c) $\underset{\sim}{\delta}^1_{2n}$ is measurable (Kunen, unpublished).

References

[1] J. W. Addison, Some consequences of the axiom of constructibility, Fund. Math. **46** (1959a), 123-135.

[2] J. W. Addison and Yiannis N. Moschovakis, Some consequences of the axiom of definable determinateness, Proc. Nat. Acad. Sci. USA, **59** (1968), 708-712.

[3] Morton Davis, Infinite games of perfect information, Advances in game theory, Ann. of Math. Study No. 52, 1964, 85-101.

[4] K. Kuratowski, Topology v. 1, Academic Press, New York & London, 1966.

[5] A. Levy, Definability in axiomatic set theory: I, in Proc. 1964 International Congress for Logic, Methodology and Philosophy of Science, Amsterdam, 1966.

[6] R. Mansfield, The theory of Σ^1_2 sets, doctoral dissertation, Stanford University, 1969.

[7] __________, Perfect subsets of definable sets of real numbers, Pacific Journal of Mathematics, (2) **35** (1970), 451-457.

[8] __________, A Souslin operation on Π^1_2, Israel Journal of Mathematics, (3) **9** (1971), 367-379.

[9] D. A. Martin, The axiom of determinateness and reduction principles in the analytical hierarchy, Bull. Amer. Math. Soc. 74 (1968), 687-689.

[10] __________, Pleasant and unpleasant consequences of determinateness, unpublished manuscript circulated in March 1970.

[11] D. A. Martin and R. M. Solovay, A basis theorem for Σ^1_3 sets of reals, Ann. of Math., 89 (1969), 138-160.

[12] Yiannis N. Moschovakis, The Suslin-Kleene theorem for countable structures, Duke Math. Journal, (2) 37 (June 1970), 341-352.

[13] __________, Determinacy and prewellorderingsof the continuum, Math Logic and Foundations of Set Theory, Edited by Y. Bar Hsillel, North Holland, Amsterdam-London, 1970, 24-62.

[14] __________, Uniformization in a playful universe, Bull. Amer. Math. Soc., to appear.

[15] __________, Descriptive set theory, a foundational approach, in preparation.

[16] J. Mycielski, On the axiom of determinateness, Fund. Math., 53 (1964), 205-224.

[17] J. Mycielski and S. Swierczkowski, On the Lebesgue measurability and the axiom of determinateness, Fund. Math., 54 (1964), 67-71.

[18] Jack H. Silver, Measurable cardinals and Δ^1_3 wellorderings, to appear.

[19] R. M. Solovay, A model of set theory in which every set is Lebesgue measurable, Ann. of Math., 92 (1970), 1-56.

[20] __________, On the cardinality of Σ^1_2 sets of reals, Foundations of Mathematics, Symposium papers commemorating the 60th birthday of Kurt Gödel, Springer-Verlag, 1966, 58-73.

[21] R. M. Solovay, Measurable cardinals and the axiom of determinateness, Lecture notes prepared in connection with the Summer Institute on Axiomatic Set Theory held at UCLA, Summer 1967.

August 1971

University of California, Los Angeles

Postscript. While this paper was being typed, we received a preprint from Martin titled "Projective sets and cardinal numbers: some questions related to the continuum problem." This appears to contain most of the results of Martin that we have listed as "unpublished" or credited them to [10].[4)]

---

4) Martin's paper will appear in the Journal of Symbolic Logic.

PARTIALLY PLAYFUL UNIVERSES

Howard Becker

Department of Mathematics
University of California
Los Angeles, CA 90024

If Γ is a pointclass, $\mathrm{Det}(\Gamma)$ means that every subset of $\mathbb{R}$ (= ${}^{\omega}\omega$ = "the reals") in Γ is determined. (See [13] for definitions.) The pointclasses we are interested in are usually the analytical classes, Σ^1_n, Π^1_n, Δ^1_n, the projective classes, $\underset{\sim}{\Sigma}^1_n$, $\underset{\sim}{\Pi}^1_n$, $\underset{\sim}{\Delta}^1_n$ (see [13] or [16, Chapter 7.8]), or (Power set of $\mathbb{R}$) $\cap L[\mathbb{R}]$. Projective determinacy, PD, is the hypothesis that all projective sets are determined.

$\mathrm{Det}(\underset{\sim}{\Delta}^1_1)$ is a theorem of ZFC (Martin [12]) and $\mathrm{Det}(\underset{\sim}{\Sigma}^1_1)$ is a theorem of ZFC + $\exists$ a Ramsey cardinal (Martin [10]), or even of ZFC + $\forall\alpha$ ($\alpha^{\#}$ exists). Beyond this point, determinacy is a new axiom, and a very strong one. There has been a lot of work done in recent years on the consequences of the axiom of projective determinacy, particularly in descriptive set theory. Much of the classical theory of Π^1_1 and Σ^1_2 has been generalized to the higher analytical pointclasses [14].

The recent interest in the consequences of PD and $\mathrm{Det}(\underset{\sim}{\Delta}^1_n)$ has led to an interest in metamathematical questions related to these axioms, and hence in models of them. A playful universe is a transitive model of ZF + PD. A partially playful universe is a transitive model of $\mathrm{ZF} + \mathrm{Det}(\underset{\sim}{\Delta}^1_n)$, for some $n \geq 2$. The topic of this paper is certain specific examples of partially playful universes.

This is a survey article. Very few of the results presented here are my own. In fact, most of them are due to Kechris and Moschovakis, together or separately. The main theorem proved at the end of this paper is due to Harrington and Kechris (unpublished).

I wish to thank Professors Kechris and Moschovakis for their help while I was working on this paper.

1. Preliminaries

A knowledge of the theory of scales is necessary in order to read this paper. Most of the results that are needed can be found in the first three sections of [8]. Sections 5 and 9 of that paper are also the source of some of the material in this paper; the theorems given there without proof are proved here. The reader is also assumed to be familiar with models of set theory, and in particular, with L.

Our notation and terminology is that of [8] and of [13].

We will always assume Dependent Choice, DC (see [8]). We do not assume the full axiom of choice, AC. All theorems stated in this paper are theorems of ZF + DC.

Any additional assumptions, including determinacy, will be explicitly stated in the hypothesis of the theorem.

We will frequently state without proof, and use, results from descriptive set theory. These are all theorems of ZF + DC. We give no references for many of these results; that is because they are "well known" theorems which have never been published. Fortunately, all these results will soon appear in print in [14].

The first thing that one must recognize in studying partially playful universes, is that we cannot prove that any exist, in ZFC. Even assuming the existence of measurable cardinals, it is not possible to prove the consistency of $\mathrm{Det}(\underset{\sim}{\Delta}^1_2)$. For Solovay [18] has shown that assuming $\mathrm{Det}(\underset{\sim}{\Delta}^1_2)$, there are models of set theory in which there exists a measurable cardinal. (A stronger result of this type is due to Green [2], who proved that $\mathrm{Det}(\underset{\sim}{\Delta}^1_2)$ implies that there is an inner model of the following theory:

$\mathrm{ZF} + \mathrm{AC} + \exists \kappa$ (κ a measurable cardinal and $\{\lambda < \kappa : \lambda$ a measurable cardinal$\}$ has normal measure 1)).

So we will be assuming that a certain amount of determinacy is true in the real world. Then assuming determinacy in V, we can construct models of determinacy. This is similar to what is done with L; if ZF is assumed true in the real world, then we can prove that $L \models \mathrm{ZF}$.

There are three basic types of partially playful universes that will be examined here. All three were originally introduced by Moschovakis. All the models will be similar to L. The models will all be of the form "the smallest transitive model of ZF + DC containing all ordinals and $P(\Sigma^1_n)$", where P is some property of the pointclass Σ^1_n. For small n ($n = 1$ or $n = 2$, depending on the type), this smallest model will indeed be L. For larger n it is not. Hence these models are analogs of L for the higher analytical pointclasses.

<u>Definition</u>. A transitive model M of ZF is <u>Σ^1_n-correct</u> if for every Σ^1_n formula $\varphi(x_1,\ldots,x_k)$ and any reals $\alpha_1,\ldots,\alpha_k$ in M, $\varphi(\alpha_1,\ldots,\alpha_k) \Leftrightarrow M \models \varphi(\alpha_1,\ldots,\alpha_k)$.

<u>Lemma 1</u>. Assume $\mathrm{Det}(\underset{\sim}{\Delta}^1_{n-1})$. If M is a Σ^1_n-correct model, then $M \models \mathrm{Det}(\underset{\sim}{\Delta}^1_{n-1})$.

<u>Proof</u>. Let S be a $\underset{\sim}{\Delta}^1_{n-1}$ set in M. Then there is an $\alpha \in M$ and there are Π^1_{n-1} formulas $\varphi(x,\alpha)$ and $\psi(x,\alpha)$ such that for $x \in M$, $x \in S \Leftrightarrow M \models \varphi(x,\alpha) \Leftrightarrow M \models \neg\psi(x,\alpha)$.

Strategies can be identified with reals, since they are functions from finite sequences of integers to integers. If σ is a strategy for I and τ is a strategy for II, $\sigma * \tau$ denotes the outcome of the game where I plays according to the strategy σ and II plays according to τ. The function $(\sigma,\tau) \mapsto \sigma * \tau$ is clearly recursive.

$M \models \forall x\ (\varphi(x,\alpha) \leftrightarrow \neg\ \psi(x,\alpha))$, so by Σ^1_n-correctness, $\forall x\ (\varphi(x,\alpha) \leftrightarrow \neg\ \psi(x,\alpha))$ is true. So these formulas define a $\underset{\sim}{\Delta}^1_{n-1}$ set S', in V. By hypothesis, either I or II has a winning strategy for S'. If I wins S', then $\exists\sigma\ \forall\tau\ \varphi(\sigma*\tau,\alpha)$. Since φ is Π^1_{n-1}, this is Σ^1_n, hence by Σ^1_n-correctness, $M \models \exists\sigma\forall\tau\ \varphi(\sigma*\tau,\alpha)$. So $M \models$ I has a winning strategy for S. On the other hand, if II wins S', then $\exists\tau\forall\sigma\ \psi\ (\sigma*\tau,\alpha)$, by the same argument this is true in M, and so $M \models$ II has a winning strategy for S. ⊣

2. The smallest Σ^1_n-correct model

We will construct the smallest Σ^1_n-correct model containing all ordinals. But first, a short digression into descriptive set theory is necessary.

The most important application of scales is in proving that determinacy implies uniformization. The following theorem (see [8], Section 3) is a fairly weak version of what can be proved:

Theorem 2. (Moschovakis; Novikoff-Kondo-Addison for $n \leq 2$). Let $n \geq 1$. Let k be the greatest even integer $< n$. Then $\mathrm{Det}(\underset{\sim}{\Delta}^1_k)$ implies that every Π^1_{n-1} set can be uniformized by a Π^1_n set.

This theorem, as stated, is not quite good enough for working with models of set theory. For suppose M is a Σ^1_n-correct model. Let $\varphi(x,y)$ be a Π^1_{n-1}-formula, hence absolute for M. The theorem, applied inside M, says that there is a Π^1_n formula $\varphi^*(x,y)$ such that $M \models \varphi^*$ uniformizes φ. But there is no guarantee that this φ^* uniformizes φ in the real world; that may require a different Π^1_n formula.

For this reason, a more absolute version of Theorem 2 is needed. Theorem 3, below, will do. It is, of course, a very unnatural way to state the theorem if one is interested only in descriptive set theory, rather than in models.

Theorem 3. Let $n \geq 1$. Let k be the greatest even integer $< n$. For every Π^1_{n-1} formula $\varphi_{n-1}(x,y)$, there is a Π^1_n formula $\varphi^*_{n-1}(x,y)$ such that

$$[ZF + DC + \mathrm{Det}(\underset{\sim}{\Delta}^1_k)] \vdash (\varphi^*_{n-1} \text{ uniformizes } \varphi_{n-1}).$$

The way to prove Theorem 3 is to go through Moschovakis' proof of Theorem 2, and observe that everything still works.

Definition. Let $\varphi_{n-1}(m,\alpha,\beta)$ be a fixed Π^1_{n-1} formula defining a Π^1_{n-1} universal relation (that is, every Π^1_{n-1} relation on $\mathbb{R}^2$ is equal to $\{(\alpha,\beta) : \varphi_{n-1}(m_0,\alpha,\beta)\}$ for some $m_0 \in \omega$). Let $\varphi^*_{n-1}(m,\alpha,\beta)$ be the Π^1_n uniformizing formula given by Theorem 3. Let

$$F^*_n(m,\alpha) = \begin{cases} \text{the unique } \beta \text{ such that } \varphi^*_{n-1}(m,\alpha,\beta), & \text{if } \exists\beta\varphi_{n-1}(m,\alpha,\beta) \\ 0 & , \text{if } \forall\beta\neg\varphi_{n-1}(m,\alpha,\beta). \end{cases}$$

F_n^* is essentially the function that uniformizes every $\underset{\sim}{\Pi}^1_{n-1}$ set. The graph of F_n^* is Π^1_n.

Lemma 4. (Moschovakis). Let $n \geq 2$. Let k be the greatest even integer $< n$. Assume $\mathrm{Det}(\underset{\sim}{\Delta}^1_k)$.

1. If M is a transitive model of ZF and M is closed under F_n^*, then M is Σ^1_n-correct.

2. If M is a transitive model of ZF + DC and M is Σ^1_n-correct, then M is closed under F_n^*.

Proof. 1. Assume M is closed under F_n^*. It is proved by induction on $i \leq n$, that M is Σ^1_i-correct. For $i = 1$ this is the well known fact that Σ^1_1 formulas are absolute for all transitive models of ZF. Assume true for $i - 1$. Without loss of generality, we may assume, by coding sequences, that all formulas contain at most one parameter from M. So consider the Π^1_{i-1} formula $\psi(\alpha,x)$, where $\alpha \in M$. It must be shown that if $\exists\beta\psi(\alpha,\beta)$, then $M \models \exists\beta\psi(\alpha,\beta)$.

Since $\varphi_{n-1}(m,\alpha,\beta)$ is Π^1_{n-1} universal and ψ is Π^1_{i-1}, $i \leq n$, there is a fixed m_0 such that $\psi(\alpha,\beta) \Longleftrightarrow \varphi_{n-1}(m_0,\alpha,\beta)$. $\exists\beta\psi(\alpha,\beta) \Rightarrow \exists\beta\varphi_{n-1}(m_0,\alpha,\beta) \Rightarrow \varphi_{n-1}(m_0,\alpha,\beta')$, where $\beta' = F_n^*(m_0,\alpha)$, $\Rightarrow \psi(\alpha,\beta')$. Since M is closed under F_n^*, $\beta' \in M$. $\alpha,\beta' \in M$, ψ is Π^1_{i-1}, so by the induction hypothesis absolute, hence $M \models \psi(\alpha,\beta')$.

2. Assume M is Σ^1_n-correct. By Lemma 1, $M \models \mathrm{Det}(\underset{\sim}{\Delta}^1_k)$. Let $\alpha \in M$, and suppose that for some m, $\exists\beta\varphi_{n-1}(m,\alpha,\beta)$. Since $M \models \mathrm{Det}(\underset{\sim}{\Delta}^1_k)$ and $M \models \mathrm{DC}$, by Theorem 3, $M \models \varphi^*_{n-1}(m,\alpha,\beta)$, for some $\beta \in M$. φ^*_{n-1} is Π^1_n, hence absolute, so $\varphi^*_{n-1}(m,\alpha,\beta)$ is true. So $\beta = F_n^*(m,\alpha)$, and $F_n^*(m,\alpha) \in M$. ⊣

In light of Lemma 4, it is clear how to define an L-like model which is the smallest Σ^1_n-correct model containing all ordinals.

Definition. Let $\mathcal{D}(X)$ = all subsets of X which are first order definable in the structure $\langle X, \in \upharpoonright X\rangle$ with parameters from X.

Definition. $M^n_0 = \emptyset$

$M^n_{\xi+1} = M^n_\xi \cup \mathcal{D}(M^n_\xi) \cup \{F_n^*(m,\alpha) : m \in \omega,\ \alpha \in \mathbb{R} \cap M^n_\xi\}$

$M^n_\lambda = \bigcup_{\xi<\lambda} M^n_\xi$, if λ is a limit ordinal

$M^n = \bigcup_{\xi \in \mathrm{Ord}} M^n_\xi$.

Let $\leq_\xi$ be a well ordering of M^n_ξ. Then $\{F_n^*(m,\alpha) : m \in \omega,\ \alpha \in \mathbb{R} \cap M^n_\xi\}$ is well ordered by the formula

$$\beta_1 \leq \beta_2 \Longleftrightarrow (\exists m)(\exists\alpha_1 \in M^n_\xi)[\beta_1 = F_n^*(m,\alpha_1) \,\&\, (\forall k < m)(\forall\alpha_2 \in M^n_\xi)(\beta_2 \neq F_n^*(k,\alpha_2)) \,\&\, (\forall\alpha_2 \in M^n_\xi)((\alpha_2 \leq_\xi \alpha_1 \,\&\, \alpha_2 \neq \alpha_1) \rightarrow \beta_2 \neq F_n^*(m,\alpha_2))].$$

Using this fact, it is possible to define a well ordering $\leq_{M^n}$, of M^n, analogous to the canonical well ordering of L. Since the details are tedious and unenlightening, we omit the formal definition.

<u>Theorem</u> 5. (Moschovakis). Let $n \geq 2$. Let k be the greatest even integer $< n$. Assume $\mathrm{Det}(\underset{\sim}{\Delta}^1_k)$.

1. M^n is a transitive model of ZF. It is closed under F^*_n and it is Σ^1_n-correct.
2. The map $\xi \mapsto M^n_\xi$ and the relation $x \in M^n_\xi$ are definable by formulas which are absolute for all Σ^1_n-correct models of ZF which are closed under F^*_n, and therefore for all Σ^1_n-correct models of ZF + DC.
3. $M^n \models \forall x \exists \xi (x \in M^n_\xi)$ (i.e. $M^n \models V = M^n$).
4. $M^n \models AC$.
5. M^n is the smallest transitive model of ZF + DC which contains all the ordinals and which is Σ^1_n-correct.
6. $M^n \models \mathrm{Det}(\underset{\sim}{\Delta}^1_k)$. If $\mathrm{Det}(\underset{\sim}{\Delta}^1_{n-1})$, then $M^n \models \mathrm{Det}(\underset{\sim}{\Delta}^1_{n-1})$.
7. $M^n \models \mathrm{Scale}(\Pi^1_i)$ and $\mathrm{Unif}(\Pi^1_i)$ for i odd, $i \leq n$,
   $M^n \models \mathrm{Scale}(\Sigma^1_i)$ and $\mathrm{Unif}(\Sigma^1_i)$ for i even, $i \leq n+1$.
8. $M^n \models GCH$.
9. $\mathbb{R} \cap M^n$ is Σ^1_{n+1} and $(\leq_{M^n} \restriction \mathbb{R})$ is a Σ^1_{n+1} well ordering and $M^n \models (\leq_{M^n} \restriction \mathbb{R})$ is Σ^1_{n+1}-good.
10. $M^n \models \mathrm{Scale}(\Sigma^1_i)$ and $\mathrm{Unif}(\Sigma^1_i)$, for $i \geq n+1$.

<u>Remark</u>. Scale (Γ) means every set in Γ admits a Γ-scale. Unif(Γ) means every relation in Γ can be uniformized by a relation in Γ. A well ordering is Γ-good if the initial segments can be coded in a Δ manner. (The L well ordering is Σ^1_2-good.) For details, see [8].

<u>Outline of proof</u>. 1. That M^n is a transitive model of ZF is proved in the same manner as that L is a transitive model of ZF. We will not bore the reader with the details. It is obviously closed under F^*_n, and so by Lemma 4, is Σ^1_n-correct.

2. The graph of F^*_n is Π^1_n. Hence the function F^*_n is absolute for Σ^1_n-correct models which are closed under F^*_n. With this additional fact, the proof used for the analogous theorem for L will work for M^n. The absoluteness for Σ^1_n-correct models of DC follows from Lemma 4.

3. follows from 1 and 2.

4. $\leq_{M^n}$ is a definable well ordering of the universe. The proof is similar to that for $\leq_L$.

5. follows from 2.

6. Lemma 1.

7. $\mathrm{Det}(\underset{\sim}{\Delta}^1_k)$ implies the scale and uniformization properties (see [8]).

8. To prove that $M^n \models GCH$, it will suffice to show that for every cardinal (of M^n) κ, and every $S \in M^n$, if $S \subset \kappa$ then $S \in M^n_\xi$ for some $\xi < (\kappa^+)^{M^n}$. We work in M^n. Hence AC is true.

Let Θ be a sufficiently large finite subset of $ZF + DC + \mathrm{Det}(\underset{\sim}{\Delta}^1_k)$. (In particular, Θ implies part 2, i.e. the map $\xi \mapsto M^n_\xi$ is absolute for Σ^1_n-correct models of Θ.) Since $\Theta + V = M^n$ is true (in M^n), by the reflection principle, there is a transitive set M' containing $S, \mathbb{R}$, and κ such that $M' \models \Theta + V = M^n$. By the downward Löwenheim-Skolem theorem, there is an elementary submodel M'' of M' such that $\kappa \subset M''$, $S \in M''$, and the cardinality of M'' is κ. Since $M'' \prec M'$, it is well founded. Let M_T be the transitive collapse of M''.

$M_T \equiv M'' \prec M'$, so $M_T \models \Theta + V = M^n$. Since the reals can effectively be identified with subsets of ω, they are all fixed by the collapsing map. So $\mathbb{R} \cap M_T = \mathbb{R} \cap M'' \prec \mathbb{R} \cap M' = \mathbb{R}$. Hence M_T is Σ^1_n-correct. By part 2 of this theorem, the absoluteness of the map $\xi \mapsto M^n_\xi$ for Σ^1_n-correct models of Θ, M_T must actually be an M^n_ξ, for some ξ.

Since the cardinality of M_T is κ, $M_T = M^n_\xi$ for some $\xi < \kappa^+$. $S \in M''$, $S \subset \kappa \subset M''$, so the transitive closure of S is in M''. Therefore the collapsing map fixes S. Hence $S \in M_T = M^n_\xi$.

9. Consider countable models of the form $N_E = \langle \omega, E \rangle$, where E is a binary relation on ω. N_E is called adequate if

a) $N_E \models (\Theta + V = M^n)$.

b) E is well founded.

c) $(\forall i)(i \in \omega^{N_E} \leftrightarrow \exists m\ (i = 2^m))$.

d) $(\forall m)(\forall k)(k < m \leftrightarrow 2^k\ E\ 2^m)$.

Truth in N_E involves quantification only over ω, so is Δ^1_1. Therefore, clearly $\{E \subset \omega \times \omega : N_E \text{ is adequate}\}$ is Π^1_1.

(N_E, a) represents α adequately if

a) N_E is adequate.

b) $a \in \omega$ & $N_E \models a \in \mathbb{R}$.

c) $(\forall m)(\forall k)(\alpha(m) = k \leftrightarrow N_E \models a(2^m) = 2^k)$.

This is also Π^1_1.

An adequate model is closed under F_n^*

iff $(\forall a)[(N_E \models a \in \mathbb{R}) \rightarrow$
$(\exists b)(\exists\alpha)(\exists\beta)((N_E,a)$ represents α adequately &
(N_E,b) represents β adequately &
$F_n^*(\alpha) = \beta)]$

iff $(\forall a)(\forall\alpha)(\forall\beta)[((N_E,a)$ represents α adequately &
$F_n^*(\alpha) = \beta) \rightarrow (\exists b)((N_E,b)$ represents β adequately$)]$.

Since the graph of F_n^* is Π_n^1, this is Δ_{n+1}^1.

$\alpha \in M^n$ iff

$(\exists E \subset \omega \times \omega)(\exists a)[(N_E,a)$ represents α adequately &
N_E is closed under $F_n^*]$

and

$\alpha \leq_{M^n} \beta$ iff
$(\exists E \subset \omega \times \omega)(\exists a)(\exists b)[(N_E,a)$ represents α adequately &
(N_E,b) represents β adequately & N_E is closed under F_n^* &
$N_E \models (a \leq_{M^n} b)]$.

This is so since N_E is Σ_n^1-correct (Lemma 4), and any Σ_n^1-correct model of $\Theta + V = M^n$ is an M_ξ^n, by the reasoning used in the proof of part 8. Thus the ordering $(\leq_{M^n})^{N_E}$ is the real $\leq_{M^n}$. So $\mathbb{R} \cap M^n$ and $(\leq_{M^n} \restriction \mathbb{R})$ are Σ_{n+1}^1. Similarly, assuming $V = M^n$,

$$\{\beta : \beta \leq_{M^n} \alpha\} = \{(\gamma)_m : m \in \omega\}$$

iff

$(\exists E \subset \omega\times\omega)(\exists a)(\exists c)[(N_E,a)$ represents α adequately &
(N_E,c) represents γ adequately & N_E is closed under F_n^* &
$N_E \models (\{\beta : \beta \leq_{M^n} a\} = \{(c)_m : m \in \omega\})]$

iff

$(\forall E \subset \omega\times\omega)(\forall a)(\forall c)[((N_E,a)$ represents α adequately &
(N_E,c) represents γ adequately & N_E is closed under $F_n^*) \rightarrow$
$N_E \models (\{\beta : \beta \leq_{M^n} a\} = \{(c)_m : m \in \omega\})]$.

So the coding of initial segments is Δ_{n+1}^1 and the well ordering is Σ_{n+1}^1-good.

10. The existence of a good Σ_{n+1}^1 well ordering of $\mathbb{R}$ implies the Scale and Uniformization properties. (See [8].) ⊣

M^2 is the smallest Σ_2^1-correct model containing all ordinals. Therefore, $M^2 = L$. This also follows directly from the definition of M^2; for Unif(Π_1^1) is a theorem of ZF + DC, hence true in L, so φ_1^* uniformizes in L, and range$(F_2^*) \subset L$.

Thus the M^n's are actually analogs of L for the higher analytical pointclasses, as promised.

It is a theorem of ZF + DC that Π^1_1 and Σ^1_2 have the scale property, and that for any n, Σ^1_n and Π^1_n cannot both have the scale property. For $n \geq 3$, the question of whether the scale property belongs to the Π class, the Σ class, or neither cannot be settled in ZFC. In Diagrams 1 and 2, below, the circled pointclasses are the classes which have the scale property, assuming V = L and PD, respectively.

Diagram 1

$V = L\ (= M^2)$

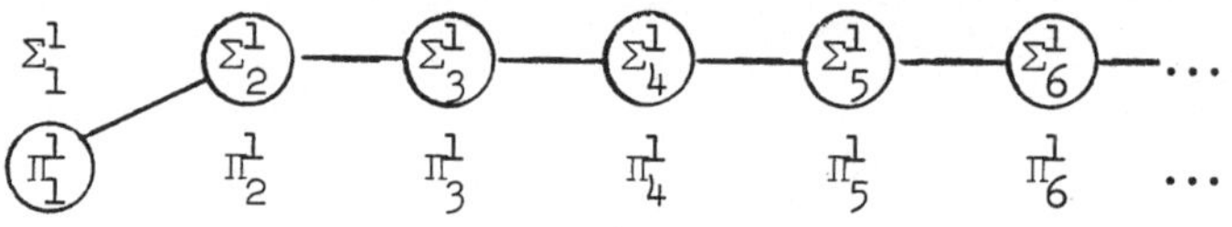

Diagram 2

PD

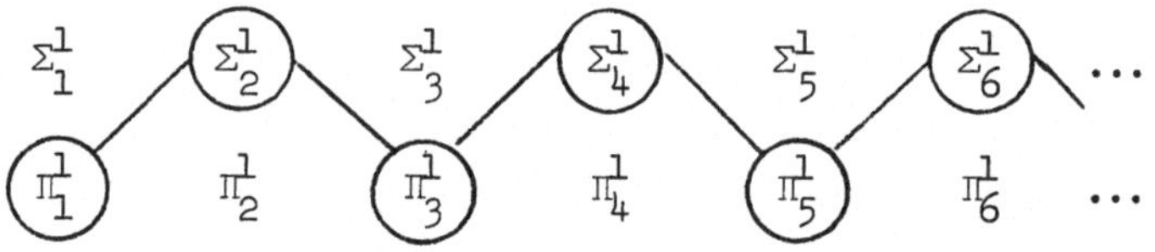

In the partially playful universes, the scale property follows the PD pattern up to a point, and then takes on the L pattern. The classes that have the scale property are shown in Diagrams 3 and 4 below, assuming $V = M^3$, $V = M^4$, $V = M^5$, and $V = M^6$. The general situation should be clear from the first five cases. The "periodicity of order 2" seen in the changing scale diagrams, is a phenomenon that occurs quite frequently in descriptive set theory (assuming determinacy).

Diagram 3

$V = M^3$
or
$V = M^4$

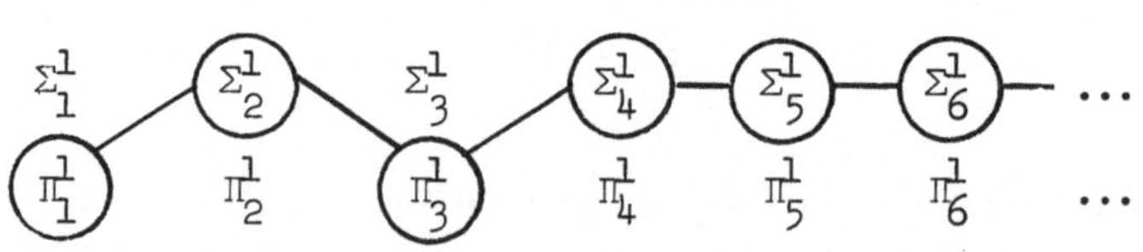

Diagram 4

$V = M^5$
or
$V = M^6$

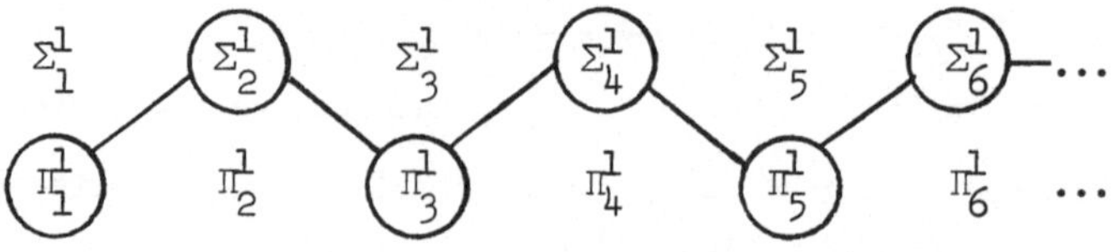

The pointclass with the scale property obviously also has the prewellordering property (see [8] for definitions and details), and this property, too, cannot be satisfied by both Π^1_n and Σ^1_n. Thus these two properties are completely determined by the $\mathrm{Det}(\underset{\sim}{\Delta}^1_k) + V = M^n$ axioms. These properties imply most of the structure theorems of descriptive set theory. The regularity properties (Lebesgue measurability, etc.) are implied by determinacy, and their negations are implied by the existence of a good well ordering (as is the case for L, with its Σ^1_2-good well ordering). Hence the axiom $\mathrm{Det}(\underset{\sim}{\Delta}^1_k) + V = M^n$ answers almost all the usual questions in descriptive set theory.

There are a few improvements that can be made in Theorem 5.

<u>Theorem</u> 6. (Martin [11]). For all even k, $\mathrm{Det}(\underset{\sim}{\Delta}^1_k) \Rightarrow \mathrm{Det}(\underset{\sim}{\Sigma}^1_k)$.

Trivially, $\mathrm{Det}(\underset{\sim}{\Sigma}^1_k) \Longleftrightarrow \mathrm{Det}(\underset{\sim}{\Pi}^1_k)$. So by part 6 of Theorem 5, $M^n \models (\mathrm{Det}(\underset{\sim}{\Sigma}^1_k) + \mathrm{Det}(\underset{\sim}{\Pi}^1_k))$.

Part 9 of Theorem 5 says that $\mathbb{R} \cap M^n$ and $(\leq_{M^n} \restriction \mathbb{R})$ are Σ^1_{n+1}. For even n, we can do better than this. If n is even then there is a Σ^1_n well ordering of $\mathbb{R} \cap M^n$. We postpone the proof (see Theorem 28). Harrington has proved that if n is even, $\mathbb{R} \cap M^n$ is Σ^1_n, but we will not give a proof of that in this paper.

$M^n \subset M^{n+1}$, since M^n is the <u>smallest</u> Σ^1_n-correct model, so

$$M^2 \subsetneq M^3 \subset M^4 \subsetneq M^5 \subset M^6 \subsetneq \cdots \quad .$$

That $M^n \neq M^{n+1}$ for even n follows from the scale diagrams. For odd n, it is still true that $M^n \neq M^{n+1}$, but this requires a different proof. It was first proved by Kechris.

<u>Theorem</u> 7. Let $n \geq 2$. Assume $\mathrm{Det}(\underset{\sim}{\Sigma}^1_n)$. Then $(\exists \alpha \in \mathbb{R})(\alpha \in M^{n+1} \setminus M^n)$.

This will be proved later. (See the remarks after Corollary 10.) Note that in proving, say $\exists \alpha \in M^4 \setminus M^3$, the extra hypothesis $\mathrm{Det}(\underset{\sim}{\Sigma}^1_3)$ is needed; $\mathrm{Det}(\underset{\sim}{\Delta}^1_2)$ is clearly not sufficient since it is consistent with $V = M^3$ (Theorem 5).

The absoluteness properties of M^n, given in part 2 of Theorem 5, are applicable only to Σ^1_n-correct models. M^n, unlike L, is not absolute for arbitrary transitive models. In fact, for every $\alpha \in \mathbb{R}$, there is a transitive model of $V = M^4$ containing α. To see this, note that the set

$$P = \{\alpha : (\forall \text{countable well founded model } M \text{ of } \Theta + V = M^4)(\alpha \notin M)\}$$

is Π^1_2. If P were not empty it would contain a Π^1_3 singleton [8]. But it can't contain a Π^1_3 singleton, since they are all in the real M^4, and so by the reflection principle, are in the complement of P.

3. <u>Indiscernibles and sharps</u>

This section requires a knowledge of the theory of indiscernibles in L. See [1, Chapter 8].

There is an analog of $0^\#$ for M^n, $0^\#_n$, which is essentially the set of formulas true in M^n when an increasing sequence of uncountable cardinals is substituted for the variables. $0_2^\#$ is the ordinary $0^\#$. $0_2^\#$ is a Π_2^1 singleton. $0_n^\#$ will turn out to be a Π_n^1 singleton. The formal definition is given below. "Formulas of the language of set theory" means first order formulas in the language with $\in$ and $=$ and no parameters. We identify formulas with their Gödel numbers.

Definition. $0_n^\#$, if it exists, is defined to be the unique set F of (Gödel numbers of) formulas of the language of set theory such that

1. F is a maximal consistent set of formulas.

2. All axioms of $ZFC + Det(\underset{\sim}{\Delta}_k^1) + V = M^n$ are in F (where k is the greatest even integer $< n$).

3. For any term t, built up from the definable Skolem functions (definable via the M^n well ordering $\leq_{M^n}$), and any formula φ of the language, the following formulas are all in F:

i) $(V_i$ is an ordinal & $V_i < V_j)$, where $i < j$

ii) $t(V_1,\ldots,V_i) <_{M^n} V_j$, where $i < j$

iii) $[t(V_1,\ldots,V_m,V_{i_1},\ldots,V_{i_k}) \leq_{M^n} V_m \rightarrow$
$t(V_1,\ldots,V_m,V_{i_1},\ldots,V_{i_k}) = t(V_1,\ldots,V_m,V_{j_1},\ldots,V_{j_k})]$,
where $m < i_1 < \ldots < i_k$, $m < j_1 < \ldots < j_k$

iv) $[V_k <_{M^n} t(V_1,\ldots,V_{k-1},V_{k+1},\ldots,V_m) \rightarrow$
$V_{k+1} \leq_{M^n} t(V_1,\ldots,V_{k-1},V_{k+1},\ldots,V_m)]$

v) $[\varphi(V_{i_1},\ldots,V_{i_m}) \leftrightarrow \varphi(V_{j_1},\ldots,V_{j_m})]$,
where $i_1 < \ldots < i_m$, $j_1 < \ldots < j_m$

4. For any well ordering $<'$ of ω, the model M generated by the ordered set $\langle\omega,<'\rangle$ of indiscernbiles, whose increasing sequences satisfy the formulas in F, is well ordered by $(\leq_{M^n})^M$.

5. $(\forall m \in \omega)(\forall \alpha,\beta \in \mathbb{R})$ $(\forall$Skolem terms $t_1,t_2)$ [if the formula $(t_1 \in \mathbb{R}$ & $t_2 \in \mathbb{R}$ & $t_2 = F_n^*(m,t_1))$ is in F, and for all $j,k \in \omega$,

$(\alpha(j) = k \leftrightarrow$ the formula $t_1(j) = k$ is in $F)$ &

$(\beta(j) = k \leftrightarrow$ the formula $t_2(j) = k$ is in $F)$, then $\beta = F_n^*(m,\alpha)$].

The only differences between the above definition and the definition of $0^\#$ are the addition of clause 5, the substitution of $Det(\underset{\sim}{\Delta}_k^1) + V = M^n$ for $V = L$ in 2, and the use of $\leq_{M^n}$ rather than $\leq_L$. 5 says that the model is closed under F_n^* (assuming it's a well founded model to begin with), and hence by Lemma 4 is Σ_n^1-correct.

From this definition it is clear that $0_n^\#$ is a Π_n^1 singleton, hence a Δ_{n+1}^1 real, if it exists.

Theorem 8. Let $n \geq 2$. Let k be the greatest even integer $< n$. Assume $\mathrm{Det}(\underset{\sim}{\Delta}^1_k)$. The following are equivalent:

a) $0^{\#}_n$ exists.

b) There is a closed unbounded class of ordinals, $\mathcal{I}$, such that

i) $\mathcal{I}$ contains all uncountable cardinals

ii) The elements of $\mathcal{I}$ are indiscernibles in M^n (i.e. for any uncountable cardinal κ, the ordinals in $\{\xi \in \mathcal{I} : \xi < \kappa\}$ are indiscernibles in M^n_κ)

iii) Every element of M^n is definable from members of $\mathcal{I}$.

c) There is an uncountable cardinal κ such that the cardinals $\geq \kappa$ are indiscernibles in M^n.

The existence of a Ramsey cardinal implies that a), b), and c) are true. The proof of this fact, and of Theorem 8, is identical to the proof of the analogous theorems for L due to Silver and Solovay (essentially [1, Theorems 8.4.8 and 8.5.2]). The assumption $\mathrm{Det}(\underset{\sim}{\Delta}^1_k)$ is needed only to ensure that M^n exists.

There is a major difference between the M^n's $(n \geq 3)$ and L, with regard to the existence of sharps. There are models in which $\mathbb{R} \cap L$ is countable but $0^{\#}$ doesn't exist, for example the Cohen extension of a countable model of $V = L$ obtained by collapsing $\aleph_1$.

Theorem 9. Let $n \geq 3$. Let k be the greatest even integer $< n$. Assume $\mathrm{Det}(\underset{\sim}{\Delta}^1_k)$. If $\mathbb{R} \cap M^n$ is countable, then $0^{\#}_n$ exists.

Proof. Let α be a real which codes the countable set $S = \mathbb{R} \cap M^n$, i.e. $S = \{(\alpha)_0, (\alpha)_1, \ldots,\}$. M^n is the smallest transitive model of ZFC containing all ordinals and closed under the function F^*_n. Since $\mathrm{range}(F^*_n) \subset \mathbb{R}$, clearly $M^n \models V = L[\mathbb{R}]$.

$$\begin{aligned} M^n &= (L[\mathbb{R}])^{M^n} \\ &= L[\mathbb{R}^{M^n}] \\ &= L[\mathbb{R} \cap M^n] \\ &= L[S] \\ &= (L[S])^{L[\alpha]} \qquad (\text{since } \alpha \text{ codes } S,\ S \in L[\alpha]). \end{aligned}$$

Solovay has proved that $\mathrm{Det}(\underset{\sim}{\Delta}^1_2)$ implies that all sharps exist [18]. We have $\mathrm{Det}(\underset{\sim}{\Delta}^1_k)$, $k \geq 2$. So $\alpha^{\#}$ exists. Therefore the class of uncountable cardinals is indiscernible in $L[\alpha]$. M^n is definable in $L[\alpha]$ by a formula whose only parameter is α, i.e. the formula $x \in M^n$ iff

$$\exists S(S \subset \mathbb{R} \ \&\ \forall \beta(\beta \in S \leftrightarrow \exists m(\beta = (\alpha)_m) \ \&\ x \in L[S]).$$

Hence the uncountable cardinals must be indiscernibles in M^n. By Theorem 8, $0^{\#}_n$ exists. ⊣

Corollary 10. Let $n \geq 3$. Let k be the greatest even integer $< n+1$. Assume $\mathrm{Det}(\underset{\sim}{\Delta}^1_k)$. If $\mathbb{R} \cap M^n$ is countable, then $0^\#_n \in M^{n+1} \setminus M^n$.

Proof. $0^\#_n$ exists by Theorem 9. It is in M^{n+1} since it is a Π^1_n singleton and M^{n+1} is Σ^1_{n+1}-correct. It is not in M^n, since if it were it would still be a Π^1_n singleton in M^n, since M^n is Σ^1_n-correct; hence it would be definable in M^n and so truth in M^n would be definable in M^n, contradicting Tarski's Theorem. ⊣

Theorem 7 follows from this corollary. For even n it has already been proved. Let n be odd. Let S be the set of $\leq_{M^n}$-least codes for each ordinal less than $\aleph_1^{M^n}$. Since the M^n well ordering is Σ^1_{n+1}-good, S is Σ^1_{n+1}. Clearly S has cardinality equal to $\aleph_1^{M^n} = \beth_1^{M^n}$, and since closed sets of ordinal codes are bounded, S can contain no perfect subset. By uniformizing the Π^1_n set which S is a projection of (using $\mathrm{Det}(\underset{\sim}{\Delta}^1_k)$), we obtain a Π^1_n set of cardinality $\beth_1^{M^n}$ which has no perfect subset. $\mathrm{Det}(\Pi^1_n)$ implies that this set must be countable. So $\mathbb{R} \cap M^n$ is also countable.

Thus PD implies that $(\forall n \geq 2)(0^\#_n \in M^{n+1} \setminus M^n)$. So not only are all the M^n distinct, but M^{n+1} knows everything about M^n. The fact that if $0^\#_n$ exists it is a Π^1_n singleton and hence in $M^{n+1} \setminus M^n$, is due to Kechris.

There is another difference between M^n and L. Although M^n is Σ^1_n-correct, we cannot conclude that all models containing M^n are Σ^1_n-correct $(n \geq 3)$. This is not so. It is clear from the proof of Theorem 9, that M^n is contained in some $L[\alpha]$. $L[\alpha] \models \neg\mathrm{Det}(\underset{\sim}{\Delta}^1_2)$, hence by Lemma 1 is not Σ^1_3-correct.

4. Consistency proofs

The two consistency results given below are corollaries of Theorem 5. As in the case with L, the proofs of these results are finitary. The notation $\mathrm{Con}(T)$ means the theory T is consistent; this is of course a statement of number theory.

Corollary 11. (Moschovakis).

$\mathrm{Con}(ZF + DC + \mathrm{Det}(\underset{\sim}{\Delta}^1_{n-1})) \rightarrow \mathrm{Con}(ZF + AC + \mathrm{Det}(\underset{\sim}{\Delta}^1_{n-1}) + \neg\mathrm{Det}(\underset{\sim}{\Delta}^1_{n+1}))$.

Proof. $M^n \models \mathrm{Det}(\underset{\sim}{\Delta}^1_{n-1})$, by part 6 of Theorem 5. By part 9, M^n has a Σ^1_{n+1}-good well ordering of $\mathbb{R}$, hence a Δ^1_{n+1}-good well ordering and this contradicts $\mathrm{Det}(\underset{\sim}{\Delta}^1_{n+1})$. ⊣

Consequently, PD cannot be proved from $\mathrm{Det}(\underset{\sim}{\Delta}^1_n)$, for any n. A better result will be given later using a different model; for odd n, $\mathrm{Det}(\underset{\sim}{\Delta}^1_n)$ does not imply $\mathrm{Det}(\underset{\sim}{\Sigma}^1_n)$. (Corollary 23, and remarks following.) Martin has proved that for even $n > 0$, $\mathrm{Det}(\underset{\sim}{\Delta}^1_n)$ does not imply $\mathrm{Det}(\underset{\sim}{\Delta}^1_{n+1})$. The only known proofs of this type of consistency result all involve the use of scales and the Second Periodicity Theorem [8, Theorem 3C-1].

Of course we also get the analog of Gödel's consistency proof:

Corollary 12. (Moschovakis).

$$\mathrm{Con}(\mathrm{ZF} + \mathrm{DC} + \mathrm{Det}(\utilde{\Delta}^1_n)) \rightarrow \mathrm{Con}(\mathrm{ZFC} + \mathrm{Det}(\utilde{\Delta}^1_n) + \mathrm{GCH}).$$

An obvious question at this point is whether it is possible to prove the independence of the continuum hypothesis, CH, from determinacy, or to prove other analogs of classical independence results obtained by forcing. There are some obstacles to doing this. It is not known when $\mathrm{Det}(\utilde{\Delta}^1_n)$ is preserved under Cohen extensions. There is no known proof of $\mathrm{Con}(\mathrm{ZFC} + \mathrm{Det}(\utilde{\Delta}^1_2)) \rightarrow \mathrm{Con}(\mathrm{ZFC} + \mathrm{Det}(\utilde{\Delta}^1_2) + \neg \mathrm{CH})$, or of anything resembling it. But this may be provable. Other analogs of classical independence results are unprovable. A classical theorem due to Cohen is $\mathrm{Con}(\mathrm{ZFC}) \rightarrow \mathrm{Con}(\mathrm{ZFC} + \mathbb{R} \cap L$ is countable). The weakest analog of this would seem to be $\mathrm{Con}(\mathrm{ZFC} + \mathrm{Det}(\utilde{\Delta}^1_2)) \rightarrow \mathrm{Con}(\mathrm{ZFC} + \mathrm{Det}(\utilde{\Delta}^1_2) + \mathbb{R} \cap M^3$ is countable).

Theorem 13. If $(\mathrm{ZFC} + \mathrm{Det}(\utilde{\Delta}^1_2) + \mathbb{R} \cap M^3$ is countable) is consistent, then the following is not provable in $\mathrm{ZFC} + \mathrm{Det}(\utilde{\Delta}^1_2) + \mathbb{R} \cap M^3$ is countable:

(*) $\mathrm{Con}(\mathrm{ZFC} + \mathrm{Det}\ (\utilde{\Delta}^1_2)) \rightarrow \mathrm{Con}(\mathrm{ZFC} + \mathrm{Det}(\utilde{\Delta}^1_2) + \mathbb{R} \cap M^3$ is countable).

Proof. By Theorem 9, $(\mathrm{ZFC} + \mathrm{Det}(\utilde{\Delta}^1_2) + \mathbb{R} \cap M^3$ is countable) implies that $0^\#_3$ exists. The existence of $0^\#_3$ implies that $M^3_{\aleph_1} \models (\mathrm{ZFC} + \mathrm{Det}(\utilde{\Delta}^1_2))$. This follows from Theorem 8.b); it is the same proof as that used for the analogous theorem for $0^\#$ and $L_{\aleph_1}$. So

$$(\mathrm{ZFC} + \mathrm{Det}(\utilde{\Delta}^1_2) + \mathbb{R} \cap M^3 \text{ is countable}) \vdash \mathrm{Con}(\mathrm{ZFC} + \mathrm{Det}(\utilde{\Delta}^1_2)).$$

If (*) was provable in $\mathrm{ZFC} + \mathrm{Det}(\utilde{\Delta}^1_2) + \mathbb{R} \cap M^3$ is countable, so would be the statement $\mathrm{Con}(\mathrm{ZFC} + \mathrm{Det}(\utilde{\Delta}^1_2) + \mathbb{R} \cap M^3$ is countable). That is,

$$(\mathrm{ZFC} + \mathrm{Det}(\utilde{\Delta}^1_2) + \mathbb{R} \cap M^3 \text{ is countable}) \vdash$$
$$\mathrm{Con}(\mathrm{ZFC} + \mathrm{Det}(\utilde{\Delta}^1_2) + \mathbb{R} \cap M^3 \text{ is countable}).$$

By Gödel's second incompleteness theorem [16, page 213], the theory $(\mathrm{ZFC} + \mathrm{Det}(\utilde{\Delta}^1_2) + \mathbb{R} \cap M^3$ is countable) must be inconsistent. Contradiction. ⊣

Since (*) is unprovable in $\mathrm{ZFC} + \mathrm{Det}(\utilde{\Delta}^1_2) + \mathbb{R} \cap M^3$ is countable, it certainly has no finitary proof.

Another classical result is Solovay's Theorem [19]:

$\mathrm{Con}(\mathrm{ZFC} + \exists$ at least 1 inaccessible cardinal) $\rightarrow$

$\mathrm{Con}(\mathrm{ZFC}$ + every uncountable OD set of reals has a perfect subset),

where OD means ordinal definable using countable sequences of ordinals as parameters. The methods of Theorem 13 can be used to show that the following weak analog of Solovay's Theorem is unprovable:

$\mathrm{Con}(\mathrm{ZFC} + \mathrm{Det}(\underset{\sim}{\Delta}^1_2) + \exists$ arbitrarily large inaccessible cardinals$) \to$

$\mathrm{Con}(\mathrm{ZFC} + \mathrm{Det}(\underset{\sim}{\Delta}^1_2) +$ every uncountable Π^1_3 set of reals has a perfect subset$)$.

Thus some classical consistency results for ZFC $(= \mathrm{ZFC} + \mathrm{Det}(\underset{\sim}{\Delta}^1_1))$ cannot be generalized to $\mathrm{ZFC} + \mathrm{Det}(\underset{\sim}{\Delta}^1_2)$, essentially since they involve proving the consistency of a large cardinal axiom, the existence of $0^{\#}_3$. There is another type of consistency result which would be nice to have, but which is, for this reason, impossible. It is the following type:

$\mathrm{Con}(\mathrm{ZFC} + \mathrm{Det}(\underset{\sim}{\Delta}^1_2)) \to \mathrm{Con}(\mathrm{ZFC} + \mathrm{Det}(\underset{\sim}{\Delta}^1_4))$.

We will later present a method of constructing models of determinacy using forcing. In particular, we will get a model of $\mathrm{ZFC} + \mathrm{Det}(\underset{\sim}{\Delta}^1_n) + 2^{\aleph_0} = \aleph_2$ (Theorem 20). As is probably clear from the above, the proof that such models exist uses some strong set theoretic axioms. The M^n's are not the right models to start out with in these forcing proofs; the ground models will be another type of partially playful universe.

5. <u>A playful universe</u>

Assuming PD, we can also define a playful universe, M^ω, following Moschovakis.

<u>Definition</u>. $M^\omega_0 = \emptyset$.

$$M^\omega_{\xi+1} = M^\omega_\xi \cup \mathcal{D}(M^\omega_\xi) \cup [\bigcup_{n\in\omega}\{F^*_n(m,\alpha) : m \in \omega,\ \alpha \in \mathbb{R} \cap M^\omega_\xi\}]$$

$$M^\omega_\lambda = \bigcup_{\xi<\lambda} M^\omega_\xi, \quad \text{if } \lambda \text{ is a limit ordinal}$$

$$M^\omega = \bigcup_{\xi\in\mathrm{Ord}} M^\omega_\xi.$$

Since M^ω is closed under F^*_n for all n, it is Σ^1_n-correct, for all n (Lemma 4). Hence by Lemma 1, $M^\omega \models \mathrm{PD}$. This implies that the scale diagram for M^ω is Diagram 2, above. $\mathbb{R} \cap M^\omega$ is an elementary substructure of $\mathbb{R}$, so descriptive set theory in M^ω is practically indistinguishable from that in the real world (assuming PD).

PD implies that there is no projective well ordering of $\mathbb{R}$. However, $\mathbb{R} \cap M^\omega$ has a hyperprojective well ordering. The rest of Theorem 5 generalizes to M^ω. In particular, $M^\omega \models \mathrm{GCH}$, which gives the consistency result:

$$\mathrm{Con}(\mathrm{ZF} + \mathrm{DC} + \mathrm{PD}) \to \mathrm{Con}(\mathrm{ZFC} + \mathrm{PD} + \mathrm{GCH}).$$

6. <u>The smallest model containing the tree</u> T^n

Let $\{\varphi_n\}_{n\in\omega}$ be a scale on A. Thus for each $\alpha \in A$, the scale assigns to α a sequence of ordinals $\varphi_0(\alpha), \varphi_1(\alpha), \ldots$. α is a sequence of integers, $\alpha(0), \alpha(1), \ldots$. So to α there corresponds a pair of sequences. A pair of sequences can be identified with a sequence of pairs. So with each scale on A, there is a tree, whose members are finite initial segments of these sequences of pairs, and which is ordered by extension. Formally,

Definition. With each κ-scale $\{\varphi_n\}_{n\in\omega}$ on A, define the associated tree T on $w \times \kappa$ by

$$T = \{((\alpha(0),\varphi_0(\alpha)),(\alpha(1),\varphi_1(\alpha)),\ldots,(\alpha(n),\varphi_n(\alpha))) : \alpha \in A,\ n \in \omega\}.$$

For each $\beta \in \mathbb{R}$, define $T(\beta)$ to be

$$\{(\xi_0,\ldots,\xi_n) : ((\beta(0),\xi_0),\ldots,(\beta(n),\xi_n)) \in T\}.$$

$T(\beta)$ is a tree on κ. A branch of a tree is an infinite sequence such that every finite initial segment of it is in the tree.

$[T]$ = the body of the tree T

= the set of all branches of T.

T is well founded if $[T] = \emptyset$.

Lemma 14. Let A be a set of reals, $\{\varphi_n\}_{n\in\omega}$ a κ-scale on A, T the associated tree. Then

$$\alpha \in A \Longleftrightarrow T(\alpha) \text{ is not well founded}$$

$$\Longleftrightarrow \exists f\ ((\alpha,f) \in [T]).$$

Proof. If $\alpha \in A$, then $(\varphi_0(\alpha),\varphi_1(\alpha),\ldots)$ is a branch of $T(\alpha)$.

Conversely, suppose $(\xi_0,\xi_1,\ldots)$ is a branch of $T(\alpha)$. By definition of associated tree, there must exist reals $\alpha_0,\alpha_1,\ldots$ in A so that for each n $((\alpha_n(0),\varphi_0(\alpha_n)),\ldots,(\alpha_n(n),\varphi_n(\alpha_n))) = ((\alpha(0),\xi_0),\ldots,(\alpha(n),\xi_n))$. This implies that $\lim_{n\to\infty}\alpha_n = \alpha$ and for $m \leq n$, $\varphi_m(\alpha_n) = \xi_m$. By the definition of scale, [8, Section 3A], $\alpha \in A$. ⊣

This is a generalization of the classical characterization of Π^1_1 in terms of well foundedness. That characterization is what is behind the Shoenfield absoluteness theorem, that models containing all countable ordinals are Σ^1_2-correct. Using the trees associated with scales, this theorem can also be generalized.

Theorem 15. (Folklore). Let T be the tree associated with a κ-scale on a set A. Let M be a transitive model of ZF + DC such that $T \in M$. Let $Q \subset \mathbb{R}^2$ and let $f : \mathbb{R}^2 \to \mathbb{R}$ be a recursive function such that $Q(\alpha,\beta) \Longleftrightarrow f(\alpha,\beta) \in A$. Then there exists a tree S on $\omega \times \kappa$ such that $S \in M$ and

1. for all α,

$$\exists\beta Q(\alpha,\beta) \Longleftrightarrow S(\alpha) \text{ is not well founded}$$

2. for all $\alpha \in M$,

$$(\exists\beta \in M)Q(\alpha,\beta) \Longleftrightarrow S(\alpha) \text{ is not well founded.}$$

Proof.

1. $\exists\beta Q(\alpha,\beta) \iff \exists\beta(f(\alpha,\beta) \in A)$

$$\iff \exists\beta\exists\gamma(f(\alpha,\beta) = \gamma \ \&\ \gamma \in A)$$

$$\iff \exists\beta\exists\gamma[\forall n R(\bar{\alpha}(n),\bar{\beta}(n),\bar{\gamma}(n)) \ \&\ \gamma \in A],$$

where R is some recursive relation on $\omega\times\omega\times\omega$

(*) $$\iff \exists\beta\exists\gamma[\forall n R(\bar{\alpha}(n),\bar{\beta}(n),\bar{\gamma}(n)) \ \&\ T(\gamma) \text{ is not well founded}].$$

The last equivalence above follows from Lemma 14. Let S' be the set

$$\{((a_0,(b_0,c_0,\xi_0)),\ldots,(a_k,(b_k,c_k,\xi_k))) : ((c_0,\xi_0),\ldots,(c_k,\xi_k)) \in T \ \&$$
$$R(\langle a_0,\ldots,a_k\rangle,\langle b_0,\ldots,b_k\rangle,\langle c_0,\ldots,c_k\rangle)\}.$$

S' is a tree on $\omega\times(\omega\times\omega\times\kappa)$. Since $T \in M$, $S' \in M$. (*) says that $S'(\alpha)$ is not well founded. So

$$\exists\beta Q(\alpha,\beta) \iff S'(\alpha) \text{ is not well founded.}$$

Let g be a bijection from $\omega\times\omega\times\kappa$ onto κ, such that $g \in M$. Let S be the tree on $\omega\times\kappa$ isomorphic to S' by g. That is,

$$S = \{((a_0,g(b_0,c_0,\xi_0)),\ldots,(a_n,g(b_n,c_n,\xi_n))) : ((a_0,b_0,c_0,\xi_0),\ldots,(a_n,b_n,c_n,\xi_n)) \in S\}.$$

Clearly $S \in M$ and

$$\exists\beta Q(\alpha,\beta) \iff S(\alpha) \text{ is not well founded.}$$

This proves 1.

2. α,g,T are in M. Let $A' = A \cap M$. A' is a set in M, namely $\{\gamma \in \mathbb{R} : T(\gamma)$ is not well founded$\}$, since well foundedness is absolute. Let $Q'(\alpha,\beta) \iff f(\alpha,\beta) \in A'$. Then $Q' = Q \cap M$. The proof of part 1 can now be carried out inside M, using A' and Q' instead of A and Q. Note that the definitions of S' and S are absolute for M. Thus $M \models [\exists\beta Q'(\alpha,\beta) \iff S(\alpha)$ is not well founded$]$, so $(\exists\beta \in M)Q(\alpha,\beta) \iff S(\alpha)$ is not well founded. ⊣

Definition. A Π^1_n set $P \subset \mathbb{R}$ is complete Π^1_n if for every $k \in \omega$ and for every Π^1_n set $A \subset \mathbb{R}^k$, there is a recursive function $f : \mathbb{R}^k \to \mathbb{R}$ such that for all $(\alpha_1,\ldots,\alpha_k) \in \mathbb{R}^k$, $(\alpha_1,\ldots,\alpha_k) \in A \iff f(\alpha_1,\ldots,\alpha_k) \in P$.

For every n, there is a complete Π^1_n set. For odd n, $\mathrm{Det}(\underset{\sim}{\Delta}^1_{n-1})$ implies that every Π^1_n set admits a Π^1_n scale. So assume $\mathrm{Det}(\underset{\sim}{\Delta}^1_{n-1})$, and let $\{\varphi^n_m\}_{m\in\omega}$ be a fixed Π^1_n scale on a complete Π^1_n set, P^n. Let T^n be the associated tree of $\{\varphi^n_m\}_{m\in\omega}$.

Corollary 16. Let n be odd. Assume $\mathrm{Det}(\underset{\sim}{\Delta}^1_{n-1})$. Any transitive model M of ZF + DC containing T^n is Σ^1_{n+1}-correct.

Proof. We prove by induction on $i \leq n+1$ that M is Σ^1_i-correct. As in the proof of Lemma 4, it is enough to show that if M is Σ^1_{i-1}-correct, $\psi(\alpha,x)$ is a Π^1_{i-1} formula, $\alpha \in M$, and $\exists\beta\psi(\alpha,\beta)$, then $M \models \exists\beta\psi(\alpha,\beta)$.

Since ψ is Π^1_{i-1} it is Π^1_n. P^n is a complete Π^1_n set, so there is a recursive function $f:\mathbb{R}^2 \to \mathbb{R}$ such that $\psi(\alpha,\beta) \Longleftrightarrow f(\alpha,\beta) \in P^n$. By Theorem 15 (with $A = P^n$, $T = T^n$, $Q = \{(\alpha,\beta) : \psi(\alpha,\beta)\}$), $\exists\beta\psi(\alpha,\beta) \Rightarrow (\exists\beta \in M)\psi(\alpha,\beta)$. By the induction hypothesis, ψ is absolute, so $M \models \exists\beta\psi(\alpha,\beta)$. ⊣

Corollary 16, together with Lemma 1, allows us to define other partially playful universes. For n odd, $L[T^n]$ is the smallest transitive model of ZF containing all ordinals and containing T^n as an element.

Theorem 17. Let n be odd. Assume $\mathrm{Det}(\underset{\sim}{\Delta}^1_{n-1})$.

1. $L[T^n]$ is a transitive model of ZFC.
2. $L[T^n]$ is Σ^1_{n+1}-correct.
3. $L[T^n] \models \mathrm{Det}(\underset{\sim}{\Delta}^1_{n-1})$. If $\mathrm{Det}(\underset{\sim}{\Delta}^1_n)$, then $L[T^n] \models \mathrm{Det}(\underset{\sim}{\Delta}^1_n)$.

Proof.

1. For any set A, $L[A] \models$ ZF, and if there is a well ordering of A in $L[A]$, then there is a definable well ordering of the universe of $L[A]$ and $L[A] \models$ AC. T^n is a tree whose elements are finite sequences of pairs of ordinals. By coding finite sequences of ordinals by single ordinals, T^n can be well ordered in $L[T^n]$.
2. Corollary 16.
3. follows from 2 and Lemma 1. ⊣

Moschovakis has proved that $T^1 \in L$ ([8], Corollary to Theorem 9C-2.). Hence $L[T^1] = L$. So the $L[T^n]$'s, like the M^n's, can be viewed as analogs of L for the higher analytical pointclasses.

Note that T^n depends on the choice of the scale $\{\varphi^n_m\}_{m\in\omega}$; there is more than one complete Π^1_n set P^n, and more than one Π^1_n scale on each such P^n. It is an open question whether $L[T^n]$ also depends on the choice of the scale (for $n \geq 3$).

1. Conjecture (Moschovakis). $L[T^n]$ is independent of the choice of scale. That is, if T^n_1 and T^n_2 are the associated trees of Π^1_n scales on complete Π^1_n sets, then $L[T^n_1] = L[T^n_2]$.
2. Conjecture (Kechris). $L[T^n]$ is the smallest transitive model M of ZFC containing all ordinals, which has the property that for any transitive model N of ZFC with $M \subset N$, N is Σ^1_{n+1}-correct.

2 implies 1. It is also an open question whether $L[T^n] \subset L[T^{n+2}]$; of course if 1 is false, the answer to this may depend on the choice of scale.

7. Forcing and the independence of CH

By Corollary 16, every Cohen extension of $L[T^n]$ is Σ^1_{n+1}-correct, hence by Lemma 1 is a model of $\mathrm{Det}(\underset{\sim}{\Delta}^1_n)$. Thus the standard proof that there exists a model of $ZFC + 2^{\aleph_0} = \aleph_2$, should give a model of $ZFC + \mathrm{Det}(\underset{\sim}{\Delta}^1_n) + 2^{\aleph_0} = \aleph_2$, if $L[T^n]$ is taken to be the ground model. The problem with this is that $L[T^n]$ might not have any Cohen extensions. After all, it is not countable. In fact T^n is uncountable, so this method may not even work for any $L_\xi[T^n]$.

As previously stated, the construction of the models requires an additional assumption. The assumption needed is:

Definable well ordering hypothesis. Let $\leq_A$ be a well ordering of a set $A \subset \mathbb{R}$ such that A and $\leq_A$ are both in $L[\mathbb{R}]$. Then A is countable.

Lemma 18. Let n be odd. Assume $\mathrm{Det}(\underset{\sim}{\Delta}^1_{n-1})$. T^n (and hence $L[T^n]$) is absolute for $L[\mathbb{R}]$.

Proof. T^n is the associated tree of a Π^1_n scale $\{\varphi^n_m\}_{m\in\omega}$ on a Π^1_n set P^n. By definition of Π^1_n scale, there is a Π^1_n relation $S \subset \omega \times \mathbb{R}^2$ such that

$$(*) \qquad \beta \in P^n \Rightarrow \forall m \forall \alpha[(\alpha \in P^n \ \&\ \varphi_m(\alpha) \leq \varphi_m(\beta)) \Longleftrightarrow S(m,\alpha,\beta)].$$

Since $L[\mathbb{R}]$ contains all reals, all analytical formulas are absolute for $L[\mathbb{R}]$. In particular, P^n and S are absolute for $L[\mathbb{R}]$. The scale $\{\varphi^n_m\}_{m\in\omega}$, that is, the sequence of functions from P^n into the ordinals, is defined absolutely from P^n and S by the formula (*). Its tree, T^n, is defined absolutely from the scale and the set P^n; see the definition of associated tree at the beginning of Section 6 to verify this. So T^n is absolute for $L[\mathbb{R}]$. ⊣

Lemma 19. Let n be odd. Assume $\mathrm{Det}(\underset{\sim}{\Delta}^1_{n-1})$ and the definable well ordering hypothesis. Then for all $m \in \omega$, $(\beth_m)^{L[T^n]}$ is countable.

Proof. The proof is by induction on m. For $\beth_0$ it is trivial. So assume it is true for $\beth_m$, in order to prove it for $\beth_{m+1}$.

Let α be a real which codes the countable ordinal $\xi = (\beth_m)^{L[T^n]}$. Consider $L[T^n,\alpha]$, the smallest model of ZF containing all ordinals and T^n and α, i.e. the relativized $L[T^n]$. Clearly $L[T^n,\alpha] \models \xi$ is countable. Let $\eta = (\beth_{m+1})^{L[T^n]}$. Since $L[T^n] \models (\mathrm{Card}(\eta) \leq \mathrm{Card}(\text{Power set of } \xi))$, and $L[T^n] \subset L[T^n,\alpha]$, $L[T^n,\alpha] \models (\mathrm{Card}(\eta) \leq \mathrm{Card}(\text{Power set of } \xi))$. So $L[T^n,\alpha] \models \mathrm{Card}(\eta) \leq \beth_1$, that is, $\mathrm{Card}((\beth_{m+1})^{L[T^n]}) \leq \mathrm{Card}((\beth_1)^{L[T^n,\alpha]})$. So it will suffice to prove $(\beth_1)^{L[T^n,\alpha]}$ is countable.

By Lemma 18, T^n is absolute for $L[\mathbb{R}]$, so $L[T^n,\alpha] \subset L[\mathbb{R}]$. $L[T^n,\alpha] \models$ (there is a well ordering of the reals), by the relativized version of Theorem 17, part 1. Since $L[T^n,\alpha] \subset L[\mathbb{R}]$, this well ordering is in $L[\mathbb{R}]$. By the definable well ordering hypothesis, $\mathbb{R} \cap L[T^n,\alpha]$ is countable. That is, $(\beth_1)^{L[T^n,\alpha]}$ is countable. ⊣

Theorem 20. (Kunen-Moschovakis). Let k be even. Assume the definable well ordering hypothesis. Then $\mathrm{Det}(\underset{\sim}{\Delta}^1_k)$ implies that there is a transitive model of $(\mathrm{ZFC} + \mathrm{Det}(\underset{\sim}{\Delta}^1_k) + 2^{\aleph_0} = \aleph_2)$.

Outline of proof. Assuming $\mathrm{Det}(\underset{\sim}{\Delta}^1_k)$, we construct a model N of $(\mathrm{ZFC} + \mathrm{Det}(\underset{\sim}{\Delta}^1_k) + 2^{\aleph_0} = \aleph_2)$ by forcing. $\mathrm{Det}(\underset{\sim}{\Delta}^1_k)$ implies that T^{k+1} and hence $L[T^{k+1}]$ exists. $L[T^{k+1}]$ is the ground model.

Now extend $L[T^{k+1}]$ to a model N, such that $L[T^{k+1}] \subset N$ and $N \models (\mathrm{ZFC} + 2^{\aleph_0} = \aleph_2)$. N is constructed in the usual way, as a Cohen extension, or a finite sequence of Cohen extensions; see [3] or [16]. (Caution: We cannot assume $L[T^{k+1}]$ satisfies GCH or even CH.) First get a generic extension that satisfies $2^{\aleph_0} = \aleph_1$ and $2^{\aleph_1} = \aleph_2$, and then add a generic $\aleph_2$-sequence of reals. We will not give a detailed proof; the reader familiar with forcing will have no trouble producing one.

By Lemma 19, the necessary generic extensions must exist, since there are only countably many sets of conditions in the ground model. Since $T^{k+1} \in L[T^{k+1}] \subset N$, by Corollary 16, N is Σ^1_{k+2}-correct. We are assuming $\mathrm{Det}(\underset{\sim}{\Delta}^1_k)$, so by Lemma 1, $N \models \mathrm{Det}(\underset{\sim}{\Delta}^1_k)$. $\dashv$

In the above theorem, $\aleph_2$ can obviously be replaced by $\aleph_m$ for any $m \geq 1$. In fact, the technique used in the above proof will work for a much larger set of $\aleph_\xi$'s, but it does have its limits. Some other analogs of classical consistency results can also be proved in this manner.

Many mathematicians who work in descriptive set theory or definability theory believe it is "obviously true" that any definable well ordered set of reals must be countable. (Every element of $L[\mathbb{R}]$ is ordinal definable from a real parameter.) But from the point of view of axiom systems, rather than intuition, the definable well ordering hypothesis is a very strong statement. PD does not imply it. The model M^ω with its hyperprojective well ordering of the reals demonstrates this. No large cardinal axiom is known to imply it. The determinacy of any pointclass with reasonable closure properties implies that every well ordering in that pointclass is countable. So $\mathrm{Det}((\text{Power set of } \mathbb{R}) \cap L[\mathbb{R}])$ does imply the definable well ordering hypothesis.

Assuming the definable well ordering hypothesis, PD implies that there is a transitive model of $(\mathrm{ZFC} + \mathrm{PD} + 2^{\aleph_0} = \aleph_2)$. The proof is the same as for Theorem 20, except that the ground model to use is $L[\tilde{T}]$, where $\tilde{T}$ is the sequence $T^1, T^3, T^5, \ldots$.

8. The Harrington-Kechris Theorem

Until quite recently, little beyond Theorem 17 was known about the $L[T^n]$'s. Then Harrington and Kechris characterized the reals in these models, assuming a lot of determinacy.

First of all, for odd n, $\mathrm{Det}(\underset{\sim}{\Sigma}^1_n)$ implies that there is a largest countable Π^1_n set, C_n, and a largest countable Σ^1_{n+1} set, C_{n+1}. (That is, C_{n+1} is countable and Σ^1_{n+1}, and if S is a countable Σ^1_{n+1} set then $S \subset C_{n+1}$, and similarly for C_n.) Thus the pointclasses which have a largest countable set are the same as those that have the scale property - Π^1_n for n odd and Σ^1_k for k even.

Actually, $\mathrm{Det}(\underset{\sim}{\Delta}^1_{n-1})$ is enough to prove the existence of a largest thin (i.e. containing no perfect subset) Π^1_n set, C_n. $\mathrm{Det}(\underset{\sim}{\Delta}^1_{n-1})$ does not imply C_n is countable. C_{n+1} is defined to be $\{\alpha : \exists\beta\,(\beta \in C_n \;\&\; \alpha \text{ is recursive in } \beta)\}$, which is clearly Σ^1_{n+1}, and which contains every countable Σ^1_{n+1} set, because Π^1_n has the uniformization property. $\mathrm{Det}(\underset{\sim}{\Sigma}^1_n)$ then implies that C_n, and hence C_{n+1}, are in fact countable. It also implies that C_{n+1} admits a Σ^1_{n+1}-good well ordering. These are theorems of Kechris and Moschovakis. A proof of them can be found in [4] or [6].

$L[\mathbb{R}]$ is, of course, the smallest transitive model of ZF containing all ordinals and all reals. We write $\mathrm{Det}(L[\mathbb{R}])$ for $\mathrm{Det}((\text{Power set of } \mathbb{R}) \cap L[\mathbb{R}])$. This assumption that every set of reals in $L[\mathbb{R}]$ is determined is far stronger than PD.

Theorem 21. (Harrington-Kechris; Kechris-Martin for $n = 3$). Let n be odd. Assume $\mathrm{Det}(L[\mathbb{R}])$. Then $\mathbb{R} \cap L[T^n] = C_{n+1}$.

Solovay [17] proved (without determinacy) that if $\mathbb{R} \cap L$ is countable, then it is the largest countable Σ^1_2 set. That is, $C_2 = \mathbb{R} \cap L$. So these C_{n+1}'s can be thought of as analogs of the constructible reals, and appropriately enough, they are the reals in the analogs of L. Since $L[T^1] = L$, for the case $n = 1$, Theorem 21 follows from Solovay's result.

A proof of this theorem will be given in the last section of this paper. First we want to point out some of its consequences.

For one thing, it implies that the reals in $L[T^n]$ are independent of the choice of the scale. This is a partial solution of the Moschovakis conjecture.

It also yields an easy proof of the main properties of a third type of partially playful universe.

9. The smallest model containing C_k

Definition. For even k, let $L^k = L[C_k]$, the smallest model of ZF containing all ordinals and containing C_k as an element.

Since $C_2 = \mathbb{R} \cap L$, $L^2 = L$. So again these are analogs of L of the sort described in Section 1.

Corollary 22. (Moschovakis). Let $k \geq 2$ be even. Assume $\mathrm{Det}(L[\mathbb{R}])$. Then $\mathbb{R} \cap L^k = C_k$.

Proof. $L^k = L[C_k]$

$= L[(\mathbb{R})^{L[T^{k-1}]}]$ (Theorem 21)

$= (L[\mathbb{R}])^{L[T^{k-1}]}$.

So $\mathbb{R} \cap L^k = \mathbb{R} \cap L[T^{k-1}] = C_k$. ⊣

Corollary 23. (Kechris). Let $k \geq 2$ be even. Assume $\mathrm{Det}(L[\mathbb{R}])$.

1. L^k is Σ^1_k-correct.
2. $L^k \models$ There exists a Σ^1_k-good well ordering of $\mathbb{R}$.
3. $L^k \models AC + GCH$.
4. $L^k \models \mathrm{Det}(\underset{\sim}{\Delta}^1_{k-1})$.
5. $L^k \models \mathrm{Scale}(\Pi^1_i)$ and $\mathrm{Unif}(\Pi^1_i)$, for i odd, $i < k$.

   $L^k \models \mathrm{Scale}(\Sigma^1_i)$ and $\mathrm{Unif}(\Sigma^1_i)$, for i even and for all $i \geq k$.
6. $L^k \models \neg \mathrm{Det}(\Sigma^1_{k-1})$.

Proof.

1. L^k has the same reals as $L[T^{k-1}]$, and $L[T^{k-1}]$ is Σ^1_k-correct by Theorem 17.

2. The reals of L^k, C_k, admits a Σ^1_k-good well ordering, in the real world. By 1, the Σ^1_k formula that defines the well ordering and the Σ^1_k and Π^1_k formulas which code initial segments are all absolute for L^k.

3. $L^k = L[C_k]$, and by 2, there is a well ordering of C_k in L^k. Hence $L^k \models$ There is a definable well ordering of the universe. So $L^k \models AC$.

A good well ordering has order type $\leq \aleph_1$; this is because the initial segments can be coded by reals, and so must be countable. This fact, plus 2, implies that $L^k \models CH$.

For $\alpha > 0$, $L^k \models 2^{\aleph_\alpha} = \aleph_{\alpha+1}$ is proved in the same manner as one proves that GCH holds in L. For $L^k = L[C_k]$ and $L^k \models (\mathrm{Card}(C_k) = \aleph_1)$, so the models of $V = L[C_k]$ of cardinality $\aleph_\alpha$ needed in the proof, can be obtained, and then the whole proof goes through.

4. Follows from 1 and Lemma 1.

5. $\mathrm{Det}(\underset{\sim}{\Delta}^1_{k-2})$ implies the scale and uniformization properties up to the kth level. The existence of a Σ^1_k-good well ordering proves it for $i > k$ (see [8]).

6. The existence of a Σ^1_k-good well ordering of $\mathbb{R}$ implies that there is an uncountable Σ^1_k set with no perfect subset (namely the set of $\leq$-least codes for each ordinal, where $\leq$ is the well ordering). By 5, $\mathrm{Unif}(\Pi^1_{k-1})$, so this Σ^1_k set is the one-to-one continuous image of a Π^1_{k-1} set, which must be uncountable and have no perfect subset. This implies $\neg\mathrm{Det}(\Pi^1_{k-1})$ and therefore $\neg\mathrm{Det}(\Sigma^1_{k-1})$. ⊣

The scaled classes for L^k are the same as those for M^k, by 5 and Theorem 5. Therefore, the descriptive set theory in these two models is virtually identical.

Both of these corollaries can be proved directly, without mentioning the $L[T^n]$'s, without the Harrington-Kechris Theorem, and without such an extreme determinacy assumption. In fact, $\mathrm{Det}(\underset{\sim}{\Delta}^1_{k-2})$ will suffice for both of the corollaries, except for part 4 of Corollary 23 which requires $\mathrm{Det}(\underset{\sim}{\Delta}^1_{k-1})$. These direct proofs can be found in Kechris [4].

Since $L^k \models (\mathrm{Det}(\underset{\sim}{\Delta}^1_{k-1}) + \neg\mathrm{Det}(\Sigma^1_{k-1}))$, we get a strengthening of Corollary 11, as claimed. In fact, with a proof of Corollary 23 that uses only $\mathrm{Det}(\underset{\sim}{\Delta}^1_{k-1})$, not $\mathrm{Det}(L[\mathbb{R}])$, we can obtain a finitary proof of

$$\mathrm{Con}(ZF + DC + \mathrm{Det}(\underset{\sim}{\Delta}^1_n)) \rightarrow \mathrm{Con}(ZFC + GCH + \mathrm{Det}(\underset{\sim}{\Delta}^1_n) + \neg\mathrm{Det}(\Sigma^1_n)),$$

for odd n $(n = k - 1)$. This is false for even n, in light of Theorem 6.

<u>Corollary</u> 24. Let n be odd. Assume $\mathrm{Det}(L[\mathbb{R}])$.

1. $L[T^n] \models$ There exists a Σ^1_{n+1}-good well ordering of $\mathbb{R}$.
2. $L[T^n] \models CH$.
3. $L[T^n] \models \mathrm{Det}(\underset{\sim}{\Delta}^1_n)$.
4. $L[T^n] \models \mathrm{Scale}(\Pi^1_i)$ and $\mathrm{Unif}(\Pi^1_i)$, for i odd, $i \leq n$.
   $L[T^n] \models \mathrm{Scale}(\Sigma^1_i)$ and $\mathrm{Unif}(\Sigma^1_i)$, for i even and for all $i \geq n+1$.
5. $L[T^n] \models \neg\,\mathrm{Det}(\Sigma^1_n)$.

<u>Proof</u>. L^{n+1} and $L[T^n]$ have the same reals. 1, 3, 4, and 5 follow directly from this fact and parts 2, 4, 5, and 6, respectively, of Corollary 23. 1 implies 2, as shown in the proof of Corollary 23. ⊣

The assumption of $\mathrm{Det}(L[\mathbb{R}])$ only shortened the proof of Corollary 23; it was not needed. For Corollary 24, however, the only known proof is the one given above, which requires the Harrington-Kechris Theorem and hence $\mathrm{Det}(L[\mathbb{R}])$.

While the reals in $L[T^n]$ are now well understood, the same cannot be said for the rest of the universe. In particular, it is still an open question whether $L[T^n] \models GCH$ (for $n \geq 3$). It is conceivable that this depends on the choice of scale.

There is another consequence of Theorem 21 of a different nature. PD implies that for even k, there is a largest meager $\underset{\sim}{\Sigma}^1_k$ set (Kechris [5] for $k \geq 2$; Baire for $k = 0$). Assuming $\mathrm{Det}(L[\mathbb{R}])$, we can get an interesting characterization of these sets, as well as a new proof of their existence.

<u>Definition</u>. For even $k \geq 2$, let $\mathfrak{m}_k$ be the set of all reals which are not L^k-generic.

Lemma 25. Let M and N be two transitive models of ZF such that $\mathbb{R} \cap M = \mathbb{R} \cap N$. For all $\alpha \in \mathbb{R}$, α is M-generic iff α is N-generic.

Thus $\mathfrak{m}_k$ is also the set of all reals which are not $L[T^{k-1}]$-generic.

Corollary 26. Let $k \geq 2$ be even. Assume Det$(L[\mathbb{R}])$. $\mathfrak{m}_k$ is the largest meager Σ^1_k set.

Proof. α is L^k-generic $\Longleftrightarrow \forall\beta[(\beta \in \mathbb{R} \cap L^k$ & β codes a dense set S_β of conditions$) \rightarrow \alpha$ meets $S_\beta]$.

By Corollary 22, $\mathbb{R} \cap L^k$ is $\mathcal{C}_k$. $\mathcal{C}_k$ is Σ^1_k, so the above formula shows that the set of L^k-generic reals is Π^1_k, and $\mathfrak{m}_k$ is Σ^1_k. $\mathcal{C}_k$ is countable, so $\mathfrak{m}_k$ is meager.

Suppose $\mathfrak{m}_k$ is not the largest meager Σ^1_k set. Then there is a meager set A such that A contains an L^k-generic real α, and A is Σ^1_k. Let $\varphi(x)$ be the Σ^1_k formula that defines A.

By Theorem 21 and Corollary 22, L^k and $L[T^{k-1}]$ have the same reals. By Lemma 25, α is $L[T^{k-1}]$-generic. The Cohen extension $(L[T^{k-1}])[\alpha]$ contains the tree T^{k-1}. By Corollary 16, it is Σ^1_k-correct. $\varphi(\alpha)$, hence $(L[T^{k-1}])[\alpha] \models \varphi(\alpha)$. So there is a condition p, p an initial segment of α, such that $p \Vdash \varphi(\hat{G})$, where $\hat{G}$ is the denotation of the generic real.

Let β be any L^k-generic real with initial segment p. Then β is $L[T^{k-1}]$-generic. Since $p \Vdash \varphi(\hat{G})$, $(L[T^{k-1}])[\beta] \models \varphi(\beta)$. Then by the same reasoning as above, $\varphi(\beta)$ is true. So $\beta \in A$. Thus A contains all elements of the set $\{\beta : \beta$ is L^k-generic & p is an initial segment of $\beta\}$. Since $\mathbb{R} \cap L^k = \mathcal{C}_k$ is countable, this set is not meager. Contradiction. $\dashv$

10. Relationship between the M^n's, L^k's, and $L[T^m]$'s

Diagram 5, below, shows the relationship between the M^n's, the L^k's, and the $L[T^m]$'s.

Diagram 5

$$\begin{array}{ccccccccccccccccc}
L[T^1] & & & & L[T^3] & & & & L[T^5] & & & & L[T^7] & & & & \\
\parallel & & & & \cup & & & & \cup & & & & \cup & & & & \\
L^2 & & \subset & & L^4 & & \subset & & L^6 & & \subset & & L^8 & & \subset & & \ldots \\
\parallel & & & & \cup & & & & \cup & & & & \cup & & & & \\
M^2 & \subset & M^3 & \subset & M^4 & \subset & M^5 & \subset & M^6 & \subset & M^7 & \subset & M^8 & \subset & M^9 & \subset & \ldots
\end{array}$$

All inclusions shown are proper; except for the top line, $L^{m+1} \subset L[T^m]$, the inclusions are proper for the reals in the model.

The inclusions $M^n \subset M^{n+1}$ have already been proved (Theorem 7). For even k, C_k is the largest countable Σ^1_k set, hence it is a countable Σ^1_{k+2} set, so trivially $C_k \subset C_{k+2}$. $L^{k+2} = L[C_{k+2}]$, so $C_k \subset L^{k+2}$. Since C_k is Σ^1_k and L^{k+2} is Σ^1_{k+2}-correct, $C_k \in L^{k+2}$. Hence $L^k = L[C_k] \subset L^{k+2}$. $\mathbb{R} \setminus C_k$ is Π^1_k, hence contains a Δ^1_{k+2} real, α [8]. Then $\{\alpha\}$ is Σ^1_{k+2}, so $\alpha \in C_{k+2} \setminus C_k$. $\mathbb{R} \cap L^k = C_k$ by Corollary 22. So $(\exists \alpha \in \mathbb{R})(\alpha \in L^{k+2} \setminus L^k)$. As previously mentioned, it is an open question whether $L[T^m] \subset L[T^{m+2}]$, for $m \geq 3$.

For even k, $M^k \subset L^k$ since M^k is the smallest Σ^1_k-correct model and L^k is Σ^1_k-correct. $L^k \subset L[T^{k-1}]$, since $L^k = (L[\mathbb{R}])^{L[T^{k-1}]}$. $M^2 = L^2 = L[T^1]$, as they are all equal to L. But for all even $k > 2$, the inclusions $M^k \subset L^k \subset L[T^{k-1}]$ are proper. To see that $L^k \neq L[T^{k-1}]$, note that if α is a real that codes the countable set C_k, then $L^k = L[C_k] \subset L[\alpha]$. $L[\alpha] \models \neg \mathrm{Det}(\underset{\sim}{\Delta}^1_2)$, so by Lemma 1 it is not Σ^1_k-correct (for $k > 2$). By Corollary 16, $T^{k-1} \notin L[\alpha]$. So $T^{k-1} \notin L^k$ and $L^k \neq L[T^{k-1}]$. Theorem 27, below, states that for $k > 2$, $0^{\#}_k \in C_k$. $0^{\#}_k \notin M^k$ (Corollary 10). Thus $M^k \neq L^k$, which justifies all of Diagram 5.

<u>Theorem</u> 27. (Solovay). Let $k \geq 4$ be even. Assume $\mathrm{Det}(L[\mathbb{R}])$. Then $0^{\#}_k \in C_k$.

<u>Proof</u>. $\mathrm{Det}(L[\mathbb{R}])$ implies the definable well ordering hypothesis, so by Lemma 19, $(\beth_2)^{L[T^{k-1}]}$ is countable. Therefore there exists an $L[T^{k-1}]$-generic map G from ω onto $\mathbb{R} \cap L[T^{k-1}]$. G is essentially a sequence of reals, so it can be coded by a real γ; that is, $\mathbb{R} \cap L[T^{k-1}] = \{(\gamma)_0, (\gamma)_1, \ldots\}$.

By Corollary 16, the Cohen extension $(L[T^{k-1}])[G]$ is Σ^1_k-correct. $\mathrm{Det}(\underset{\sim}{\Delta}^1_2)$ implies that all sharps exist. So $\gamma^{\#}$ exists. This is a Σ^1_3 statement involving γ, and $\gamma \in (L[T^{k-1}])[G]$ which is Σ^1_k-correct, $k \geq 4$. Hence $(L[T^{k-1}])[G] \models \gamma^{\#}$ exists. That means, $(L[T^{k-1}])[G] \models$ (The uncountable cardinals are indiscernibles in $L[\gamma]$.).

$\mathbb{R} \cap L[T^{k-1}]$ is clearly definable in $L[\gamma]$ by a formula whose only parameter is γ. So $(L[\mathbb{R}])^{L[T^{k-1}]}$ is also definable by such a formula. Since $L[T^{k-1}]$ is Σ^1_k-correct, so is $(L[\mathbb{R}])^{L[T^{k-1}]}$; by Theorem 5, part 2, M^k is a definable inner model in $L[\mathbb{R}]^{L[T^{k-1}]}$. Putting this all together, we have that M^k is definable in $L[\gamma]$ by a formula whose only parameter is γ. Therefore, $(L[T^{k-1}])[G] \models$ (The uncountable cardinals are indiscernibles of M^k.).

The uncountable cardinals of $(L[T^{k-1}])[G]$ are precisely the cardinals $> \beth_1$ of $L[T^{k-1}]$; this is so because the forcing conditions satisfy the $\beth_1$-antichain condition (see [3]). So $L[T^{k-1}] \models$ (The cardinals $> \beth_1$ are indiscernibles of M^k). (Note that M^k is absolute for $L[T^{k-1}]$.) $L[T^{k-1}] \models \mathrm{Det}(\underset{\sim}{\Delta}^1_{k-2})$, by Theorem 17.

By Theorem 8, applied inside $L[T^{k-1}]$, $L[T^{k-1}] \models 0_k^\#$ exists. $0_k^\#$ is a Π_k^1-singleton and $L[T^{k-1}]$ is Σ_k^1-correct, so $(0_k^\#)^{L[T^{k-1}]}$ is the real $0_k^\#$. This proves that $0_k^\# \in L[T^{k-1}]$. By the Harrington-Kechris Theorem, $0_k^\# \in \mathcal{C}_k$. $\dashv$

Theorem 27 was originally proved by Solovay by an entirely different method. His proof does not require the assumption of $\mathrm{Det}(L[\mathbb{R}])$, but uses the additional hypothesis that there exists a measurable cardinal.

Remark. It is implicit in the above proof that for odd $n > 1$, $L[T^n] \models$

(*) (The cardinals $> \beth_1$ are indiscernibles for $L[\mathbb{R}]$.).

The existence of a Ramsey cardinal implies (*) and it is probably true in V. This reveals another difference between $L[T^n]$, $n > 1$, and $L[T^1] = L$. $L \models V = L[\mathbb{R}]$, so trivially $L \models \neg (*)$.

It is possible to define one type of partially playful universe inside another. By Theorem 5, part 2, for even k, $(M^k)^{M^k} = (M^k)^{L^k} = (M^k)^{L[T^{k-1}]} = M^k$. Since M^k is Σ_k^1-correct and $\mathbb{R} \cap M^k \subset \mathbb{R} \cap L^k = \mathcal{C}_k$, $M^k \models \forall\alpha\ (\alpha \in \mathcal{C}_k)$. So $(L^k)^{M^k} = (L[\mathbb{R}])^{M^k} = M^k$. Similarly, $(L^k)^{L^k} = (L[\mathbb{R}])^{L^k} = L^k$, and $(L^k)^{L[T^{k-1}]} = (L[\mathbb{R}])^{L[T^{k-1}]} = L^k$. Of course $\mathcal{C}_k$ is not countable in any of these models; the proof that $\mathcal{C}_k$ is countable requires $\mathrm{Det}(\underset{\sim}{\Sigma}_{k-1}^1)$, and the models M^k, L^k, and $L[T^{k-1}]$ are not quite that playful. That leaves one interesting case.

Let $n \geq 2$ and let m be the greatest odd integer $\leq n$. Since $M^n \models \mathrm{Det}(\underset{\sim}{\Delta}_{n-1}^1)$, $L[T^m]$ can be defined in M^n. (This is not the real T^m; it is the set which satisfies the definition of T^m in M^n.) By Theorem 17, part 2, applied inside M^n, $M^n \models (L[T^m]$ is Σ_{m+1}^1-correct). Since M^n is Σ_n^1-correct and $m+1 \geq n$, $(L[T^m])^{M^n}$ is really Σ_n^1-correct. But M^n is the <u>smallest</u> Σ_n^1-correct model. Hence $M^n = (L[T^m])^{M^n}$. Thus $M^n \models V = L[T^m]$, which is particularly interesting in light of the fact that $L[T^m] \models V \neq L[T^m]$, as will be shown below.

$$\begin{aligned} M^n &= (L[T^m])^{M^n} \\ &= (L[(T^m)^{M^n}])^{M^n} \\ &= L[(T^m)^{M^n}], \text{ by the absoluteness of constructibility.} \end{aligned}$$

This shows that the M^n's are all models of the form L[A], where A is a set of ordinals. (Recall, T^m is a set of finite sequences of ordered pairs of ordinals.) No characterization of $(T^m)^{M^n}$ is known which doesn't mention M^n, so the old definition of M^n appears to be necessary.

Let $k \geq 2$ be even. $L[T^{k-1}]$ can be defined inside itself and inside L^k, too. By Lemma 18, $L[T^{k-1}]$ is absolute for $L[\mathbb{R}]$. Since $(L[\mathbb{R}])^{L[T^{k-1}]} = L^k$, we have that

$$(L[T^{k-1}])^{L[T^{k-1}]} = (L[T^{k-1}])^{L^k} \subset L^k \subsetneq L[T^{k-1}].$$

In fact, it is equal to L^k. To prove this it will suffice to show that $C_k \subset (L[T^{k-1}])^{L[T^{k-1}]}$. We will prove this later (Corollary 33); it falls out of the proof of Theorem 21.

Having postponed one proof, we now take up another proof that we had previously put off. Theorem 5 states that $M^n \models$ ($\mathbb{R}$ admits a Σ^1_{n+1} well ordering). We said that for even k, Σ^1_{k+1} could be improved to Σ^1_k. We are now in a position to prove it.

Let $\leq_{L^k}$ be the well ordering of the reals of L^k given by Corollary 23, part 2. Since $M^k \subset L^k$, clearly $\leq_{L^k} \restriction (\mathbb{R} \cap M^k)$ is a well ordering of $\mathbb{R} \cap M^k$. L^k and M^k are both Σ^1_k-correct, so M^k is Σ^1_k-correct with respect to L^k. Therefore the Σ^1_k formula that defines $\leq_{L^k}$ in L^k is absolute for M^k, and so defines $\leq_{L^k} \restriction (\mathbb{R} \cap M^k)$ in M^k. We have just proved

<u>Theorem</u> 28. Let $k \geq 2$ be even. Assume Det($L[\mathbb{R}]$). Then $M^k \models$ ($\mathbb{R}$ admits a Σ^1_k well ordering).

Of course this implies that $M^k \models$ ($\mathbb{R}$ admits a Δ^1_k well ordering). The hypothesis Det($L[\mathbb{R}]$) was used in our proof of Corollary 23, but as we pointed out before, is not really necessary. Det($\underset{\sim}{\Delta}^1_{k-2}$) is sufficient to prove Theorem 28.

This completes our discussion of the three types of partially playful universes, the M^n's, L^k's, and $L[T^m]$'s. There is a fourth type of partially playful universe which has been studied, but which we will not consider in this paper. This is the $L[Q_n]$'s of Kechris, Martin, and Solovay; see [6] for details.

<u>Definition</u>. An nth <u>level nice analog of</u> L is a transitive model M of ZF which has the following three properties:

a) $\mathbb{R} \cap M$ is Σ^1_n
b) M is Σ^1_n-correct
c) $\exists\ \alpha \in (\mathbb{R} \setminus M)$.

For even k, the M^k's, L^k's, and $L[T^{k-1}]$'s are all k^{th} level nice analogs of L. (The M^n's are not nice for odd n.) These models have most of the important properties of L, relative to the analytical hierarchy, shifted up $k - 2$ levels. While we have discussed extensively nice analogs at the even levels, we have never mentioned any nice analogs at the odd levels. As the following theorem shows, this is not due to an oversight.

<u>Theorem</u> 29. (Kechris [5]). Let n be odd. Assume Det($\underset{\sim}{\Delta}^1_{n-1}$). There does not exist an nth level nice analog of L.

11. <u>Proof of the Harrington-Kechris Theorem</u>

The rest of this paper consists of the proof of the Harrington-Kechris Theorem (Theorem 21). We adopt the following convention for the rest of the paper:

Notation. n always denotes an odd integer. Other integer variables m,k,... may be either odd or even.

Definition. Let T be a tree on $\omega\times\kappa$. Then p[T] is the projection of the body of T, $\{\alpha \in \mathbb{R} : \exists f((\alpha,f) \in [T])\}$.

Lemma 30. Assume $\mathrm{Det}(\underset{\sim}{\Delta}^1_{n-1})$. Let B be a Σ^1_{n+1} set. There is a tree $S \in L[T^n]$ such that $B = p[S]$.

Proof. Let $Q \subset \mathbb{R}^2$ be a Π^1_n set such that $B(\alpha) \iff \exists\beta Q(\alpha,\beta)$. T^n is by definition, the tree associated with a scale on a complete Π^1_n set P^n. Since P^n is complete Π^1_n, there is a recursive function f such that $Q(\alpha,\beta) \iff f(\alpha,\beta) \in P^n$. By Theorem 15 (with $A = P^n$, $T = T^n$, $M = L[T^n]$), there is a tree S in $L[T^n]$ such that for all α, $\exists\beta Q(\alpha,\beta) \iff S(\alpha)$ is not well founded. Hence $B(\alpha) \iff S(\alpha)$ is not well founded, which means that $B = p[S]$. ⊣

Theorem 31. (Mansfield [9]). Let T be a tree on $\omega\times\kappa$ and let $B = p[T]$. If B contains an element not in L[T], then B has a nonempty perfect subset.

Mansfield's original proof uses forcing. A forcing-free proof, due to Solovay, can be found in [8, Theorem 11A-1].

Corollary 32. Assume $\mathrm{Det}(\underset{\sim}{\Delta}^1_{n-1})$. Then $C_{n+1} \subset L[T^n]$.

Proof. C_n is Π^1_n, hence Σ^1_{n+1}, so by Lemma 30, there is a tree $S \in L[T^n]$ such that $C_n = p[S]$. $\mathrm{Det}(\underset{\sim}{\Delta}^1_{n-1})$ implies that C_n has no perfect subset. By Theorem 31, $C_n \subset L[S]$. So $C_n \subset L[T^n]$. And since $C_{n+1} = \{\alpha : \exists\beta(\beta \in C_n \;\&\; \alpha \text{ is recursive in } \beta)\}$, clearly $C_{n+1} \subset L[T^n]$. ⊣

Corollary 32 is half of Theorem 21. Before proving the other half, we will tie up the one remaining loose end. We said (remarks prior to Theorem 28) that $C_{n+1} \subset (L[T^n])^{L[T^n]}$, which implies that $(L[T^n])^{L[T^n]} = (L[T^n])^{L^{n+1}} = L^{n+1}$. This follows from Mansfield's Theorem, by essentially the same proof as Corollary 32.

Corollary 33. Assume $\mathrm{Det}(\underset{\sim}{\Delta}^1_{n-1})$. Then $C_{n+1} \subset (L[T^n])^{L[T^n]}$.

Proof. $C_n \subset C_{n+1}$, and $C_{n+1} \subset L[T^n]$ by Corollary 32. By Theorem 17, $L[T^n] \models \mathrm{Det}(\underset{\sim}{\Delta}^1_{n-1})$ and $L[T^n]$ is Σ^1_{n+1}-correct. Since C_n is Π^1_n and $C_n \subset L[T^n]$, $C_n \in L[T^n]$ and $L[T^n] \models (C_n \text{ is } \Pi^1_n)$. By Lemma 30, applied inside $L[T^n]$, there is a tree $S \in (L[T^n])^{L[T^n]}$ such that $C_n = p[S]$. $\mathrm{Det}(\underset{\sim}{\Delta}^1_{n-1})$ implies that C_n has no perfect subset. By Theorem 31, $C_n \subset L[S]$. So $C_n \subset (L[T^n])^{L[T^n]}$. And since $C_{n+1} = \{\alpha : \exists\beta(\beta \in C_n \;\&\; \alpha \text{ is recursive in } \beta)\}$, clearly $C_{n+1} \subset (L[T^n])^{L[T^n]}$. ⊣

We have proved $C_{n+1} \subset \mathbb{R} \cap L[T^n]$; the reverse inclusion, $\mathbb{R} \cap L[T^n] \subset C_{n+1}$, remains to be proved. Since C_{n+1} is the largest countable Σ^1_{n+1} set, it will suffice to prove

1. $\mathbb{R} \cap L[T^n]$ is countable
2. $\mathbb{R} \cap L[T^n]$ is Σ^1_{n+1}.

Theorem 34. Assume $\mathrm{Det}(L[\mathbb{R}])$. $\mathbb{R} \cap L[T^n]$ is countable.

Proof. $\mathrm{Det}(L[\mathbb{R}])$ implies the definable well ordering hypothesis. By Lemma 19, $(\beth_1)^{L[T^n]}$ is countable. ⊣

AD (= the axiom of determinacy) is the proposition that every set of reals is determined. AD contradicts the axiom of choice, but is believed to be consistent with ZF + DC.

Lemma 35. The following are equivalent:

a) $\mathrm{Det}(L[\mathbb{R}])$ (= Det(Power set of $\mathbb{R} \cap L[\mathbb{R}]$))

b) $L[\mathbb{R}] \models AD$.

Proof. Strategies can be identified with reals. ⊣

Since $L[T^n]$ is absolute for $L[\mathbb{R}]$ (Lemma 18), $\mathbb{R} \cap L[T^n]$ is Σ^1_{n+1} iff $L[\mathbb{R}] \models (\mathbb{R} \cap L[T^n]$ is $\Sigma^1_{n+1})$. This fact, plus Lemma 35, shows that the following are equivalent:

a) $\mathrm{Det}(L[\mathbb{R}]) \Rightarrow \mathbb{R} \cap L[T^n]$ is Σ^1_{n+1}

b) $L[\mathbb{R}] \models AD \Rightarrow L[\mathbb{R}] \models (\mathbb{R} \cap L[T^n]$ is $\Sigma^1_{n+1})$.

By working inside $L[\mathbb{R}]$, to prove the remaining part of the Harrington-Kechris Theorem, it is enough to prove that

$$(ZF + DC + AD) \vdash (\mathbb{R} \cap L[T^n] \text{ is } \Sigma^1_{n+1}).$$

(Note: $L[\mathbb{R}] \models DC$.) So we now proceed to prove that $\mathbb{R} \cap L[T^n]$ is Σ^1_{n+1}, using AD + DC.

Definition. Let φ be a norm on $\mathbb{R}$. $|\alpha|$ denotes the ordinal $\varphi(\alpha)$. Let $R(\xi_0,\ldots,\xi_m,\beta_0,\ldots,\beta_k)$ be a relation on ordinals and reals. Let Γ be a pointclass. R is Γ in the codes (with respect to φ) if the relation

$$R^*(\alpha_0,\ldots,\alpha_m,\beta_0,\ldots,\beta_k) \Longleftrightarrow R(|\alpha_0|,\ldots,|\alpha_m|,\beta_0,\ldots,\beta_k)$$

is in Γ. A relation $R(\alpha,\bar{\beta})$ is invariant on α if $(|\alpha_1| = |\alpha_2| \;\&\; R(\alpha_1,\bar{\beta})) \Rightarrow R(\alpha_2,\bar{\beta})$.

Lemma 36. Let $\leq$ be a Δ^1_m prewellordering of $\mathbb{R}$. If $P(\alpha,\bar{\beta})$ is Σ^1_{m+1} and invariant on α, then there is a Π^1_m relation $Q(\alpha,\bar{\beta},\gamma)$ which is invariant on α and such that $P(\alpha,\bar{\beta}) \Longleftrightarrow \exists\gamma Q(\alpha,\bar{\beta},\gamma)$.

Proof. $P(\alpha,\beta) \Longleftrightarrow \exists\alpha'[|\alpha'| = |\alpha| \;\&\; P(\alpha',\bar{\beta})]$

$\Longleftrightarrow \exists\alpha'\exists\delta[|\alpha'| = |\alpha| \;\&\; R(\alpha',\bar{\beta},\delta)]$, where R is some Π^1_m relation

$\Longleftrightarrow \exists\gamma[|(\gamma)_0| = |\alpha| \;\&\; R((\gamma)_0,\bar{\beta},(\gamma)_1)]$.

Let $Q(\alpha,\bar{\beta},\gamma) \Longleftrightarrow [|(\gamma)_0| = |\alpha| \;\&\; R((\gamma)_0,\bar{\beta},(\gamma)_1)]$. ⊣

Computing the complexity of the set $\mathbb{R} \cap L[T^n]$ requires some techniques for computing the complexity of sets whose definition involves ordinals. Specifically, if $R(\xi,\beta)$ is Γ in the codes, what can be said about the sets of reals $(\exists \xi < \kappa)R(\xi,\beta)$ and $(\forall \xi < \kappa)R(\xi,\beta)$, where κ is the length of the norm? The next theorem answers this question in about the nicest way imaginable.

<u>Theorem</u> 37. (Harrington-Kechris; Kechris-Martin for n = 3). Let n be odd. Assume $\mathrm{Det}(\underset{\sim}{\Sigma}^1_n)$. Let $\leq$ be a Δ^1_n prewellordering of $\mathbb{R}$, and $\varphi : \mathbb{R} \to \kappa$ be the associated norm. Let $P(\xi,\bar{\beta})$ be Σ^1_{n+1} in the codes. Then

$$R(\bar{\beta}) \iff (\exists \xi < \kappa)P(\xi,\bar{\beta})$$

and

$$S(\bar{\beta}) \iff (\forall \xi < \kappa)P(\xi,\bar{\beta})$$

are both Σ^1_{n+1}.

To prove this, we need a lemma from descriptive set theory.

<u>Lemma</u> 38. Let n be odd. Assume $\mathrm{Det}(\underset{\sim}{\Delta}^1_{n-1})$. Let $\tau \in \mathbb{R}$. Suppose $F : \mathbb{R} \to \mathbb{R}$ has a $\Delta^1_n(\tau)$ graph and $\leq$ is a $\Delta^1_n(\tau)$ prewellordering of $\mathbb{R}$. Then there is a compact perfect set $K \subset \mathbb{R}$ such that (the real that codes) the tree of finite sequences whose branches form K, $T(K)$, is $\Delta^1_n(\tau)$, and there is an ordinal ξ such that $\alpha \in K \Rightarrow |F(\alpha)| = \xi$.

Usually when citing facts from descriptive set theory we have omitted the proofs. However, Lemma 38 is a new result, so we will sketch the proof here. The results about Baire category which are needed for this proof can be found in Kechris [5]. The proofs in that paper actually use $\mathrm{Det}(\underset{\sim}{\Sigma}^1_{n+1})$; but they can be improved to only $\mathrm{Det}(\underset{\sim}{\Delta}^1_{n-1})$ using the techniques of [7].

<u>Proof of Lemma</u> 38. Let $\alpha \leq_F \beta$ iff $F(\alpha) \leq F(\beta)$. Then $\leq_F$ is also a $\Delta^1_n(\tau)$ prewellordering of $\mathbb{R}$. $\mathrm{Det}(\underset{\sim}{\Delta}^1_{n-1})$ implies that $\leq_F$ and its initial segments have the property of Baire, as subsets of $\mathbb{R}^2$. By the Kuratowski-Ulam Theorem (the category version of Fubini's Theorem; see [15]), at least one of the equivalence classes of this prewellordering must be nonmeager. Let E be the $\leq_F$-least nonmeager equivalence class. Then for all $\alpha,\beta \in E$, $|F(\alpha)| = |F(\beta)|$. Let $\xi = F(\alpha)$, for some $\alpha \in E$; it will suffice to find a compact perfect $K \subset E$ such that $T(K)$ is $\Delta^1_n(\tau)$.

$E_\alpha = \{\beta : \beta \leq_F \alpha \;\&\; \alpha \leq_F \beta\}$ = the equivalence class of α. Since $\leq_F$ is a $\Delta^1_n(\tau)$ prewellordering, $\{(\alpha,\beta) : \beta \in E_\alpha\}$ is $\Delta^1_n(\tau)$. Let N_k denote the kth basic open neighborhood.

$$\alpha \in E \iff (\exists k)[(\mathbb{R} - E_\alpha) \cap N_k \text{ is meager}] \;\&\; (\forall \beta)[(\beta \leq_F \alpha \;\&\; \neg\alpha \leq_F \beta) \to E_\beta \text{ is meager}].$$

The first conjunct is equivalent to E_α being nonmeager, due to the property of Baire. By [5, Theorem 2.2.5], if $G \subset \mathbb{R}^2$ is Σ^1_n, and $G_\alpha = \{\beta : G(\alpha,\beta)\}$, then $\{\alpha : G_\alpha \text{ is not meager}\}$ is Σ^1_n, assuming $\mathrm{Det}(\underset{\sim}{\Delta}^1_{n-1})$. Therefore, E is $\Pi^1_n(\tau)$.

Assuming $\mathrm{Det}(\underset{\sim}{\Delta}^1_{n-1})$, for odd n, if a Π^1_n set is nonmeager, then player I has a Δ^1_n winning strategy for the Banach-Mazur game associated with that set [5, Corollary 4.2.2]. Using the relativized version of this theorem, I has a $\Delta^1_n(\tau)$ winning strategy σ in the Banach-Mazur game associated with E.

For any $\alpha \in {}^{\omega}2$, let $\sigma^*[\alpha]$ denote the real $\sigma(\emptyset) \frown \alpha(0) \frown \sigma(\alpha(0)) \frown \alpha(1) \frown \sigma(\alpha(0),\alpha(1)) \frown \alpha(2) \frown \sigma(\alpha(0),\alpha(1),\alpha(2)) \frown \ldots$.

That is, $\sigma^*[\alpha]$ is the outcome of the Banach-Mazur game in which I plays according to the strategy σ and II plays the single integer $\alpha(n-1)$ on his nth move. The function $f : {}^{\omega}2 \to \mathbb{R}$ defined by $f(\alpha) = \sigma^*[\alpha]$ is continuous and one-to-one. So it is a homeomorphism. Let $K = \{f(\alpha) : \alpha \in {}^{\omega}2\}$. ${}^{\omega}2$ is compact and perfect, so K is. Since σ is a winning strategy for I for E, $K \subset E$.

$T(K) = \{s \in \mathrm{Seq} : (\exists m)(\exists k_0,\ldots,k_m)[(\forall i \leq m)(k_i = 0 \vee k_i = 1)$ & (s is an initial segment of the finite sequence

$$(k_0 \frown \sigma(k_0) \frown k_1 \frown \sigma(k_0,k_1) \frown \ldots \frown k_m \frown \sigma(k_0,k_1,\ldots,k_m))]\}.$$

σ is a $\Delta^1_n(\tau)$ real, so $T(K)$ is $\Delta^1_n(\tau)$. ⊣

Proof of Theorem 37.

$$\begin{aligned} R(\bar\beta) &\iff (\exists \xi < \kappa)P(\xi,\bar\beta) \\ &\iff \exists\alpha\, P^*(\alpha,\bar\beta). \end{aligned}$$

P^* is Σ^1_{n+1}, so R is too.

$$\begin{aligned} S(\bar\beta) &\iff (\forall \xi < \kappa)P(\xi,\bar\beta) \\ &\iff \forall\alpha\, P^*(\alpha,\bar\beta), \text{ where } P^* \text{ is } \Sigma^1_{n+1} \text{ and invariant on } \alpha \\ &\iff \forall\alpha\exists\gamma\, Q(\alpha,\bar\beta,\gamma), \text{ where } Q \text{ is } \Pi^1_n \text{ and invariant on } \alpha \text{ (Lemma 36)}. \end{aligned}$$

Consider the game $G(\bar\beta)$ defined as follows: I plays a real which codes, in some recursive manner, two reals, τ and δ. II plays a real α.

$$\text{I wins iff } (\exists\gamma \in \Delta^1_n(\tau,\delta))Q(\alpha,\bar\beta,\gamma).$$

$$\text{II wins iff } (\forall\gamma \in \Delta^1_n(\tau,\delta))\neg Q(\alpha,\bar\beta,\gamma).$$

Let $W(\bar\beta,x) \iff x$ is in the payoff set for player I for $G(\bar\beta)$ (i.e. I wins the game $G(\bar\beta)$ if x is the outcome).

$\mathrm{Det}(\underset{\sim}{\Delta}^1_{n-1})$ implies that Π^1_n (n odd) is closed under quantification of the form $\exists y \in \Delta^1_n(z)$. So the definition of the payoff set for I for $G(\bar\beta)$ is Π^1_n, with parameters $\tau,\delta,\alpha,\bar\beta$. τ,δ, and α are uniformly recursive in x, so

(*) $\qquad W$ is Π^1_n.

Claim. $\forall\alpha\exists\gamma\, Q(\alpha,\bar\beta,\gamma)$ iff I has a winning strategy for $G(\bar\beta)$.

Assuming the claim,

$$S(\bar{\beta}) \Longleftrightarrow \forall\alpha\exists\gamma\, Q(\alpha,\bar{\beta},\gamma)$$

$\Longleftrightarrow$ I has a winning strategy for $G(\bar{\beta})$

$\Longleftrightarrow \exists\sigma\forall\rho\, W(\bar{\beta},\sigma*\rho)$ (where $\sigma*\rho$ denotes the outcome of the game where I plays via the strategy σ and II plays via the strategy ρ).

By (*), this is Σ^1_{n+1}, which proves the theorem.

Proof of Claim. ($\Leftarrow$)

I has a winning strategy for $G(\bar{\beta})$

$\Rightarrow (\forall\alpha$ played by II$)(\exists\gamma\in\Delta^1_n(\tau,\delta))Q(\alpha,\bar{\beta},\gamma)$

$\Rightarrow \forall\alpha\exists\gamma\, Q(\alpha,\bar{\beta},\gamma)$.

($\Rightarrow$)

Assume I does not have a winning strategy for $G(\bar{\beta})$. By (*), the payoff set for player I for $G(\bar{\beta})$ is $\Pi^1_n(\bar{\beta})$. So $G(\bar{\beta})$ is determined. Let τ be a winning strategy for II for $G(\bar{\beta})$.

Consider the function

$$F(\delta) = \text{subsequence of } ([\langle\tau,\delta\rangle]*\tau) \text{ with odd coordinates.}$$

That is, F maps δ to the real played by II in the game where II plays according to the strategy τ (i.e. the strategy coded by the real τ) and I plays $\langle\tau,\delta\rangle$, the real which codes the pair of reals (τ,δ).

F is recursive in τ, hence has a $\Delta^1_n(\tau)$ graph. Using Lemma 38, pick a K and ξ such that K is compact, perfect, and coded by a $\Delta^1_n(\tau)$ tree $T(K)$, and $(\forall\delta\in K)(|F(\delta)| = \xi)$. Let δ_0 be a fixed element of K. Let $\alpha_0 = F(\delta_0)$.

Since K is compact perfect, there is a canonical homeomorphism $h : K \to {}^{\omega}2$ between K and the Cantor set. Since $T(K)$ is $\Delta^1_n(\tau)$ the graph of h is $\Delta^1_n(\tau)$. There is a bijection b between ${}^{\omega}2$ and the reals $({}^{\omega}\omega)$ which has Δ^1_1 graph. Let $B = b\circ h$. Thus $B : K \to \mathbb{R}$ is a bijection with $\Delta^1_n(\tau)$ graph.

Let γ' be an arbitrary real. Let $\delta = B^{-1}(\gamma')$. Then $\gamma'\in\Delta^1_n(\tau,\delta)$. Let $\alpha = F(\delta)$. That means α is the real played by II, when I plays $\langle\tau,\delta\rangle$ and II plays according to the strategy τ. Since τ is a winning strategy for II, by definition of the payoff set,

$$(\forall\gamma\in\Delta^1_n(\tau,\delta))\neg Q(\alpha,\bar{\beta},\gamma).$$

Hence $\neg Q(\alpha,\bar{\beta},\gamma')$. $\alpha = F(\delta)$, $\alpha_0 = F(\delta_0)$. δ and δ_0 are both in K, so $|\alpha| = |\alpha_0| = \xi$. Since Q is invariant, $\neg Q(\alpha_0,\bar{\beta},\gamma')$. γ' is arbitrary, so $\forall\gamma\neg Q(\alpha_0,\bar{\beta},\gamma)$, and hence $\exists\alpha\forall\gamma\neg Q(\alpha,\bar{\beta},\gamma)$. This proves the claim, which proves the theorem. $\dashv$

The proof of Theorem 37 given above is a modification of the original Harrington-Kechris proof which is due to Moschovakis.

Corollary 39. Assume $Det(\underset{\sim}{\Sigma}^1_n)$. Let $\leq$ be a Δ^1_n prewellordering of $\mathbb{R}$, and $\varphi : \mathbb{R} \to \kappa$ be the associated norm. Let $P(\xi,\bar{\eta},\bar{\beta})$ be Σ^1_{n+1} in the codes (where $\xi,\bar{\eta}$ are ordinal variables). Then

$$R(\bar{\eta},\bar{\beta}) \Leftrightarrow (\exists \xi < \kappa)P(\xi,\bar{\eta},\bar{\beta})$$

and

$$S(\bar{\eta},\bar{\beta}) \Leftrightarrow (\forall \xi < \kappa)P(\xi,\bar{\eta},\bar{\beta})$$

are both Σ^1_{n+1} in the codes.

Proof. Let $P^*(\xi,\alpha_1,\ldots,\alpha_m,\bar{\beta}) \Leftrightarrow P(\xi,|\alpha_1|,\ldots,|\alpha_m|,\bar{\beta})$ and apply Theorem 37 to P^*. ⊣

Theorem 40. (Moschovakis [13, Theorem 3]). Assume AD. Let $m \geq 1$. Let $\varphi : \mathbb{R} \to \kappa$ be a Δ^1_m norm. Every relation on κ is $\underset{\sim}{\Delta}^1_m$ in the codes, with respect to φ.

Definition. Let $\varphi : \mathbb{R} \to \kappa$ be a Δ^1_m norm. Let $P \subset \kappa$, $P^* \subset \mathbb{R}$ such that $P^*(\alpha) \Leftrightarrow P(|\alpha|)$, and let $\mathcal{E} \in \mathbb{R}$ be a $\underset{\sim}{\Delta}^1_m$ code for the set P^*. Then $\mathcal{E}$ is also called the code for P, with respect to φ, and $X_{\mathcal{E}}$ denotes P.

Theorem 40 then says that for every $P \subset \kappa$, there is an $\mathcal{E}$ such that $P = X_{\mathcal{E}}$. The $\mathcal{E}$ is not unique.

Definition. $\underset{\sim}{\delta}^1_m = \sup\{\xi : \xi$ is the length of a $\underset{\sim}{\Delta}^1_m$ prewellordering of $\mathbb{R}\}$.

Theorem 41. (Kechris - Moschovakis). Assume AD. $\underset{\sim}{\delta}^1_n$ is a regular cardinal and there is an ordinal λ_n such that

1. $\underset{\sim}{\delta}^1_n = \lambda_n^+$
2. There exists a Δ^1_n norm on $\mathbb{R}$ of length λ_n.

That $\underset{\sim}{\delta}^1_n$ is a regular cardinal is Theorem 6 of [13]. 1 is Theorem 2B-5 of Section 2 of [4]; 2 is implicit in the proof of that theorem.

Notation. For odd n, let λ_n be the ordinal of Theorem 41, and let ψ^n be a fixed Δ^1_n norm on $\mathbb{R}$ of length λ_n.

Since there is no Δ^1_n prewellordering of length $\underset{\sim}{\delta}^1_n$, a different method of coding subsets of $\underset{\sim}{\delta}^1_n$ than that used above on κ, is needed. This new coding is defined below.

Code subsets of λ_n and of $\lambda_n \times \lambda_n$ with respect to the norm ψ^n, as described in Theorem 40 and the definition that follows it.

Let $A \subset \underset{\sim}{\delta}^1_n$. Let

$A^* = \{\mathcal{E} \in \mathbb{R} : X_{\mathcal{E}} \subset \lambda_n \times \lambda_n$ is a wellordering of a subset of λ_n, and its order type is in $A\}$.

<u>Definition</u>. $A \subset \underset{\sim}{\delta}^1_n$ is called Γ <u>in the codes</u> if A^* is in Γ.

Let Θ be a sufficiently large finite subset of ZF + DC.

<u>Lemma</u> 42. Let κ be an uncountable regular cardinal. Let $A \subset \kappa$. For all $\alpha \in \mathbb{R}$, $\alpha \in L[A] \Leftrightarrow (\exists \xi < \kappa)(\exists \eta < \xi)(\alpha \in L_\xi[A \cap \eta]$ & $L_\xi[A \cap \eta] \models \Theta)$.

The proof is a fairly straightforward collapsing argument, which we omit.

<u>Theorem</u> 43. Assume AD. If $A \subset \underset{\sim}{\delta}^1_n$ is Δ^1_{n+1} in the codes, then $P(\alpha) \Leftrightarrow \alpha \in (\mathbb{R} \cap L[A])$ is Σ^1_{n+1}.

<u>Proof</u>. By Theorem 41, $\underset{\sim}{\delta}^1_n$ is a regular cardinal and it is obviously uncountable. By Lemma 42,

$$P(\alpha) \Leftrightarrow (\exists \xi < \underset{\sim}{\delta}^1_n)(\exists \eta < \xi)(\alpha \in L_\xi[A \cap \eta] \;\&\; L_\xi[A \cap \eta] \models \Theta).$$

Since the cardinality of $L_\xi[A \cap \eta]$ is less than $\underset{\sim}{\delta}^1_n$, and $\underset{\sim}{\delta}^1_n = \lambda_n^+$ (Theorem 41), $L_\xi[A \cap \eta]$ is isomorphic to a structure $\langle M,E\rangle$, where M is a subset of the ordinal λ_n, and E is a binary relation on λ_n. Therefore

$P(\alpha) \Leftrightarrow$

$(\exists M)(\exists E)(\exists \zeta_1 < \lambda_n)(\exists \zeta_2 < \lambda_n)[M \subset \lambda_n$ & $E \subset M \times M$ & $\zeta_1, \zeta_2 \in M$ & $\langle M,E\rangle \models (\Theta + V = L[\zeta_1])$ & E is wellfounded & $\langle M,E\rangle \models (\zeta_2$ is an ordinal & $\zeta_1 \subset \zeta_2)$ & (if Π is the collapsing map of $\langle M,E\rangle$, then $\Pi(\zeta_1) = A \cap \Pi(\zeta_2)$ & $\alpha \in \Pi[M])]$.

It must be shown that this definition is Σ^1_{n+1}.

M and E are subsets of λ_n and $\lambda_n \times \lambda_n$, respectively, so by Theorem 40, M and E are $\underset{\sim}{\Delta}^1_n$ in the codes. Therefore the existential quantification $\exists M$ is equivalent to $\exists \beta \in \mathbb{R}$ (i.e. $\exists$ a $\underset{\sim}{\Delta}^1_n$ code), and similarly for $\exists E$. $(\exists \zeta_1 < \lambda_n)$ and $(\exists \zeta_2 < \lambda_n)$ are the type of quantifier which is dealt with in Corollary 39. So in order to show that P is Σ^1_{n+1}, all that needs to be proved is that the formula in brackets is Σ^1_{n+1} in the codes.

Truth in $\langle M,E\rangle$ involves quantification only over M, that is, over ordinals less than λ_n. $(\langle M,E\rangle \models \xi_1 \in \xi_2)$ iff $(\xi_1, \xi_2) \in E$, which is Δ^1_n in the codes, and an induction on the length of formulas, using Corollary 39, shows that $\langle M,E\rangle \models \varphi$ is Σ^1_{n+1} in the codes, for any formula φ.

$\alpha \in \Pi[M] \Leftrightarrow$

$(\exists \xi_\alpha < \lambda_n)(\forall m)(\forall k)(\exists \xi_m < \lambda_n)(\exists \xi_k < \lambda_n)[\xi_\alpha \in M$ & $\xi_m \in M$ & $\xi_k \in M$ & $\langle M,E\rangle \models (\xi_m$ is m & ξ_k is k & ξ_α is a real) & $(\alpha(m) = k \leftrightarrow \langle M,E\rangle \models \xi_\alpha(\xi_m) = \xi_k)]$.

Again using Corollary 39, and the fact that truth is Σ^1_{n+1} in the codes, $\alpha \in \Pi[M]$ is Σ^1_{n+1} in the codes.

$\Pi(\zeta_1) = A \cap \Pi(\zeta_2) \Leftrightarrow$

(*) $$(\forall \xi < \lambda_n)[(\xi, \zeta_2) \in E \rightarrow ((\xi, \zeta_1) \in E \longleftrightarrow \Pi(\zeta_1) \in A)] .$$

$\Pi(\zeta_1) \in A \iff$

$\exists \mathcal{E} \exists F\,(F\colon (\mathrm{Ord} \upharpoonright \zeta_1)^M \to \lambda_n$ & F is an order preserving bijection from ζ_1 onto $|X_{\mathcal{E}}|$ & $\mathcal{E} \in A^*) \iff$

$\forall \mathcal{E} \forall F\,[(F\colon (\mathrm{Ord} \upharpoonright \zeta_1)^M \to \lambda_n$ & F is an order preserving bijection from ζ_1 onto $|X_{\mathcal{E}}|) \to \mathcal{E} \in A^*]$,

where A^* is the set of reals used in coding A.

By hypothesis, A is Δ^1_{n+1} in the codes, so A^* is Δ^1_{n+1}. As above, the quantifiers $\exists F$ and $\forall F$ are equivalent to real quantifiers. So the formula $\Pi(\zeta_1) \in A$ is Δ^1_{n+1} in the codes. This fact, plus Corollary 39, applied to the formula (*), shows that $\Pi(\zeta_1) = A \cap \Pi(\zeta_2)$ is Σ^1_{n+1} in the codes.

E is well founded $\iff$

$\neg (\exists F)(F\colon \omega \to M \;\&\; (\forall m)(\forall k)(m < k \to (F(k),F(m)) \in E))$.

$\exists F$ is a real quantifier, so this is easily Σ^1_{n+1} in the codes. ⊣

T^n is a set whose members are finite sequences of ordered pairs of ordinals. By coding sequences, T^n can effectively be identified with a set of ordinals, A^n. Let $\langle a_0,\xi_0,\ldots,a_k,\xi_k\rangle$ denote the ordinal that codes $((a_0,\xi_0),\ldots,(a_k,\xi_k))$. $L[A^n] = L[T^n]$, and since T^n is the associated tree of a Π^1_n scale, clearly all ordinals in T^n, and hence in A^n, are less than $\underset{\sim}{\delta}^1_n$. So by Theorem 43, to prove that $\mathbb{R} \cap L[T^n]$ is a Σ^1_{n+1} set, all that remains to be shown is that A^n is Δ^1_{n+1} in the codes. The next theorem fills in this last gap in the proof of the Harrington-Kechris Theorem (Theorem 21).

<u>Theorem</u> 44. Assume AD. Let $\{\varphi_m\}_{m\in\omega}$ be a Π^1_n scale on a Π^1_n set P. Let T be its associated tree. Let

$$A = \{\langle a_0,\xi_0,\ldots,a_k,\xi_k\rangle : ((a_0,\xi_0),\ldots,(a_k,\xi_k)) \in T\}.$$

Then A is Δ^1_{n+1} in the codes.

<u>Remark</u>. Codings are as before, via the same norm, ψ^n (introduced after Theorem 41). The norms φ_m are not used in the coding.

<u>Proof</u>. Let $U(m,\mathcal{E},\alpha) \iff$

$\alpha \in P$ & $\mathcal{E}$ is a code & $(X_{\mathcal{E}} \subset \lambda_n \times \lambda_n$ is a well ordering of order type $\varphi_m(\alpha))$.

<u>Claim</u>. U is Δ^1_{n+1}.

<u>Proof of claim</u>. $U(m,\mathcal{E},\alpha) \iff$

$\alpha \in P$ & $\mathcal{E}$ is a code & $X_{\mathcal{E}}$ is a wellordering &

$\exists \rho\,[\rho$ codes a $\underset{\sim}{\Delta}^1_n$ relation $\underset{\sim}{\Delta}\rho\,(\delta,\beta)$ &

$\forall\delta\forall\beta(\underset{\sim}{\Delta}\rho\,(\delta,\beta) \to (|\delta| \in \mathrm{Field}(X_{\mathcal{E}})$ & $\varphi_m(\beta) < \varphi_m(\alpha)))$ &

$\forall\delta\forall\delta'\forall\beta\forall\beta'((\varphi_m(\beta) = \varphi_m(\beta') < \varphi_m(\alpha)$ &

$|\delta| = |\delta'|$ & $\underset{\sim}{\Delta}\rho\,(\delta,\beta)) \to \underset{\sim}{\Delta}\rho\,(\delta',\beta'))$ &

$$\forall\delta\forall\delta'\forall\beta\forall\beta'((\underset{\sim}{\Delta}\rho(\delta,\beta)\ \&\ \underset{\sim}{\Delta}\rho(\delta',\beta'))\rightarrow$$
$$(|\delta| = |\delta'| \longleftrightarrow \varphi_m(\beta) = \varphi_m(\beta'))\ \&$$
$$\forall\delta(|\delta| \in \text{Field}(X_{\mathcal{E}}) \rightarrow \exists\beta\ (\varphi_m(\beta) < \varphi_m(\alpha)\ \&\ \underset{\sim}{\Delta}\rho(\delta,\beta)))\ \&$$
$$\forall\beta\ (\varphi_m(\beta) < \varphi_m(\alpha) \rightarrow \exists\delta(\underset{\sim}{\Delta}\rho(\delta,\beta)))\ \&$$
$$\forall\delta\forall\delta'\forall\beta\forall\beta'((\underset{\sim}{\Delta}\rho(\delta,\beta)\ \&\ \underset{\sim}{\Delta}\rho(\delta',\beta'))\rightarrow$$
$$((|\delta|,|\delta'|) \in X_{\mathcal{E}} \longleftrightarrow \varphi_m(\beta) \le \varphi_m(\beta')))]$$

So the above formula is Σ^1_{n+1}. That it is actually equivalent to U, follows from the fact that the unique relation $\underset{\sim}{\Delta}\rho$ with the properties described in the above formula, is in fact $\underset{\sim}{\Delta}^1_n$; this is a consequence of Theorem 40.

So U is Σ^1_{n+1}. To see that U is also Π^1_{n+1}, note that

$\neg U(m,\mathcal{E},\alpha) \Longleftrightarrow$

$\alpha \notin P \vee \mathcal{E}$ is not a code $\vee X_{\mathcal{E}}$ is not a wellordering $\vee [\alpha \in P\ \&\ \mathcal{E}$ is a code &
$X_{\mathcal{E}}$ is a wellordering & $(\exists\beta(\varphi_m(\beta) < \varphi_m(\alpha)\ \&\ U(m,\mathcal{E},\beta)) \vee$
$\exists\beta(\beta \in P\ \&\ \varphi_m(\alpha) < \varphi_m(\beta)\ \&\ U(m,\mathcal{E},\beta)))]$.

This proves the claim.

The ordinal

$\langle a_0,\xi_0,\ldots,a_k,\xi_k\rangle = \langle a_0,|X_{\mathcal{E}_0}|,\ldots,a_k,|X_{\mathcal{E}_k}|\rangle$ (where $\mathcal{E}_0,\ldots,\mathcal{E}_k$ are reals) is in A

iff $(\exists\alpha \in P)[(\forall m \le k)(\alpha(m) = a_m\ \&\ U(m,\mathcal{E}_m,\alpha))]$

iff $(\forall\alpha_0,\ldots,\alpha_k \in P)[(\forall m \le k)U(m,\mathcal{E}_m,\alpha_m) \rightarrow$
$\exists\alpha(\ \varphi_0(\alpha) \le \varphi_0(\alpha_0)\ \&\ (\forall m \le k)(\alpha(m) = a_m\ \&\ \varphi_m(\alpha) \le \varphi_m(\alpha_m)))]$.

U is Δ^1_{n+1} and the norms φ_m are all Δ^1_n; so the first formula is Σ^1_{n+1} and the second is Π^1_{n+1}. ⊣

References

[1] F. R. Drake, Set Theory; An Introduction to Large Cardinals, North-Holland, Amsterdam, 1974.

[2] J. T. Green, Ph.D. Thesis, University of California, Berkeley, to appear.

[3] T. J. Jech, Lectures in Set Theory with Particular Emphasis on the Method of Forcing, Lecture Notes in Math., 217, Springer, Berlin, 1971.

[4] A. S. Kechris, Ph.D. Thesis, University of California, Los Angeles, 1972.

[5] A. S. Kechris, Measure and category in effective descriptive set theory, Ann. Math. Logic, 5 (1972/73), 337-384.

[6] A. S. Kechris, The theory of countable analytical sets, Trans. Amer. Math. Soc., 202 (1975), 259-298.

[7] A. S. Kechris, On a notion of smallness for subsets of the Baire space, Trans. Amer. Math. Soc., 229 (1977), 191-207.

[8] A. S. Kechris and Y. N. Moschovakis, Notes on the Theory of Scales, Circulated multilithed manuscript, 1971. Reprinted in this volume.

[9] R. Mansfield, Perfect subsets of definable sets of real numbers, Pacific J. Math., 35 (1970), 451-457.

[10] D. A. Martin, Measurable cardinals and analytic games, Fundamenta Mathematicae, 66 (1970), 287-291.

[11] D. A. Martin, $\underset{\sim}{\Delta}^1_{2n}$ determinacy implies $\underset{\sim}{\Sigma}^1_{2n}$ determinacy, Circulated notes, 1973.

[12] D. A. Martin, Borel determinacy, Ann. Math., 102 (1975), 363-371.

[13] Y. N. Moschovakis, Determinacy and Prewellorderings of the Continuum, Math. Logic and Foundations of Set Theory (Proc. Internat. Colloq., Jerusalem, 1968), North-Holland, Amsterdam, 1970, 24-62.

[14] Y. N. Moschovakis, Descriptive Set Theory, to appear.

[15] J. C. Oxtoby, Measure and Category, Springer, New York, 1971.

[16] J. R. Shoenfield, Mathematical Logic, Addison-Wesley, Reading, Mass., 1961.

[17] R. M. Solovay, On the Cardinality of Σ^1_2 sets of reals, Foundations of Mathematics, Symposium papers commemorating the 60th birthday of Kurt Gödel, Springer-Verlag, 1966, 58-73.

[18] R. M. Solovay, Measurable Cardinals and the Axiom of Determinateness, Lecture notes prepared in connection with the Summer Institute on Axiomatic Set Theory held at UCLA, Summer, 1967.

[19] R. M. Solovay, A model of set theory in which every set is Lebesgue measurable, Ann. Math., 92 (1970), 1-56.

AD AND PROJECTIVE ORDINALS

Alexander S. Kechris[1]
Department of Mathematics
California Institute of Technology
Pasadena, California 91125

This is an unpolished exposition of some work in the theory of projective ordinals under the hypothesis of definable determinacy. This is understood here as the hypothesis that every set of reals in $L[\mathbb{R}]$ is determined. Since the projective ordinals are absolute between the real world and $L[\mathbb{R}]$ we carry this study entirely <u>within</u> $L[\mathbb{R}]$. Thus we will use the full Axiom of Determinacy (AD) together with ZF + DC (DC is of course the only choice principle that is preserved under this transition to $L[\mathbb{R}]$).

Starting with the work of Martin, Moschovakis, and Solovay a decade ago, the exciting and unexpected possibility was discovered that one could calculate <u>precisely</u> the projective ordinals in terms of the aleph function. Indeed $\underset{\sim}{\delta}^1_1 = \omega_1$ is a classical result and Martin computed that $\underset{\sim}{\delta}^1_2 = \omega_2$, $\underset{\sim}{\delta}^1_3 = \omega_{\omega+1}$ and $\underset{\sim}{\delta}^1_4 = \omega_{\omega+2}$ (the last independently also due to Kunen). The computation of $\underset{\sim}{\delta}^1_5$ and the higher $\underset{\sim}{\delta}^1_n$'s is now the central problem of this theory. Kunen in 1971 has originated a major program towards achieving that goal, developing along the way some very important and powerful techniques. Part of his work is presented in the later sections of this survey and in Solovay's paper, "A Δ^1_3 coding of the subsets of ω_ω", in this volume. We are planning to present the rest in a sequel paper, along with other more recent advances in this area.

Work in descriptive set theory over the last ten years has resulted in a basically complete understanding of the analytical sets of the 3rd and 4th level of the analytical hierarchy, fully analogous to that provided by the classical effective theory for the first two. Moreover, recent results in this subject show, in our opinion, that a complete structure theory for all analytical sets at level 5 and beyond is essentially reduced to the problem of the precise calculation of the $\underset{\sim}{\delta}^1_n$'s for $n \geq 5$.

---

[1]Preparation for this paper was partially supported by NSF Grant MCS 76-17254. The author would like to thank R. M. Solovay for many interesting and helpful discussions on the topics presented in this paper.

Remark. Since some of the results we state below need actually only weaker forms of AD (like PD, etc.) we have put explicitly in the statements of the theorems the set theoretical assumptions which are used to establish them, beyond ZF + DC.

1. Definitions and the general picture

Definition. For all $n \geq 1$, let

$$\underset{\sim}{\delta}^1_n = \sup\{\xi : \xi \text{ is the length of a } \underset{\sim}{\Delta}^1_n \text{ prewellordering of } \mathcal{R} \ (= \omega^\omega)\}.$$

The following facts are known, granting AD:

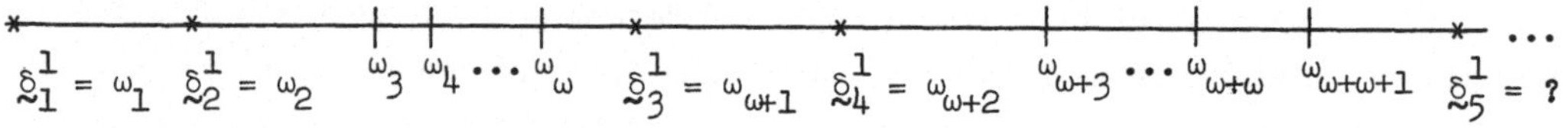

and in general for $n \geq 0$,

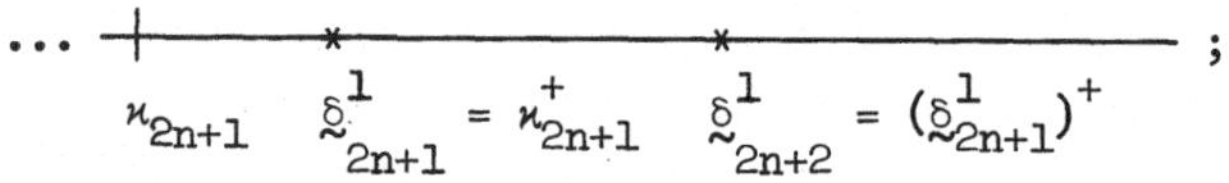

1. All the $\underset{\sim}{\delta}^1_n$ are cardinals.
2. $\underset{\sim}{\delta}^1_{2n+2} = (\underset{\sim}{\delta}^1_{2n+1})^+$.
3. $\underset{\sim}{\delta}^1_{2n+1} = \varkappa^+_{2n+1}$, where $\varkappa_{2n+1}$ is a cardinal of cofinality ω.
4. All $\underset{\sim}{\delta}^1_n$ are regular. (Note: without choice this does not follow from the fact that they are successor cardinals.) In fact:
5. All $\underset{\sim}{\delta}^1_n$ are measurable.

Proofs of these and other results will be given in the sequel.

2. For all n, $\underset{\sim}{\delta}^1_n$ is a cardinal

Definition. Let $F \subseteq \mathcal{R}$, $\leq$ a prewellordering on F, and let $\varphi : F \xrightarrow{\text{onto}} \xi =$ length $(\leq)$, be the canonical norm associated with it (i.e. $\alpha \leq \beta \Leftrightarrow \varphi(\alpha) \leq \varphi(\beta)$). If $f : \xi \to P(\mathcal{R})$, (where $P(\mathcal{R})$ = power set of $\mathcal{R}$), put

$$\text{Code } (f; \leq) = \{(\alpha,\beta) : \alpha \in F \ \& \ \beta \in f(\varphi(\alpha))\}.$$

If $A \subseteq \xi^n$, put

$$\text{Code } (A; \leq) = \{(\alpha_1 \dots \alpha_n) \in F^n : (\varphi(\alpha_1) \dots \varphi(\alpha_n)) \in A\}.$$

Definition. If $f : \xi \to P(\mathcal{R})$, we call $g : \xi \to P(\mathcal{R})$ a choice subfunction of f if

1. $\forall \eta < \xi$, $g(\eta) \subseteq f(\eta)$.
2. $\forall \eta < \xi \ [f(\eta) \neq \emptyset \Rightarrow g(\eta) \neq \emptyset]$.

Theorem 2.1. [AD] (The Coding Lemma; Moschovakis 70). Let $\leq$ be a $\underset{\sim}{\Delta}^1_n$ prewellordering of a subset of $\mathbb{R}$ with length ξ. Then every function $f : \xi \to P(\mathbb{R})$ has a choice subfunction g such that Code $(g;\leq)$ is $\underset{\sim}{\Sigma}^1_n$.

Sketch of proof. Let $G \subseteq \mathbb{R} \times \mathbb{R} \times \mathbb{R}$ be universal for the $\underset{\sim}{\Sigma}^1_n$ subsets of $\mathbb{R} \times \mathbb{R}$; α is a code for a $\underset{\sim}{\Sigma}^1_n$ set $Q \subseteq \mathbb{R} \times \mathbb{R}$ if $Q = G_\alpha = \{(\beta,\gamma) : G(\alpha,\beta,\gamma)\}$. Given $f : \xi \to P(\mathbb{R})$, let for all $\eta < \xi$, $f_\eta : \xi \to P(\mathbb{R})$, be the restriction of f to η, defined to be $\emptyset$ outside η. Suppose there is an f with no good choice subfunction, where g is good if Code $(g;\leq)$ is $\underset{\sim}{\Sigma}^1_n$. Let $\eta_0 \leq \xi$ be least such that f_{η_0} has no good choice subfunction. Clearly η_0 is limit. Consider the following game: Player I plays $\alpha \in \mathbb{R}$, player II plays $\beta \in \mathbb{R}$, and II wins if whenever α codes a good choice subfunction of f_η for some $\eta < \eta_0$, then β codes a good choice subfunction of f_θ, where $\eta < \theta < \eta_0$.

Case I: I has a winning strategy. Then for each β there is an $\eta(\beta) < \eta_0$ and a good choice subfunction $g_{\eta(\beta)}$ of $f_{\eta(\beta)}$, with code given by I's strategy applied to β. Notice that $\sup\{\eta(\beta) : \beta \in \mathbb{R}\}$ must be bounded below η_0 (otherwise the union of the $g_{\eta(\beta)}$ will be a good choice subfunction of f_{η_0}). But since η_0 was chosen least, II can easily beat I's winning strategy. Hence Case I never occurs.

Case II: II has a winning strategy. Using the recursion theorem we can find a partial continuous function h such that for all $w \in$ Field $(\leq)$ with $\varphi(w) < \eta_0$, $h(w)$ is defined and is a $\underset{\sim}{\Sigma}^1_n$-code for a good choice subfunction $g_{\theta(w)}$ of $f_{\theta(w)}$ with $\varphi(w) < \theta(w) < \eta_0$. But then if

$$g_{\eta_0}(\eta) = \bigcup_{\substack{w \in \text{Field}(\leq) \\ \varphi(w) < \eta_0}} g_{\theta(w)}(\eta),$$

g_{η_0} is a choice subfunction of f_{η_0} and $(\alpha,\beta) \in \text{Code}(g_{\eta_0};\leq) \Leftrightarrow \exists w[w \in \text{Field}(\leq)$ & $\varphi(w) < \eta_0$ & $G(h(w),\alpha,\beta)]$, hence g_{η_0} is good, contradicting the choice of η_0. ⊣

Corollary. [AD] For every $A \subseteq \xi^n$, Code $(A;\leq)$ is $\underset{\sim}{\Delta}^1_n$.

Proof. Let us take $n = 1$ for notational implicity. Let α_0, α_1 be distinct reals, and define $f : \xi \to P(\mathbb{R})$ by $f(\eta) = \begin{cases} \{\alpha_0\}, & \text{if } \eta \in A \\ \{\alpha_1\}, & \text{if } \eta \notin A \end{cases}$. Then the only choice subfunction of f is f itself. Hence Cod $(f;\leq)$ is $\underset{\sim}{\Sigma}^1_n$ by the coding Lemma, and

$$\alpha \in \text{Code}(A;\leq) \Leftrightarrow (\alpha,\alpha_0) \in \text{Code}(f;\leq)$$
$$\Leftrightarrow \alpha \in \text{Field}(\leq) \;\&\; (\alpha,\alpha_1) \notin \text{Code}(f;\leq),$$

hence Code $(A;\leq)$ is $\underset{\sim}{\Delta}^1_n$. ⊣

Theorem 2.2. [AD] (Moschovakis 70). For all $n \geq 1$, $\underset{\sim}{\delta}^1_n$ is a cardinal.

Proof. If not, let $f : \xi \to \underset{\sim}{\delta}^1_n$ be 1-1 and onto, where $\xi < \underset{\sim}{\delta}^1_n$. There is a prewellordering $\leq$ of $\mathcal{R}$ of length ξ which is $\underset{\sim}{\Delta}^1_n$. Let $<^*$ be defined on ξ by

$$\eta <^* \theta \Leftrightarrow f(\eta) < f(\theta).$$

Then by the corollary, Code $(<^*;\leq)$ is $\underset{\sim}{\Delta}^1_n$. But then Code$(<^*;\leq)$ is a prewellordering on $\mathcal{R}$ in $\underset{\sim}{\Delta}^1_n$ with length $\underset{\sim}{\delta}^1_n$, contradiction. ⊣

3. The $\underset{\sim}{\delta}^1_n$'s are successor cardinals

Definitions. Let $A \subseteq \mathcal{R}$. A norm on A is a map $\varphi : A \to$ Ordinals. The length of φ is the length of the prewellordering induced by φ on A, i.e. $\alpha \leq^{\varphi} \beta \Leftrightarrow \varphi(\alpha) \leq \varphi(\beta)$.

If $\underset{\sim}{\Gamma} = \underset{\sim}{\Pi}^1_n$ or $\underset{\sim}{\Sigma}^1_n$ and $A \in \underset{\sim}{\Gamma}$, then φ is a $\underset{\sim}{\Gamma}$-norm if the following relations are in $\underset{\sim}{\Gamma}$:

$$\alpha \leq^*_{\varphi} \beta \Leftrightarrow \alpha \in A \;\&\; [\beta \notin A \vee \varphi(\alpha) \leq \varphi(\beta)],$$

$$\alpha <^*_{\varphi} \beta \Leftrightarrow \alpha \in A \;\&\; [\beta \notin A \vee \varphi(\alpha) < \varphi(\beta)].$$

If every set in $\underset{\sim}{\Gamma}$ has a $\underset{\sim}{\Gamma}$-norm, we say that $\underset{\sim}{\Gamma}$ has the prewellordering property.

Theorem 3.1. [PD] (The Prewellordering Theorem; Martin 68, Moschovakis (see Addison-Moschovakis 68)). For all $n \geq 0$, $\underset{\sim}{\Pi}^1_{2n+1}$ and $\underset{\sim}{\Sigma}^1_{2n+2}$ have the prewellordering property (and $\underset{\sim}{\Sigma}^1_{2n+1}$, $\underset{\sim}{\Pi}^1_{2n+2}$ do not have the prewellordering property).

Definition. Let $A, B \subseteq \mathcal{R}$. We say that A is reducible to B if there is a total continuous function $f : \mathcal{R} \to \mathcal{R}$ such that $\alpha \in A \Leftrightarrow f(\alpha) \in B$. If $\underset{\sim}{\Gamma} = \underset{\sim}{\Sigma}^1_n$ or $\underset{\sim}{\Pi}^1_n$, a set $A \subseteq \mathcal{R}$ is called $\underset{\sim}{\Gamma}$-complete if $A \in \underset{\sim}{\Gamma}$ and every set $B \in \underset{\sim}{\Gamma}$ is reducible to A.

Wadge's Lemma. [AD] If $A, B \subseteq \mathcal{R}$, then either A is reducible to B or B is reducible to $\mathcal{R} - A$.

Proof. Consider the game in which I plays α, II plays β and II wins iff $\alpha \in A \Leftrightarrow \beta \in B$. ⊣

By this lemma every set in $\underset{\sim}{\Gamma} - \underset{\sim}{\Delta}$ (where $\underset{\sim}{\Gamma} = \underset{\sim}{\Sigma}^1_n$ or $\underset{\sim}{\Pi}^1_n$ and $\underset{\sim}{\Delta} = \underset{\sim}{\Delta}^1_n$) is $\underset{\sim}{\Gamma}$-complete.

Theorem 3.2 [PD] (Moschovakis 70). If φ is a $\underset{\sim}{\Pi}^1_{2n+1}$-norm on a $\underset{\sim}{\Pi}^1_{2n+1}$-complete set then

$$\text{length}(\varphi) = \underset{\sim}{\delta}^1_{2n+1}.$$

Definition. A scale on a set $A \subseteq \mathcal{R}$ is a sequence of norms $\{\varphi_n\}_{n\in\omega}$ on A such that for every sequence $\{\alpha_i\}_{i\in\omega}$ of members of A, if

1. $\lim_{i\to\infty} \alpha_i = \alpha$

and

2. For each n there is an ordinal λ_n such that $\varphi_n(\alpha_i) = \lambda_n$, for all large enough i, then $\alpha \in A$ and for all n, $\varphi_n(\alpha) \leq \lambda_n$.

The scale $\{\varphi_n\}_{n\in\omega}$ is a λ-scale if length $(\varphi_n) \leq \lambda$, $\forall n$.

If $\underset{\sim}{\Gamma} = \underset{\sim}{\Sigma}^1_n$ or $\underset{\sim}{\Pi}^1_n$, we call $\{\varphi_n\}_{n\in\omega}$ a Γ-scale if the two relations

$$S(n,\alpha,\beta) \Longleftrightarrow \alpha \leq^*_{\varphi_n} \beta$$

$$T(n,\alpha,\beta) \Longleftrightarrow \alpha <^*_{\varphi_n} \beta$$

are in $\underset{\sim}{\Gamma}$.

Theorem 3.3 [PD] (The Scale Theorem, Moschovakis 71) For $n \geq 0$, every $\underset{\sim}{\Pi}^1_{2n+1}$ $(\underset{\sim}{\Sigma}^1_{2n+2})$ set admits a $\underset{\sim}{\Pi}^1_{2n+1}$ $(\underset{\sim}{\Sigma}^1_{2n+2})$-scale.

A tree on a set X is a set of finite sequences of members of X closed under initial segments. We will consider many times trees on $\omega \times \lambda$ or $\omega \times \omega \times \lambda$, etc. where λ is an ordinal. Thus if T is a tree on $\omega \times \lambda$, its members are of the form

$((k_0,\xi_0), (k_1,\xi_1),\ldots,(k_n,\xi_n))$, where $k_i \in \omega$ and $\xi_i < \lambda$ for all $i \leq n$.

We will sometimes find it convenient to represent elements of such a T by pairs of tuples of the form

$$((k_0 \ldots k_n),\ (\xi_0 \ldots \xi_n)).$$

An (infinite) branch of a tree T on X is a sequence $f \in X^\omega$ such that for all n, $f \upharpoonright n \in T$, where $f \upharpoonright n = (f(0),\ldots,f(n-1))$. A branch of a tree on $\omega \times \lambda$ is thus a sequence $g \in (\omega \times \lambda)^\omega$, but we will represent it by the unique pair $(\alpha,f) \in \omega^\omega \times \lambda^\omega$ such that for all n,

$$g(n) = (\alpha(n),f(n)).$$

If J is a tree on $\omega \times \lambda$ put

$$[J] = \{(\alpha,f) : \alpha \in \mathcal{R},\ f \in \lambda^\omega \text{ and } \forall n\ (\alpha \upharpoonright n,\ f \upharpoonright n) \in J\}$$

(i.e. $[J]$ is the set of branches of J), and put $p[J] = \{\alpha \in \mathcal{R} : \exists f \in \lambda^\omega (\alpha,f) \in [J]\}$.

If $\{\varphi_n\}_{n\in\omega}$ is a λ-scale on a set $A \subseteq \mathcal{R}$, the tree associated with this scale is the tree on $\omega \times \lambda$ defined by

$$((k_0 \ldots k_n),(\xi_0 \ldots \xi_n)) \in T \Longleftrightarrow \exists \alpha \in A \text{ such that } \forall i \leq n (\alpha(i) = k_i \text{ and } \varphi_i(\alpha) = \xi_i).$$

Claim: For A, T as above, $p[T] = A$.

Proof. $A \subseteq p[T]$ is obvious. Let $\alpha \in p[T]$. Find $f \in \lambda^\omega$ such that $(\alpha,f) \in [T]$. Then for all n, $(\alpha \restriction n, f \restriction n) \in T$, i.e. there is a sequence of reals $\{\alpha_n\}_{n\in\omega}$ such that $\forall n\ (\alpha_n \in A)$ and

$$(\alpha \restriction n, f \restriction n) = (\alpha_n \restriction n, (\varphi_0(\alpha_n),\ldots,\varphi_{n-1}(\alpha_n))).$$

Then $\alpha_n \to \alpha$ and $\varphi_n(\alpha_i) = f(n)$ for all $i > n$, hence $\alpha \in A$. ⊣

Definition. If T is a tree on $\omega \times \lambda$, put

$$T(\alpha) = \{s \in \lambda^{<\omega} : (\alpha \restriction \text{length}(s), s) \in T\}.$$

Clearly for each α, $T(\alpha)$ is a tree on λ, and if T is the tree associated with a scale on A as above, then by the claim we have

$$\alpha \in A \Leftrightarrow T(\alpha) \text{ has a infinite branch.}$$

Definition. A set $A \subseteq \mathcal{R}$ is λ-Souslin (where λ is an ordinal) if there is a tree T on $\omega \times \lambda$ such that $A = p[T]$.

The next result is classical for $\underset{\sim}{\Sigma}^1_1$, is due to Shoenfield 61 for $\underset{\sim}{\Sigma}^1_2$, to Martin-Solovay 67 for $\underset{\sim}{\Sigma}^1_3$ (see also Mansfield 71) and to Moschovakis 71 in general.

Theorem 3.4 [PD] (i) For each $n \geq 0$, every $\underset{\sim}{\Sigma}^1_{2n+2}$ set is $\underset{\sim}{\delta}^1_{2n+1}$-Souslin. (ii) For each $n \geq 0$, every $\underset{\sim}{\Sigma}^1_{2n+1}$ set is κ_{2n+1}-Souslin, where κ_{2n+1} is a cardinal $< \underset{\sim}{\delta}^1_{2n+1}$.

Proof. Let $\langle , \rangle : \omega \times \kappa \xrightarrow[\text{onto}]{1-1} \kappa$ be a coding of pairs by ordinals less than a cardinal κ, with decoding functions $(\ )_0$ and $(\ )_1$. Then if $B \subseteq \mathcal{R} \times \mathcal{R}$ is κ-Souslin, let T be a tree on $\omega \times \omega \times \kappa$ such that $(\alpha,\beta) \in B \Leftrightarrow \exists f \in \kappa^\omega\ \forall n\ (\alpha \restriction n, \beta \restriction n, f \restriction n) \in T$. Let $((k_0,\xi_0),\ldots,(k_n,\xi_n)) \in T' \Leftrightarrow ((k_0,(\xi_0)_0,(\xi_0)_1),\ldots,(k_n,(\xi_n)_0,(\xi_n)_1)) \in T$. Then clearly $\exists \beta\, B(\alpha,\beta) \Leftrightarrow \alpha \in p[T']$, hence $\{\alpha : \exists \beta\, B(\alpha,\beta)\}$ is also κ-Souslin. So to prove (i) it is enough to show that every $\underset{\sim}{\Pi}^1_{2n+1}$ set is $\underset{\sim}{\delta}^1_{2n+1}$-Souslin. But this is obvious by the previous results and the evident fact that a $\underset{\sim}{\Pi}^1_{2n+1}$-scale on a $\underset{\sim}{\Pi}^1_{2n+1}$ set must be a $\underset{\sim}{\delta}^1_{2n+1}$-scale.

Proof of (ii). By the closure of κ-Souslin sets under real existential quantification it is enough to show that every $\underset{\sim}{\Pi}^1_{2n}$ set is κ-Souslin for some fixed $\kappa < \underset{\sim}{\delta}^1_{2n+1}$. The least such κ is the required κ_{2n+1}. Let A be a complete $\underset{\sim}{\Pi}^1_{2n}$ set, $\{\varphi_n\}_{n\in\omega}$ a $\underset{\sim}{\Pi}^1_{2n+1}$-scale on A. Since A is $\underset{\sim}{\Delta}^1_{2n+1}$, length $(\varphi_m) < \underset{\sim}{\delta}^1_{2n+1}$ for all m. Since any ω-sequence of $\underset{\sim}{\Delta}^1_{2n+1}$ prewellorderings can be put together to yield a new $\underset{\sim}{\Delta}^1_{2n+1}$ prewellordering of length at least the supremum of the lenghts of the original prewellorderings, cof $(\underset{\sim}{\delta}^1_{2n+1}) > \omega$. Hence there is a $\kappa < \underset{\sim}{\delta}^1_{2n+1}$ such that length $(\varphi_m) < \kappa$, for all m. Hence, by passing from the scale to its associated tree, A is κ-Souslin. ⊣

Definition. A set of reals is λ-Borel (λ an ordinal) if it belongs to the smallest class of sets of reals containing the open sets and closed under complements and wellordered unions of length $< \lambda$. This class is denoted by $\mathcal{B}_\lambda$.

Theorem 3.5. (Separation of $\varkappa$-Souslin sets; Lusin for $\varkappa = \omega$; see Martin 7?) If $A, B \subseteq \mathbb{R}$ are $\varkappa$-Souslin and $A \cap B = \emptyset$ then there is a $\varkappa^+$-Borel set C which separates them, i.e. $A \subseteq C$ and $C \cap B = \emptyset$.

Proof. Let $A = p[T]$, $B = p[S]$, where T and S are trees on $\omega \times \varkappa$. Define a tree U on $\omega \times \varkappa \times \varkappa$ by

$$(s,u,v) \in U \Leftrightarrow (s,u) \in T \ \&\ (s,v) \in S.$$

Since $A \cap B = \emptyset$, U is well founded, i.e. has no infinite branches. We will define a function on U

$$(s,u,v) \longmapsto C_{s,u,v} \subseteq \mathbb{R},$$

by induction on U, such that each $C_{s,u,v} \in \mathcal{B}_{\varkappa^+}$ and $C_{s,u,v}$ separates $A_{s,u}$ and $B_{s,v}$, where

$$A_{s,u} = \{\alpha \supseteq s : \exists f \supseteq u((\alpha,f) \in [T])\}$$

and

$$B_{s,v} = \{\alpha \supseteq s : \exists f \supseteq v((\alpha,f) \in [S])\}.$$

We can take then $C = C_{\emptyset,\emptyset,\emptyset}$.

Note that $A_{s,u} = \bigcup_{n,\xi} A_{\widehat{sn},\widehat{u\xi}}$ and $B_{s,v} = \bigcup_{m,\eta} B_{\widehat{sm},\widehat{v\eta}}$. So it is enough to define $D_{n,\xi,m,\eta} \in \mathcal{B}_{\varkappa^+}$ such that $D_{n,\xi,m,\eta}$ separates $A_{\widehat{sn},\widehat{u\xi}}$ from $B_{\widehat{sm},\widehat{v\eta}}$, since we can then take

$$C_{s,u,v} = \bigcup_{n,\xi} \bigcap_{m,\eta} D_{n,\xi,m,\eta}$$

Assume we have defined all $C_{\widehat{sn},\widehat{u\xi},\widehat{v\eta}}$, when $(\widehat{sn},\widehat{u\xi},\widehat{v\eta}) \in U$.

Case I: $n = m$ and $(\widehat{sn},\widehat{u\xi},\widehat{v\eta}) \in U$.

Then take

$$D_{n,\xi,m,\eta} = C_{\widehat{sn},\widehat{u\xi},\widehat{v\eta}}.$$

Case II: $n = m$ and $(\widehat{sn},\widehat{u\xi},\widehat{v\eta}) \notin U$. Then either $A_{\widehat{sn},\widehat{u\xi}} = \emptyset$ or $B_{\widehat{sn},\widehat{v\eta}} = \emptyset$, so they can be trivially separated.

Case III: $n \neq m$. Then $A_{\widehat{sn},\widehat{u\xi}}$ and $B_{\widehat{sm},\widehat{v\eta}}$ can be separated by disjoint open neighborhoods. ⊣

Theorem 3.6 (Generalized Souslin Theorem; see Martin 7?) If A, $\mathbb{R} - A$ are $\varkappa$-Souslin, then $A \in \mathcal{B}_{\varkappa^+}$. ⊣

Theorem 3.7 [AD] ($\subseteq$ Martin 7? ; $\supseteq$ Moschovakis 71). For all $n \geq 0$, $\mathcal{B}_{\underset{\sim}{\delta}^1_{2n+1}} = \underset{\sim}{\Delta}^1_{2n+1}$.

Proof. That $\underset{\sim}{\Delta}^1_{2n+1} \subseteq \mathcal{B}_{\underset{\sim}{\delta}^1_{2n+1}}$ follows from the fact that each $\underset{\sim}{\Delta}^1_{2n+1}$ set is $\varkappa_{2n+1}$-Souslin for some $\varkappa_{2n+1} < \underset{\sim}{\delta}^1_{2n+1}$ (Theorem 3.4 (ii)). For the other direction, it is enough to show that $\underset{\sim}{\Delta}^1_{2n+1}$ is closed under unions of length $< \underset{\sim}{\delta}^1_{2n+1}$. If not, let $\theta < \underset{\sim}{\delta}^1_{2n+1}$ be least such that for some sequence $\{A_\xi\}_{\xi<\theta}$ of $\underset{\sim}{\Delta}^1_{2n+1}$ sets, $\bigcup_{\xi<\theta} A_\xi = A \notin \underset{\sim}{\Delta}^1_{2n+1}$. Clearly θ is an uncountable cardinal, and we may assume $A_\xi \subseteq A_\eta$ if $\xi \leq \eta < \theta$ and $A_\lambda = \bigcup_{\xi<\lambda} A_\xi$ if $\lambda = \bigcup\lambda < \theta$. Let $\leq$ be a $\underset{\sim}{\Delta}^1_{2n+1}$ prewellordering of $\mathcal{R}$ such that length $(\leq) = \theta$. Let $f : \theta \to P(\mathcal{R})$ be given by $f(\xi) = \{\varepsilon : \varepsilon$ is a $\underset{\sim}{\Delta}^1_{2n+1}$ code of $A_\xi\}$, where ε is a $\underset{\sim}{\Delta}^1_{2n+1}$-code if $\varepsilon = \langle \varepsilon_0, \varepsilon_1 \rangle$ and the $\underset{\sim}{\Pi}^1_{2n+1}$ set coded by ε_0 equals the $\underset{\sim}{\Sigma}^1_{2n+1}$ set coded by ε_1. Denote by $\underset{\sim}{\Delta}_\varepsilon$ the $\underset{\sim}{\Delta}^1_{2n+1}$ set coded by ε, if ε is a $\underset{\sim}{\Delta}^1_{2n+1}$-code.

Let g be a choice subfunction of f such that Code $(g; \leq)$ is $\underset{\sim}{\Sigma}^1_{2n+1}$. Then

$$\alpha \in A \Leftrightarrow \exists \beta[(w,\beta) \in \text{Code } (g; \leq) \;\&\; \alpha \in \underset{\sim}{\Delta}_\beta].$$

Hence A is $\underset{\sim}{\Sigma}^1_{2n+1}$. By Wadge's Lemma A is $\underset{\sim}{\Sigma}^1_{2n+1}$-complete. For $\alpha \in A$, let $\psi(\alpha)$ = the unique $\xi < \theta$ such that $\alpha \in A_{\xi+1} - A_\xi$.

Claim. ψ is a $\underset{\sim}{\Sigma}^1_{2n+1}$-norm (which is a contradiction since it implies that $\underset{\sim}{\Sigma}^1_{2n+1}$ has the prewellordering property).

Proof of Claim. We have $\alpha \leq^*_\psi \beta \Leftrightarrow \exists \xi < \theta[\alpha \in (A_{\xi+1} - A_\xi) \;\&\; \beta \notin A_\xi]$ and $\alpha <^*_\psi \beta \Leftrightarrow \exists \xi < \theta[\alpha \in (A_{\xi+1} - A_\xi) \;\&\; \beta \notin A_{\xi+1}]$. So $\leq^*_\psi$ and $<^*_\psi$ are both unions of $< \underset{\sim}{\delta}^1_{2n+1}$ $\underset{\sim}{\Delta}^1_{2n+1}$ sets, hence as before they are $\underset{\sim}{\Sigma}^1_{2n+1}$. ⊣

Definition. If J is a tree on a set X and $u \in X^{<\omega}$ = set of finite sequences from X, then $J_u = \{v \in X^{<\omega} : u^\frown v \in J\}$.

Notation. $|J| < \xi$ means J is wellfounded and has rank $< \xi$. (Put also $|\emptyset| = -1$.)

Theorem 3.8 (Sierpinski for $\varkappa = \omega$). If $A \subseteq \mathcal{R}$ is $\varkappa$-Souslin, then $A \in \mathcal{B}_{\varkappa^{++}}$.

Proof. Let $A = p[T]$, where T is a tree on $\omega \times \varkappa$. For each $\xi < \varkappa^+$ and $u \in \varkappa^{<\omega}$ put

$$A^\xi_u = \{\alpha : |T(\alpha)_u| < \xi\}.$$

Then if length $(u) = n$,

$$A^0_u = \{\alpha : (\alpha \restriction n, u) \notin T\}$$

$$A^{\xi+1}_u = A^\xi_u \cup \bigcap_{\eta<\lambda} A^\xi_{u^\frown\eta}$$

$$A^\lambda_u = \bigcup_{\xi<\lambda} A^\xi_u, \quad \text{if } \lambda = \bigcup\lambda > 0.$$

Thus $A^{\xi}_{u} \in \mathcal{B}_{\kappa^+}$ for all u and ξ. But

$$\begin{aligned}\alpha \notin A &\Leftrightarrow \alpha \notin P[T]\\ &\Leftrightarrow T(\alpha) \text{ is well founded}\\ &\Leftrightarrow \exists \xi < \kappa^+(|T(\alpha)| < \xi)\\ &\Leftrightarrow \exists \xi < \kappa^+(\alpha \in A^{\xi}_{\emptyset})\end{aligned}$$

⊣

Theorem 3.9. (Martin 7?) If A is κ-Souslin and $\mathrm{cof}(\kappa) > \omega$, then $A \in \mathcal{B}_{\kappa^+}$.

Proof. Let $A = p[T]$, where T is a tree on $\omega \times \kappa$. Then $\alpha \in A \Leftrightarrow T(\alpha)$ is not well founded $\Leftrightarrow \exists \xi < \kappa(T^{\xi}(\alpha)$ is not well founded), where $T^{\xi} = T$ restricted to ordinals $< \xi$. Now apply 3.8. ⊣

Theorem 3.10 (Kechris 74). For all n, $\underset{\sim}{\delta}^1_{2n+1} = (\kappa_{2n+1})^+$, where κ_{2n+1} is a cardinal of cofinality ω.

Proof. Let (by 3.4) κ_{2n+1} = least κ such that every $\underset{\sim}{\Sigma}^1_{2n+1}$ set is κ-Souslin. If $(\kappa_{2n+1})^{++} \le \underset{\sim}{\delta}^1_{2n+1}$, then every $\underset{\sim}{\Sigma}^1_{2n+1}$ set is in $\mathcal{B}_{(\kappa_{2n+1})^{++}} \subseteq \mathcal{B}_{\underset{\sim}{\delta}^1_{2n+1}} = \underset{\sim}{\Delta}^1_{2n+1}$, a contradiction. By 3.4, $\kappa_{2n+1} < \underset{\sim}{\delta}^1_{2n+1}$, hence $(\kappa_{2n+1})^+ = \delta^1_{2n+1}$. If $\mathrm{cof}(\kappa_{2n+1}) > \omega$, then by 3.9 every $\underset{\sim}{\Sigma}^1_{2n+1}$ set is in $\mathcal{B}_{(\kappa_{2n+1})^+} = \mathcal{B}_{\underset{\sim}{\delta}^1_{2n+1}} = \underset{\sim}{\Delta}^1_{2n+1}$, contradiction. ⊣

Theorem 3.11 [Kunen 71a , Martin 7 ?] If $\prec \subseteq \mathcal{R} \times \mathcal{R}$ is wellfounded and κ-Souslin, then $|\prec| < \kappa^+$.

Proof. Let $\alpha \prec \beta \Leftrightarrow \exists f \in \kappa^{\omega}((\alpha,\beta,f) \in [T])$, where T is a tree on $\omega\times\omega\times\kappa$. Put

$$T_{\prec} = \{(\alpha_0,\alpha_1,\ldots,\alpha_n) : \alpha_0 \succ \alpha_1 \succ \ldots \succ \alpha_n\}.$$

By induction one easily checks that for each $\alpha \in$ Field $(\prec)$ and each $\alpha_0 \ldots \alpha_n$ such that $\alpha_0 \succ \alpha_1 \succ \ldots \succ \alpha_n \succ \alpha$ we have

$$|\alpha|_{\prec} = |(\alpha_0 \ldots \alpha_n,\alpha)|_{T_{\prec}} .$$

Hence $|\prec| \le |T_{\prec}|$.

Let S consist of all sequences of the form

$$s = ((s_1,t_1,u_1),\ldots,(s_n,t_n,u_n)),$$

where $s_i = t_{i+1}$ for all $i < n$ and $(s_i,t_i,u_i) \in T$, for all $i \le n$. Thus $s_i,t_i \in \omega^{<\omega}$, $u_i \in \kappa^{<\omega}$. For s,s' as above, define

$$s \succ^* s' \Leftrightarrow \text{length } (s) < \text{length } (s') \text{ and for all } i < \text{length } (s),$$
$$(s'_i,t'_i,u'_i \text{ properly extend } s_i,t_i,u_i).$$

For any α,β such that $\alpha < \beta$ let $h_{\alpha,\beta}$ be the leftmost branch of $T(\alpha,\beta)$. Now define $f : T_< \to S$ by

$$f(\alpha_0 \dots \alpha_n) = ((\alpha_1 \upharpoonright n, \alpha_0 \upharpoonright n, h_{\alpha_1,\alpha_0} \upharpoonright n), \dots, (\alpha_n \upharpoonright n, \alpha_{n-1} \upharpoonright n, h_{\alpha_n,\alpha_{n-1}} \upharpoonright n)).$$

Clearly f embeds in an order preserving way $T_<$ into $(S,<^*)$. It only remains to show $<^*$ is wellfounded.

If not, let $s_0 >^* s_1 >^* s_2 >^* \dots$, where $s_n = ((s_1^n, t_1^n, u_1^n), \dots, (s_{k_n}^n, t_{k_n}^n, u_{k_n}^n))$. Then $k_n \to \infty$, $t_1^n \to \alpha_0$, $s_1^n = t_2^n \to \alpha_1$, $s_2^n = t_3^n \to \alpha_2, \dots$ and $u_1^n \to f_1$, $u_2^n \to f_2, \dots$, where for all n, $(\alpha_{n+1}, \alpha_n, f_{n+1}) \in [T]$. Hence $\alpha_0 > \alpha_1 > \alpha_2 > \dots$, a contradiction. ⊣

<u>Theorem</u> 3.12 [AD] (Kunen 71a , Martin 7?). For all $n \geq 0$, $(\underset{\sim}{\delta}^1_{2n+1})^+ = \underset{\sim}{\delta}^1_{2n+2}$.

<u>Proof</u>. If φ is a $\underset{\sim}{\Pi}^1_{2n+1}$-norm on a $\underset{\sim}{\Pi}^1_{2n+1}$-complete set, then length $(\varphi) = \underset{\sim}{\delta}^1_{2n+1}$. Since its associated prewellordering is $\underset{\sim}{\Delta}^1_{2n+2}$, we have $\underset{\sim}{\delta}^1_{2n+1} < \underset{\sim}{\delta}^1_{2n+2}$. Hence $\underset{\sim}{\delta}^1_{2n+2} \geq (\underset{\sim}{\delta}^1_{2n+1})^+$. Since every $\underset{\sim}{\Sigma}^1_{2n+2}$ relation is $\underset{\sim}{\delta}^1_{2n+1}$-Souslin, we have by 3.11 that $(\underset{\sim}{\delta}^1_{2n+1})^+ \geq \underset{\sim}{\delta}^1_{2n+2}$. ⊣

<u>Theorem</u> 3.13 [AD] (Moschovakis 70 for odd n, Kechris 74 for even n). For all n, $\underset{\sim}{\delta}^1_n < \underset{\sim}{\delta}^1_{n+1}$.

<u>Proof</u>. The theorem for n odd follows from 3.12. Suppose $\underset{\sim}{\delta}^1_{2m} = \underset{\sim}{\delta}^1_{2m+1}$. Then $\underset{\sim}{\delta}^1_{2m+1} = (\varkappa_{2m+1})^+ = \underset{\sim}{\delta}^1_{2m} = (\underset{\sim}{\delta}^1_{2m-1})^+$, hence $\varkappa_{2m+1} = \underset{\sim}{\delta}^1_{2m-1}$ which is a contradiction since cofinalify $(\underset{\sim}{\delta}^1_{2m-1}) > \omega$. ⊣

<u>Theorem</u> 3.14 [AD] (Moschovakis 70 , for odd n; Kunen 71a , Martin 7? , for all n). For all n, $\underset{\sim}{\delta}^1_n = \sup\{\xi : \xi$ is the length of a $\underset{\sim}{\Sigma}^1_n$ wellfounded relation$\}$.

<u>Proof</u>. By 3.4, 3.11 and 3.12. ⊣

4. <u>The $\underset{\sim}{\delta}^1_n$'s are regular</u>

<u>Theorem</u> 4.1 [AD] (Moschovakis 70 for odd n; Kunen 71a , for all n]. For all n, $\underset{\sim}{\delta}^1_n$ is regular.

<u>Proof</u>. Assume not, and let $f : \lambda \to \underset{\sim}{\delta}^1_n$ be a cofinal map, with $\lambda < \underset{\sim}{\delta}^1_n$. Let $\leq$ be a $\underset{\sim}{\Delta}^1_n$ prewellordering of $\mathcal{R}$ of length λ with corresponding norm φ. Let $g(\xi) = \{\alpha : \alpha$ is a $\underset{\sim}{\Sigma}^1_n$ code of a $\underset{\sim}{\Sigma}^1_n$ well founded relation of length $f(\xi)\}$. Let g' be a choice subfunction of g such that Code $(g';\leq)$ is $\underset{\sim}{\Sigma}^1_n$.

Let $W \subseteq \mathcal{R} \times \mathcal{R} \times \mathcal{R}$ be $\underset{\sim}{\Sigma}^1_n$ universal, and put

$(\alpha,\beta,\gamma) \prec (\alpha',\beta',\gamma') \Leftrightarrow [\alpha = \alpha',\ \beta = \beta',\ (\alpha,\beta) \in \text{Code}\ (g';\leq) \text{ and } (\gamma,\gamma') \in W_\beta]$.

Clearly $\prec$ is $\underset{\sim}{\Sigma}^1_n$ and wellfounded. But for any $\xi < \lambda$, if α is such that $\varphi(\alpha) = \xi$, then for any fixed $\beta \in g'(\xi)$ the map

$$\gamma \mapsto (\alpha,\beta,\gamma)$$

embeds W_β into $\prec$. So $|\prec| \geq |W_\beta| = f(\xi)$, hence $|\prec| = \underset{\sim}{\delta}^1_n$, a contradiction. ⊣

5. The $\underset{\sim}{\delta}^1_n$'s are measurable

Theorem 5.1 [AD] [Solovay for $n = 1$ (see 67), 2; Martin 71 for odd n; Kunen 71b in general]. For all n, $\underset{\sim}{\delta}^1_n$ is measurable.

Proof. Let $W \subseteq \mathcal{R}^3$ be universal $\underset{\sim}{\Sigma}^1_n$ and let

$$S = \{\alpha : W_\alpha \text{ is wellfounded binary relation}\}.$$

For $\alpha \in S$, let $|\alpha| = \text{length}\ (W_\alpha)$. Hence

$$\underset{\sim}{\delta}^1_n = \sup\{|\alpha| : \alpha \in S\}.$$

For $A \subseteq \underset{\sim}{\delta}^1_n$, consider the following game G^A first used (for $n = 1$ and with a different coding) by Solovay in his original proof that ω_1 is measurable. I plays α, II plays β, and II wins iff $\{[\exists i((\alpha)_i \notin S \text{ or } (\beta)_i \notin S)$ and if i_0 is the least such i, then $(\alpha)_{i_0} \notin S]$ or $[\forall i((\alpha)_i \in S \text{ and } (\beta)_i \in S)$ and

$$\sup\{|(\alpha)_0|,|(\beta)_0|,|(\alpha)_1|,|(\beta)_1|,\ldots\} = \sup_i\{|(\alpha)_i|,|(\beta)_i|\} \in A]\}.$$

Here we think of a real α as coding an ω-sequence of reals $\{(\alpha)_i\}_{i\in\omega}$, where $(\alpha)_i(m) = \alpha(p_i^{m+1})$ ($p_i = i^{\text{th}}$ prime).

Now define $U \subseteq \text{power}\ (\underset{\sim}{\delta}^1_n)$ by

$$A \in U \Leftrightarrow \text{II has a winning strategy in } G^A.$$

We show that U is a $\underset{\sim}{\delta}^1_n$-additive measure on $\underset{\sim}{\delta}^1_n$.

Lemma 1. $A \in U$ and $B \supseteq A \Rightarrow B \in U$.

Proof. Trivial.

Lemma 2. $A, B \in U \Rightarrow A \cap B \in U$.

Proof. Given reals α,β, let $\alpha \oplus \beta$ be a real such that $(\alpha \oplus \beta)_{2n} = (\alpha)_n$ and $(\alpha \oplus \beta)_{2n+1} = (\beta)_n$ for all n. Suppose II has a winning strategy τ in G^A, and a winning strategy σ in G^B. To win $G^{A\cap B}$, given a move α of I, II simultaneously builds reals β', β'' such that β' is the result of τ against $\alpha \oplus \beta''$, and β'' is the result of σ against $\alpha \oplus \beta'$. II's actual play is then $\beta' \oplus \beta''$. If there is i_0 such that $\forall j \leq i_0((\alpha)_j \in S)$ and $(\beta' \oplus \beta'')_{i_0} \notin S$ we are led to a contradiction. If $\forall i((\alpha)_i \in S$ and $(\beta' \oplus \beta'')_i \in S)$ then $\sup_i\{|(\alpha)_i|,|(\beta' \oplus \beta'')_i|\} = \sup\{|(\alpha\oplus\beta')_i|,|(\beta'')_i|\} = \sup\{|(\alpha\oplus\beta'')_i|,|(\beta)_i|\} \in A \cap B$.

Lemma 3. U contains no bounded sets.

Proof. If A is bounded, then since $\sup\{|\alpha| : \alpha \in S\} = \underset{\sim}{\delta}^1_n$, player I can easily win G^A.

Lemma 4. U is an ultrafilter.

Proof. We must show that $A \notin U \Rightarrow (\underset{\sim}{\delta}^1_n - A) \in U$. But if II has no winning strategy in G^A, then II can essentially follow I's winning strategy in G^A to win $G^{(\underset{\sim}{\delta}^1_n - A)}$.

Lemma 5. U is $\underset{\sim}{\delta}^1_n$-additive.

Proof. Let $\{A_\xi\}_{\xi<\eta<\underset{\sim}{\delta}^1_n}$ be a sequence of $< \underset{\sim}{\delta}^1_n$ members of U. It suffices to show $\bigcap_{\xi<\eta} A_\xi \neq \emptyset$.

Let $\leq$ be a $\underset{\sim}{\Delta}^1_n$ prewellordering of $\mathcal{R}$ of length η with associated norm φ. For $\xi < \eta$ let

$$f(\xi) = \{\tau : \tau \text{ is a winning strategy for II in } G^{A_\xi}\}.$$

Let g be a choice subfunction of f such that Code $(g;\leq)$ is $\underset{\sim}{\Sigma}^1_n$, say in $\Sigma^1_n(y)$.

Claim. For each $m \geq 0$, there is a function $f_m : S^{m+1} \to S$ such that for all $\alpha^0 \ldots \alpha^m \in S$, for all α with $(\alpha)_i = \alpha^i$ if $i \leq m$ and for all $\tau \in \bigcup_{\xi<\eta} g(\xi)$,

$$|f_m(\alpha^0 \ldots \alpha^m)| \geq |(\tau[\alpha])_m|,$$

where $\tau[\alpha]$ = II's play when I plays α and II follows τ.

Proof of Claim. Given $\alpha^0 \ldots \alpha^m \in S$, consider the following wellfounded relation:

$$\langle\alpha,x,\tau,z\rangle \prec_{\alpha^0\ldots\alpha^m} \langle\alpha',x',\tau',z'\rangle \Leftrightarrow \alpha = \alpha' \;\&\; x = x' \;\&\; \tau = \tau' \;\&\forall i \leq m((\alpha)_i = \alpha^i) \;\&$$
$$(x,\tau) \in \text{Code } (g;\leq) \;\&\; (z,z') \in W_{(\tau[\alpha])_m}.$$

Then $\prec_{\alpha^0\ldots\alpha^m}$ is $\Sigma^1_n(\alpha^0 \ldots \alpha^m, y)$ with length $\geq |(\tau[\alpha])_m|$ for any τ, α as above. Clearly one can find a continuous f_m such that $f_m(\alpha^0 \ldots \alpha^m)$ is a $\underset{\sim}{\Sigma}^1_n$-code for $\prec_{\alpha^0\ldots\alpha^m}$, proving the claim.

Now let $\alpha^0 \in S$ and define inductively

$$\alpha^{m+1} = f_m(\alpha^0 \ldots \alpha^m).$$

Let $\theta = \sup\{|\alpha^m| : m \in \omega|$. Given $\xi < \eta$, let I play α such that $\forall i((\alpha)_i = \alpha^i)$, and let II play using a strategy τ from $g(\xi)$, producing a real $\beta = \tau[\alpha]$. Then $\forall i((\beta)_i \in S)$ and $\sup\{|(\alpha)_i|, |(\beta)_i|\} = \sup\{|(\alpha)_i|\} = \theta$. Since II's strategy τ was winning, $\theta \in A_\xi$. Hence $\theta \in \bigcap_{\xi<\eta} A_\xi$, proving the lemma. ⊣

6. Calculating $\underset{\sim}{\delta}^1_n$ for $n \leq 4$.

Theorem 6.1 (Classical) $\underset{\sim}{\delta}^1_1 = \omega_1$.

Proof. Every $\underset{\sim}{\Sigma}^1_1$ set is ω-Souslin. So every $\underset{\sim}{\Sigma}^1_1$ wellfounded relation has length $< \omega_1$.

Theorem 6.2 [AD] (Martin 7?) $\underset{\sim}{\delta}^1_2 = \omega_2$.

Proof. Obvious, since $\underset{\sim}{\delta}^1_2 = (\underset{\sim}{\delta}^1_1)^+$. ⊣

Theorem 6.3 $[\forall\alpha\,(\alpha^\#$ exists)] (Martin-Solovay 67). Every $\underset{\sim}{\Sigma}^1_3$ set is ω_ω-Souslin.

Proof. We will show that every Π^1_2 set admits a Δ^1_3-scale $\{\varphi_n\}_{n\in\omega}$ such that for all n, length $(\varphi_n) < \omega_\omega$. Note that it suffices to prove that the Π^1_2 set $\mathcal{R}^\# = \{\alpha^\# : \alpha \in \mathcal{R}\}$ admits such a scale. Because if $\{\varphi^*_n\}_{n\in\omega}$ is such a scale on $\mathcal{R}^\#$, put $\varphi_n(\alpha) = \langle\varphi^*_0(\alpha^\#),\alpha^\#(0),\varphi^*_1(\alpha),\alpha^\#(1),\ldots,\varphi^*_n(\alpha^\#),\alpha^\#(n)\rangle$, where $\langle\ \rangle$ refers to the ordinal of the 2n-tuple under the lexicographical order. Then $\{\varphi^*_n\}_{n\in\omega}$ is a Δ^1_3-scale when restricted to any Π^1_2 set A: To show this (assuming $\{\varphi^*_n\}_{n\in\omega}$ has the right properties), let $\alpha_i \in A$, $\alpha_i \to \alpha$, and $\varphi_n(\alpha_i) = \lambda_n, \forall i \geq n$. This implies that $\alpha^\#_i \to \beta$ for some β. Since each $\varphi^*_n(\alpha^\#_i)$ is eventually constant, $\beta = \bar{\alpha}^\#$ for some $\bar{\alpha}$. Since there is a recursive f such that $f(\beta^\#) = \beta$ for all β, $\bar{\alpha} = \alpha$. To see that $\alpha \in A$, note that for some Π^1_2 formula φ

$$\alpha \in A \Longleftrightarrow L[\alpha] \models \varphi(\alpha)$$
$$\Longleftrightarrow \alpha^\#(n_0) = 0,$$

where n_0 is a Gödel number for $\varphi(\dot{\alpha})$ ($\dot{\alpha}$ is the constant symbol denoting α). Hence since for all i, $\alpha_i \in A$, we have $\alpha^\#_i(n_0) = 0 \therefore \alpha^\#(n_0) = 0$, i.e. $\alpha \in A$.

Finally, to see that $\varphi_n(\alpha) \leq \lambda_n$, pick k large enough so that $\forall i \geq k$, $\forall p \leq n$, $\varphi^\#_p(\alpha_i)$ is constant and $\alpha^\#_i \restriction (n+1) = \alpha^\# \restriction (n+1)$. Then

$$\varphi_n(\alpha) = \langle\varphi^*_0(\alpha^\#),\alpha^\#(0),\ldots,\varphi^*_n(\alpha^\#),\alpha^\#(n)\rangle$$
$$\leq \langle\varphi^\#_0(\alpha^\#_k),\alpha^\#_k(0),\ldots,\varphi^*_n(\alpha^\#_k),\alpha^\#_k(n)\rangle = \lambda_n.$$

Since the sharp operation is Δ^1_3, it is clear that $\{\varphi_n\}_{n\in\omega}$ is a Δ^1_3-scale.

So we must produce a Δ^1_3-scale on $\mathcal{R}^\#$ whose norms have length $< \omega_\omega$. Let $\tau'_0,\tau'_1,\tau'_2\ldots$ be a recursive enumeration of all definable Skolem functions or terms (with variables) in the theory $ZF + V = L[\dot{\alpha}] + \dot{\alpha} \in \mathcal{R}$ (where $\dot{\alpha}$ is a constant symbol) and put $\tau_n = \text{rank}(\tau'_n)$ so that τ_n takes only ordinal values. Say $\tau_n = \tau_n\ (v_1 \ldots v_{k_n})$. Then define

$$\varphi^*_n(\alpha^\#) = \tau_n^{L[\alpha]}(\omega_1 \ldots \omega_{k_n}).$$

Note that

$$\varphi_n^*(\alpha^{\#}) < \varphi_n^*(\beta^{\#}) \Leftrightarrow \tau_n^{L[\alpha]}(\omega_1 \ldots \omega_{k_n}) < \tau_n^{L[\beta]}(\omega_1 \ldots \omega_{k_n})$$
$$\Leftrightarrow L[\alpha,\beta] \models \psi(\alpha,\beta,\omega_1 \ldots \omega_{k_n})$$
$$\Leftrightarrow \langle\alpha,\beta\rangle^{\#}(m) = 0,$$

where m is obtained recursively from n. Hence $\{\varphi_n^*\}_{n\in\omega}$ is a Δ_3^1-scale, if it is a scale.

We also have length $(\varphi_n^*) < \omega_\omega$ since in fact $\tau_n^{L[\alpha]}(\omega_1 \ldots \omega_{k_n}) < \omega_{k_n+1}$ for all α (because every cardinal is an indiscernible for every $L[\alpha]$).

To show $\{\varphi_n^*\}_{n\in\omega}$ is a scale, let $\alpha_i^{\#} \in \mathfrak{R}^{\#}$, $\alpha_i^{\#} \to \beta$ and $\varphi_n^*(\alpha_i^{\#}) = \lambda_n$ for $i > n$. Note that

$$\beta \in \mathfrak{R}^{\#} \Leftrightarrow P(\beta) \ \& \ \Gamma(\beta,\omega_1) \text{ is wellfounded,}$$

where P is Π_1^0 expressing "β is a set of Gödel numbers of formulas satisfying the syntactical conditions for a remarkable character relative to some real α", and $\Gamma(\beta,\xi)$ is the model of $ZF + V = L[\dot{\alpha}]$ generated by ξ indiscernibles on the basis of β.

Since $P(\alpha_i^{\#})$ for all i, we know that $P(\beta)$ holds.

Let $\mathcal{J}^{\alpha}$ = class of Silver indiscernibles for $L[\alpha]$. Let $C = \bigcap_{i,j}(\mathcal{J}^{\langle\alpha_i,\alpha_j\rangle} \cap \omega_1)$. Thus C is closed unbounded in ω_1. Let $\{c_\xi\}_{\xi<\omega_1}$ be its increasing enumeration.

Since for any fixed n, $\tau_n^{L[\alpha_i]}(\omega_1 \ldots \omega_{k_n})$ becomes eventually constant, the same is true of $\tau_n^{L[\alpha_i]}(c_{\xi_1} \ldots c_{\xi_{k_n}})$ for all $\xi_1 < \ldots < \xi_{k_n} < \omega_1$. So define $f : ORD^{\Gamma(\beta,\omega_1)} \to ORD$ by $f(\tau_n^{\Gamma(\beta,\omega_1)}(i_{\xi_1} \ldots i_{\xi_{k_n}}))$ = eventual value of $\tau_n^{L[\alpha_i]}(c_{\xi_1} \ldots c_{\xi_{k_n}})$, where $I = \{i_\xi : \xi < \omega_1\}$ is a generating set of indiscernibles for $\Gamma(\beta,\omega_1)$.

<u>Claim.</u> f is well defined and order preserving.

<u>Proof.</u> Suppose $\tau_n^{\Gamma(\beta,\omega_1)}(i_{\xi_1} \ldots i_{\xi_{k_n}}) = \tau_m^{\Gamma(\beta,\omega_1)}(i_{\xi'_1} \ldots i_{\xi'_{k_m}})$ where $\xi_1 < \ldots < \xi_{k_n}$ and $\xi'_1 < \ldots < \xi'_{k_m}$. Then there is a ψ such that $\Gamma(\beta,\omega_1) \models \psi(i_{\eta_1} \ldots i_{\eta_\ell}) \Leftrightarrow \tau_n(i_{\xi_1} \ldots i_{\xi_{k_n}}) = \tau_m(i_{\xi'_1} \ldots i_{\xi'_{k_m}})$, where $\eta_1 < \ldots < \eta_\ell$ is $\{\xi_1 \ldots \xi_{k_n}, \xi'_1 \ldots \xi'_{k_n}\}$ written in increasing order.

Thus $\beta([\psi(v_1 \ldots v_\ell)]) = 0$. Since $\alpha_i^{\#} \to \beta$, the eventual value of $\alpha_i^{\#}([\psi(v_1 \ldots v_\ell)])$ is 0. Hence eventually $\tau_n^{L[\alpha_i]}(c_{\xi_1} \ldots c_{\xi_{k_n}}) = \tau_m^{L[\alpha_i]}(c_{\xi'_1} \ldots c_{\xi'_{k_m}})$, so f is well defined. Similarly f is order preserving.

Hence $\Gamma(\beta,\omega_1)$ is well founded, so $\beta \in \mathfrak{R}^{\#}$. So $\beta = \alpha^{\#}$, where $\alpha_i \to \alpha$. Thus $\Gamma(\beta,\omega_1) = L_{\omega_1}[\alpha]$. Let $\{i_\xi^{\alpha} : \xi < \omega_1\}$ be the increasing enumeration of the Silver indiscernibles of $L_{\omega_1}[\alpha]$. Let $C^* = \{\xi < \omega_1 : c_\xi = i_\xi^{\alpha} = \xi\}$. Then C^* is closed unbounded, and since f is order preserving, $\forall c_{\xi_1} < \ldots < c_{\xi_{k_n}}$ in C^* we have

$$\tau_n^{L[\alpha]}(c_{\xi_1}\ldots c_{\xi_{k_n}}) = \tau_n^{L[\alpha]}(i^{\alpha}_{\xi_1}\ldots i^{\alpha}_{\xi_{k_n}}) \leq f(\tau_n^{L[\alpha]}(i^{\alpha}_{\xi_1}\ldots i^{n}_{\xi_{k_n}}))$$

$$= \text{eventual value of } \tau_n^{L[\alpha_i]}(c_{\xi_1}\ldots c_{\xi_{k_n}}).$$

Thus $\tau_n^{L[\alpha]}(\omega_1\ldots\omega_{k_n}) \leq$ eventual value of $\tau_n^{L[\alpha_i]}(\omega_1\ldots\omega_{k_n}) = \lambda_n$, i.e. $\varphi_n^*(\alpha^{\#}) \leq \lambda_n$. Hence $\{\varphi_n^*\}_{n\in\omega}$ is a scale. ⊣

Theorem 6.4 [AD] (Martin 7?). $\underset{\sim}{\delta}^1_3 = \omega_{\omega+1}$.

Proof. We have $\underset{\sim}{\delta}^1_3 = \varkappa_3^+$, where $\varkappa_3 > \omega$ is a cardinal of cofinality ω and $\varkappa_3$ is the least cardinal such that every $\underset{\sim}{\Sigma}^1_3$ set is $\varkappa_3$-Souslin. Hence $\varkappa_3 \leq \omega_\omega$, so $\varkappa_3 = \omega_\omega$, since countable choice implies that ω_ω is the second cardinal of cofinality ω. Hence $\underset{\sim}{\delta}^1_3 = \omega_{\omega+1}$. ⊣

Corollary 6.5 [AD] (Kunen 71a , Martin 7?). $\underset{\sim}{\delta}^1_4 = \omega_{\omega+2}$. ⊣

7. The closed unbounded measure on ω_1.

Theorem 7.1 [AD] (Solovay 67 for n = 1, Moschovakis 70 in general). Let $P \in \underset{\sim}{\Pi}^1_{2n+1} - \underset{\sim}{\Delta}^1_{2n+1}$, and let φ be a $\underset{\sim}{\Pi}^1_{2n+1}$ norm on P with associated prewellordering $\leq$. Then for every $A \subseteq \underset{\sim}{\delta}^1_{2n+1}$, Code $(A;\leq) \in \underset{\sim}{\Pi}^1_{2n+1}$.

Proof. By the Coding Lemma we know that for all $\xi < \underset{\sim}{\delta}^1_{2n+1}$, Code $(A\cap\xi;\leq) \in \underset{\sim}{\Delta}^1_{2n+1}$. Consider the following game: I plays w, II plays α, and II wins iff [$w \in P \Rightarrow \alpha$ is a $\underset{\sim}{\Delta}^1_{2n+1}$-code of a set $\underset{\sim}{\Delta}_\alpha$ such that

$$\text{Code}(A\cap(\varphi(w)+1);\leq) \subseteq \underset{\sim}{\Delta}_\alpha \subseteq \text{Code}(A;\leq)].$$

If I has a winning strategy σ, then $\{\sigma(\alpha) : \alpha \in \mathcal{R}\} = Q$ is a $\underset{\sim}{\Sigma}^1_1$ subset of P, hence by boundedness, $\xi = \sup\{\varphi(w) : w \in Q\} < \underset{\sim}{\delta}^1_{2n+1}$. So II can easily beat this strategy by playing a $\underset{\sim}{\Delta}^1_{2n+1}$-code of Code $(A\cap(\xi+1);\leq)$.

Hence II has a winning strategy τ, and

$$w \in \text{Code}(A;\leq) \Leftrightarrow w \in P \;\&\; w \in \underset{\sim}{\Delta}_{\tau(w)},$$

so Code $(A;\leq)$ is $\underset{\sim}{\Pi}^1_{2n+1}$. ⊣

Theorem 7.2 [AD] (Solovay 67). For every $A \subseteq \omega_1$, $\exists\alpha\in\mathcal{R}\,(A \in L[\alpha])$.

Proof. Let WO = $\{\alpha : \alpha$ codes a wellordering of $\omega\}$ and for $\alpha \in$ WO, let $|\alpha|$ = the ordinal coded by α. For $A \subseteq \omega_1$, let Code(A) = $\{\alpha : |\alpha| \in A\}$. By the above, for every $A \subseteq \omega_1$, Code (A) $\in \underset{\sim}{\Pi}^1_1$. We will now show (in ZF + DC) that for any A with Code (A) $\in \underset{\sim}{\Sigma}^1_2$, $\exists\alpha\,(A \in L[\alpha])$.

Let $P \in \Sigma^1_2$ be such that $\alpha \in \text{Code}(A) \Leftrightarrow P(\beta_0,\alpha)$ for some β_0. Then $\xi \in A \Leftrightarrow \exists\alpha(P(\beta_0,\alpha) \;\&\; |\alpha| = \xi)$.

Case I: For some γ, $\omega_1^{L[\gamma]} = \omega_1$.

Then $\xi \in A \Leftrightarrow L[\gamma,\beta_0] \models \exists\alpha(P(\beta_0,\alpha)\ \&\ |\alpha| = \xi)$, so $A \in L[\gamma,\beta_0]$.

Case II: For all $\gamma, \omega_1^{L[\gamma]} < \omega_1$.

Then if C_ξ is the notion of forcing which collapses ξ to ω (for $\xi < \omega_1$), there are C_ξ-generic over $L[\beta_0]$ sets (since $(\text{power}(\xi))^{L[\beta_0]}$ is countable). Hence

$$\xi \in A \Leftrightarrow \exists\alpha\ (P(\beta_0,\alpha)\ \&\ |\alpha| = \xi)$$

$$\Leftrightarrow \text{For all } C_\xi\text{-generic over } L[\beta_0]G,$$

$$L[\beta_0,G] \models \exists\alpha(P(\beta_0,\alpha)\ \&\ |\alpha| = \xi)$$

$$\Leftrightarrow \emptyset \mid \vdash^{L[\beta_0]}_{C_\xi} \varphi(\hat{\beta}_0,\hat{\xi}),$$

for some formula φ.

Since forcing is definable in $L[\beta_0]$, this shows that $A \in L[\beta_0]$. ⊣

Theorem 7.3 [AD] (Solovay 67). There is a unique normal measure μ on ω_1, namely

$$\mu(A) = 1 \Leftrightarrow A \text{ contains a closed unbounded set.}$$

Proof. It suffices to show that for all $A \subseteq \omega_1$, either A or $\omega_1 - A$ contains a cub set. Given $A \subseteq \omega_1$, let α be such that $A \in L[\alpha]$. Then $A = \tau^{L[\alpha]}(i^\alpha_{\eta_1} \ldots i^\alpha_{\eta_k}, i^\alpha_{\xi_1} \ldots i^\alpha_{\xi_m})$ for some $\eta_1 < \ldots < \eta_k < \omega_1 \le \xi_1 \ldots < \xi_m$ and some term τ. But then $C = \{i^\alpha_\eta : \eta_k < \eta < \omega_1\}$ is closed unbounded, and either $C \subseteq A$ or $C \subseteq \omega_1 - A$. ⊣

8. Uniform indiscernibles and the ω_n's for $n \le \omega$.

Definition. An ordinal u is a uniform indiscernible if $\forall\alpha \in \mathbb{R}\ (u \in \mathcal{I}^\alpha)$, where $\mathcal{I}^\alpha = \{i^\alpha_\xi\}_{\xi\in\mathrm{ORD}}$ is the class of Silver indiscernibles for $L[\alpha]$. Let

$$U = \{u_\xi\}_{\xi \in \mathrm{ORD}}$$

be the increasing enumeration of the uniform indiscernibles.

Clearly $u_1 = \omega_1$, U is closed unbounded and every cardinal is in U. Hence $u_\xi \le \omega_\xi$.

Theorem 8.1 [AD] (Martin 7 ?) $u_\omega = \omega_\omega$.

Proof. In the proof of 6.3 we could have used

$$\varphi^*_n(\alpha^\#) = \tau^{L[\alpha]}_n\ (u_1 \ldots u_{k_n}),$$

hence $\utilde{\Sigma}^1_3$ sets are u_ω-Souslin. Hence as in 6.4

$$(u_\omega)^+ = \utilde{\delta}^1_3 = (\omega_\omega)^+$$

hence $\omega_\omega = u_\omega$. ⊣

Lemma A $[\forall\alpha(\alpha^{\#}$ exists)]. For every ordinal ξ there is a real α, a term τ, and ordinals $\eta_1 < \dots < \eta_m$ such that $\xi = \tau^{L[\alpha]}(u_{\eta_1} \dots u_{\eta_m})$.

Proof. By induction on ξ. Clear if $\xi < \aleph_1$. So assume true for all $\xi' < \xi$, and let $\xi \notin U$. Then for some α, $\xi \notin \mathcal{I}^{\alpha}$. Thus

$$\xi = \sigma^{L[\alpha]}(i^{\alpha}_{\theta_1} \dots i^{\alpha}_{\theta_m}, i^{\alpha}_{\theta_{m+1}} \dots i^{\alpha}_{\theta_k})$$

for some term σ and some

$$i^{\alpha}_{\theta_1} < \dots < i^{\alpha}_{\theta_m} < \xi < i^{\alpha}_{\theta_{m+1}} < \dots < i^{\alpha}_{\theta_k}.$$

Thus

$$\xi = \sigma^{L[\alpha]}(i^{\alpha}_{\theta_1} \dots i^{\alpha}_{\theta_m}, \vec{\aleph}),$$

where $\vec{\aleph}$ is a sequence of large enough cardinals. Now for all $j \leq m, i^{\alpha}_{\theta_j}$ can be defined in some $L[\beta]$ using uniform indiscernibles (by induction hypothesis) hence

$$\xi = \tau^{L[\alpha,\beta]}(u_{\rho_1} \dots u_{\rho_n}, \vec{\aleph}),$$

for some $\rho_1 < \dots < \rho_n$ and we are done. $\dashv$

Definition. For λ an ordinal, let

$$(\lambda,\alpha)^+ = \text{first element of } \mathcal{I}^{\alpha} > \lambda,$$

$$(\lambda^+)^{\alpha} = \text{first cardinal in } L[\alpha] > \lambda.$$

Lemma B $[\forall\alpha(\alpha^{\#}$ exists)]. For all ξ,

$$u_{\xi+1} = \sup_{\alpha\in\mathcal{R}} (u_\xi,\alpha)^+ = \sup_{\alpha\in\mathcal{R}} (u_\xi{}^+)^{\alpha}$$

Proof. Clearly

$$u_{\xi+1} \geq \sup_{\alpha\in\mathcal{R}} (u_\xi^+)^{\alpha} \geq \sup_{\alpha\in\mathcal{R}} (u_\xi^+)^{\alpha^{\#}}$$

$$\geq \sup_{\alpha\in\mathcal{R}} (u_\xi,\alpha)^+ > u_\xi,$$

so it suffices to show that $\sup_{\alpha\in\mathcal{R}}(u_\xi,\alpha)^+$ is a uniform indiscernible. Given $\beta \in \mathcal{R}$, $\lambda = \sup_{\alpha\in\mathcal{R}}(u_\xi,\alpha)^+ = \sup_{\alpha \geq_T \beta^{\#}}(u_\xi,\alpha)^+$ (where $\geq_T$ is Turing reducibility). Hence λ is a sup of members of $\mathcal{I}^{\beta}$, so $\lambda \in \mathcal{I}^{\beta}$. Hence $\lambda \in U$. $\dashv$

Lemma C. $[\forall\alpha\ (\alpha^{\#}$ exists)]. If $\xi < u_{\theta+1}$, then there exist $\theta_1 < \dots < \theta_n \leq \theta$, $\alpha \in \mathcal{R}$ and a term τ such that

$$\xi = \tau^{L[\alpha]}(u_{\theta_1} \dots u_{\theta_n}).$$

Proof. We can assume inductively that $u_\theta \leq \xi < u_{\theta+1}$. Then

$$\xi = \sigma^{L[\beta]}(u_{\theta_1} \dots u_{\theta_{n-1}}, u_{\eta_1} \dots u_{\eta_k}),$$

where

$$u_{\theta_1} < \dots < u_{\theta_{n-1}} \leq u_\theta < u_{\eta_1} < \dots < u_{\eta_k},$$

by Lemma A. Find γ such that

$$\xi < (u_\theta, \gamma)^+ < u_{\theta+1}$$

and let $\lambda_1 < \dots < \lambda_k$ be the first k cardinals above u_θ in $L[\alpha]$, where $\alpha = \langle \beta^\#, \gamma^\# \rangle$. Then $\lambda_1 \dots \lambda_k \in \mathscr{I}^\beta$ and $(u_\theta, \gamma)^+ < \lambda_1$. Hence $\xi = \sigma^{L[\beta]}(u_{\theta_1} \dots u_{\theta_{n-1}}, \lambda_1 \dots \lambda_k)$ and since $\lambda_1 \dots \lambda_k$ are definable in $L[\alpha]$ from u_θ, we have for some term τ

$$\xi = \tau^{L[\alpha]}(u_{\theta_1} \dots u_{\theta_{n-1}}, u_\theta).$$

⊣

<u>Theorem</u> 8.2 $[\forall \alpha\ (\alpha^\#$ exists)] (Solovay). For all ξ, cofinality $(u_{\xi+1})$ = cofinality (u_2).

<u>Proof</u>. Define $f : u_2 \to u_{\xi+1}$ by $f(\tau^{L[\alpha]}(u_1)) = \tau^{L[\alpha]}(u_\xi)$. By indiscernibility, f is well defined and order preserving. Since $\sup_{\alpha \in \mathcal{R}}(u_\xi, \alpha)^+ = u_{\xi+1}$, f is cofinal.

⊣

<u>Theorem</u> 8.3 [AD] (Martin 7 ?). For all $2 \leq n < \omega$, cofinality $(\omega_n) = \omega_2$.

<u>Proof</u>. $u_\omega = \omega_\omega$, hence $\omega_n = u_{k_n+1}$ for some k_n.

⊣

<u>Theorem</u> 8.4 [AD] (Kunen, Solovay). For all $1 \leq n \leq \omega$, $u_n = \omega_n$.

<u>Proof</u>. Let μ be the closed unbounded measure on ω_1.

Suppose that for all n, we can prove

$$u_n^{\omega_1} / \mu \cong u_{n+1}.$$

If for some k, $\overline{\overline{u_k}} = \overline{\overline{u_{k+1}}}$ then

$$\overline{\overline{u_{k+2}}} = \overline{\overline{(u_{k+1}^{\omega_1}/\mu)}} = \overline{\overline{(u_k^{\omega_1}/\mu)}} = \overline{\overline{u_{k+1}}}$$

and similarly $\overline{\overline{u_{k+1}}} = \overline{\overline{u_{k+n}}}$ for all n. Hence $u_\omega < \omega_\omega$, a contradiction. Hence it is enough to show that $\forall n\ (u_n^{\omega_1}/\mu \cong u_{n+1})$.

<u>Lemma</u>. $\forall n,\ u_n^{\omega_1} \subseteq \tilde{L} \stackrel{\text{def}}{=} \bigcup_{\alpha \in \mathcal{R}} L[\alpha]$.

<u>Proof</u>. We prove by induction on $\xi < u_\omega$ that $\xi^{\omega_1} \subseteq \tilde{L}$. If for some α, ξ is not a cardinal in $L[\alpha]$, then this is obvious by induction hypothesis. If ξ is a cardinal in all $L[\alpha]$'s then it is a uniform indiscernible, hence has cofinality = cofinality (u_2), so it is enough to show that $u_2 = \omega_2$.

To see that $u_2 = \omega_2$: Clearly $u_2 \leq \omega_2$. Suppose $u_2 < \omega_2$. Let $A \subseteq \omega_1 \times \omega_1$ be a well ordering of ω_1 with order type u_2. Then for some $\alpha \in \mathcal{R}$, $A \in L[\alpha]$. Hence the order type of A is $< (\omega_1^+)^\alpha < u_2$, contradiction. Hence $u_2 = \omega_2$.

Now to complete the proof of 8.4, let $f \in u_n^{\omega_1}$, $n \geq 1$.. Then there is a real β such that for some term τ and μ-almost all ξ,

$$f(\xi) = \tau^{L[\beta]}(\xi; u_1 \dots u_{n-1}).$$

Indeed, since $f \in \tilde{L}$ we have by Lemma C that for some $\alpha \in \mathbb{R}$ and for some term σ

$$f(\xi) = \sigma^{L[\alpha]}(\xi; u_1 \dots u_{n-1}, u_n) < u_n,$$

so for all $\xi \in \mathcal{J}^\alpha \cap \omega_1$,

$$\begin{aligned} f(\xi) &= \sigma^{L[\alpha]}(\xi; u_1 \dots u_{n-1}, (u_{n-1}, \alpha)^+) \\ &= \tau^{L[\beta]}(\xi; u_1 \dots u_{n-1}) \end{aligned}$$

for some term τ and $\beta = \alpha^\#$.

Let $[f]_\mu$ be the equivalence class of f in $u_n^{\omega_1}/\mu$ and define

$$j([f]_\mu) = \tau^{L[\beta]}(u_1; u_2 \dots u_n) < u_{n+1}.$$

Clearly j is well defined and order preserving by indiscernibility.

Let $\theta < u_{n+1}$. Find τ, α such that $\theta = \tau^{L[\alpha]}(u_1 \dots u_n)$. Now define $f \in u_n^{\omega_1}$ by

$$f(\xi) = \tau^{L[\alpha]}(\xi; u_1 \dots u_{n-1}).$$

Then $j([f]_\mu) = \theta$, so j is onto.

Hence $j : u_n^{\omega_1}/\mu \cong u_{n+1}$. ⊣

To summarize: From AD,

$$\underset{\sim}{\delta}_1^1 = \omega_1 = u_1, \quad \text{measurable}$$

$$\underset{\sim}{\delta}_2^1 = \omega_2 = u_2, \quad \text{measurable}$$

For $n \geq 3$, $\omega_n = u_n$ singular, cofinality $= \omega_2$

$$\underset{\sim}{\delta}_3^1 = \omega_{\omega+1}, \quad \text{measurable}$$

$$\underset{\sim}{\delta}_4^1 = \omega_{\omega+2}, \quad \text{measurable.}$$

Proposition 8.5 [AD]. For $n \geq 3$, $\underset{\sim}{\delta}_n^1 = u_{\underset{\sim}{\delta}_n^1}$.

Proof. If $\underset{\sim}{\delta}_n^1 < u_{\underset{\sim}{\delta}_n^1}$, then since $\underset{\sim}{\delta}_n^1$ is a cardinal, $\underset{\sim}{\delta}_n^1 = u_\xi$ for some $\xi < \underset{\sim}{\delta}_n^1$. Now ξ cannot be a successor (otherwise $\text{cof}(\underset{\sim}{\delta}_n^1) = \omega_2$). Hence ξ is limit, hence $\underset{\sim}{\delta}_n^1$ is singular, contradiction. ⊣

Basic open problem. Compute $\underset{\sim}{\delta}_5^1$.

9. Back to the real world

For a moment we interrupt the development of the theory of projective ordinals in the context of ZF + DC + AD, to see what is the picture of these ordinals in a context with Choice and Projective Determinacy only.

Theorem 9.1.

0) $\underset{\sim}{\delta}_1^1 = \omega_1$.

1) $\underset{\sim}{\delta}_2^1 \le \omega_2$ (Martin, 7 ?)

$[\forall\alpha\ (\alpha^\#\text{ exists})]\ \underset{\sim}{\delta}_2^1 = u_2$ (Martin 7 ?)

2) $[AC + \forall\alpha\ (\alpha^\#\text{ exists})]\ \underset{\sim}{\delta}_3^1 \le \omega_3$ (Martin 7 ?)

3) $[AC + PD]\ \underset{\sim}{\delta}_4^1 \le \omega_4$ (Kunen 71a , Martin 7 ?)

4) $[PD]\ \underset{\sim}{\delta}_{2n+2}^1 \le (\underset{\sim}{\delta}_{2n+1}^1)^+$ (Kunen 71a , Martin 7 ?)

5) $[PD]$ For all n, $\underset{\sim}{\delta}_n^1 < \underset{\sim}{\delta}_{n+1}^1$ (Kechris 74 , Moschovakis 70)

Proof. 1) That $\underset{\sim}{\delta}_2^1 \le \omega_2$ follows from the Kunen-Martin theorem and the fact that every $\underset{\sim}{\Sigma}_2^1$ set is ω_1-Souslin. That $\underset{\sim}{\delta}_2^1 \le u_2$ follows from the fact that if $A \in \Sigma_2^1(\alpha)$, then $A = p[T]$, where $T \in L[\alpha]$ is a tree on $\omega \times \omega_1$. By the proof of the Kunen-Martin theorem the length of a $\Sigma_2^1(\alpha)$ wellfounded relation is $<(\omega_1^+)^\alpha < u_2$, hence $\underset{\sim}{\delta}_2^1 \le u_2$.

To see that $u_2 \le \underset{\sim}{\delta}_2^1$: For every $\alpha \in \mathcal{R}$ and for every $\xi < (\omega_1^+)^\alpha$ we can find a term τ such that

$$\xi = \tau^{L[\alpha]}(i^\alpha_{\xi_1} \dots i^\alpha_{\xi_n}, \omega_1, \vec{\aleph}),$$

for some $\xi_1 < \dots < \xi_n < \omega_1 < \vec{\aleph}$. Coding the $\xi_1 \dots \xi_n$ by reals we can easily find a $\Pi_1^1(\alpha^\#)$ prewellordering of reals of length $> (\omega_1^+)^\alpha$, so $u_2 = \sup_\alpha(\omega_1^+)^\alpha \le \underset{\sim}{\delta}_2^1$.

2) To prove $\underset{\sim}{\delta}_3^1 \le \omega_3$: We know that every $\underset{\sim}{\Sigma}_3^1$ set is u_ω-Souslin by the Martin-Solovay theorem. Hence $\underset{\sim}{\delta}_3^1 \le (u_\omega)^+$. Since $\forall n \le 2$, $\mathrm{cof}(u_n) = \mathrm{cof}(u_2)$, we must have $u_n < \omega_3$. But ω_3 is regular, hence $u_\omega < \omega_3$. Hence $\underset{\sim}{\delta}_3^1 \le \omega_3$.

3,4) These follow from the fact that every $\underset{\sim}{\Sigma}_{2n+2}^1$ set is $\underset{\sim}{\delta}_{2n+1}^1$-Souslin.

5) To prove $\underset{\sim}{\delta}_{2n+1}^1 < \underset{\sim}{\delta}_{2n+2}^1$ use the fact that a $\underset{\sim}{\Pi}_{2n+1}^1$ norm on a complete $\underset{\sim}{\Pi}_{2n+1}^1$ set has length exactly $\underset{\sim}{\delta}_{2n+1}^1$.

To prove that $\underset{\sim}{\delta}_{2n}^1 < \underset{\sim}{\delta}_{2n+1}^1$, prove first that $\underset{\sim}{\delta}_{2n+1}^1 = \sup\{\xi : \xi$ is the length of a $\underset{\sim}{\Sigma}_{2n+1}^1$ wellfounded relation$\}$ using the recursion theorem (see Moschovakis 70). Then let $W \subseteq \mathcal{R} \times \mathcal{R} \times \mathcal{R}$ be universal $\underset{\sim}{\Sigma}_{2n}^1$, and let

$$(\alpha,\beta) \prec (\alpha',\gamma) \Leftrightarrow \alpha = \alpha' \ \&\ W_\alpha \text{ is wellfounded } \&\ W(\alpha,\beta,\gamma).$$

Then $\prec$ is $\underset{\sim}{\Delta}_{2n+1}^1$ and dominates every $\underset{\sim}{\Sigma}_{2n}^1$ wellfounded relation. Hence $\underset{\sim}{\delta}_{2n}^1 < \underset{\sim}{\delta}_{2n+1}^1$.

⊣

Theorem 9.2 (Sierpinski for $\kappa = \omega$). If A is κ-Souslin, then A is the union of κ^+ sets in $\mathcal{B}_{\kappa^+}$.

Proof. Let $\alpha \in A \Leftrightarrow T(\alpha)$ not wellfounded, T a tree on $\omega \times \kappa$. For $\alpha \in A$ let

$$\psi(\alpha) = \sup\{|T(\alpha)_u| : u \in \kappa^{<\omega} \ \& \ T(\alpha)_u \text{ is wellfounded}\} < \kappa^+.$$

For $\xi < \kappa^+$, Let $B_\xi = \{\alpha \in A : \psi(\alpha) \leq \xi\}$. Since $A = \bigcup_{\xi<\kappa^+} B_\xi$, it is enough to show each $B_\xi \in \mathcal{B}_{\kappa^+}$.

For $\xi < \kappa^+$,

$$\alpha \in B_\xi \Leftrightarrow T(\alpha) \text{ is not wellfounded } \& \ \forall u[T(\alpha)_u \text{ wellfounded} \Rightarrow |T(\alpha)_u| \leq \xi],$$

therefore

$$\alpha \notin B_\xi \Leftrightarrow T(\alpha) \text{ is wellfounded} \vee \exists u[T(\alpha)_u \text{ is wellfounded } \& \ |T(\alpha)_u| > \xi]$$

$$\Leftrightarrow |T(\alpha)| < \xi + 1 \vee \exists u[|T(\alpha)_u| = \xi + 1].$$

Since by the proof of 3.8

$$A_u^\xi = \{\alpha : |T(\alpha)_u| < \xi\} \in \mathcal{B}_{\kappa^+},$$

we are done. ⊣

Theorem 9.3. 1) Every $\underset{\sim}{\Sigma}^1_2$ set is the union of $\aleph_1$ Borel sets (Sierpinski).
2) [AC + $\forall\alpha$ ($\alpha^\#$ exists)]. Every $\underset{\sim}{\Sigma}^1_3$ set is the union of $\aleph_2$ Borel sets. (Martin 7 ?).
3) [AC + PD] Every $\underset{\sim}{\Sigma}^1_4$ set is the union of $\aleph_3$ Borel sets. (Martin 7?).

Proof. 1) Every $\underset{\sim}{\Pi}^1_1$ set is the union of $\aleph_1$ Borel sets. Hence every $\underset{\sim}{\Sigma}^1_2$ set if the union of $\aleph_1 \underset{\sim}{\Sigma}^1_1$ sets. So it suffices to show that the $\underset{\sim}{\Sigma}^1_1$ sets are unions of $\aleph_1$ Borel sets. This follows from 9.2. (This proof uses AC. This can be avoided by using the uniformization theorem for $\underset{\sim}{\Pi}^1_1$ sets.)
2) Every $\underset{\sim}{\Sigma}^1_3$ set is u_ω-Souslin. Since $\mathrm{cof}(u_n) \leq \aleph_2$ for $n < \omega$, we have $u_\omega < \aleph_3$. So $\underset{\sim}{\Sigma}^1_3$ sets are $\aleph_2$-Souslin. If A is $\underset{\sim}{\Sigma}^1_3$ then for some tree T on $\omega \times \aleph_2$

$$\alpha \in A \Leftrightarrow T(\alpha) \text{ not wellfounded}$$

$$\Leftrightarrow \exists \xi < \aleph_2 \ (T^\xi(\alpha) \text{ not wellfounded}) \quad (\text{see } 3.9).$$

So A is the union of $\aleph_2$ many $\aleph_1$-Souslin sets. By the same argument, each $\aleph_1$-Souslin is the union of $\aleph_1$ many ω-Souslin (i.e. $\underset{\sim}{\Sigma}^1_1$) sets and we are done.
3) Similar, using the fact that every $\underset{\sim}{\Sigma}^1_4$ set is $\underset{\sim}{\delta}^1_3$-Souslin, and the fact that $\underset{\sim}{\delta}^1_3 \leq \aleph_3$. ⊣

Basic open problem. Is it true that (from any reasonable hypotheses and AC): $\underset{\sim}{\delta}^1_n \leq \aleph_n$, for $n \geq 5$?

10. Infinite exponent partition relations and the singular measures μ_λ.

Definition. If α,β,γ are ordinals with $\gamma \leq \beta \leq \alpha$, we put

$$\alpha \rightarrow (\beta)^\gamma$$

iff for every $X \subseteq \alpha^\gamma \uparrow = \{f \in \alpha^\gamma : f \text{ increasing}\}$ there is an $H \subseteq \alpha$ of order type β such that either $H^\gamma \uparrow \subseteq X$ or $H^\gamma \uparrow \subseteq \neg X$.

Remark. ZFC $\vdash \neg \exists \kappa\ (\kappa \rightarrow (\omega)^\omega)$.

Definition. Let κ be a regular cardinal, and let λ be a regular cardinal $< \kappa$. The filter μ_λ is the collection of all subsets of κ which contain a λ-closed unbounded set. ($A \subseteq \kappa$ is λ-closed if every increasing λ-sequence from A has its limit in A.)

Theorem 10.1 (Kleinberg 70).

1) If κ is a regular uncountable cardinal, $\lambda < \kappa$ a regular cardinal, and $\kappa \rightarrow (\kappa)^{\lambda+\lambda}$, then μ_λ is a normal measure.

2) If κ is a regular uncountable cardinal with $< \kappa$ many regular cardinals below κ, and $\forall \xi < \kappa\ (\kappa \rightarrow (\kappa)^\xi)$, then the normal measures on κ are exactly the μ_λ for λ regular $< \kappa$.

Proof of 2) from 1). By 1) we know each that μ_λ is a normal measure. Let μ be another normal measure. For λ regular $< \kappa$ let

$$E_\lambda = \{\xi < \kappa : \text{cofinality } (\xi) = \lambda\}.$$

Then the E_λ's are pairwise disjoint, and

$$\bigcup_{\substack{\lambda \text{ regular} \\ \lambda < \kappa}} E_\lambda = \{\xi < \kappa : \xi \text{ limit ordinal}\}.$$

Since there are $< \kappa$ regular cardinals below κ we can find a regular $\lambda_0 < \kappa$ such that

$$E_{\lambda_0} \in \mu.$$

Suppose $\mu \neq \mu_{\lambda_0}$. Then we can find a λ_0-closed unbounded A such that

$$B = \kappa - A \in \mu.$$

For $\xi \in B \cap E_{\lambda_0}$, let

$$g(\xi) = \sup(A \cap \xi).$$

Then $g(\xi) < \xi$ for all $\xi \in B \cap E_{\lambda_0}$, hence g is μ - a.e. constant, which contradicts the unboundedness of A. So $\mu = \mu_{\lambda_0}$.

Proof of 1). Assume $\kappa \rightarrow (\kappa)^{\lambda+\lambda}$.

To show that μ_λ is a normal measure, let $f : \kappa \to \kappa$ be pressing down. Consider $X \subseteq \kappa^{\lambda+\lambda}\uparrow$ given by

$$G \in X \Leftrightarrow f(\sup_{\alpha<\lambda} G(\alpha)) = f(\sup_{\alpha<\lambda} G(\lambda+\alpha)).$$

Let $H \subseteq \kappa$ be homogeneous for this partition, with card $(H) = \kappa$.

Suppose $H^{\lambda+\lambda}\uparrow \subseteq \neg X$. Let C be the set of limits of increasing λ sequences from H. Then C is λ-closed unbounded and for $\xi, \eta \in C$,

$$\xi < \eta \Rightarrow f(\xi) \neq f(\eta),$$

i.e. f is 1-1 on C. Now we inductively define an increasing λ-sequence $\{\gamma_\eta\}_{\eta<\lambda}$ of elements from C as follows:

$$\gamma_0 = \text{least member of } C$$
$$\gamma_\eta = \text{least element } \gamma \text{ of } C \text{ greater than all } \gamma_\theta \text{ for } \theta < \eta$$

which satisfies:

$$\forall\delta\ (\delta > \gamma \text{ and } \delta \in C \Rightarrow \forall\theta < \eta[f(\delta) > \gamma_\theta].$$

Then if $\gamma = \text{limit}_{\eta<\lambda}\, \gamma_\eta \in C$ we have $f(\gamma) < \gamma$, hence for some $\eta < \lambda$, $f(\gamma) < \gamma_\eta$, hence $\gamma \leq \gamma_{\eta+1}$, a contradiction.

Hence $H^{\lambda+\lambda}\uparrow \subseteq X$. Then if $\xi < \eta$ are both in C, we can find $G \in H^{\lambda+\lambda}\uparrow$ such that

$$\xi = \sup_{\alpha<\lambda} (G(\alpha)),\ \eta = \sup_{\alpha<\lambda} G(\lambda+\alpha)$$

and hence $f(\xi) = f(\eta)$. So f is constant on C, proving normality for μ_λ.

To see now that μ_λ is an ultrafilter, look at the characteristic functions of subsets of κ (which are of course pressing down). To see that μ_λ is κ-additive, let $\{A_\theta\}_{\theta<\rho<\kappa} \subseteq \mu_\lambda$ and suppose $\bigcap_{\theta<\rho} A_\theta = \emptyset$, towards a contradiction. Then $\kappa = \bigcup_{\theta<\rho} (\kappa - A_\theta)$, so consider

$$f(\xi) = \begin{cases} \text{least } \theta < \rho \text{ such that } \xi \notin A_\theta, & \text{if } \xi \geq \rho \\ 0 & \text{otherwise} \end{cases}$$

Then for some $\theta < \rho$, $\{\xi : f(\xi) = \theta\}$ contains a λ-closed unbounded set, i.e. $\mu_\lambda(\kappa - A_\theta) = 1$, a contradiction. ⊣

11. Countable exponent partition relations for $\underset{\sim}{\delta}^1_n$, n odd.

We present first in an abstract form Martin's method for proving infinite exponent partition relations from AD. It is a modification of Solovay's technique used in the proof of 5.1.

Lemma 11.1. (Martin) Let $\kappa > \omega$ be a regular cardinal, $\lambda \leq \kappa$ an ordinal.

Assume:

1) There is $\{C_\xi\}_{\xi<\omega\cdot\lambda}$, with $C_\xi \subseteq \mathcal{R}$, and for each $\xi < \omega\cdot\lambda$ a map $\varepsilon \to f^\xi(\varepsilon)$ from C_ξ into κ such that if $C = \bigcap_{\xi<\omega\cdot\lambda} C_\xi$ and for $\varepsilon \in C$ we let

$f_\varepsilon(\xi) = f^\xi(\varepsilon)$ then $\varepsilon \mapsto f_\varepsilon$ maps C onto ${}^{\omega\cdot\lambda}\varkappa$.

2) There are $\{C_{\xi,\theta}\}_{\xi<\omega\cdot\lambda,\theta<\varkappa}$ such that

$$C_{\xi,\theta} \subseteq \bigcap_{\xi'\leq\xi} C_{\xi'}$$

and if $\sigma : \mathcal{R} \to \mathcal{R}$ is continuous and $\sigma[\bigcap_{\xi'\leq\xi} C_{\xi'}] \subseteq C_\xi$ then for all $\xi < \omega\cdot\lambda$ and $\theta < \varkappa$ we have

$$G^\sigma(\xi,\theta) =^{\text{def}} \sup\{f^\xi(\sigma(\varepsilon)) + 1 : \varepsilon \in C_{\xi,\theta}\} < \varkappa.$$

3) If $f \in {}^{\omega\cdot\lambda}\varkappa\uparrow$ then there is $\varepsilon \in C$ such that $f_\varepsilon = f$ and $\varepsilon \in C_{\xi,f(\xi)}$, $\forall \xi < \omega\cdot\lambda$.

Then:

$$AD + DC \Rightarrow \varkappa \to (\varkappa)^\lambda.$$

<u>Proof</u>. Let $A \subseteq \varkappa^\lambda\uparrow$ and consider the following game:

I $\quad$ II

$\varepsilon^I \quad \varepsilon^{II} \quad$ II wins iff

(1) $\exists \xi < \omega\cdot\lambda\ (\varepsilon^I \notin C_\xi \vee \varepsilon^{II} \notin C_\xi)$ and if ξ_0 is the least such ξ then $\varepsilon^I \notin C_{\xi_0}$.

or

(2) $\forall \xi < \omega\cdot\lambda(\varepsilon^I \in C_\xi \wedge \varepsilon^{II} \in C_\xi)$ and $\langle \sup_n\{f_{\varepsilon^I}(\omega\cdot\theta+n), f_{\varepsilon^{II}}(\omega\cdot\theta+n)\}\rangle_{\theta<\lambda} \in A$.

Without loss of generality we can assume that II has a winning strategy σ. Then by (1) $\sigma[\bigcap_{\xi'\leq\xi} C_{\xi'}] \subseteq C_\xi$, $\forall\xi < \omega\cdot\lambda$. So by 2) above $G^\sigma(\xi,\theta) < \varkappa$. By the regularity of $\varkappa$ let $D \subseteq \varkappa$ be closed unbounded such that

$$\rho \in D \Rightarrow \forall\xi < \omega\cdot\lambda\ \forall\theta < \varkappa\ [\xi < \rho \wedge \theta < \rho \Rightarrow G^\sigma(\xi,\theta) < \rho].$$

Let

$$D\not\vdash^\lambda = \{g \in D^\lambda\uparrow : \exists f \in {}^{\omega\cdot\lambda}\varkappa\uparrow\ \forall\theta < \lambda(g(\theta) = \sup_n f(\omega\cdot\theta+n)\}.$$

We claim that $D\not\vdash^\lambda \subseteq A$, which completes the proof since then $H^\lambda\uparrow \subseteq A$, where $H = \{\delta_{\xi+\omega} : \xi < \varkappa\}$, where $\{\delta_\nu\}_{\nu<\varkappa}$ is the increasing enumeration of D. Let $g \in D\not\vdash^\lambda$ and let $f \in {}^{\omega\cdot\lambda}\varkappa\uparrow$ be such that $g(\theta) = \sup_n f(\omega\cdot\theta+n)$. Then by 3) above find $\varepsilon \in C$ such that $f_\varepsilon = f$ and $\varepsilon \in C_{\xi,f(\xi)}$. Then for all $\theta < \lambda$ and all $n \in \omega$:

$$f^{\omega\cdot\theta+n}(\sigma(\varepsilon)) < G^\sigma(\omega\cdot\theta+n,\ f(\omega\cdot\theta+n)) < g(\theta),$$

since

$$\omega\cdot\theta + n \leq f(\omega\cdot\theta+n) < g(\theta) \in D.$$

So for all $\theta < \lambda$

$$\sup_n\{f_\varepsilon(\omega\cdot\theta+n),\ f_{\sigma(\varepsilon)}(\omega\cdot\theta+n)\} = g(\theta),$$

therefore $g \in A$. $\dashv$

Theorem 11.2 [AD] (Martin 71). For any $n \geq 0$,

$$\underset{\sim}{\delta}^1_{2n+1} \to (\underset{\sim}{\delta}^1_{2n+1})^\lambda, \quad \forall \lambda < \omega_1.$$

Proof. Fix $t : \omega \cdot \lambda \xrightarrow[\text{onto}]{1\text{-}1} \omega$. For a real α, set $\alpha_\xi = (\alpha)_i$, where $t(\xi) = i$. Let also W be a complete $\underset{\sim}{\Pi}^1_{2n+1}$ set, φ a $\underset{\sim}{\Pi}^1_{2n+1}$-norm on W with range $\underset{\sim}{\delta}^1_{2n+1}$ and for $\alpha \in W$, write $|\alpha| = \varphi(\alpha)$.

Define now for $\xi < \omega \cdot \lambda$,

$$C_\xi = \{\alpha : \alpha_\xi \in W\}$$

and for $\alpha \in C_\xi$,

$$f^\xi(\alpha) = |\alpha_\xi|.$$

Finally let for $\xi < \omega \cdot \lambda$, $\theta < \underset{\sim}{\delta}^1_{2n+1}$:

$$C_{\xi,\theta} = \{\alpha : \forall \xi' \leq \xi \exists \eta' \leq \theta (\alpha_{\xi'} \in W \wedge |\alpha_{\xi'}| \leq \eta')\} .$$

Now obviously properties 1), 3) of Lemma 11.1 are satisfied so it is enough to verify 2). For that notice that $C_{\xi,\theta} \in \underset{\sim}{\Delta}^1_{2n+1}$ so that if σ is continuous then $\sigma[C_{\xi,\theta}]$ is $\underset{\sim}{\Sigma}^1_{2n+1}$. If also $\sigma[C_{\xi,\theta}] \subseteq C_\xi$ then $\{\alpha_\xi : \alpha \in \sigma[C_{\xi,\theta}]\}$ is a $\underset{\sim}{\Sigma}^1_{2n+1}$ subset of W, so by boundedness

$$G^\sigma(\xi,\theta) = \sup\{|\alpha_\xi| + 1 : \alpha \in \sigma[C_{\xi,\theta}]\} < \underset{\sim}{\delta}^1_{2n+1}$$

and we are done. ⊣

12. $\omega_1 \to (\omega_1)^{\omega_1}$.

Definition. For $C \subseteq \varkappa$, put

$$C\!\!\not{p} = C\!\!\not{p}^\varkappa = \{f \in C^\varkappa\!\uparrow : \exists g \in \varkappa^\varkappa\!\uparrow \forall \xi, f(\xi) = \sup_n g(\omega \cdot \xi + n)\}.$$

It is not hard to check that

$$\varkappa \to (\varkappa)^\varkappa \Leftrightarrow \forall X \subseteq \varkappa^\varkappa\!\uparrow \exists C (C \text{ is closed unbounded on } \varkappa \text{ and } C\!\!\not{p} \subseteq X \text{ or } C\!\!\not{p} \subseteq \neg X).$$

Theorem 12.1 [AD] (Martin; see Martin-Paris 71).

$$\omega_1 \to (\omega_1)^{\omega_1}.$$

Proof. We will apply Lemma 11.1 again. Let $\tau_0, \tau_1, \tau_2, \ldots$ be a recursive enumeration of terms in the language of $ZF + V = L[\dot{\alpha}]$, which take only ordinal values, as in the proof of 6.3. For $\xi < \omega_1$ put, using again the notation of 6.3:

$$C_\xi = \{\varepsilon : \varepsilon = \widehat{n\varepsilon'} = \langle n, \varepsilon'(0), \varepsilon'(1), \ldots \rangle \ \&\ P(\varepsilon') \ \&\ \text{the wellfounded part of } \Gamma(\xi + \omega, \varepsilon') \text{ has an ordinal of order type } \xi \text{ (denoted also by } \xi) \text{ and } \tau_n^{\Gamma(\xi+\omega,\varepsilon')}(\xi) \text{ also belongs to the wellfounded part}\}.$$

Then

$$\varepsilon \in C \Leftrightarrow \forall \xi < \omega_1 (\varepsilon \in C_\xi) \Leftrightarrow \varepsilon = \widehat{n\alpha^\#}, \text{ for some } n,\alpha.$$

For $\varepsilon \in C_\xi$, let also

$$f^\xi(\varepsilon) = \tau_n^{\Gamma(\xi+\omega,\varepsilon')}(\xi).$$

Then if $\varepsilon \in C$, say $\varepsilon = \widehat{n\alpha^\#}$, we have for all $\xi < \omega_1$:

$$f_\varepsilon(\xi) = \tau_n^{L[\alpha]}(\xi).$$

Finally put for $\xi < \omega_1$, $\theta < \omega_1$

$$C_{\xi,\theta} = \{\varepsilon : \forall \xi' \leq \xi \exists \eta' \leq \theta(\varepsilon \in C_{\xi'} \wedge \text{ if } \varepsilon = \widehat{n\varepsilon'}, \text{ then } \tau_n^{\Gamma(\xi'+\omega,\varepsilon')}(\xi') \leq \eta')\}.$$

Clearly conditions 1), 3) or 11.1 are satisfied. To verify also condition 2) note first that each $C_{\xi,\theta}$ is Borel. So if σ is continuous $\sigma[C_{\xi,\theta}]$ is Σ_1^1. If moreover $\sigma[C_{\xi,\theta}] \subseteq C_\xi$ then an easy boundedness argument shows that $G^\sigma(\xi,\theta) < \omega_1$ and we are done. ⊣

13. The Martin-Paris theorem

Definition. Let κ be an uncountable cardinal, μ a normal measure on κ, and assume $\kappa^\kappa/\mu \cong \kappa^+$. For each $f \in \kappa^\kappa$, let $f(\kappa) = [f]$ (thus $\kappa^+ = \{f(\kappa) : f \in \kappa^\kappa\}$). A μ as above is canonical if it has the following selection property: If $\pi < \kappa^+$, and $\{\xi_\lambda\}_{\lambda<\pi}$ is a π-sequence of ordinals $< \kappa^+$, then there is a sequence $\{f_\lambda\}_{\lambda<\pi} \subseteq \kappa^\kappa$ such that $f_\lambda(\kappa) = \xi_\lambda$.

Note that if such a measure exists, then κ^+ is regular.

Theorem 13.1 [AD] (Solovay). The measure μ_ω on ω_1 is canonical.

Proof. If $f \in \omega_1^{\omega_1}$, we can find τ, α such that

$$\forall \eta < \omega_1,\ f(\eta) = \tau^{L[\alpha]}(\eta).$$

It is easy to check that $f(\omega_1) = \tau^{L[\alpha]}(\omega_1)$. So in particular

$$\omega_1^{\omega_1}/\mu_\omega \cong \omega_2.$$

Now let $\{\xi_\lambda\}_{\lambda<\pi}$ be a sequence of ordinals less than ω_2. Without loss of generality $\pi = \omega_1$, and

$$\{\xi_\lambda : \lambda < \omega_1\} = \xi < \omega_2.$$

Define the following prewellordering on ω_1:

$$\lambda \lesssim \lambda' \Rightarrow \xi_\lambda \leq \xi_{\lambda'}.$$

By the proof of Solovay's theorem 7.2, find τ,α such that

$$\lesssim\ = \tau^{L[\alpha]}(\omega_1).$$

For some term σ then,

$$\xi_\lambda = \sigma^{L[\alpha]}(\lambda,\omega_1), \forall\lambda < \omega_1.$$

Take $f_\lambda(\eta) = \sigma^{L[\alpha]}(\lambda,\eta)$. $\dashv$

Notation. Let $\varkappa$ be an uncountable cardinal carrying a canonical measure μ. Let $\varkappa \leq \pi < \varkappa^+$, and fix $h : \varkappa \xrightarrow[\text{onto}]{1\text{-}1} \pi$. For any $\theta < \varkappa$, let $\bar{\theta},\rho_\theta$ be such that

$$\rho_\theta : \bar{\theta} \xrightarrow[\text{onto}]{1\text{-}1} h[\theta] = \{h(\xi) : \xi < \theta\},$$

with ρ_θ order preserving. Then consider the normal (i.e. increasing and continuous) function $\chi : \varkappa \to \varkappa$ such that $\chi_{\theta+1} - \chi_\theta = \bar{\theta}$ (where $\chi_\theta = \chi(\theta)$). For $\lambda \in h[\theta]$ let

$$\chi_{\theta,\lambda} = \chi_\theta + \rho_\theta^{-1}(\lambda).$$

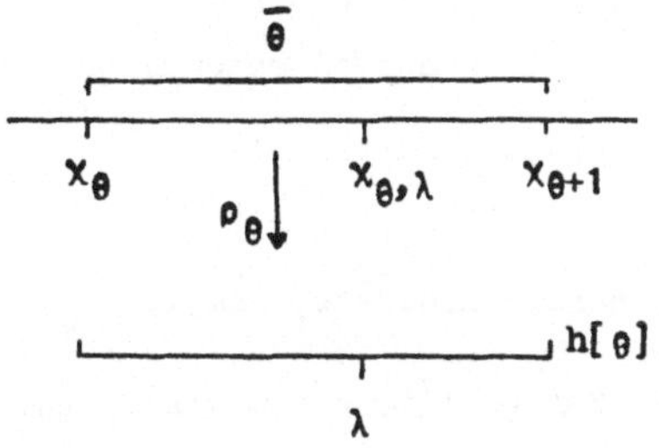

Then $\lambda < \lambda' \in h[\theta] \Rightarrow \chi_{\theta,\lambda} < \chi_{\theta,\lambda'}$.

Thus to each $\lambda < \pi$ we can assign an increasing

$$\psi_\lambda : \varkappa \to \varkappa,$$

(actually ψ_λ is defined from a point on), where

$$\psi_\lambda(\theta) = \chi_{\theta,\lambda}.$$

Thus $\lambda < \lambda' \Rightarrow \psi_\lambda(\theta) < \psi_{\lambda'}(\theta)$.

Now let $f \in \varkappa^\varkappa\uparrow$. For $\lambda < \pi$ let

$$f^{[\lambda]}(\theta) = f(\psi_\lambda(\theta)).$$

Thus $f^{[\lambda]} \in \varkappa^\varkappa\uparrow$. Now let

$$\tilde{f}^\pi \in (\varkappa^+)^\pi\uparrow$$

be defined by

$$\tilde{f}^\pi(\lambda) = f^{[\lambda]}(\varkappa).$$

(Recall that for a function $g \in \varkappa^\varkappa$, $g(\varkappa)$ denotes the image of g in the ultrapower $\varkappa^\varkappa/\mu$.) Since $\lambda < \lambda' \Rightarrow$ for all θ from a point on, $\psi_\lambda(\theta) \underset{\sim}{<} \psi_{\lambda'}(\theta)$, we see that $\lambda < \lambda' \Rightarrow \{\theta : f^{[\lambda]}(\theta) < f^{[\lambda']}(\theta)\}$ has measure 1, so $\tilde{f}^\pi$ is indeed increasing.

For $A \subseteq \varkappa$, let

$$(A^{\varkappa}\uparrow)^{\tilde{\pi}} = \{f^{\tilde{\pi}} : f \in A^{\varkappa}\uparrow\}$$

and let

$$A^{*} = \{f(\varkappa) : f \in A^{\varkappa}\}.$$

Note that if A is unbounded in $\varkappa$, then A^{*} is unbounded in $\varkappa^{+}$. (If $f \in \varkappa^{\varkappa}$, define $g \in A^{\varkappa}$ by $g(\xi)$ = least member of $A > f(\xi)$; then $g(\varkappa) < f(\varkappa)$).

Theorem 13.2 [AD] (Martin-Paris 71). Let $\varkappa$ be an uncountable cardinal, μ a canonical measure on $\varkappa, \varkappa \leq \pi < \varkappa^{+}$, $A \subseteq \varkappa$ unbounbed. Then

$$(A^{*} - (\varkappa + 1))^{\pi}\uparrow \subseteq (A^{\varkappa}\uparrow)^{\tilde{\pi}}.$$

Corollary 13.3 [AD] (Martin-Paris 71). Let $\varkappa$ be an uncountable cardinal carrying a canonical measure μ. If $\varkappa \rightarrow (\varkappa)^{\varkappa}$, then $\forall \pi < \varkappa^{+}$, $\varkappa^{+} \rightarrow (\varkappa^{+})^{\pi}$. Hence for any regular $\lambda < \varkappa^{+}$, μ_{λ} is a normal measure on $\varkappa^{+}$.

Corollary 13.4 [AD] (Martin-Paris 71)

1) $\forall \pi < \omega_2, \omega_2 \rightarrow (\omega_2)^{\pi}$.

2) ω_2 has exactly two normal measures, namely $\mu_{\omega}, \mu_{\omega_1}$.

Proof of 13.3. Let $X \subseteq (\varkappa^{+})^{\pi}\uparrow$, $\varkappa \leq \pi < \varkappa^{+}$. Put $\tilde{\pi}X = \{f \in \varkappa^{\varkappa}\uparrow : f^{\tilde{\pi}} \in X\}$. Let $H \subseteq \varkappa$ have cardinality $\varkappa$ such that, say, $H^{\varkappa}\uparrow \subseteq \tilde{\pi}X$. Then by 13.2

$$(H^{*} - (\varkappa + 1))^{\pi}\uparrow \subseteq (H^{\varkappa}\uparrow)^{\tilde{\pi}} \subseteq (\tilde{\pi}(X))^{\tilde{\pi}} \subseteq X.$$

⊣

Proof of 13.2. Let $f \in (A^{*} - (\varkappa + 1))^{\pi}\uparrow$. Then find $\{f_{\lambda}\}_{\lambda < \pi} \subseteq \varkappa^{\varkappa}$ such that

$$f_{\lambda}(\varkappa) = f(\lambda).$$

We want to find $G \in A^{\varkappa}\uparrow$ such that

$$\forall \lambda < \pi,\ G^{\tilde{\pi}}(\lambda) = f_{\lambda}(\varkappa), \quad \text{i.e.}$$

$$\forall \lambda < \pi,\ G^{[\lambda]}(\varkappa) = f_{\lambda}(\varkappa), \ \text{i.e.}$$

$$\forall \lambda < \pi,\ G(x_{\theta,\lambda}) = f_{\lambda}(\theta) \text{ for } \mu\text{-almost all } \theta.$$

For that it is enough to have for μ-almost all θ,

$$\forall \lambda \in h[\theta],\ G(x_{\theta,\lambda}) = f_{\lambda}(\theta).$$

To prove this we need the following

Lemma. There is a set C of μ-measure 1 such that if $\theta \in C$:

i) $\forall \lambda \in h[\theta], (f_{\lambda}(\theta) \in A)$

and

ii) $\forall \lambda < \lambda' \in h[\theta](f_{\lambda}(\theta) < f_{\lambda'}(\theta))$.

Proof of lemma. For fixed ξ, η, let $C_{\xi,\eta} = \{\theta : f_{h(\xi)}(\theta) \in A$ and $h(\xi) < h(\eta) \Rightarrow f_{h(\xi)}(\theta) < f_{h(\eta)}(\theta)\}$. Then each $C_{\xi,\eta} \in \mu$ (since for all $\lambda < \pi$,

$f_\lambda(\kappa) \in A^* \Rightarrow f_\lambda$ is μ-equivalent to some element of A^κ, hence $\{\theta : f_\lambda(\theta) \in A\} \in \mu$).

Now let $C = \{\theta : \forall \xi, \eta < \theta\ (\theta \in C_{\xi,\eta})\}$. Then C has μ-measure 1 and has the required properties. So the proof of the lemma is complete.

To finish the proof of 13.2: Let θ_0 be large enough so that $0 \in h[\theta_0]$. Now define

$$G(\chi_{\theta,\lambda}) = \begin{cases} f_\lambda(\theta), & \text{if } \theta \in C - \theta_0, \lambda \in h[\theta] \text{ and} \\ & \quad f_0(\theta) > \sup_{\zeta < \chi_\theta} G(\zeta) \\ \text{least member of } A, & \text{otherwise.} \\ \text{greater than all} & \\ G(\zeta) \text{ for } \zeta < \chi_{\theta,\lambda} & \end{cases}$$

Claim. For μ-almost all θ, the first definition occurs.

Proof. Since χ is normal, for μ-almost all $\theta, \chi_\theta = \theta = \sup_{\zeta < \chi_\theta} G(\zeta)$. Since $f_0(\kappa) > \kappa$, for μ-almost all θ, $f_0(\theta) > \theta = \sup_{\zeta < \chi_\theta} G(\zeta)$, which proves the claim.

By the properties of C given in the lemma, $G \in A^\kappa\uparrow$ and is as desired. ⊣

Theorem 13.5 [AD] (Martin-Paris 71). Let κ be an uncountable cardinal, μ a normal measure on κ, $\kappa_2 \cong \kappa^\kappa/\mu$. For each $\pi < \kappa^+$ there is a map $f \to \tilde{f}^\pi$ sending $\kappa^\kappa\uparrow$ into $\kappa_2^\pi\uparrow$, such that if $A \subseteq \kappa$, $F \in (A^* - (\kappa+1))^\pi\uparrow$, and $\exists G \in (\kappa^\kappa)^\pi$ such that

$$[G(\lambda)]_\mu = F(\lambda),$$

then there is an $f \in A^\kappa\uparrow$ with $\tilde{f}^\pi = F$.

Corollary 13.6. In the notation of 13.5, if $\kappa \to (\kappa)^\kappa$, then $\kappa_2 \to (\kappa_2)^\pi$, $\forall \pi < \omega_1$.

Proof of theorem. For $\pi \geq \kappa$ see the proof of 13.2. Assume $\pi < \kappa$. Define a normal function χ on κ by $\chi_{\theta+1} - \chi_\theta = \pi$. Let $\chi_{\theta,\lambda} = \chi_\theta + \lambda$ for $\lambda < \pi$. For $f \in \kappa^\kappa$ let

$$f^{[\lambda]}(\theta) = f(\chi_{\theta,\lambda}),$$

and let $\tilde{f}^\pi = \{f^{[\lambda]}\}_{\lambda<\pi}$. Now repeat the proof of 13.2. ⊣

For the corollary just notice that if $A \subseteq \kappa$ has cardinality κ then $A^* - (\kappa+1)$ has order type κ_2.

14. The measure μ_ω on $\underset{\sim}{\delta}^1_n$, n odd.

Definition. Let κ be a cardinal. If W is a wellordering of (a subset of) κ, let for $\xi < \kappa$

$$H_W(\xi) = |W \restriction \xi|,$$

where $W \restriction \xi = W \cap (\xi \times \xi)$. Clearly $H_W : \kappa \to \kappa$.

Theorem 14.1. Let κ be an uncountable cardinal, μ a normal measure on κ. Then for any wellordering W on κ,

$$[H_W]_\mu = |W|.$$

In particular, $\{[H_W]_\mu : W \text{ a wellordering on } \kappa\} = \kappa^+$.

Proof. Notice first that $\{[H_W]_\mu : W \text{ a wellordering on } \kappa\}$ is an initial segment of ordinals. Because if $F : \kappa \to \kappa$ is such that $[F]_\mu < [H_W]_\mu$, then for μ-almost all ξ, $F(\xi) < |W \restriction \xi|$. Hence there is a map $\xi \mapsto \xi^* < \xi$ such that for μ-almost all ξ, $F(\xi) = |W^{\xi^*} \restriction \xi|$ (where for any wellordering W, W^x = initial segment of W determined by x). So by normality, find ξ_0 such that $F(\xi) = |W^{\xi_0} \restriction \xi|$ μ-almost everywhere. Hence $[F]_\mu = [H_{W^{\xi_0}}]_\mu$.

To prove the theorem it is enough to show that for W, V wellorderings on κ,

$$[H_W]_\mu < [H_V]_\mu \Leftrightarrow |W| < |V|.$$

If $H_W(\xi) < H_V(\xi)$ μ- a.e., let

$$F_\xi : W \restriction \xi \xrightarrow{\text{isomorphism}} V^{\xi^*} \restriction \xi, \text{ for } \mu\text{-almost all } \xi,$$

with $\xi^* < \xi$. Then by normality, there is ξ_0 such that

$$F_\xi : W \restriction \xi \xrightarrow{\text{isomorphism}} V^{\xi_0} \restriction \xi,$$

μ-almost everywhere. For each $\eta \in \text{Field}(W)$, let

$$f_\eta(\xi) = F_\xi(\eta) < \xi,$$

so by normality $f_\eta(\xi) = g(\eta)$ for μ-a.a. ξ (i.e. $g(\eta)$ is the constant value assumed μ - a.e. by f_η). Then $g : W \xrightarrow{\text{embd}} V^{\xi_0}$, so $|W| < |V|$. Similarly $H_W(\xi) \leq H_V(\xi)$ μ- a.e. $\Rightarrow |W| \leq |V|$. ⊣

Corollary 14.2. Let κ be an uncountable cardinal, μ a normal measure on κ. Then the following are equivalent:

1) μ is canonical.

2) $\kappa^\kappa/\mu \cong \kappa^+$ and κ^+ is regular.

3) Every $F : \kappa \to \kappa$ is μ - a.e. equal to some H_W, W a wellordering on κ, and κ^+ is regular.

Proof. 1) $\Rightarrow$ 2) by definition.

2) $\Leftrightarrow$ 3) by 14.1.

2) $\Rightarrow$ 1) Enough to show the selection property, and since κ^+ is regular, it is enough to show that if $\rho < \kappa^+$ then we can find $\{f_\lambda\}_{\lambda<\rho}$, where $f_\lambda : \kappa \to \kappa$ such that $[f_\lambda]_\mu = \lambda$. Pick a wellordering W of κ with $|W| = \rho$. For each $\lambda < \rho$ let $\lambda^* < \kappa$ be such that

$$|W^{\lambda^*}| = \lambda.$$

Then let $f_\lambda = H_{W^{\lambda^*}}$ ⊣

Theorem 14.3 [AD] (Kunen 71c). For each $n \geq 0$, the measure μ_ω on $\underset{\sim}{\delta}^1_{2n+1}$ is canonical.

Proof. We need some lemmas first.

Lemma A. There is a $\underset{\sim}{\Pi}^1_{2n+1}$ set $G \subseteq \mathcal{R}$ and a $\underset{\sim}{\delta}^1_{2n+1}$-scale $\{\varphi_m\}_{m\in\omega}$ on G such that if we put $\psi(\alpha) = \sup_m \varphi_m(\alpha)$, then

1) $\varphi_m(\alpha) < \psi(\alpha)$, $\forall \alpha \in G$, $\forall m \in \omega$.

2) $\{\psi(\alpha) : \alpha \in G\} \in \mu_\omega$.

3) If $A \subseteq G$ is $\underset{\sim}{\Sigma}^1_{2n+1}$, then $\sup_{\alpha \in A} \psi(\alpha) < \underset{\sim}{\delta}^1_{2n+1}$.

Proof. Let W be a $\underset{\sim}{\Pi}^1_{2n+1}$-complete set of reals, $\{\chi_m\}_{m\in\omega}$ a $\underset{\sim}{\Pi}^1_{2n+1}$-scale on W, where the range of each χ_m is included in $\underset{\sim}{\delta}^1_{2n+1}$. Put for $\alpha \in W$,

$$\overline{\chi}_m(\alpha) = \chi_0(\alpha) + \chi_1(\alpha) + \ldots + \chi_m(\alpha).$$

Then $\{\overline{\chi}_m\}_{m\in\omega}$ is a $\underset{\sim}{\delta}^1_{2n+1}$-scale on W and for all $m, \alpha \in W$, $\overline{\chi}_m(\alpha) < \sup_m \overline{\chi}_m(\alpha)$ (we can clearly assume here without loss of generality that always $\chi_m(\alpha) > 0$). Consider now

$$G = \{\alpha : \forall i, (\alpha)_i \in W\}$$

and for $\alpha \in G$, $m \in \omega$ define

$$\varphi_m(\alpha) = \overline{\chi}_{(m)_0}((\alpha)_{(m)_1}),$$

where $m \mapsto ((m)_0,(m)_1)$ is a 1-1 correspondence between ω and $\omega \times \omega$. Clearly $\{\varphi_m\}_{m\in\omega}$ is a $\underset{\sim}{\delta}^1_{2n+1}$-scale on G and if $\psi(\alpha) = \sup_m \varphi_m(\alpha)$ then properties 1), 3) are satisfied.

Claim. $\{\psi(\alpha) : \alpha \in G\}$ is ω-closed unbounded.

Proof. Clearly it is unbounded. Let

$$\psi(\alpha^0) < \psi(\alpha^1) < \ldots \to \lambda$$

where $\alpha^0, \alpha^1, \ldots \in G$. Let $\alpha \in G$ be such that

$$(\alpha)_i = (\alpha^{(i)_0})_{(i)_1}.$$

Then

$$\varphi_m(\alpha) = \overline{\chi}_{(m)_0}(\alpha_{(m)_1}) = \overline{\chi}_{(n)_0}((\alpha^{(m)_{1,0}})_{(m)_{1,1}}) < \psi(\alpha^{(m)_{1,0}}) < \lambda,$$

where $(m)_{i,j} = ((m)_i)_j$.

If $\theta < \lambda$, find j large enough and $\varkappa$ such that $\varphi_\varkappa(\alpha^j) > \theta$. Then if m is such that

$$(m)_0 = (k)_0,\ (m)_{1,0} = j,\ (m)_{1,1} = (k)_1,$$

we have $\varphi_m(\alpha) = \overline{\chi}_{(k)_0}((\alpha^j)_{(k)_1}) > \theta$.

Hence $\psi(\alpha) = \lambda$. ⊣

Lemma B. There is a tree U on $\omega\times\varkappa_{2n+1}$ such that $\sup_{\alpha\in\mathcal{R}}\{|U(\alpha)| : U(\alpha)$ is wellfounded$\} = \underset{\sim}{\delta}^1_{2n+1}$.

Proof. Let S be a $\underset{\sim}{\Sigma}^1_{2n+1}$-complete set of reals and let U be a tree on $\omega\times\varkappa_{2n+1}$ such that $p[U] = S$. ⊣

To prove now the theorem: Let $F : \underset{\sim}{\delta}^1_{2n+1} \to \underset{\sim}{\delta}^1_{2n+1}$ be given, and consider the following game: I plays α, II plays β, and II wins iff

$$\alpha \in G \Rightarrow U(\beta) \text{ is wellfounded and } |U(\beta)| > F(\psi(\alpha)).$$

If I has a winning strategy then by Lemma A.3) and Lemma B we get a contradiction. So assume σ_0 is a winning strategy for II. Let T be the tree on $\omega\times\underset{\sim}{\delta}^1_{2n+1}$ coming from the scale $\{\varphi_n\}_{n\in\omega}$ on G (thus $G = p[T]$). Then for all α

(*) $$T(\alpha) \text{ not wellfounded} \Rightarrow U(\sigma_0[\alpha]) \text{ is wellfounded and}$$
$$F(\psi(\alpha)) < |U(\sigma_0[\alpha])|.$$

Let $\sigma[\alpha] = \beta \Leftrightarrow \forall n\ (\sigma\restriction n, \alpha\restriction n, \beta\restriction n) \in S$, S a tree on $\omega\times\omega\times\omega$. Let R be the tree on $\omega\times\omega\times\omega\times\underset{\sim}{\delta}^1_{2n+1}\times\varkappa_{2n+1}$ defined by

$$(s,a,b,u,v) \in R \Leftrightarrow (s,a,b) \in S \ \&\ (b,v) \in U \ \&\ (a,u) \in T.$$

Then $R(\sigma_0)$ is wellfounded by (*). Suppose $\alpha \in G \ \&\ \psi(\alpha) = \xi > \varkappa_{2n+1}$. Let $f \in \xi^\omega$ be defined by $f(n) = \varphi_n(\alpha)$. Thus if $\beta = \sigma_0[\alpha]$, for any $v \in U(\beta)$ we have

$$(\alpha\restriction \ell h(v), \beta\restriction \ell h(v), f\restriction \ell h(v), v) \in R(\sigma_0)\restriction\xi = \{(a,b,u,v) \in R(\sigma_0) : u \in \xi^{<\omega}\}$$

where $\ell h(\alpha_0 \ldots a_{m-1}) = m$. Hence

$$F(\xi) < |U(\sigma_0[\alpha])| \leq |R(\sigma_0)\restriction\xi|.$$

If we let $W(\sigma_0)$ be the Kleene-Brouwer wellordering of $R(\sigma_0)$, viewed as a wellordering of $\underset{\sim}{\delta}^1_{2n+1}$ (after identifying $\omega^{<\omega}\times\omega^{<\omega}\times(\underset{\sim}{\delta}^1_{2n+1})^{<\omega}\times(\varkappa_{2n+1})^{<\omega}$ with $\underset{\sim}{\delta}^1_{2n+1}$), then we have by the above

$$F(\xi) < H_{W(\sigma_0)}(\xi)\ \ \mu_\omega\text{-a.e.}$$

Hence $F(\xi) = H_{W(\sigma_0)^{\xi_0}}(\xi)$ μ_ω-a.e. for some ξ_0. So μ_ω is canonical. ⊣

15. <u>The measures μ_λ, with $\lambda > \omega$, on $\underset{\sim}{\delta}^1_n$, n odd.</u>

Lemma 15.1 [AD]. There is a relation $W \subseteq \mathcal{R}\times\underset{\sim}{\delta}^1_{2n+1}\times\underset{\sim}{\delta}^1_{2n+1}$, with the following properties:

1) If $W_\varepsilon(\xi,\eta) \Leftrightarrow W(\varepsilon,\xi,\eta)$, then for every $F : \underset{\sim}{\delta}^1_{2n+1} \to \underset{\sim}{\delta}^1_{2n+1}$ there is $\varepsilon_0 \in \mathcal{R}_{\xi_0}$ and $\xi_0 < \underset{\sim}{\delta}^1_{2n+1}$ such that W_{ε_0} is a wellordering and $F(\xi) = |(W_{\varepsilon_0})^{\xi_0}\restriction\xi|$ μ_ω-a.e. In particular, $\sup\{|W_{\varepsilon_0}| : W_{\varepsilon_0}$ is a wellordering$\} = \underset{\sim}{\delta}^1_{2n+2}$.

2) If $W_{\xi,\eta} = \{\varepsilon : W(\varepsilon,\xi,\eta)\}$ then for each $\xi,\eta, W_{\xi,\eta}$ is an open set of reals. In particular for each $\xi_0,\xi,\theta < \underset{\sim}{\delta}^1_{2n+1}$, $\{\varepsilon : |(W_\varepsilon)^{\xi_0}\restriction\xi| < \theta\} \in \underset{\sim}{\Delta}^1_{2n+1}$.

3) Let $P \subseteq \mathcal{R}$ be a $\underset{\sim}{\Pi}^1_{2n+1}$-complete set and φ a $\underset{\sim}{\Pi}^1_{2n+1}$-norm on it, $\varphi : P \xrightarrow{\text{onto}} \underset{\sim}{\delta}^1_{2n+1}$. Then there are $\underset{\sim}{\Pi}^1_{2n+1}$, $\underset{\sim}{\Sigma}^1_{2n+1}$ relations Q,S resp. such that

$$\alpha,\beta \in P \Rightarrow [W(\varepsilon,\varphi(\alpha),\varphi(\beta)) \Leftrightarrow Q(\varepsilon,\alpha,\beta) \Leftrightarrow S(\varepsilon,\alpha,\beta)].$$

Proof. By the proof of 14.3 there is a tree T on $\omega \times \underset{\sim}{\delta}^1_{2n+1}$ such that if $C : \underset{\sim}{\delta}^1_{2n+1} \xrightarrow[\text{onto}]{\text{1-1}} (\underset{\sim}{\delta}^1_{2n+1})^{<\omega}$ and we put

$$W(\varepsilon,\xi,\eta) \Leftrightarrow C(\xi),\ C(\eta) \in T(\varepsilon) \ \&\ C(\xi) <_{KB} C(\eta),$$

where $<_{KB}$ is the Kleene-Brouwer ordering then 1) is satisfied. Also 2) is obvious since if $W(\varepsilon,\xi,\eta)$ holds and $n \geq \ell h(C(\xi)), \ell h(C(\eta))$ then for all δ with $\delta \restriction n = \varepsilon \restriction n$ we have $W(\delta,\xi,\eta)$. (To compute that $P^{\theta}_{\xi_0,\xi} = \{\varepsilon : |(W_\varepsilon)^{\xi_0} \restriction \xi| < \theta\}$ is $\underset{\sim}{\Delta}^1_{2n+1}$ proceed by induction on θ.)

Finally to check 3): By the proof of 7.1 we can find $Q^* \subseteq \omega \times \mathcal{R}^2$ and $S^* \subseteq \omega \times \mathcal{R}^2$ in $\underset{\sim}{\Pi}^1_{2n+1}, \underset{\sim}{\Sigma}^1_{2n+1}$ resp. such that (identifying below $\omega^{<\omega}$ with ω):

$$\alpha,\beta \in P \Rightarrow [(u,C(\varphi(\alpha))),\ (u,C(\varphi(\beta))) \in T \wedge C(\varphi(\alpha)) <_{KB} C(\varphi(\beta))$$
$$\Leftrightarrow Q^*(u,\alpha,\beta) \Leftrightarrow S^*(u,\alpha,\beta)].$$

Let then

$$Q(\varepsilon,\alpha,\beta) \Leftrightarrow \exists n\, Q^*(\varepsilon \restriction n,\alpha,\beta)$$
$$S(\varepsilon,\alpha,\beta) \Leftrightarrow \exists n\, S^*(\varepsilon \restriction n,\alpha,\beta).$$

Then for $\alpha,\beta \in P$,

$$Q(\varepsilon,\alpha,\beta) \Leftrightarrow \exists n[(\ \varepsilon \restriction n\ ,\ C(\varphi(\alpha))),(\ \varepsilon \restriction n\ ,\ C(\varphi(\beta)) \in T$$
$$\wedge\ C(\varphi(\alpha)) <_{KB} C(\varphi(\beta))]$$
$$\Leftrightarrow\ C(\varphi(\alpha)),\ C(\varphi(\beta)) \in T(\varepsilon)$$
$$\wedge\ C(\varphi(\alpha)) <_{KB} C(\varphi(\beta))$$
$$\Leftrightarrow\ W(\varepsilon,\varphi(\alpha),\ \varphi(\beta))$$

and similarly for S. ⊣

Lemma 15.2 [AD]. Let $n \geq 0$ be given and assume $\lambda > \omega$ is a regular cardinal $< \underset{\sim}{\delta}^1_{2n+1}$. Then there is a function $F : \underset{\sim}{\delta}^1_{2n+1} \to \underset{\sim}{\delta}^1_{2n+1}$ such that for every wellordering U on $\underset{\sim}{\delta}^1_{2n+1}$ there is a wellordering V on $\underset{\sim}{\delta}^1_{2n+1}$ with $|V| > |U|$ and a cub set $C \subseteq \underset{\sim}{\delta}^1_{2n+1}$ such that $H_V(\xi) < F(\xi)$, $\forall \xi \in E_\lambda \cap C$. In particular if μ_λ is a normal measure, $H_U < F\ \mu_\lambda$ - a.e. (Recall that $E_\lambda = \{\xi : \mathrm{cof}(\xi) = \lambda\}$).

Proof. Let W be as in the Lemma 15.1. Put for $\xi \in E_\lambda$

$$F(\xi) = \sup\{|W_\varepsilon \restriction \xi| + 1 : \varepsilon \text{ is such that } \forall \eta < \xi\ (|W_\varepsilon \restriction \eta| < \xi)\}$$

and let $F(\xi) = 0$ if $\xi \notin E_\lambda$. If now $\xi \in E_\lambda$ i.e. $\mathrm{cof}(\xi) = \lambda > \omega$ then

$$\forall \eta < \xi (W_\varepsilon \restriction \eta \text{ is wellfounded} \Rightarrow W_\varepsilon \restriction \xi \text{ is wellfounded}),$$

so $F(\xi) < \underset{\sim}{\delta}^1_{2n+1}$, by boundedness and 15.1. Now given a wellordering U find ε_0 such that $|U| < |W_{\varepsilon_0}|$ and then find a closed unbounded C such that

$$\xi \in C \Rightarrow \forall \eta < \xi(|W_{\varepsilon_0} \upharpoonright \eta| < \xi).$$

Then $\xi \in E_\lambda \cap C \Rightarrow F(\xi) > |W_{\varepsilon_0} \upharpoonright \xi|$ and we are done. $\dashv$

Theorem 15.3 [AD] (Kunen 71c). If μ_λ, $\lambda > \omega$, is a normal measure on $\underset{\sim}{\delta}^1_{2n+1}$ then $\underset{\sim}{\delta}^1_{2n+1}{}^{\underset{\sim}{\delta}^1_{2n+1}}/\mu_\lambda > (\underset{\sim}{\delta}^1_{2n+1})^+ = \underset{\sim}{\delta}^1_{2n+2}$.

Proof. If F is as in 15.2 then by 14.1 $[F]_{\mu_\lambda} \geq \underset{\sim}{\delta}^1_{2n+2}$. $\dashv$

From Kunen's result (see Solovay's paper, "A Δ^1_3 coding of the subsets of ω_ω," this volume) that $\underset{\sim}{\delta}^1_3 \to (\underset{\sim}{\delta}^1_3)^\lambda$, $\forall \lambda < \underset{\sim}{\delta}^1_3$ it follows that the conclusion of 15.3 holds for regular $\omega < \lambda < \underset{\sim}{\delta}^1_3$ i.e. for $\lambda = \omega_1, \omega_2$. It also holds for $\lambda = \omega_1$ for any $\underset{\sim}{\delta}^1_{2n+1}$, $n > 0$, by the remarks following 16.1.

16. Countable exponent partition relations on $\underset{\sim}{\delta}^1_n$, n even.

Theorem 16.1 [AD] (Kunen 71c). For all $n \geq 1$, $\underset{\sim}{\delta}^1_{2n} \to (\underset{\sim}{\delta}^1_{2n})^\lambda, \forall \lambda < \omega_1$.

Proof. Fix $\lambda < \omega_1$ and a map $t : \omega \cdot \lambda \xrightarrow[\text{onto}]{1-1} \omega$. Let $P \subseteq \mathcal{R}$ be $\underset{\sim}{\Pi}^1_{2n+1}$-complete, φ a $\underset{\sim}{\Pi}^1_{2n+1}$-norm on P with range $\underset{\sim}{\delta}^1_{2n+1}$ and put $|\alpha| = \varphi(\alpha)$. For any real α and any $\xi < \omega \cdot \lambda$ let $\alpha_\xi = (\alpha)_i$, where $t(\xi) = i$.

Fix $X \subseteq (\delta^1_{2n+1})^\lambda \uparrow$. Then consider the game

I	II	II wins iff
ε^I, α^I	$\varepsilon^{II}, \alpha^{II}$	1) $\exists \xi < \omega \cdot \lambda$ such that α^I_ξ or $\alpha^{II}_\xi \notin P$ and for the least such ξ, say ξ_0, $\alpha^I_{\xi_0} \notin P$.
	or	2) $\forall \xi < \omega \cdot \lambda\ (\alpha^I_\xi, \alpha^{II}_\xi \in P) \wedge$ for some $\langle \zeta, \xi \rangle \in \underset{\sim}{\delta}^1_{2n+1} \times (\omega \cdot \lambda)$ $(W_{\varepsilon^I_\xi})^{\lvert\alpha^I_\xi\rvert} \upharpoonright \zeta$ or $(W_{\varepsilon^{II}_\xi})^{\lvert\alpha^{II}_\xi\rvert} \upharpoonright \zeta$ is not wellordered and for the lexicographically least such $\langle \zeta_0, \xi_0 \rangle$, $(W_{\varepsilon^I_{\xi_0}})^{\lvert\alpha^I_{\xi_0}\rvert} \upharpoonright \zeta_0$ is not wellordered.
	or	3) both 1) and 2) fail and letting $F_{I,\xi}(\zeta) = \lvert(W_{\varepsilon^I_\xi})^{\lvert\alpha^I_\xi\rvert} \upharpoonright \zeta\rvert$ and similarly for II we have $\langle \sup_n \{[F_{I,\omega\cdot\theta+n}], [F_{II,\omega\cdot\theta+n}]\}\rangle_{\theta<\lambda} \in X$.

Here W is as in 15.1 and $[F] = [F]_{\mu_\omega}$.

Assume without loss of generality that II has a winning strategy σ. Put

$$(\varepsilon_\sigma^{II},\alpha_\sigma^{II}) = \sigma[\varepsilon^I,\alpha^I].$$

Let then for $\xi < \omega\cdot\lambda$; $\zeta,\eta,\iota < \underset{\sim}{\delta}_{2n+1}^1$:

$$\Theta(\xi,\zeta,\eta,\iota) = \sup\{|(W_{(\varepsilon_\sigma^{II})_\xi})^{|(\alpha_\sigma^{II})_\xi|} \restriction \zeta| + 1:$$

$$\varepsilon^I,\alpha^I \text{ are such that } \forall\bar\xi < \omega\cdot\lambda\ (\alpha_{\bar\xi}^I \in P \wedge |\alpha_{\bar\xi}^I| < \iota)$$

and for all

$$\langle\zeta',\xi'\rangle \leq_{\ell ex}\langle\zeta,\xi\rangle$$

$$|(W_{\varepsilon_{\xi'}^I})^{|\alpha_{\xi'}^I|}\restriction\zeta'| < \eta\}.$$

By 15.1. $\Theta(\xi;\zeta,\eta,\iota) < \underset{\sim}{\delta}_{2n+1}^1$ so we can find a cub C such that

$$\rho \in C \Rightarrow \forall\xi < \omega\cdot\lambda\ \forall\zeta,\eta,\iota < \rho(\Theta(\xi,\zeta,\eta,\iota) < \rho).$$

Put $H = (C^* \cap E_\omega) - (\underset{\sim}{\delta}_{2n+1}^1)$ where C^* = image of C under the embedding generated by the ultrapower relative to μ_ω. We will show that if $f \in H^{\not\,\lambda}$ then $f \in X$.

Fix such an f and then find $g \in (\underset{\sim}{\delta}_{2n+2}^1)^{\omega\cdot\lambda}\uparrow$ such that

$$\lim_{n<\omega} g(\omega\cdot\theta + n) = f(\theta),\ \forall\theta < \lambda.$$

Then find ε^I,α^I such that $W_{\varepsilon_\xi^I}$ is wellordered and $\alpha_\xi^I \in P$ for all $\xi < \omega\cdot\lambda$ and such that $[F_{I,\omega\cdot\theta+n}] = g(\omega\cdot\theta+n)$. Put $\varepsilon^{II} = \varepsilon_\sigma^{II},\alpha^{II} = \alpha_\sigma^{II}$. Then $\alpha^{II} \in P$ for all $\xi < \omega\cdot\lambda$ and $(W_{\varepsilon_\xi^{II}})^{|\alpha_\xi^{II}|}$ is wellordered for all $\xi < \omega\cdot\lambda$, so it is enough to show that

$$[F_{II,\omega\cdot\theta+n}] < f(\theta),\ \forall\theta < \lambda,\ \forall n \in \omega.$$

Pick $F \in C^{\underset{\sim}{\delta}_{2n+1}^1}$ such that $[F] = f(\theta)$. Then we have to show that

$$F_{II,\omega\cdot\theta+n}(\zeta) < F(\zeta)\ \ \mu_\omega\text{-a.e.}$$

Let $I \in \mu_\omega$ be such that

$\zeta \in I \Rightarrow$ 1) $F(\zeta) > \zeta$

2) $F(\zeta) > \iota > \sup\{|\alpha_\xi^I| : \xi < \omega\cdot\lambda\}$ (for some ι)

3) $F_{I,\xi'}(\zeta') < \zeta, \forall\xi' < \omega\cdot\lambda, \forall\zeta' < \zeta$

4) $F_{I,\omega\cdot\theta+n+1}(\zeta) < F(\zeta)$

5) $F_{I,\xi'}(\zeta) < F_{I,\xi''}(\zeta), \forall\xi' < \xi'' < \omega\cdot\lambda.$

Then we claim that for $\zeta \in I$,

$$F_{II,\omega\cdot\theta+n}(\zeta) = |(W_{\varepsilon^{II}_{\omega\theta+n}})^{|\alpha^{II}_{\omega\theta+n}|} \restriction \zeta| < F(\zeta),$$

which completes the proof. Since $F(\zeta) \in C$ we have

$$\forall \xi_1 < \omega\lambda\ \forall \zeta_1, \eta_1, z_i < F(\zeta)(\Theta(\xi_1, \zeta_1, \eta_1, z_1) < F(\zeta)),$$

so since $\zeta < F(\zeta)$, $z < F(\zeta)$ we only have to show that for some $\eta < F(\zeta)$ and all $\langle \zeta', \xi' \rangle \leq_{\ell ex} \langle \zeta, \omega\theta+n \rangle$ we have $F_{I,\xi}(\zeta') < \eta$. Take

$$\eta = \max\{F_{I,\omega\theta+n+1}(\zeta), \zeta\} < F(\zeta).$$

Let $\langle \zeta', \xi' \rangle \leq_{\ell ex} \langle \zeta, \omega\theta+n \rangle$. Then we have:

<u>Case</u> 1. $\zeta' < \zeta$: Then $F_{I,\xi'}(\zeta') < \zeta \leq \eta$.

<u>Case</u> 2. $\zeta' = \zeta$ and $\xi' \leq \omega\theta+n$: Then $F_{I,\xi'}(\zeta) < F_{I,\omega\theta+n+1}(\zeta) \leq \eta < F(\zeta)$ and we are done. ⊣

Kechris has recently shown that for all $n \geq 2$, $\underset{\sim}{\delta}^1_n \to (\underset{\sim}{\delta}^1_n)^\lambda$, $\forall \lambda < \omega_2$. Thus μ_{ω_1} is a normal measure for all $\underset{\sim}{\delta}^1_n$, $n \geq 2$.

17. <u>The measure</u> μ_ω <u>on</u> $\underset{\sim}{\delta}^1_n$, n <u>even</u>.

<u>Theorem</u> 17.1 [AD] (Kunen 71c). For all $n \geq 0$, μ_ω is a normal measure on $\underset{\sim}{\delta}^1_{2n+2}$ and is generated by the sets of the form $(C^* \cap E_\omega) - \underset{\sim}{\delta}^1_{2n+1}$, where $C \subseteq \underset{\sim}{\delta}^1_{2n+1}$ is closed unbounded and C^* is the image of C under the embedding generated by the ultrapower relative to μ_ω on $\underset{\sim}{\delta}^1_{2n+1}$.

<u>Proof</u>. By 10.1 and Section 16 μ_ω is a normal measure on $\underset{\sim}{\delta}^1_{2n+1}$. To prove the extra statement we show that if $f : \underset{\sim}{\delta}^1_{2n+2} \to \underset{\sim}{\delta}^1_{2n+2}$ is pressing down there is a set as above on which f is constant. For that consider the partition of $(\underset{\sim}{\delta}^1_{2n+2})^{\omega+\omega}\uparrow$ as in 10.1. Then by the proof of 16.1 there is a closed unbounded $C \subseteq \underset{\sim}{\delta}^1_{2n+1}$ such that if $p \in (C^* \cap E_\omega)^{\omega+\omega}\uparrow$ and $p(\omega) > \lim_{n<\omega} p(n)$ and $p(0) > \underset{\sim}{\delta}^1_{2n+1}$ then

$$f(\sup_n (p(n))) = f(\sup_n (p(\omega+n))).$$

Let $D \subseteq C$ be the set of all limit points of C and $E \subseteq D$ the set of all limit points of D. Then both D, E are cub and $E^* \cap E_\omega \subseteq D^* \cap E_\omega \subseteq C^* \cap E_\omega$, while every point of E^* is a limit point of D^*, which in turn is a limit point of C^*. So if $\theta < \eta$ are in $(E^* \cap E_\omega) - \underset{\sim}{\delta}^1_{2n+1}$, find

$$\theta_0 < \theta_1 < \cdots \to \theta < \eta_0 < \eta_1 < \cdots \to \eta$$

$\theta_i, \eta_i \in D^* - \underset{\sim}{\delta}^1_{2n+1}$. Put $p(n) = \omega^{th}$ element of C^* above θ_n and $p(\omega+n) =$ the ωth element of C^* above η_n. Then $p(n) \to \theta$ since $\theta_n \leq p(n) \leq \theta_{n+1}$ and similarly $p(\omega+n) \to \eta$. Since $p \in (C^* \cap E_\omega)^{\omega+\omega}\uparrow$ and $p(\omega) > \lim_n p(n) = \theta$ and $p(0) > \underset{\sim}{\delta}^1_{2n+1}$ we have $f(\sup_n(p(n))) = f(\theta) = f(\sup_n(p(\omega+n))) = f(\eta)$ $\therefore$ f is constant on $(E^* \cap E_\omega) - \underset{\sim}{\delta}^1_{2n+1}$ and we are done. ⊣

Theorem 17.2 [AD] (Kunen 71 c). If $F : \underset{\sim}{\delta}^1_{2n+2} \to \underset{\sim}{\delta}^1_{2n+2}$, there is $J : \underset{\sim}{\delta}^1_{2n+1} \to \underset{\sim}{\delta}^1_{2n+1}$ such that $F(\xi) \le J^*(\xi)$, for all ξ in $(\underset{\sim}{\delta}^1_{2n+1}, \underset{\sim}{\delta}^1_{2n+2})$, where again $J^* : \underset{\sim}{\delta}^1_{2n+2} \to \underset{\sim}{\delta}^1_{2n+2}$ is the image of J under the embedding generated by μ_ω on $\underset{\sim}{\delta}^1_{2n+1}$.

Proof. In the notation of 16.1 consider the game

I	II	II wins if
ε^I, α^I	ε^{II}	1) $\alpha^I \notin P$ or
		2) $\alpha^I \in P$ and either for some ξ, $(W_{\varepsilon^I})^{\lvert\alpha^I\rvert} \restriction \xi$ or $W_{\varepsilon^{II}} \restriction \xi$ is not a wellordering and for the least such, say ξ_0, $(W_{\varepsilon^I})^{\lvert\alpha^I\rvert} \restriction \xi_0$ is not a wellordering or for all ξ, $(W_{\varepsilon^I})^{\lvert\alpha^I\rvert} \restriction \xi$ and $W_{\varepsilon^{II}} \restriction \xi$ are wellorderings and if $f_I(\xi) = \lvert(W_{\varepsilon^I})^{\lvert\alpha^I\rvert} \restriction \xi\rvert$ and $f_{II}(\xi) = \lvert W_{\varepsilon^{II}} \restriction \xi\rvert$ then $F([f_I]) < [f_{II}]$, where $[f] = [f]_{\mu_\omega}$.

Claim. I does not have a winning strategy.

Proof. Suppose he had one τ. Then we will show that there is $K : \underset{\sim}{\delta}^1_{2n+1} \to \underset{\sim}{\delta}^1_{2n+1}$ such that if II plays correctly so that f_{II} is defined then if f_I is produced following τ, then $[f_I] < [K]$. If II then "plays f_{II}" such that $[f_{II}] > \sup\{F(\theta) : \theta < [K]\}$ we immediately have a contradiction. To define K let $(\varepsilon^I_\tau, \alpha^I_\tau) = \tau\,[\varepsilon^{II}]$. Then let

$$K(\xi) = \sup\{\lvert(W_{\varepsilon^I_\tau})^{\lvert\alpha^I_\tau\rvert} \restriction \xi\rvert + 1 : \varepsilon^{II} \text{ is such that}$$

$$\forall \eta < \xi\ (\lvert W_{\varepsilon^{II}} \restriction \eta\rvert < \xi)\}.$$

Then $K(\xi) < \underset{\sim}{\delta}^1_{2n+1}$ by boundedness, since if $W_{\varepsilon^{II}} \restriction \eta$ is wellordered for all $\eta < \xi$ then $(W_{\varepsilon^I})^{\lvert\alpha^I_\tau\rvert} \restriction \xi$ must be wellordered by the rules of the game and the fact that τ is a winning strategy for I. Suppose now II "plays f_{II}" and find C cub in $\underset{\sim}{\delta}^1_{2n+1}$ such that

$$\xi \in C \Rightarrow \forall\, \eta < \xi\ (f_{II}(\eta) < \xi).$$

Then if I produces following his strategy f_I we have $f_I(\xi) < K(\xi)$ for all $\xi \in C$ ∴ $[f_I] < [K]$ and the proof of the claim is complete.

So II has a winning strategy σ. Then put for $\xi < \theta < \underset{\sim}{\delta}^1_{2n+1}$:

$$J'(\xi,\theta) = \sup\{|W_{\varepsilon^{II}_\sigma} \upharpoonright \xi| + 1 : \alpha^I, \varepsilon^I \text{ are such that } |\alpha^I| < \xi \text{ and } |(W_{\varepsilon^I})^{|\alpha^I|} \upharpoonright \xi| \leq \theta\},$$

where as usual $\varepsilon^{II}_\sigma = \sigma[\alpha^I, \varepsilon^I]$. By boundedness again $J'(\xi,\theta) < \underset{\sim}{\delta}^1_{2n+1}$. So put for $\theta < \underset{\sim}{\delta}^1_{2n+1}$

$$J(\theta) = \sup_{\xi<\theta} J'(\xi,\theta) < \underset{\sim}{\delta}^1_{2n+1}.$$

Now we want to prove that

$$F(\xi') < J^*(\xi'), \forall \xi' \in (\underset{\sim}{\delta}^1_{2n+1}, \underset{\sim}{\delta}^1_{2n+2}).$$

Fix such a ξ' and find α^I, ε^I such that $[f_I] = \xi'$. Thus we have to show that $F([f_I]) < J^*([f_I])$. Since $F([f_I]) < [f_{II}]$ it is enough to check that $[f_{II}] \leq J^*([f_I])$ i.e. $f_{II}(\xi) \leq J(f_I(\xi))$ μ_ω-a.e. Let $\xi > |\alpha^I|$ and $f_I(\xi) > \xi$ (this happens μ_ω - a.e.). Put $\theta = f_I(\xi) > \xi$. Then

$$J(f_I(\xi)) = J(\theta) \geq J'(\xi,\theta) > f_{II}(\xi).$$

⊣

<u>Theorem</u> 17.3 [AD] (Kunen 71c). For all $n \geq 1$, $\underset{\sim}{\delta}^1_{2n+2}{}^{\underset{\sim}{\delta}^1_{2n+2}}/\mu_\omega = (\underset{\sim}{\delta}^1_{2n+2})^+$ and $\mathrm{cof}((\underset{\sim}{\delta}^1_{2n+2})^+) = \underset{\sim}{\delta}^1_{2n+2}$.

<u>Proof</u>. Let $F : \underset{\sim}{\delta}^1_{2n+2} \to \underset{\sim}{\delta}^1_{2n+2}$. Find $J : \underset{\sim}{\delta}^1_{2n+1} \to \underset{\sim}{\delta}^1_{2n+1}$ such that for $\xi > \underset{\sim}{\delta}^1_{2n+1}$, $F(\xi) < J^*(\xi)$. Let W be a wellordering on $\underset{\sim}{\delta}^1_{2n+1}$ such that $J(\xi) = H_W(\xi)\mu_\omega$ - a.e., say $J(\xi) = H_W(\xi)$, $\forall \xi \in I \in \mu_\omega$. Then $J^* = (H_W)^* = H_{W^*}$ on I^* ∴ $F < H_{W^*}$ on $I^* \cap E_\omega$ ∴ $[F]_{\mu_\omega} < [H_{W^*}]_{\mu_\omega} = |W^*| < (\underset{\sim}{\delta}^1_{2n+2})^+$. So $\underset{\sim}{\delta}^1_{2n+2}{}^{\underset{\sim}{\delta}^1_{2n+2}}/\mu_\omega = (\underset{\sim}{\delta}^1_{2n+2})^+$.

Given now $\xi < \underset{\sim}{\delta}^1_{2n+2}$ find $f : \underset{\sim}{\delta}^1_{2n+1} \to \underset{\sim}{\delta}^1_{2n+1}$ such that $[f]_{\mu_\omega} = \xi$. Then $f^* : \underset{\sim}{\delta}^1_{2n+2} \to \underset{\sim}{\delta}^1_{2n+2}$. Put

$$g(\xi) = [f^*]_{\mu_\omega} < (\underset{\sim}{\delta}^1_{2n+2})^+.$$

It is easy to see that g is well defined. Moreover g is cofinal by the preceding fact and 17.2.

⊣

<u>Corollary</u> 17.4 [AD] (Kunen 71d). For all $n \geq 1$, $\underset{\sim}{\delta}^1_{2n} \not\to (\underset{\sim}{\delta}^1_{2n})^{\underset{\sim}{\delta}^1_{2n}}$.

<u>Proof</u>. By 17.3 and 13.6.

⊣

18. Some singular cardinals

Let A,B be two transitive classes, $i : A \to B$ a Δ_0-elementary embedding i.e. for every Δ_0 φ, $A \models \varphi(a_1 \ldots a_n) \Leftrightarrow B \models \varphi(ia_1 \ldots ia_n)$. Put $i(A) = \bigcup\{ia : a \in A\}$. If A is closed under transitive closure, $i(A)$ is transitive and is therefore equal to the smallest transitive class containing the range of i.

[Proof. If $x \in y \in ia$, put $b = TC(a)$. Then $A \models \forall y \in a \forall x \in y\ (x \in b)$ $\therefore$ $B \models \forall y \in ia\ \forall x \in y\ (x \in ib)$ $\therefore$ $x \in ib$.]

Put $HWO = \{a : TC(a) \text{ is wellorderable}\}$ and for each ordinal $\varkappa$ let $Fn(HWO,\varkappa) = HWO^{\varkappa} \cap HWO = \{F \in HWO^{\varkappa} : \text{range}(F) \in HWO\}$. If $\mathfrak{w}$ is a countably complete ultrafilter on $\varkappa$, let $Fn(HWO,\varkappa)/\mathfrak{w}$ be the usual ultrapower. By Los' theorem

$$Fn(HWO,\varkappa)/\mathfrak{w} \models \varphi([F_1] \ldots [F_n]) \Rightarrow$$
$$\{\xi \in \varkappa : HWO \models \varphi(F_1(\xi) \ldots F_n(\xi))\} \in \mathfrak{w},$$

for $\varphi \in \Delta_0$. So $Fn(HWO,\varkappa)/\mathfrak{w}$ is wellfounded and extensional so it can be collapsed to a transitive class $Ult(HWO,\mathfrak{w})$. Let

$$i^{\mathfrak{w}} : HWO \to Ult(HWO,\mathfrak{w})$$

be the usual embedding which by the above is Δ_0-elementary. Also

$$Ult(HWO,\mathfrak{w}) = i^{\mathfrak{w}}(HWO) \subseteq HWO.$$

[Proof. For $Ult(HWO,\mathfrak{w}) = i^{\mathfrak{w}}(HWO)$: If $[F] \in Fn(HWO,\varkappa)/\mathfrak{w}$ and $z = \text{range}(F)$, then $\{\xi < \varkappa : HWO \models F(\xi) \in z\} \in \mathfrak{w}$ $\therefore$ $Fn(HWO,\varkappa)/\mathfrak{w} \models [F] \in [\xi \mapsto z]$ $\therefore$ $Ult(HWO,\mathfrak{w}) \models [F] \in i^{\mathfrak{w}}(z)$. (We identify here $[F]$ with its collapse.)

For $i^{\mathfrak{w}}(HWO) \subseteq HWO$: Enough to show $\{[G] : [G] \in [F]\}$ is wellorderable. Let $<$ be a wellordering of $TC(\text{range }(F))$. For $[G], [G'] \in F$ let

$$[G] < [G'] \Rightarrow \{x : G(x) < G(x')\} \in \mathfrak{w};$$

then $<$ is a wellordering on $\{[G] : [G] \in [F]\}$.]

Lemma 18.1 (Kunen 71d). Let $\varkappa \leq \lambda$ be two cardinals such that $cof(\lambda) = \varkappa$. Let $\mathfrak{w}$ be a countably complete uniform ultrafilter on $\varkappa$. Then $i^{\mathfrak{w}}(\lambda) \geq \lambda^+$.

Proof. Fix $\lambda \leq \gamma < \lambda^+$. To show $i^{\mathfrak{w}}(\lambda) > \gamma$. Let R be a wellordering of λ of type γ. Let $F : \varkappa \xrightarrow{cof} \lambda$. Then $i^{\mathfrak{w}}(\lambda) > [F]_{\mathfrak{w}} \geq \sup\{i^{\mathfrak{w}}(\xi) : \xi < \lambda\}$. Now $|R \restriction \xi| < \lambda$, $\forall \xi < \lambda$, therefore in particular $|R \restriction F(\eta)| < \lambda$, $\forall \eta < \varkappa$. So $|i^{\mathfrak{w}}(R) \restriction [F]_{\mathfrak{w}}| < i^{\mathfrak{w}}(\lambda)$. But also $|i^{\mathfrak{w}}(R) \restriction [F]_{\mathfrak{w}}| \geq |i^{\mathfrak{w}}(R) \restriction \{i^{\mathfrak{w}}(\xi) : \xi < \lambda\}| = |R| = \gamma$ $\therefore$ $i^{\mathfrak{w}}(\lambda) > \gamma$. ⊣

Theorem 18.2 [AD] (Kunen 71d). Let $n \geq 0$, $\delta^1_{2n+1} = \omega_{\rho+1}$ and $\delta^1_{2n+2} = \omega_{\rho+2}$. Then for each $k \geq 2$

1) $cof(\omega_{\rho+k}) = \omega_{\rho+2}$.

2) There is $h : \mathrm{Ult}(\mathrm{HWO},\mathcal{U}) \to \mathrm{HWO}$ Δ_0-elementary, such that $\omega_{\rho+k} = \sup\{h(\xi) : \xi < \omega_{\rho+2}\}$, where $\mathcal{U} = \mu_\omega$ on $\omega_{\rho+1}$.

Proof. Clearly 2) $\Rightarrow$ 1). We prove now 2) by induction on k. It is trivial for $k = 2$, with h = identity. So assume $k > 2$ and it holds for all $2 \leq k' < k$. If $\mathcal{V} = \mu_\omega$ on $\omega_{\rho+2}$ then by Lemma 18.1, $i^{\mathcal{V}}(\omega_{\rho+k-1}) \geq \omega_{\rho+k}$.

Case I. $i^{\mathcal{V}}(\omega_{\rho+k'}) = \omega_{\rho+k}$, for some $2 \leq k' < k$.

By induction hypothesis let h satisfy 2) for $\omega_{\rho+k'}$, i.e. $\omega_{\rho+k'} = \sup\{h(\xi) : \xi < \omega_{\rho+2}\}$. Define $h^{00} : \mathrm{Ult}(\mathrm{HWO},\mathcal{U}) \to \mathrm{HWO}$ by

$$h^{00}([F]_{\mathcal{U}}) = [h \cdot i^{\mathcal{U}}F]_{\mathcal{V}}.$$

First we check that h^{00} is well defined. Assume $F = G$ on $I \in \mathcal{U}$. Then $i^{\mathcal{U}}F = i^{\mathcal{U}}G$ on $i^{\mathcal{U}}I \in \mathcal{V}$ so $h \cdot i^{\mathcal{U}}F = h \cdot i^{\mathcal{U}}G$ $\mathcal{V}$- a.e. It is equally routine to check that h^{00} is Δ_0-elementary. We shall prove now that

$$\sup\{h^{00}(\xi) : \xi < \omega_{\rho+2}\} = i^{\mathcal{V}}(\omega_{\rho+k'}) = \omega_{\rho+k},$$

which will complete the proof in this case.

First notice that $h^{00}(\xi) < i^{\mathcal{V}}(\omega_{\rho+k'})$ if $\xi < \omega_{\rho+2}$, since if $[F]_{\mathcal{U}} = \xi$, where $F : \omega_{\rho+1} \to \omega_{\rho+1}$, clearly $h \cdot i^{\mathcal{U}}F : \omega_{\rho+2} \to \omega_{\rho+k'}$. Conversely, if $\theta < i^{\mathcal{V}}(\omega_{\rho+k'})$ then for some $G : \omega_{\rho+2} \to \omega_{\rho+k'}$, $\theta \leq [G]_{\mathcal{V}}$. Let $F' : \omega_{\rho+2} \to \omega_{\rho+2}$ be given by

$$F'(\eta) = \text{least } \zeta < \omega_{\rho+2} \text{ such that } G(\eta) \leq h(\zeta).$$

Then $G \leq h \cdot F'$ everywhere. By 17.2 let $F : \omega_{\rho+1} \to \omega_{\rho+1}$ be such that $F' \leq i^{\mathcal{U}}F$ $\mathcal{V}$-a.e. Then $G \leq h \cdot i^{\mathcal{U}}F$ $\mathcal{V}$-a.e. so

$$\theta \leq [G]_{\mathcal{V}} \leq [h \circ i^{\mathcal{V}}F]_{\mathcal{V}} = h^{00}([F]_{\mathcal{V}}) = h^{00}(\xi),$$

where $\xi < \omega_{\rho+2}$ and we are done.

Case II. $i^{\mathcal{V}}(\omega_{\rho+k'-1}) < \omega_{\rho+k} < i^{\mathcal{V}}(\omega_{\rho+k'})$, for some $3 \leq k' < k$.

Then find $F : \omega_{\rho+2} \to \omega_{\rho+k'}$ such that $[F]_{\mathcal{V}} = \omega_{\rho+k}$. Then F is cofinal in $\omega_{\rho+k'}$ since otherwise $[F]_{\mathcal{V}} < i^{\mathcal{V}}(\xi)$ for some $\xi < \omega_{\rho+k'}$. But $\mathrm{card}(\xi) \leq \omega_{\rho+k'-1}$ $\therefore$ $\mathrm{card}(i^{\mathcal{V}}(\xi)) \leq \mathrm{card}(i^{\mathcal{V}}(\omega_{\rho+k'-1})) < \omega_{\rho+k}$ $\therefore$ $i^{\mathcal{V}}(\xi) < \omega_{\rho+k}$, a contradiction. Put

$$h^0 = i^{\mathcal{V}} \circ h,$$

where h comes from our induction hypothesis for $\omega_{\rho+k'}$. Clearly $h^0 : \mathrm{Ult}(\mathrm{HWO},\mathcal{U}) \to \mathrm{HWO}$ is Δ_0-elementary. We will show that h^0 works for $\omega_{\rho+k}$.

Notice first that

$$\sup\{h^0(\alpha) : \alpha < \omega_{\rho+2}\} = [h \restriction \omega_{\rho+2}]_{\mathcal{V}}.$$

Indeed, $h^0(\xi) = i^{\mathcal{V}}(h(\xi)) < [h \restriction \omega_{\rho+2}]_{\mathcal{V}}$, since $h \restriction \omega_{\rho+2}$ is cofinal in $\omega_{\rho+k'}$. On the other hand if $[F]_{\mathcal{V}} < [h \restriction \omega_{\rho+2}]_{\mathcal{V}}$ then $F(\xi) < h(\xi)$ $\mathcal{V}$- a.e., so restricting to ξ's of cofinality ω if we put

$$T(\xi) = \text{least } \eta < \xi \text{ such that } F(\xi) < h(\eta),$$

then $T(\xi)$ is pressing down $\mathcal{V}$ - a.e., because h is continuous at limits of cofinality ω, so $T(\xi) = \theta < \omega_{\rho+2}$ $\mathcal{V}$ - a.e. Then $F(\xi) < h(\theta)$ $\mathcal{V}$ - a.e. i.e. $[f]_{\mathcal{V}} < i^{\mathcal{V}}(h(\theta))$ and we are done.

So enough to show

$$[h \upharpoonright \omega_{\rho+2}]_{\mathcal{V}} = \omega_{\rho+k} = [F]_{\mathcal{V}}.$$

Now clearly $[h \upharpoonright \omega_{\rho+2}]_{\mathcal{V}} \leq [F]_{\mathcal{V}}$, since F is unbounded in $\omega_{\rho+k'}$. So it suffices to prove that

$$[F]_{\mathcal{V}} = \omega_{\rho+k} \leq [h \upharpoonright \omega_{\rho+2}]_{\mathcal{V}}.$$

For that it is again enough to show that

$$([h \upharpoonright \omega_{\rho+2}]_{\mathcal{V}})^+ \geq i^{\mathcal{V}}(\omega_{\rho+k'}).$$

Fix $\gamma < i^{\mathcal{V}}(\omega_{\rho+k'})$. Then $\gamma = [H]_{\mathcal{V}}$, where $H : \omega_{\rho+2} \to \omega_{\rho+k'}$, so $\gamma \leq [h \cdot F]_{\mathcal{V}}$ for some $F : \omega_{\rho+2} \to \omega_{\rho+2}$. Find $H_W : \omega_{\rho+1} \to \omega_{\rho+1}$ such that $[F]_{\mathcal{V}} \leq [i^{\mathcal{U}} H_W]_{\mathcal{V}} = [H_{i^{\mathcal{U}}W}]_{\mathcal{V}}$. Then $\gamma \leq [h \circ H_{i^{\mathcal{U}}W}] = [h \circ H_V]_{\mathcal{V}}$, where $V = i^{\mathcal{U}}W$ is a wellordering on $\omega_{\rho+2}$. Fix $g \in \text{Ult}(HWO,\mathcal{U})$ such that for some $I \in \mathcal{V}$, $I \in \text{Ult}(HWO,\mathcal{U})$, $g(\xi) : H_{\mathcal{V}}(\xi) \xrightarrow{1-1} \xi$, $\forall \xi \in I$. [To see these exist find $P(\xi) : |W \upharpoonright \xi| \xrightarrow{1-1} \xi$, $\forall \xi \in X \in \mathcal{U}$ $\therefore$ $i^{\mathcal{U}}P(\xi) : |V \upharpoonright \xi| \xrightarrow{1-1} \xi$, $\forall \xi \in i^{\mathcal{U}}X \in \mathcal{V}$. Put $g = i^{\mathcal{U}}P$, $I = i^{\mathcal{U}}X$]. Since $g \in \text{Ult}(HWO,\mathcal{U})$,

$$\xi \in I \Rightarrow h(g(\xi)) : h \circ H_V(\xi) \xrightarrow{1-1} h(\xi)$$

$$\therefore \quad [h \circ g]_{\mathcal{V}} : [h \circ H_V]_{\mathcal{V}} \xrightarrow{1-1} [h \upharpoonright \omega_{\rho+2}]_{\mathcal{V}}$$

$$\therefore \quad \gamma \leq [h \circ H_V]_{\mathcal{V}} < ([h \upharpoonright \omega_{\rho+2}]_{\mathcal{V}})^+ .$$

⊣

References

J. W. Addison and Y. N. Moschovakis

[1968] Some consequences of the axiom of definable determinateness, Proc. Nat. Acad. Sci., USA 59 (1968), 708-712.

A. S. Kechris

[1974] On projective ordinals, J. Symb. Logic, 39 (1974), 269-282.

E. M. Kleinberg

[1970] Strong partition properties for infinite cardinals, J. Symb. Logic, 35 (1970), 410-428.

K. Kunen

[1971a] A remark on Moschovakis' Uniformization theorem, circulated note, March 1971.

[1971b] Measurability of $\underset{\sim}{\delta}_n^1$, circulated note, April 1971.

[1971c] Some singular cardinals, circulated note, September 1971.

[1971d] Some more singular cardinals, circulated note, September 1971.

R. Mansfield

[1971] A Souslin operation on Π^1_2, Israel J. Math., 9 (1971), 367-379.

D. A. Martin

[1968] The axiom of determinateness and reduction principles in the analytical hierarchy, Bull. Amer. Math. Soc., 74 (1968), 687-689.

[1971] Determinateness implies many cardinals are measurable, circulated note, May 1971.

[197?] Projective sets and cardinal numbers: Some questions related to the continuum problem, J. Symb. Logic, to appear.

D. A. Martin and J. B. Paris

[1971] AD $\Rightarrow \exists$ exactly 2 normal measures on ω_2, circulated note, March 1971.

D. A. Martin and R. Solovay

[1967] A basis theorem for Σ^1_3 sets of reals, Ann. Math., 89 (1969), 138-160.

Y. N. Moschovakis

[1970] Determinacy and prewellorderings of the continuum, Mathematical Logic and Foundations of Set Theory, Y. Bar-Hillel, Ed., North Holland, Amsterdam and London, 1970, 24-62.

[1971] Uniformization in a playful universe, Bull. Amer. Math. Soc., 77 (1971), 731-736.

J. R. Shoenfield

[1961] The problem of predicativity, Essays on the Foundations of Mathematics, Magnes Press, Hebrew University, Jerusalem 1961, 132-139.

R. M. Solovay

[1967] Measurable cardinals and the axiom of determinateness, Lecture notes prepared in connection with the Summer Institute on Axiomatic Set Theory, UCLA, 1967.

A Δ^1_3 CODING OF THE SUBSETS OF ω_ω

Robert M. Solovay

Department of Mathematics
California Institute of Technology
Pasadena, CA 91125

and

Department of Mathematics
University of California
Berkeley, CA 94720

Introduction. We present Kunen's [3] analysis of the measures on ω_ω, from AD + DC, together with some light-face refinements that follow by mixing Kunen's techniques with a theorem of Kechris-Martin. Our purpose in this introduction is (a) to list some applications of the Kunen method (to be presented in Section B); (b) to give an overview of the more technical results to be proved; (c) to give some idea of the motivation behind the technicalities that follow.

The principal results of type (a) are as follows (AD + DC is assumed throughout):

Let $\lambda < \underset{\sim}{\delta}^1_3$. Then $\underset{\sim}{\delta}^1_3 \to (\underset{\sim}{\delta}^1_3)^\lambda$. This should be compared with the result of Martin that $\underset{\sim}{\delta}^1_1 \to (\underset{\sim}{\delta}^1_1)^{\underset{\sim}{\delta}^1_1}$. It is still open whether or not $\underset{\sim}{\delta}^1_3 \to (\underset{\sim}{\delta}^1_3)^{\underset{\sim}{\delta}^1_3}$. Kunen's proof uses a highly detailed analysis at level 3, and it is not known how to generalize his work to odd n's greater than 3. In particular, it is open whether $\underset{\sim}{\delta}^1_5 \to (\underset{\sim}{\delta}^1_5)^\lambda$ for all $\lambda < \underset{\sim}{\delta}^1_5$, though we have already seen in [1] that $\underset{\sim}{\delta}^1_5 \to (\underset{\sim}{\delta}^1_5)^\lambda$ for countable λ.

Note that by earlier work in [1], it follows that there are exactly three normal measures on $\underset{\sim}{\delta}^1_3$ (concentrating on points of cofinality ω_0, ω_1, and ω_2 respectively).

(b) Our proof will be based on a Δ^1_3 encoding of the subsets of ω_ω, (cf. B4). To state what this means recall that the theory of sharps together with the fact that the ω_i's $1 \leq i < \omega$ are precisely the first ω uniform indiscernibles gives a natural Δ^1_3 encoding of the ordinals $< \omega_\omega$. We shall produce a Δ^1_3 set, C, of codes for subsets of ω_ω so that the relation: the ordinal coded by x lies in the set coded by y is Δ^1_3.

To get such a coding we need a concrete way of generating all subsets of ω_ω. We prove that every non-empty subset of ω_ω is the countable union of simple sets. Here a simple set is a subset of some ω_m which is the 1 - 1 image, by some function constructible from a real, of one of a countable sequence of standard sets, $A^j_{m,k}$. [This is slightly stronger than what we prove below (owing to a different definition of "simple"); it is not hard to prove the stronger claim with Kunen's method.]

By an ingenious reduction, (cf. B3) this is reduced to an analysis of measures on ω_ω. Some further simple reductions (which take place in B1 and B2) reduce

*The author would like to express his thanks to the Sherman Fairchild Distinguished Scholars Program at Caltech for its generous support during the academic year 1976-1977. Thanks also to Greg Ennis for the conscientious painstaking work of transferring a series of lectures to the printed page.

the problem to the following: We are given integers $m \geq 0$, $k \geq 1$. Each closed subset C of ω_1 determines in a canonical manner a subset $\tilde{C}$ of ω_{m+1} (the image of C under a suitable elementary embedding of some $L[z]$ mapping ω_1 into ω_{m+1}.) In this way the sets $\tilde{C}^k$ form a filter base for a filter $\mathfrak{F}$ on ω_{m+1}^k. Our problem is to characterize the ultrafilters that extend $\mathfrak{F}$. The main result proved in part A shows that there are only finitely many such ultrafilters. They live on disjoint sets $A^1_{m,k},\ldots,A^{\ell}_{m,k}$ (where union is the set of k-tuples of limit ordinals less than ω_{m+1}) and the decomposition is Δ^1_3 in the codes.

The proof is a refinement of the Martin-Paris analysis of normal measures on ω_2, which was presented in [1]. We replace the study of k-tuples of ordinals less than ω_{m+1} by the study of k-tuples of functions from ω_1^m into ω_1. We analyse these k-tuples carefully enough so as to be able to imitate the Martin-Paris proof. Of course in our more general context the technicalities will be considerably greater.

We emphasize that we work in ZF + AD + DC in the following. Any unexplained notation is as in [1].

A. Classification of tuples of ordinals.

A.1. W is the canonical normal measure on ω_1. W_n is the product measure $W \times \cdots \times W$ (n times). If A is a set, A^n is the n-fold cartesian power. $A^{[n]}$, for A linearly ordered, is the set of strictly increasing n-tuples from A.

A.2. $\tilde{L} = \bigcup\{L[x] : x \in \mathbb{R}\}$. For each $x \in \mathbb{R}$, we have an elementary embedding

$$i_{W_n} : L[x] \to L[x],$$

given by the transitive realization of the ultrapower. These maps piece together to give a map

$$i_{W_n} : \tilde{L} \to \tilde{L}.$$

(Note that this last map is not elementary since ω_1 is definable in $\tilde{L}$ but $i_{W_n}(\omega_1) \neq \omega_1$.)

Note that W_n gives $\omega_1^{[n]}$ measure 1 (and we usually view W_n as a measure on $\omega_1^{[n]}$). If $F : \omega_1^{[n]} \to L[x]$, then F determines (via transitive realization of the ultrapower) an element $[F] \in L[x]$.

Let $\pi_i : \omega_1^{[n]} \to \omega_1$, be given by $\pi_i(\alpha_1,\ldots,\alpha_n) = \alpha_i$.

<u>Lemma</u>: $[\pi_i] = \omega_i$. $i_{W_n}(\omega_1) = \omega_{n+1}$.

<u>Proof</u>: Clearly $\{\vec{\alpha} \in \omega_1^{[n]} : L[z^{\#}] \models \alpha_i$ is an indiscernible for $L[z]\}$ has W_n-measure 1. Thus $[\pi_i]$ is a uniform indiscernible. So clearly is $i_{W_n}(\omega_1)$. Since clearly

$$[\pi_1] < \cdots < [\pi_n] < i_{W_n}(\omega_1),$$

it suffices to show that if $F : \omega_1^{[n]} \to \omega_1$ is such that $[F]$ is a uniform indiscernible, then $[F] = [\pi_i]$ for some i.

Now $F \in L[z]$, for some $z \in \mathbb{R}$ since $F \subseteq L_{\omega_1}$. By increasing the Turing degree of z, we may suppose F is definable in $L[z]$ from z, ω_1. Pick i minimal such that $[F] \leq [\pi_i]$. (The case when $[F] > [\pi_n]$ is handled similarly.) If $[F] = [\pi_i]$, we are done. Otherwise, for almost all $\vec{\alpha}$, $F(\vec{\alpha})$ is definable in $L[z]$ from z, and indiscernibles for $L[z]$ distinct from $F(\vec{\alpha})$. It follows that $F(\vec{\alpha})$ is a.e. not an indiscernible for $L[z]$. Whence $[F]$ is not a uniform indiscernible. ⊣

Corollary: Let $\lambda < \omega_{m+1}$. Let $F : \omega_1^{[m]} \to \omega_1$ such that $[F] = \lambda$. Let $\tilde{F} : \omega_{m+1}^{[m]} \to \omega_{m+1}$ be $i_{W_m}(F)$. Then $\lambda = \tilde{F}(\omega_1,\ldots,\omega_m)$.

Proof: By Los' theorem, this amounts to

$$F(\vec{\alpha}) = F(\pi_1(\vec{\alpha}),\ldots,\pi_m(\vec{\alpha})).$$ ⊣

A.3. We now state the theorem which is the main goal of Section A.

Theorem (Kunen [3]). Let $m \geq 0$. Let $k \geq 1$. Let $X_{m,k} = \{\langle\lambda_1,\ldots,\lambda_k\rangle :$ For $1 \leq i \leq k$, λ_i is a limit ordinal $< \omega_{m+1}\}$. We shall construct a decomposition:

$$X_{m,k} = A^1_{m,k} \cup \cdots \cup A^{\ell}_{m,k} \quad \text{(disjoint union)}$$

with the following properties:

(1) Let C be a cub subset of ω_1. Let $\tilde{C} = i_{W_m}(C)$. Then $\tilde{C}^k \cap A^i_{m,k} \neq \emptyset$.

(2) Sets of the form $\tilde{C}^k \cap A^i_{m,k}$ are the basis for an ultrafilter on $A^i_{m,k}$. (This ultrafilter is clearly countably additive since the intersection of countably many cub's is cub.)

(3) Let $h : \omega_1 \to \omega_1$ be normal (i.e. strictly increasing and continuous). Let $\tilde{h} : \omega_{m+1} \to \omega_{m+1}$ be $i_{W_m}(h)$. If $\vec{\alpha} \in A^i_{m,k}$, then

$$\langle\tilde{h}(\alpha_1),\ldots,\tilde{h}(\alpha_k)\rangle \in A^i_{m,k}.$$

We shall see that properties (1) through (3) characterize the partition of $X_{m,k}$ into the $A^i_{m,k}$'s.

A.4. For certain applications we need that our partition is Δ^1_3 in the codes. A code for an ordinal $< \omega_\omega$ is a pair $z = \langle n, x^{\#}\rangle$, where $x \in \mathbb{R}$. If z is a code as above then we put

$$|z| = \tau_n^{L[x]}(\omega_1 \cdots \omega_{k_n})$$

where $\tau_0, \tau_1, \ldots$ is recursive enumeration of all the terms in the language of $L[x]$ (x occuring as a constant symbol) taking always ordinal values.

The following result is an effective version (Δ^1_3 replacing $\underset{\sim}{\Delta}^1_3$) of a theorem of Kunen [3].

Theorem: (1) Let $\ell(m,k)$ be the number of pieces into which $X_{m,k}$ is decomposed. (Cf. Theorem A.3.) Then the map $\langle m,k\rangle \to \ell(m,k)$ is Δ^1_3.

(2) The following relation, $\Phi(k,m,j,x)$ is Δ^1_3: (Here k,m,j are in ω, and x in $\mathcal{R}$) $k \geq 1$; $m \geq 0$; $1 \leq j \leq \ell(m,k)$; for $1 \leq i \leq k$, $(x)_i$ codes a limit ordinal $\lambda_i < \omega_{m+1}$; $\langle \lambda_1, \ldots, \lambda_k\rangle \in A^j_{m,k}$.

A.5. Let HF be the class of hereditarily finite sets ($HF = L_\omega$ and there is a canonical well-ordering of HF of type ω.)

The decomposition of $X_{m,k}$ will arise in the following way. We will define a map $\Psi_{m,k} : X_{m,k} \to HF$. The range of $\Psi_{m,k}$ will be finite, say $\{x_1, \ldots, x_\ell\}$, where $x_1 < \cdots < x_\ell$ with respect to the canonical well-ordering of HF. Then $A^j_{m,k} = \Psi^{-1}_{m,k}(\{x_j\})$.

The map $\Psi_{m,k}$ will have the following properties:

(α) Range $(\Psi_{m,k})$ is finite.

(β) $\underset{\sim}{\Psi}_{m,k}$ is an invariant: If $h : \omega_1 \to \omega_1$ is normal, $\tilde{h} = i_{W_m}(h)$, $\vec{\alpha} \in X_{m,k}$ and $\beta_i = \tilde{h}(\alpha_i)$ for $1 \leq i \leq k$, then

$$\Psi_{m,k}(\vec{\alpha}) = \Psi_{m,k}(\vec{\beta})$$

(γ) $\Psi_{m,k}$ is Δ^1_3 in the codes: The following relation is Δ^1_3: $m \geq 0$, $k \geq 1$, for $1 \leq i \leq k$, $(x)_i$ codes a limit ordinal $\lambda_i < \omega_{m+1}$ and

$$\Psi_{m,k}(\vec{\lambda}) = s.$$

(Here $s \in HF$, $x \in \mathcal{R}$, $m,k \in \omega$.)

A.6. The case $m = 0$ is trivial but atypical. First we describe the invariant $\Psi_{o,k}$:

$$\Psi_{o,k}(\langle \lambda_1, \ldots, \lambda_k\rangle) = \{\langle i,j\rangle : \lambda_i < \lambda_j\}.$$

It is evident that $\Psi_{o,k}$ has the properties (α), (β), and (γ) of A.5.

A moment's thought will show that we can effectively tell which $s \in HF$ lie in range $(\Psi_{o,k})$. Thus the portion of A.4 that relates to $m = 0$ is evident from (γ) holding for $\Psi_{o,k}$ (uniformly in k).

Finally we verify Theorem A.3 for $X_{o,k}$ with respect to the decomposition induced by $\Psi_{o,k}$. Let $A^j_{o,k}$ be a component of this decomposition. Then there is an integer r, with $1 \leq r \leq k$, integers $s_1,\dots,s_r$ with $1 \leq s_i \leq k$, and integers n_i, $1 \leq i \leq k$, with $1 \leq n_i \leq r$ such that for $\vec{\alpha} \in A^j_{o,k}$ $\alpha_i = \alpha_{s_{n_i}}$;

$\alpha_{s_1} < \alpha_{s_2} < \cdots < \alpha_{s_r}$.

By means of the map $A^j_{o,k} \cong \mathrm{Lim}^{[r]} : \langle \alpha_1,\dots,\alpha_n\rangle \to \langle \alpha_{s_1},\dots,\alpha_{s_r}\rangle$, the claims of Theorem A.3 for the case $m=0$ reduce to the following well-known fact: sets of the form $C^{[r]}$, C cub, are a basis for W_r.

A.7. We turn now to the definition of $\Psi_{m,k}$ for $m > 0$. Our strategy will be as follows. We define three simpler invariant functions, $\Psi^1_{m,k}$, $\Psi^2_{m,k}$, $\Psi^3_{m,k}$ which satisfy (α), (β), (γ) of A.5. We then put

$$\Psi_{m,k}(\vec{\alpha}) = \langle \Psi^1_{m,k}(\vec{\alpha}),\dots,\Psi^3_{m,k}(\vec{\alpha})\rangle.$$

$\Psi_{m,k}$ will then inherit the properties (α), (β) and (γ) from the $\Psi^i_{m,k}$'s.

We now fix $m \geq 1$, $k \geq 1$. Let $\vec{\lambda} \in X_{m,k}$. Pick $F_1,\dots,F_k$ mapping $\omega_1^{[m]}$ into ω_1 such that $[F_i] = \lambda_i$. Pick $z \in \mathbb{R}$ such that $\langle F_i : 1 \leq i \leq k\rangle$ is definable from z, ω_1 in $L[z]$. Let I_z be the set of canonical indiscernibles for $L[z]$ which are less than ω_1.

We pick $\vec{\gamma} \in I_z^{[2m]}$. Then $\Psi^1_{m,k}(\vec{\lambda}) = \{\langle i,\vec{r},j,\vec{s}\rangle : 1 \leq i,j, \leq k;$ $1 \leq r_1 < \cdots < r_m \leq 2m;\ 1 \leq s_1 < \cdots < s_m \leq 2m;$

$$F_i(\gamma_{r_1},\dots,\gamma_{r_m}) < F_j(\gamma_{s_1},\dots,\gamma_{s_m})\}.$$

It is necessary to see $\Psi^1_{m,k}$ depends only on $\vec{\lambda}$ and not on the choices of the F's, z, and γ's. Evidently the choice of the γ's are irrelevant so long as they are chosen in I_z. If $\vec{F}'$, z' are different choices, we can find E cub so that

(a) F_i and F_i' agree on $E^{[m]}$ $(1 \leq i \leq k)$,

(b) $E \subseteq I_z \cap I_{z'}$.

But now if we take $\vec{\gamma} \in E^{[2m]}$, we will get the same value of Ψ^1 from the primed choices as from the unprimed ones.

How about properties (α) through (γ)? Properties (α) and (γ) are evident. As for property (β) note that $[h \circ F_i] = \tilde{h}(\lambda_i)$. Since h is order-preserving, (β) is now clear for Ψ^1.

A.8. Let D be a closed unbounded set. D' is the set of limit points of D. Define a function $H^D_i : D'^{[m]} \to \omega_1$:

$$H_i^D(\vec{\gamma}) = \sup\{F_j(\vec{\delta}) : \vec{\delta} \in D^{[m]} \text{ and } F_j(\vec{\delta}) < F_i(\vec{\gamma})\}.$$

Evidently if E is a cub $\subseteq D$, $[H_i^E] \leq [H_i^D]$, since for $\vec{\gamma} \in E'^{[m]}$, the sup is over a smaller set for $H_i^E(\vec{\gamma})$ than for $H_i^D(\vec{\gamma})$.

Definition: i is of type II if for some cub D, $[H_i^D] < [F_i]$.

We put $\Psi^2_{m,k}(\vec{\lambda}) = \{i : i$ is of type II$\}$.

As usual we must verify independence of the choice of F's. But if F_j and F'_j agree on $E^{[m]}$, E cub, and for some D, $[H_i^D] < [F_i]$, then by the remark of the paragraph preceding the definition, $[H_i^{D\cap E}] < [F_i]$. But then $H_i^{D\cap E} = H_i'^{D\cap E}$ (since F_j's and F'_j's look alike on $E^{[m]}$) so $[H_i'^{D\cap E}] < F'_i]$. So i is of type II with respect to the F''s (if it is with respect to the F's).

Property (α) is evident. Since if h is normal, it preserves the notion of limit and non-limit, and since $[h \circ F_i] = \tilde{h}(\lambda_i)$, property (β) is clear.

Our proof of perperty (γ) for Ψ^2 will be preceded by two lemmas.

Lemma 1. Let $\vec{\lambda}$, $\vec{F}$, z be as above. Let D, E be cub subsets of ω_1 with $E \subseteq D \subseteq I_z$. Suppose that for some $\vec{\delta} \in E'^{[m]}$, $H_i^E(\vec{\delta}) < F_i(\vec{\delta})$. Then for every $\vec{\gamma} \in D'^{[m]}$, $H_i^D(\vec{\gamma}) < F_i(\vec{\gamma})$.

Proof: Let $\xi = H_i^E(\vec{\delta})$. Let $\theta_1,\ldots,\theta_n$ be ordinals of I_z such that $\xi = \tau^{L[z]}(\theta_1,\ldots,\theta_n)$. Pick $\theta'_1,\ldots,\theta'_n$ in E so that

(a) $\theta_i \leq \theta'_i$ $(1 \leq i \leq n)$

(b) $\vec{\theta}$, $\vec{\delta}$ and $\vec{\theta}'$, $\vec{\delta}$ are similarly ordered

(c) If $\alpha_1,\alpha_2 \in \{0,\theta'_1,\ldots,\theta'_n,\delta_1,\ldots,\delta_m\}$ with $\alpha_1 < \alpha_2$, then there at least m members of E strictly between α_1 and α_2.

(There is no difficulty doing this since the δ_i's are limit points of E.)

Let $\xi' = \tau^{L[z]}(\theta'_1,\ldots,\theta'_n)$. By (a) $\xi \leq \xi'$. By (b) and $E \subseteq I_z$, $\xi' < F_i(\vec{\delta})$.

Next select $\theta^*_1,\ldots,\theta^*_n$ in D with

(d) $\vec{\theta}^*$, $\vec{\gamma}$ similarly ordered to $\vec{\theta}'$, $\vec{\delta}$.

(This is possible since the γ_i's are in D' and $\vec{\gamma}$ is similarly ordered to $\vec{\delta}$.) Put $\xi^* = \tau^{L[z]}(\vec{\theta}^*)$. By (d) and the fact that $E \subseteq D \subseteq I_z$, we have $\xi^* < F_i(\vec{\gamma})$. Thus to prove $H_i^D(\vec{\gamma}) < F_i(\vec{\gamma})$ it suffices to show $H_i^D(\vec{\gamma}) \leq \xi^*$.

Deny this towards a contradiction. Then for some $\vec{\eta} \in D^{[m]}$, j, we have $\xi^* < F_j(\vec{\eta}) < F_i(\vec{\gamma})$. By (c), we can choose $\vec{\eta}' \in E^{[m]}$ so that

(e) $\vec{\theta}^*$, $\vec{\eta}$, $\vec{\gamma}$ are ordered similarly to $\vec{\theta}'$, $\vec{\eta}'$, $\vec{\delta}$.

But then by (e) and $E \subseteq D \subseteq I_z$,

(f) $\xi' < F_j(\vec{\eta}') < F_i(\vec{\delta})$.

But this contradicts the fact that $H_i^E(\delta) \leq \xi'$. ⊣

Lemma 2: Let $D \subseteq I_z$ be cub. Then if i is of type I, $H_i^D(\vec{\gamma}) = F_i(\vec{\gamma})$ for all $\vec{\gamma} \in D'^{[m]}$. If i is of type II, $H_i^D(\vec{\gamma}) < F_i(\vec{\gamma})$ for all $\vec{\gamma} \in D'^{[m]}$.

Proof: Suppose that $F_i(\vec{\gamma}) > H_i^D(\vec{\gamma})$ for one $\vec{\gamma} \in D'^{[m]}$. (Note that $F_i(\vec{\gamma}) \geq H_i^D(\vec{\gamma})$ is evident from the definitions.) Apply Lemma 1 with $D = E$. Then $F_i(\vec{\gamma}) > H_i^D(\vec{\gamma})$ for every $\vec{\gamma} \in D'^{[m]}$. Whence $[F_i] > [H_i^D]$, and i is of type II. This proves our first claim.

Next suppose i of of type II. Then for some cub E, $[H_i^E] < [F_i]$. Replacing E, if need be, by $E \cap D$, But then Lemma 1 guarantees $H_i^D(\vec{\gamma}) < F_i(\vec{\gamma})$ for all $\vec{\gamma} \in D'^{[m]}$ (since the $\vec{\delta}$ needed for Lemma 1 is guaranteed by $[H_i^E] < [F_i]$.) ⊣

It is now easy to prove Ψ^2 is "Δ_3^1 in the codes". If $(x)_i$ codes λ_i, we can take x itself for our z. The criterion of Lemma 2 can be checked in $L[z^{\#}]$, hence recursively in $z^{\#\#}$, hence Δ_3^1 in z. This proves Ψ^2 satisfies (γ).

A.9. Definition. $\widetilde{cf^L}(\alpha) = \min\{cf^{L[z]}(\alpha) : z \in \mathbb{R}\}$. Evidently $\widetilde{cf^L}(\alpha) \leq \alpha$. $\widetilde{cf^L}(\alpha)$ is regular in each $L[z]$. So if α is a limit ordinal, $\widetilde{cf^L}(\alpha)$ is either ω or a uniform indiscernible.

It follows that $\widetilde{cf^L}(\lambda_i) = \omega_j$ for some j with $o \leq j \leq m$. We put

$$\Psi^3_{m,k}(\vec{\lambda}) = \{\langle i,j\rangle : i \text{ is of type II and } cf(\lambda_i) = \omega_j\}.$$

Evidently Ψ^3 satisfies (α). If $h : \omega_1 \to \omega_1$ is normal, then so is $\tilde{h} : \omega_{m+1} \to \omega_{m+1}$. If g maps ω_m cofinally into λ_i, $\tilde{h}\circ g$ maps ω_m cofinally into $\tilde{h}(\lambda_i)$. If g is order preserving, so is $\tilde{h}\circ g$. It follows $\widetilde{cf^L}(\tilde{h}(\lambda_i)) = \widetilde{cf^L}(\lambda_i)$ and Ψ^3 satisfies (β).

We need the following lemmas to show Ψ^3 satisfies (γ).

Lemma 3: Let $z \in \mathbb{R}$. Let λ be an infinite ordinal which is regular in $L[z]$ but not an indiscernible of $L[z]$. Then λ is cofinal with ω in $L[z^{\#}]$.

Proof: We may as well assume $\lambda > \omega$. Let $\gamma = \sup\{\xi < \lambda : \xi = \omega$ or ξ is an indiscernible for $L[z]\}$. Then $\omega \leq \gamma \leq \lambda$. If $\gamma = \lambda$, then γ must be an indiscernible for $L[z]$ (since the class of indiscernibles is closed and $\gamma > \omega$). This contradicts our assumption on λ. So $\gamma < \lambda$.

Let $\theta_1, \theta_2, \ldots,$ be the first ω indiscernibles for $L[z]$ greater than λ. Then every ordinal $\xi < \lambda$ is definable in $L[z]$ from ordinals $\leq \gamma$ and some of the θ_i's.

We set $S_i = \{\eta < \lambda : \eta$ is definable in $L[z]$ from ordinals in $\gamma \cup \{\gamma\} \cup \{\theta_1, \ldots, \theta_i\}\}$. Then $S_i \in L[z]$ and in $L[z]$ S_i has power γ. Since λ is regular in $L[z]$, $\xi_i = \sup(S_i)$ is less than λ. By the preceding paragraph, the

ξ_i's are cofinal in λ. But clearly the sequence ξ_i is definable from λ in $L[z^\#]$. ⊣

Lemma 4: Let $\lambda = \tau^{L[z]}(\omega_1,\ldots,\omega_m)$, λ a limit ordinal. Then $cf^{L[z^\#]}(\lambda) = cf^{\tilde{L}}(\lambda)$.

Proof: If $cf^{L[z]}(\lambda) = \omega_i$, for $o \leq i \leq m$, then evidently $\omega_i = cf^{L[z^\#]}(\lambda) = cf^{\tilde{L}}(\lambda)$. If not, $cf^{L[z]}(\lambda)$ is not an indiscernible in $L[z]$ (since it is definable in $L[z]$ from $\omega_1,\ldots,\omega_m$ but is distinct from each of $\omega_1,\ldots,\omega_m$). Whence by Lemma 3, $cf^{L[z^\#]}(cf^{L[z]}(\lambda)) = \omega$. But then clearly $cf^{L[z^\#]}(\lambda) = cf^{\tilde{L}}(\lambda) = \omega$. ⊣

Lemma 4 makes it evident that Ψ^3 is Δ^1_3 in the codes. For again if $(x)_i$ codes λ_i, we can take x as our z and compute $cf^{\tilde{L}}(\lambda_i)$ recursively from $x^{\#\#}$.

A.10. Now as promised in A.5 we put

$$\Psi_{m,k}(\vec{\lambda}) = \langle \Psi^1_{m,k}(\vec{\lambda}), \Psi^2_{m,k}(\vec{\lambda}), \Psi^3_{m,k}(\vec{\lambda})\rangle$$

for $\vec{\lambda} \in X_{m,k}$. Thus the range of $\Psi_{m,k}$ is a finite subset of HF, say $\{s_1,\ldots,s_{\ell(m,k)}\}$, where $s_1 < \cdots < s_{\ell(m,k)}$ in the canonical well ordering of HF. Let $A^j_{m,k} = \Psi^{-1}_{m,k}(\{s_j\})$.

The proof of Theorem A.4.2 will follow from A.4.1 and the fact that Ψ^1, Ψ^2, Ψ^3 all satisfy property (γ) of A.5. To prove Theorem A.4.1 we need the Kechris-Martin Theorem.

A.11. Theorem (Kechris-Martin [2]). Let $A \subseteq \omega_\omega$ be nonempty and Π^1_3 in the codes. Then

$$\exists\, x \in \Delta^1_3(|x| \in A).$$

We need the following corollaries:

Lemma 1: Let $R \subseteq \omega^n_\omega$ be Δ^1_3 in the codes.
(a) $\neg R$ is Δ^1_3 in the codes.
(b) If $S(\eta_1 \cdots \eta_{n-1}) \Leftrightarrow \exists\, \xi < \omega_\omega R(\xi,\vec{\eta})$, then S is Δ^1_3 in the codes.
(c) If $T(\eta_1 \cdots \eta_{n-1}) \Leftrightarrow \forall\, \xi < \omega_\omega R(\xi,\vec{\eta})$, then T is Δ^1_3 in the codes.

Proof: (a) is obvious since the set of codes is Δ^1_3.

(b) Let $R^+(x_1 \cdots x_n) \Leftrightarrow \forall i \leq n$ (x_i codes an ordinal $< \omega_\omega$) and $R(|x_1| \cdots |x_n|)$. Then $S^+(x_2 \cdots x_n) \Leftrightarrow \exists\, x_1$ (x_1 codes an ordinal and $R^+(x_1,x_2,\ldots,x_n)$). So S is Σ^1_3 in the codes. Also, by the Kechris-Martin Theorem, $S^+(x_2 \cdots x_n) \Leftrightarrow \exists\, x_1 \in \Delta^1_3(x_2 \cdots x_n)$ [x_1 codes an ordinal and $R^+(x_1 \cdots x_n)$] so S is Π^1_3 in the codes.

(c) Follows from (a) and (b). ⊣

Lemma 2: Let $R \subseteq \omega_\omega^n$ be Δ_3^1 in the codes, $R \neq \emptyset$. Then $\exists x \in \Delta_3^1$ $[(x)_1 \cdots (x)_n$ code ordinals $< \omega_\omega$ & $R(|(x)_1| \cdots |(x)_n|)]$.

Proof: By induction on n. $n = 1$ is the Kechris-Martin Theorem. If $n > 1$, let $S(\eta_1 \cdots \eta_{n-1}) \Leftrightarrow \exists \xi < \omega_\omega R(\eta_1,\ldots,\eta_{n-1},\xi)$. By Lemma 1b, S is Δ_3^1 in the codes, so by inductive hypothesis let $x \in \Delta_3^1$ be such that $S(|(x)_1| \cdots |(x)_{n-1}|)$. Then $\{\xi : R(|(x)_1| \cdots |(x)_{n-1}|,\xi)\}$ is $\Delta_3^1(x)$ in the codes, hence Δ_3^1 in the codes, so $\exists y \in \Delta_3^1$ with $R(|(x)_1| \cdots |(x)_{n-1}|,|y|)$. Let $z \in \Delta_3^1$ be such that $(z)_i = (x)_i$ if $i < n$ and $(z)_n = y$. ⊣

Lemma 3: $\{\langle m,k,s\rangle : s \in \text{Range } \Psi_{m,k}\}$ is Δ_3^1.

Proof: $s \in \text{Range } \Psi_{m,k} \Leftrightarrow \exists \vec{x}[|\vec{x}| \in X_{m,k}$ & $\Psi_{m,k}(|\vec{x}|) = s]$. Thus since the components of $\Psi_{m,k}$ all satisfy (γ), this is Σ_3^1. But by Lemma 2 we can replace the existential quantifiers by $\exists x \in \Delta_3^1$, hence it's Π_3^1. ⊣

Now Theorem A.4.1 follows, since

$$\ell(m,k) = \text{card } (\text{Range } \Psi_{m,k}),$$

so ℓ is clearly a Δ_3^1 function. Theorem A.4.2 follows easily from this.

A.12. It remains to prove that our partition $X_{m,k} = A_{m,k}^1 \cup \cdots \cup A_{m,k}^\ell$ has the properties 1, 2 of A.3. (Property 3 holds since Ψ^1,Ψ^2,Ψ^3 satisfy (β)).

Let $\text{Norm} = \{h : \omega_1 \to \omega_1 : h \text{ normal}\}$.

The following is a variant of Martin's partition theorem (12.1 in [1]).

Lemma: Let $\Phi : \text{Norm} \to \{0,1\}$ be such that $\Phi(h)$ depends only on $h \mid \text{Lim}$. Then $\exists j \in \{0,1\}$, $\exists$ cub C such that if $h \in \text{Norm}$, $h : \omega_1 \to C$, then $\Phi(h) = j$.

Proof: Let $\Psi : P(\omega_1) \to \{0,1\}$ be given by $\Psi(A) = 0$ iff $\exists h \in \text{Norm}[\Phi(h) = 0$ and

$$A = \{h(\omega\cdot\xi + \omega) : \xi < \omega_1\}].$$

By Martin (Theorem 12.1 in [1]) $\exists$ cub C and a $j \in \{0,1\}$ such that if $A \in C^{\not\,\omega_1}$, then $\Psi(A) = j$. Say $j = 0$. Given $h \in \text{Norm}$, $h : \omega_1 \to C$, let

$$A = \{h(\omega\cdot\xi + \omega) : \xi < \omega_1\}.$$

Thus $A \in C$, so $\Psi(A) = 0$. Thus for some $h' \in \text{Norm}$, with $\Phi(h') = 0$ we have

$$A = \{h'(\omega\cdot\xi + \omega) : \xi < \omega_1\}.$$

Since h, h' are both increasing, we must have $\forall \xi < \omega_1(h(\omega\cdot\xi + \omega) = h'(\omega\cdot\xi + \omega))$. But the closure of $\{\omega\cdot\xi + \omega : \xi < \omega_1\}$ is the set of limit ordinals

$< \omega_1$, so since h, h′ are continuous they must agree at all limit ordinals. This proves the case $j = 0$, and the case $j = 1$ is easier. ⊣

A.13. Fix $m > 0$, $k \geq 1$, $1 \leq j \leq \ell(m,k)$ such that if $\langle \lambda_1 \cdots \lambda_k \rangle \in A^j_{m,k}$ then at least one $\lambda_i \geq \omega_1$. Note that if $\Psi_{m,k}(A^j_{m,k}) = s$ we can tell using s alone whether $\exists\, i\ \lambda_i \geq \omega_1$, since if $[F_i] = \lambda_i$ then $\lambda_i < \omega_1$ iff F_i is constant a.e. (W_m), which can be determined from $\Psi^1_{m,k}$. So the property "$\exists\, i(\lambda_i \geq \omega_1)$" is true of all tuples $\langle \lambda_1 \cdots \lambda_k \rangle$ in $A^j_{m,k}$ if it's true of one tuple. (Note also that if $A^j_{m,k}$ is a subset of ω_1^k then we're done by the $m = 0$ case.)

<u>Lemma</u>: $\exists\, G_1 \cdots G_k : \omega_1^{[m]} \to \omega_1$ such that

(i) $\Psi_{m,k}([G_1] \cdots [G_k]) = s$

(ii) If $\vec{\gamma} \in \omega^k_{m+1}$, $\Psi(\vec{\gamma}) = s$, $C \subseteq \omega_1$ is cub and $\vec{\gamma} \in \tilde{C}'$, then there is a normal $h : \omega_1 \to C$ such that

$$\langle \tilde{h}([G_1]), \ldots, \tilde{h}([G_k]) \rangle = \vec{\gamma}.$$

Granting the lemma, we can prove 1, 2 of A.3 as follows:

(1) Given $C \subseteq \omega_1$ cub, let $h : \omega_1 \to C$ be normal. Then $\langle \tilde{h}([G_1]), \ldots, \tilde{h}([G_k]) \rangle \in \tilde{C}^k$. By (γ), $\langle \tilde{h}([G_1]) \cdots \tilde{h}([G_k]) \rangle \in A^j_{m,k}$, so $\tilde{C}^k \cap A^j_{m,k} \neq \emptyset$.

(2) Let $B \subseteq A^j_{m,k}$. We want a cub $C \subseteq \omega_1$ such that either

$$\tilde{C}^k \cap A^j_{m,k} \subseteq B$$

or

$$\tilde{C}^k \cap A^j_{m,k} \cap B = \emptyset.$$

Define $\Phi : \mathrm{Norm} \to \{0,1\}$ by

$$\Phi(h) = \begin{cases} 0 & \text{if } \langle \tilde{h}([G_1]) \cdots \tilde{h}([G_k]) \rangle \in B \\ 1 & \text{otherwise} \end{cases}$$

Since $[G_i] \in \mathrm{Lim}$, $G_i(a) \in \mathrm{Lim}$ a.e. Hence since $\langle \tilde{h}([G_1]) \cdots \tilde{h}([G_k]) \rangle = \langle [h \circ G_1], \ldots, [h \circ G_k] \rangle$ we have $h \mid \mathrm{Lim} = h' \mid \mathrm{Lim} \Rightarrow \forall\, i([h \circ G_1] = [h' \circ G_i])$, so $\Phi(h) = \Phi(h')$.

Hence we can apply Lemma A.12 to get a $j \in \{0,1\}$ and a cub C such that for all normal $h : \omega_1 \to C$, $\Phi(h) = j$.

If $\vec{\gamma} \in \tilde{C}'^k \cap A^j_{m,k}$ then $\exists\, h : \omega_1 \to C$ normal with $\vec{\gamma} = \langle \tilde{h}([G_1]) \cdots \tilde{h}([G_k]) \rangle$. So $\vec{\gamma} \in B$ always (never) if $\Phi(h) = 0$ ($\Phi(h) = 1$) for all $h : \omega_1 \to C$. Thus either $\tilde{C}'^k \cap A^j_{m,k} \subseteq B$ or $\tilde{C}'^k \cap A^j_{m,k} \cap B = \emptyset$. ⊣

Thus Kunen's theorem is done once the lemma is proved.

A.14. Proof of lemma: We have a fixed invariant $s, \lambda_1 \cdots \lambda_k$ with $\Psi(\vec{\lambda}) = s$, where at least one λ_i is $\geq \omega_1$. Consider all $m+1$ tuples $\langle i, \alpha_1 \cdots \alpha_m \rangle$ such that $1 \leq i \leq k$, $\alpha_1 < \cdots < \alpha_m < \omega_i$. We define an equivalence relation $\sim$ on these tuples as follows: Given two such tuples $\langle i, \alpha_1 \cdots \alpha_m \rangle$, $\langle j, \beta_1 \cdots \beta_m \rangle$, let $1 \leq r_1 < \cdots < r_m \leq 2m$, $1 \leq s_1 < \cdots < s_m \leq 2m$ be such that $r_1 \cdots r_m, s_1 \cdots s_m$ are ordered similarly to $\alpha_1 \cdots \alpha_m$, $\beta_1 \cdots \beta_m$. Then put $\langle i, \vec{\alpha} \rangle \sim \langle j, \vec{\beta} \rangle$ iff $\langle i, r_1 \cdots r_m, j, s_1 \cdots s_m \rangle \notin \Psi^1(\vec{\lambda})$ and $\langle i, s_1 \cdots s_m, j, r_1 \cdots r_m \rangle \notin \Psi^1(\vec{\lambda})$.

Let S be the set of $\sim$-equivalence classes. Linearly order S by $[i, \alpha_1 \cdots \alpha_m] <_S [j, \beta_1 \cdots \beta_m]$ iff $\langle i, r_1 \cdots r_m, j, s_1 \cdots s_m \rangle \in \Psi^1(\vec{\lambda})$, where $[\ ]$ denotes equivalence class and $r_1 \cdots r_m, s_1 \cdots s_m$ are as above. Note that $\langle S, <_S \rangle$ depends only on s, not on $\vec{\lambda}$.

Let $[F_i] = \lambda_i$ for $i \leq k$, and let $z \in \mathbb{R}$ be such that $F_1 \cdots F_k$ are definable from ω_1 in $L[z]$ and $cf^{L[z]}(\lambda_i) = cf^{\tilde{L}}(\lambda_i)$ for $i \leq k$. Let I_z be the indiscernibles for $L[z] < \omega_1$. Let $h : \omega_1 \to I_z$ enumerate I_z in increasing order. Define

$$h^* : \{1, \ldots, k\} \times \omega_1^{[m]} \to \omega_1$$

by $h^*(i, \alpha_1 \cdots \alpha_m) = F_i(h(\alpha_1) \cdots h(\alpha_m))$.

<u>Claim 1</u>: $\sim$ is an equivalence relation.

<u>Proof</u>: $h^*(i, \alpha_1 \cdots \alpha_m) = h^*(j, \beta_1 \cdots \beta_m)$ iff $\langle i, \alpha_1 \cdots \alpha_m \rangle \sim \langle j, \beta_1 \cdots \beta_m \rangle$, by definition of Ψ^1. ⊣

<u>Claim 2</u>: $<_S$ is well defined.

<u>Proof</u>: h^* induces a 1 - 1 order preserving map from S into ω_1 (call this induced map h^* also). ⊣

<u>Claim 3</u>: S is order isomorphic to ω_1.

<u>Proof</u>: Some λ_i is $\geq \omega_1$, hence F_i is not constant a.e. Hence $F_i[I_z^{[m]}]$ has power ω_1, and $h^*[S] \supseteq F_i[I_z^{[m]}]$. ⊣

For $[i, \alpha_1 \cdots \alpha_m] = x \in S$, say x is of type I (II) if i is of type I (II) (i.e. if $i \notin (\in) \Psi^2(\vec{\lambda})$).

<u>Claim 4</u>: The type of $[i, \alpha_1 \cdots \alpha_m]$ does not depend on a choice of representative.

<u>Proof</u>: Let $\gamma_1 \cdots \gamma_m \in \text{Lim}$. Then $h^*([i, \vec{\gamma}])$ is not a limit point of $h^*[S] \Leftrightarrow \sup\{h^*(j, \vec{\delta}) : h^*(j, \vec{\delta}) < h^*(i, \vec{\gamma})\} < h^*(i, \vec{\gamma}) \Leftrightarrow \sup\{F_j(h(\delta_1) \cdots h(\delta_m)) :$

$F_j(h(\delta_1) \cdots h(\delta_m)) < F_i(\underrightarrow{h(\gamma_1) \cdots h(\gamma_m)})\} < F_i(h(\gamma_1) \cdots h(\gamma_m)) \Leftrightarrow \forall \vec{\gamma} \in \mathrm{Lim}^{[m]}$ $\sup\{F_j(\overline{h(\vec{\delta})}) : F_j(\overline{h(\vec{\delta})}) < F_i(\overline{h(\vec{\gamma})})\} < F_i(\overline{h(\vec{\gamma})})$ (by Lemma A.8.2) $\Leftrightarrow$ i is of type II. ⊣

<u>Definition</u>: Let $[i,\alpha_1 \cdots \alpha_m] \in S$ be of type II. Then

$$\mathrm{cf}([i,\alpha_1 \cdots \alpha_m]) = \begin{cases} \omega & \text{if } \mathrm{cf}^{\tilde{L}}(\lambda_i) = \omega \\ \alpha_j & \text{if } \mathrm{cf}^{\tilde{L}}(\lambda_i) = \omega_j,\ j > 0. \end{cases}$$

<u>Claim 5</u>: $\mathrm{cf}([i,\alpha_1 \cdots \alpha_m])$ does not depend on choice of representative.

<u>Proof</u>: Let $[i,\alpha_1 \cdots \alpha_m] = [j,\beta_1 \cdots \beta_m]$. Let $r_1 \cdots r_m, s_1 \cdots s_m$ be integers in $1,\ldots,2m$ such that $\vec{r}, \vec{s}$ and $\vec{\alpha}, \vec{\beta}$ are similarly ordered. Then by our choice of z, $\mathrm{cf}^{L[z]}(F_i(\omega_{r_1} \cdots \omega_{r_m})) = \omega_n$ for some n, $0 \leq n \leq 2m$. Since $F_i(\omega_{r_1} \cdots \omega_{r_m}) = F_j(\omega_{s_1} \cdots \omega_{s_m})$, $\mathrm{cf}^{L[z]}(F_j(\omega_{s_1} \cdots \omega_{s_m})) = \omega_n$. If $n = 0$, we're done. Otherwise, if $\mathrm{cf}^{L[z]}(F_i(\omega_{r_1} \cdots \omega_{r_m})) = \omega_{r_k}$ and $\mathrm{cf}^{L[z]}(F_j(\omega_{s_1} \cdots \omega_{s_m})) = \omega_{s_\ell}$, then $\omega_{r_k} = \omega_{s_\ell}$, hence by the similar ordering of $\vec{r}, \vec{s}$, and $\vec{\alpha}, \vec{\beta}$ we have $\alpha_k = \beta_\ell$. So cf is well defined. ⊣

Now let $T = S \cup \{\langle x,\alpha\rangle : x \in S,\ x$ of type II, $\exists \vec{\gamma} \in \mathrm{Lim}^{[m]} \exists i(x = [i,\vec{\gamma}])$ and $\alpha < \mathrm{cf}(x)\}$. We define $<_T$ on T as follows: for $y_1, y_2 \in T$, $y_1 <_T y_2$ iff

(a) $(y_1 = x_1$ or $y_1 = \langle x_1,\alpha_1\rangle)$ & $(y_2 = x_2$ or $y_2 = \langle x_2,\alpha_2\rangle)$ <u>and</u> $x_1 <_S x_2$

or (b) $y_1 = \langle x,\alpha_1\rangle$ and $[y_2 = x$ or $(y_2 = \langle x,\alpha_2\rangle$ and $\alpha_1 < \alpha_2)]$.

Note that $\langle T,<_T\rangle$ depends only on s, not on $\vec{\lambda}$.

<u>Claim 6</u>: T is order isomorphic with ω_1.

<u>Proof</u>: T is obtained as follows: Start with S. For certain x in S, we adjoin a countable wellordered set of elements less than x and greater than any predecessor of x in S. So T is well-ordered, its initial segments are countable, and the power of T is ω_1. ⊣

So we can identify T with ω_1. Now we can define the functions G_i as stated in the lemma:

$$G_i(\gamma_1 \cdots \gamma_m) = [i,\gamma_1 \cdots \gamma_m]$$

where we view $[i,\gamma_1 \cdots \gamma_m]$ as an element of ω_1 under isomorphism $T \cong \omega_1$. Thus $G_i : \omega_1^{[m]} \to \omega_1$.

Claim 7: $\Psi^1([G_1] \cdots [G_k]) = \Psi^1(\vec{\lambda})$.

Proof: Let $\vec{\gamma} \in I_z^{[2m]}$. Then $\Psi^1([\vec{G}]) = \{\langle i, r_1 \cdots r_m, j, s_1 \cdots s_m\rangle :$ $1 \le r_1 < \cdots < r_m \le 2m,\ 1 \le s_1 < \cdots < s_m \le 2m,\ 1 \le i,\ j \le k$ and $G_i(\gamma_{r_1} \cdots \gamma_{r_m}) < G_j(\gamma_{s_1} \cdots \gamma_{s_m})\} = \{\langle i, r_1 \cdots r_m, j, s_1 \cdots s_m\rangle : ---$ and $[i, \gamma_{r_1} \cdots \gamma_{r_m}] <_s [j, \gamma_{s_1} \cdots \gamma_{s_m}]\} = \Psi^1(\vec{\lambda})$. ⊣

Claim 8: $\Psi^2([G_1] \cdots [G_k]) = \Psi^2(\vec{\lambda})$.

Proof: Must show i is of type I w.r.t. $\vec{\lambda}$ iff i is of type II w.r.t. $[G_1] \cdots [G_k]$. Suppose i is of type I w.r.t. $\vec{\lambda}$. Then $\forall \vec{\gamma} \in \mathrm{Lim}^{[m]}$,

$$\sup\{F_j(\overline{h(\vec{\delta})}) : F_j(\overline{h(\vec{\delta})}) < F_i(\overline{h(\vec{\gamma})})\} = F_i(\overline{h(\vec{\gamma})}),$$

which by definition of h^* implies that $\forall \vec{\gamma} \in \mathrm{Lim}^{[m]}$,

$$\sup\{h^*([j,\vec{\delta}]) : h^*([j,\vec{\delta}]) < h^*([i,\vec{\gamma}])\} = h^*([i,\vec{\gamma}]).$$

Suppose $\exists \vec{\gamma} \in \mathrm{Lim}^{[m]}$ such that

$$\sup\{[j,\vec{\delta}] : [j,\vec{\delta}] < [i,\vec{\gamma}]\} < [i,\vec{\gamma}].$$

Then since i is not of type II, by the way S is embedded in T we must have

$$\sup\{[j,\vec{\delta}] : [j,\vec{\delta}] < [i,\vec{\gamma}]\} \le [k,\vec{\beta}] < [i,\vec{\gamma}]$$

for some $[k,\vec{\beta}] \in S$. This yields a contradiction by applying h^*. Hence $\forall \vec{\gamma} \in \mathrm{Lim}^{[m]}$,

$$\sup\{[j,\vec{\delta}] : [j,\vec{\delta}] < [i,\vec{\gamma}]\} = [i,\vec{\gamma}],$$

so i is of type I w.r.t. $[G_1] \cdots [G_k]$, by Lemma A.8.2.

Clearly (by the definition of T), if i is of type II w.r.t $\vec{\lambda}$, then $\forall \vec{\gamma} \in \mathrm{Lim}^{[m]}$,

$$\sup\{[j,\vec{\delta}] : [j,\vec{\delta}] < \langle [i,\vec{\gamma}]\} < \langle [i,\vec{\gamma}], 0\rangle < [i,\vec{\gamma}].$$

Hence i is of type II w.r.t. $[G_1] \cdots [G_k]$. ⊣

Claim 9: $\Psi^3([G_1] \cdots [G_k]) = \Psi^3(\vec{\lambda})$.

Proof: Suppose i is of type II, with $\mathrm{cf}[i, \gamma_1 \cdots \gamma_m] = \gamma_j$. Then by the construction of T, $[i, \gamma_1 \cdots \gamma_m]$ is cofinal with γ_j in T. The construction can be done within L, hence

$$\mathrm{cf}^L G_i(\gamma_1 \cdots \gamma_m) = \mathrm{cf}^L(\gamma_j).$$

Hence $cf^L[G_i] = cf^L[\pi_j] = \omega_j$. The case $cf = \omega_o$ is similar. ⊣

Hence $\Psi([G_1] \cdots [G_k]) = s$ as desired.

Now to prove part (ii) of the lemma, suppose $C \subseteq \omega_1$ is cub with $\lambda_i \in \tilde{C}'$. We will find a normal $h^{***} : \omega_1 \to C$ such that $\tilde{h}^{***}([G_i]) = \lambda_i$.

For the remainder of the proof of the Lemma "definable in $L[z]$" will mean "definable from z,ω_1 in $L[z]$".

We have $z \in \mathfrak{R}$ with $F_1 \cdots F_k$ definable from z,ω_1 and $cf^{L[z]}(\lambda_i) = cf^{\tilde{L}}(\lambda_i)$. By increasing the Turing degree of z if necessary, we may assume C is also definable from ω_1 in $L[z]$. Since $\lambda_i \in \tilde{C}'$, we have $F_i(\gamma_1 \cdots \gamma_m) \in C'$ a.e. Since $C \in L[z]$, by indiscernibility we get

$$\forall \gamma_1 < \cdots < \gamma_m \; F_i(h(\gamma_1) \cdots h(\gamma_m)) \in C' .$$

So $h^* : S \to C'$.

h^* is definable in $L[z^\#]$. Let $x = [i,\gamma_1 \cdots \gamma_m]$ be of type II, all $\gamma_i \in \text{Lim}$. Let $g(x) =$ least map in $L[z]$ of $cf^{L[z]}(h^*(x))$ into $h^*(x)$ continuously, order preservingly, cofinally, and with range contained in

$$C \cap \{\xi : \xi > \sup(h^*(S) \cap h^*(x))\}.$$

Note that since $h^*(x)$ is not a limit point of $h^*(S)$, such a map exists.

Now define $h^{**} : T \to C$ as follows: if $x \in S$, $h^{**}(x) = h^*(x)$. For $\langle x,\alpha\rangle \in T - S$,

$$h^{**}(\langle x,\alpha\rangle) = \begin{cases} g(x)(\alpha) & \text{if } cf(h^*(x)) = \omega \\ g(x)(h(\alpha)) & \text{otherwise.} \end{cases}$$

(Note that $\alpha < cf\ x \Rightarrow h(\alpha) < cf\ h^*(x)$).

It is an easy exercise to prove that h^{**} is order preserving (using the fact that for all α, $g(x)(\alpha) > \sup(h^*(S) \cap h^*(x))$). Also $h^{**}(G_i(\alpha_1 \cdots \alpha_m)) = h^{**}([i,\alpha_1 \cdots \alpha_m]) = F_i(h(\alpha_1) \cdots h(\alpha_m))$, for all $\vec{\alpha} \in \omega_1^{[m]}$. So since $h(\alpha) = \alpha$ a.e. we have $\tilde{h}^{**}([G_i]) = \lambda_i$. But h^{**} is not normal. To get our desired normal h^{***} we need the following easy result:

<u>Fact</u>: Let $A \subseteq \omega_1$ have order type ω_1, $B =$ closure of A in ω_1. Let h_A, h_B be the enumerations in order of A, B resp. Suppose λ is a limit ordinal with $h_A(\lambda)$ a limit point of A. Then $h_A(\lambda) = h_B(\lambda)$.

Thus let h^{***} be the enumeration of the closure of $h^{**}(T)$. Since C is closed, $h^{***} : \omega_1 \to C$. Now if $\gamma_1 < \cdots < \gamma_m$ are limit ordinals, then $[i,\gamma_1 \cdots \gamma_m]$ is a limit point of T, and $h^{**}([i,\gamma_1 \cdots \gamma_m])$ is a limit point of $h^{**}[T]$ (since if $x = [i,\gamma_1 \cdots \gamma_m]$ is of type II and $cf^{L[z]}(F_i(\omega_1 \cdots \omega_m)) = \omega_j$, then $cf^{L[z]}(h^*(x)) = h(\alpha_j)$). So by the fact, $h^{***}([i,\gamma_1 \cdots \gamma_m]) = h^{**}([i,\gamma_1 \cdots \gamma_m])$, so $\tilde{h}^{***}([\vec{G}_i]) = \vec{\lambda}$. ⊣

B. Applications

B.1. Kunen's theorem gives us a partition $(\omega_{m+1} \cap \mathrm{Lim})^k = A^1_{m,k} \cup \cdots \cup A^\ell_{m,k}$ such that each piece carries a canonical measure $V^j_{m,k}$. In this section we prove the following:

Theorem (Kunen [3]). Let V be a (countably additive) measure on ω_ω. Then $\exists\, h : \omega_\omega \to \omega^k_{m+1}$ for some m,k, such that

(1) $h \in \tilde{L}$.

(2) h is $1-1$ a.e. (V).

(3) $h_*(V) = V^j_{m,k}$ for some j.

(4) $\exists\, g : \omega^k_{m+1} \to \omega_\omega$, $g \in \tilde{L}$, such that g is $1-1$ a.e. $(V^j_{m,k})$ and $g = h^{-1}$ a.e.

B.2. By countable additivity of V we may assume V is on ω_n for some $1 \leq n < \omega$. We define the restricted ultrapower of OR in the same way as the ordinary ultrapower except we only consider functions $f : \omega_n \to \mathrm{OR}$ which lie in $\tilde{L}$. Given such an f, $[f]_{\tilde{L}}$ is its image in the restricted ultrapower.

Lemma: Let $z \in \mathbb{R}$. $L[z] \models \psi([h_1]_{\tilde{L}} \cdots [h_m]_{\tilde{L}})$ iff $L[z] \models \psi(h_1(\gamma) \cdots h_m(\gamma))$ a.e. (V).

Proof: Usual Łoś proof works. ⊣

Recall the following (Lemma C in §8 of [1]).

Lemma: Let $\alpha \in \mathrm{OR}$. Then there are uniform indiscernibles $\gamma_1 \cdots \gamma_k \leq \alpha$ and a real z such that α is definable in $L[z]$ from $\gamma_1 \cdots \gamma_k$.

Proof of the theorem in B.1: Let $z \in \mathbb{R}$ and $[f_1]_{\tilde{L}} < \cdots < [f_k]_{\tilde{L}} \leq [\mathrm{id}]_{\tilde{L}}$ uniform indiscernibles such that $[\mathrm{id}]_{\tilde{L}} = t^{L[z]}([f_1]_{\tilde{L}} \cdots [f_k]_{\tilde{L}})$. Let $h : \omega_n \to \omega^k_n$ be

$$h(\alpha) = \langle f_1(\alpha) \cdots f_k(\alpha) \rangle.$$

We have a reverse map $g : \omega^k_n \to \omega_n$ given by $g(\gamma_1 \cdots \gamma_k) = t^{L[z]}(\gamma_1 \cdots \gamma_k)$.

Clearly $g \circ h = \mathrm{id}$ a.e.

Pick j such that $A^j_{m,k} \in h_*(V)$ (here of course $m = n-1$). A unique such j exists since there are finitely many $A^j_{m,k}$'s, all disjoint. Let $C \subseteq \omega_1$ be cub. Since $[f_i]$ is a uniform indiscernible for $i = 1 \cdots k$, we have $f_i(\alpha) \in \tilde{C}$ a.e. (V). Thus $\tilde{C}^k \in h_*(V)$. So $h_*(V) = V^j_{m,k}$.

That $h \circ g = \mathrm{id}$ a.e. $(V^j_{m,k})$ follows from the following general fact: given (two valued) measure spaces (X,U), (X',U'), $h : X \to X'$ such that $h_*(U) = U'$, and

$g : X' \to X$ such that $g \circ h = id$ a.e. (U), then $h \circ g = id$ a.e. (U′).
(Proof: if $A \in U$ and $g(h(x)) = x$ for $x \in A$, then let $B = h[A] \in U'$. On B, g is clearly equal to h^{-1}). ⊣

B.3 Definition. $S \subseteq \omega_\omega$ is simple if $\exists$ m,j,k, C a cub subset of ω_1, and an $F : \omega_{m+1}^k \to \omega_\omega$ such that

(1) F is 1 - 1 on $A^j_{m,k} \cap \tilde{C}^k$
(2) $S = F[A^j_{m,k} \cap \tilde{C}^k]$
(3) $F \in \tilde{L}$.

Theorem (Kunen [3]): If $A \subseteq \omega_\omega$, then A is a countable union of simple sets.

We need the following; where $\Theta = \sup\{\xi : \xi$ is a surjective image of $\mathbb{R}\}$.

Lemma: If $\mathcal{I}$ is a proper σ-ideal on λ, where $\lambda < \Theta$, then there is a countably additive ultrafilter $\mathcal{U}$ on λ such that $A \in \mathcal{I} \Rightarrow A \notin U$.

Proof: By Moschovakis [4] let $h : \mathbb{R} \xrightarrow{\text{onto}} P(\lambda)$. So there is a $h_1 : \mathbb{R} \xrightarrow{\text{onto}} \mathcal{I}$. Let $g : \mathcal{D} \to \lambda$ (where $\mathcal{D}$ = Turing degrees) be given by

$$g(d) = \text{least } \alpha < \lambda[\alpha \notin \bigcup\{h(x) : x \leq_T d\}],$$

where $\mathcal{D}$ = Turing degrees. Then if $\mathcal{U}_0$ is the Martin measure on $\mathcal{D}$, $g_*(\mathcal{U}_0) = \mathcal{U}$ is as desired. ⊣

Proof of theorem: Fix $A \subseteq \omega_\omega$. Let $\mathcal{I} = \{B \subseteq \bigcup_i A_i$: where A_i are simple, and either $A_i \subseteq A$ or $A_i \cap A = \emptyset$ for $i < \omega\}$. $\mathcal{I}$ is a σ-ideal. If $\mathcal{I}$ is not proper, then $A \in \mathcal{I}$ and we are done.

So assume $\mathcal{I}$ is proper, towards a contradiction. By the lemma, let $\mathcal{U}$ be a countably additive ultrafilter such that $B \in \mathcal{I} \Rightarrow B \notin \mathcal{U}$. By Theorem B.1 there are functions $H : \omega_\omega \to (\omega_{m+1})^k$ and $G : (\omega_{m+1})^k \to \omega_\omega$, both in $\tilde{L}$, which demonstrate that $\mathcal{U}$ is equivalent to $V^j_{m,k}$.

Case I: $A \in \mathcal{U}$.

Then $H[A] \in V^j_{m,k}$. So $\exists$ cub $C \subseteq \omega_1$ such that $H[A] \supseteq A^j_{m,k} \cap \tilde{C}^k$. Hence $A \supseteq G[A^j_{m,k} \cap \tilde{C}^k]$ and we may assume G is 1 - 1 on $A^j_{m,k} \cap \tilde{C}^k$. Hence $X = G[A^j_{m,k} \cap \tilde{C}^k]$ is simple, so $X \in \mathcal{I}$. Hence $X \notin \mathcal{U}$, a contradiction to $H_*(V^j_{m,k}) = \mathcal{U}$.

Case II: $A \notin \mathcal{U}$.

Proof is similar to case I. ⊣

B.4. Theorem (Kunen [3], effectivized). There is a Δ^1_3 coding of subsets of ω_ω, i.e. there is a Δ^1_3 $C_\omega \subseteq \mathbb{R}$ and a map $C_\omega \to P(\omega_\omega)$ taking ε to X_ε, such that $\{X_\varepsilon : \varepsilon \in C_\omega\} = P(\omega_\omega)$, and the relation "w codes an ordinal $< \omega_\omega$ & $\varepsilon \in C_\omega$ & $|w| \in X_\varepsilon$" is Δ^1_3.

Proof: Let $\varepsilon \in C^*_\omega \Leftrightarrow \varepsilon = \langle m,k,j,\ulcorner\varphi\urcorner,\alpha,\ulcorner\psi\urcorner\rangle$ & $\{\xi < \omega_1 : L[\alpha] \models \varphi(\xi)\} = C$ is cub & $\{(\vec{\xi},\eta) \in \omega_\omega)^{k+1} : L[\alpha] \models \psi(\omega_1 \cdots \omega_r, \vec{\xi}, \eta)\} = F$ is a function from ω^k_{m+1} into ω_ω which is 1 - 1 on $A^j_{m,k} \cap \tilde{C}^k$.

For $\varepsilon \in C^*_\omega$, let $X^*_\varepsilon = F[A^j_{m,k} \cap \tilde{C}^k]$. Put $\varepsilon \in C_\omega \Leftrightarrow \forall i(\varepsilon_i \in C^*_\omega)$. Let

$$X_\varepsilon = \bigcup_i X^*_{\varepsilon_i}$$

for $\varepsilon \in C_\omega$.

Using the Kechris-Martin Theorem and its collaries (cf. A.11), it can be seen that this coding is Δ^1_3. $\dashv$

B.5. Corollary. (Kunen [3]). If $\mathcal{U}$ is an ultrafilter on ω_ω then $i^{\mathcal{U}}(\underset{\sim}{\delta}^1_3) = \underset{\sim}{\delta}^1_3$.

Proof: We must show that if $f : \omega_\omega \to \underset{\sim}{\delta}^1_3$, then $[f]_{\mathcal{U}} < \underset{\sim}{\delta}^1_3$. It's enough to consider $f : \omega_\omega \to \omega_\omega$ (since for $\omega_\omega \leq \lambda < \delta^1_3$,

$$\overline{\overline{\lambda^{\omega_\omega}/\mathcal{U}}} = \overline{\overline{\omega_\omega^{\omega_\omega}/\mathcal{U}}}).$$

Furthermore, it is enough to show that $i^{V^j_{m,k}}(\omega_\omega) < \underset{\sim}{\delta}^1_3$. To see this: Define $\delta \prec \varepsilon \Leftrightarrow \delta,\varepsilon \in C_\omega$ & X_δ, X_ε "are" functions from $A^j_{m,k} \to \omega_\omega$ &

$$[X_\delta]_{V^j_{m,k}} < [X_\varepsilon]_{V^j_{m,k}}.$$

This last inequality is expressed by

$$\exists C \subseteq \omega_1 \text{ cub } [\forall \xi \in \tilde{C}^k \cap A^j_{m,k}(X_\delta(\vec{\xi}) < X_\varepsilon(\vec{\xi}))].$$

Thus $\prec$ is $\underset{\sim}{\Sigma}^1_3$, so it is bounded below $\underset{\sim}{\delta}^1_3$. $\dashv$

B.6. Theorem. (Kunen [3]). For all $\lambda < \underset{\sim}{\delta}^1_3$, $\underset{\sim}{\delta}^1_3 \to (\underset{\sim}{\delta}^1_3)^\lambda$.

Proof: Assume $\lambda \geq \omega_\omega$. Fix $t : \omega \cdot \lambda \to \omega_\omega$ 1 - 1 and onto. For $\rho < \omega_\omega$, define

$$(\varepsilon)_\rho = \begin{cases} \{\langle\eta,\eta'\rangle : \langle\rho,\eta,\eta'\rangle \in X_\varepsilon\} & \text{if } \varepsilon \in C_\omega \\ \emptyset & \text{otherwise.} \end{cases}$$

For $\xi < \omega \cdot \lambda$, let $\varepsilon_\xi = (\varepsilon)_\rho$ where $t(\xi) = \rho$. If ε_ξ "is" a well-ordering of ω_ω, let $|\varepsilon_\xi|$ be its length.

Fix $A \subseteq (\underset{\sim}{\delta}^1_3)^\lambda \uparrow$, and consider the game in which I plays α, II plays β, and II wins if either

(1) For some $\xi < \omega \cdot \lambda$, α_ξ or β_ξ is not a w.o. of ω_ω, and if ξ_0 = least

such ξ, then α_{ξ_0} is not a w.o.,

or (2) For all $\xi < \omega \cdot \lambda$, α_ξ and β_ξ are w.o.'s of ω_ω, and

$$\{\sup_n(\max\{|\alpha_{\omega\cdot\theta+n}|, |\beta_{\omega\cdot\theta+n}|\})\}_{\theta<\lambda} \in A.$$

Assume II has a winning strategy σ. Then for some $\xi < \lambda$, $\eta < \underset{\sim}{\delta}^1_3$ let

$$\Theta\ (\xi,\eta) = \sup\{|\sigma[\alpha]_\xi| + 1 : \forall \xi' \leq \xi\ (\alpha_{\xi'} \text{ is a w.o. of } \omega_\omega) \text{ and}$$
$$\sup\{|\alpha_{\xi'}| : \xi' \leq \xi\} < \eta\}.$$

<u>Claim</u>: $\Theta\ (\xi,\eta) < \underset{\sim}{\delta}^1_3$.

<u>Proof</u>: It is easy to construct a $\underset{\sim}{\Sigma}^1_3$ wellfounded relation on $\mathbb{R}$ of length $\geq \Theta\ (\xi,\eta)$.

The rest of the proof is as in 1 1.1 of [1]. ⊣

<u>References</u>

[1] A. S. Kechris, AD and projective ordinals, this volume.

[2] A. S. Kechris and D. A. Martin, On the theory of Π^1_3 sets of reals, Bull. Amer. Math. Soc., <u>84</u> (1978), 149-151.

[3] K. Kunen, On $\underset{\sim}{\delta}^1_5$, mimeographed notes, August 1971.

[4] Y. N. Moschovakis, Determinacy and prewellorderings of the continuum, Math. Logic and Foundations of set theory, Ed. by Y. Bar Hillel, North Holland, Amsterdam-London, 1970, 24-62.

WADGE DEGREES AND DESCRIPTIVE SET THEORY

Robert Van Wesep

Department of Mathematics
California Institute of Technology
Pasadena, California 91125

and

Department of Mathematics
Dartmouth College
Hanover, New Hampshire 03755

The work to be presented here is taken principally from the three sources Steel [1977], Van Wesep [1977], and Wadge, listed in the bibliography. There is so far nothing published on this subject except the Van Wesep paper in the JSL.

In Sections 1, 2, and 3 we provide a general picture of the Wadge degrees. In Section 4 we prove some results of Steel concerning functions from the Turing degrees to $\aleph_1$ modulo the Martin measure and apply them to a computation of the length of the Wadge ordering of $\underset{\sim}{\Delta}^1_n$ sets. Then in Section 5 we prove some results about the separation, reduction, and prewellordering properties for suitable classes of sets of reals.

We work throughout in ZF + DC + AD, but the reader will be able to determine when and how the determinateness assumption may be relaxed in proving corresponding results about restricted classes of sets of reals.

1. Definitions

Definition: Baire space = ${}^{\omega}\omega$. An element of ${}^{\omega}\omega$ is a real. An interval of Baire is a set $[s] = \{\alpha \in {}^{\omega}\omega \mid s \subseteq \alpha\}$, where $s \in {}^{\underset{\smile}{\omega}}\omega = \bigcup_{n\in\omega} {}^{n}\omega$.

All our topological notions are with respect to the Baire topology, which is defined by taking the intervals of Baire as the basic open sets.

Definition: For $A,B \subseteq {}^{\omega}\omega$, $A \leq_w B$ iff there is a continuous $f : {}^{\omega}\omega \to {}^{\omega}\omega$ such that $A = f^{-1}(B)$.

Clearly $\leq_w$ is reflexive and transitive.

Definition: $A \equiv_w B$ iff $A \leq_w B$ and $B \leq_w A$. A w-degree (Wadge degree) is an equivalence class of $\equiv_w$.

Consider the game $G_w(A,B)$:

I	II	
$\alpha(0)$		I plays $\alpha(0)$, II passes or plays, I plays $\alpha(1)$, II passes or plays, etc. II's plays in order are $\beta(0),\beta(1),\ldots$. II wins iff he plays infinitely often and $\alpha \in A \Leftrightarrow \beta \in B$.
$\alpha(1)$		
$\vdots$		
$\alpha(n_0)$	$\beta(0)$	
$\alpha(n_0+1)$		
$\vdots$		
$\alpha(n_1)$	$\beta(1)$	
$\vdots$		
α	β	

It is easy to see that $A \leq_w B \Leftrightarrow$ II wins (i.e., has a winning strategy in) $G_w(A,B)$.

<u>Definition</u>: Let $G_\ell(A,B)$ be the following game:

I	II	
$\alpha(0)$	$\beta(0)$	I plays $\alpha(0)$, II plays $\beta(0)$, I plays $\alpha(1)$, II plays $\beta(1)$, etc. II wins iff $\alpha \in A \Leftrightarrow \beta \in B$.
$\alpha(1)$	$\beta(1)$	
$\vdots$	$\vdots$	
α	β	

<u>Definition</u>: $A \leq_\ell B$ iff II wins $G_\ell(A,B)$. $A \equiv_\ell B$ iff $A \leq_\ell B$ and $B \leq_\ell A$. An <u>ℓ-degree (Lipschitz degree)</u> is an equivalence class of $\equiv_\ell$.

Of course, $A \leq_\ell B \Rightarrow A \leq_w B$.

<u>2. The Lipschitz ordering</u>

<u>Lemma 2.1</u>: (Wadge's Lemma). For $A,B \subseteq {}^\omega\omega$, either $A \leq_\ell B$ or $B \leq_\ell {\sim}A$, <u>a fortiori</u>, either $A \leq_w B$ or $B \leq_w {\sim}A$.

<u>Proof</u>: Immediate from the determinateness of $G_\ell(A,B)$. $\dashv$

We shall generally use upper case Roman letters for sets of reals, and lower case Roman letters for their (ℓ- or w-) degrees.

<u>Definition</u>: If Γ is any class of sets of reals, then $\check{\Gamma} = \{A \subseteq {}^\omega\omega \mid {\sim}A \in \Gamma\}$ is the <u>dual</u> class to Γ.

Wadge's Lemma gives us the following information about the Lipschitz ordering:

1) If a is a selfdual ℓ-degree and b is any ℓ-degree, then $b <_\ell a$, $b = a$, or $b >_\ell a$.

2) If a is a nonselfdual ℓ-degree, and b is any ℓ-degree, then

(i) $b <_\ell a$ and $b <_\ell \check{a}$,

or (ii) $b = a$,

or (iii) $b = \check{a}$,

or (iv) $b >_\ell a$ and $b >_\ell \check{a}$.

<u>Definition</u>: Let $A_i \subseteq {}^\omega\omega$, $i \in \omega$. Then $\mathrm{join}_{i\in\omega} A_i = \{ \langle i\rangle^\frown\alpha \mid i \in \omega, \alpha \in A_i\}$. For $A,B \subseteq {}^\omega\omega$, A join $B = \{ \langle i\rangle^\frown\alpha \mid n \in \omega, n \text{ even}, \alpha \in A\} \cup \{ \langle n\rangle^\frown\alpha \mid n \in \omega, n \text{ odd}, \alpha \in B\}$.

<u>A picture of the lower ℓ-degrees.</u>

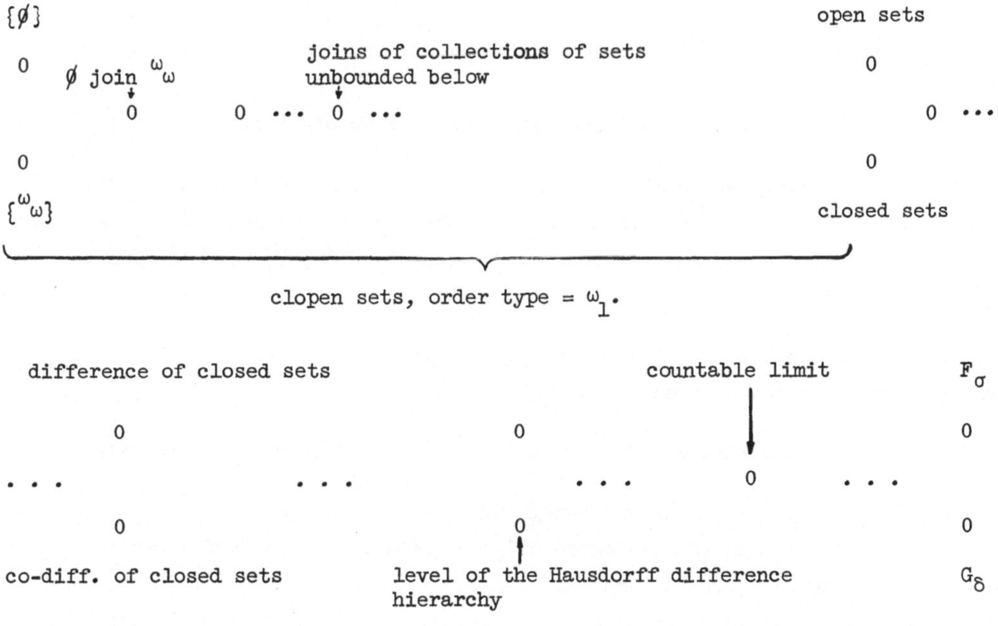

<u>Definition</u>: For $A \subseteq {}^\omega\omega$, $s \in {}^{\mathcal{G}}\omega$, $s^\frown A = \{ s^\frown\alpha \mid \alpha \in A\}$, $A_s = \{\alpha \mid s^\frown\alpha \in A\}$.

<u>Theorem 2.2.</u> (Martin [1973]). The relation $\leq_\ell$ is well-founded.

<u>Proof</u>: Suppose not. By DC there are sets A_n, $n \in \omega$, such that for any n, $A_{n+1} <_\ell A_n$. It is easy to see that I wins each of the games $G_\ell(A_n, A_{n+1})$ and $G_\ell(A_n, {\sim}A_{n+1})$, say by the strategies f_1^n and f_0^n, respectively.

To any $\alpha \in {}^\omega 2$ assign the sequence $\langle f^n_{\alpha(n)} \mid n \in \omega\rangle$ of strategies and the sequence $\langle \gamma^n_\alpha \mid n \in \omega\rangle$ of reals, as indicated in the following diagram:

$$\begin{array}{ccccccc}
 & f^n_{\alpha(n)} & & & & f^1_{\alpha(1)} & f^0_{\alpha(0)} \\
\cdots & A_{n+1} & A_n & \cdots & A_2 & A_1 & A_0 \\
 & \gamma^{n+1}_\alpha(0) & \gamma^n_\alpha(0) & & \gamma^2_\alpha(0) & \gamma^1_\alpha(0) & \gamma^0_\alpha(0) \\
 & \gamma^{n+1}_\alpha(1) & \gamma^n_\alpha(1) & & \gamma^2_\alpha(1) & \gamma^1_\alpha(1) & \gamma^0_\alpha(1) \\
 & \vdots & \vdots & & \vdots & \vdots & \vdots
\end{array}$$

where for each n, $\gamma^n_\alpha(0)$ is just the first move according to the strategy $f^n_{\alpha(n)}$, and $\gamma^n_\alpha(k)$ is the response of $f^n_{\alpha(n)}$ to $\gamma^{n+1}_\alpha \restriction k$.

Now if α and α' are in ${}^\omega 2$, and $n \in \omega$ is such that $(\forall m \geq n)\alpha(m) = \alpha'(m)$, then $\gamma^n_\alpha = \gamma^n_{\alpha'}$. For $n \in \omega$, define T^n to be the set of $\beta \in {}^\omega 2$ such that $\gamma^n_{s\frown\beta} \in A_n$ for any $s \in {}^n 2$. Then for any s in ${}^n 2$ we have

$$T^0_s = \begin{cases} T^n & \text{if } s \text{ contains an even number of 1's.} \\ \sim T^n & \text{if } s \text{ contains an odd number of 1's.} \end{cases}$$

Now no T^n is either meager or comeager, because its two details of rank one are complementary. But if T^0 is not meager, then (by the consequence of AD that all sets of reals have the Baire property) for some $s \in {}^{<\omega}2$, $T^0_s = T^{lh(s)}$ is comeager, a contradiction. ⊣

Definition: We let $ord_\ell(a)$ be the order type of the ℓ-degrees below a (having first coalesced each degree and its dual).

Some of the properties of ℓ-degrees may be inferred from their ordinals.

Lemma 2.3: Suppose a is an ℓ-degree.

(1) If $a \neq \check{a}$, then the ℓ-degree of joins of a set in a and a set in $\check{a}$ is the minimum ℓ-degree above a and $\check{a}$.

(2) If $a = \check{a}$, then the ℓ-degree of $\langle 0 \rangle \frown A$ for some $A \in a$ is the minimum ℓ-degree above a.

(3) If $ord_\ell(a)$ is a limit ordinal of cofinality ω, then a is the degree of joins of ω-sequences of sets unbounded below a. So $a = \check{a}$.

(4) If $ord_\ell(a)$ is a limit ordinal of uncountable cofinality, then $a \neq \check{a}$.

Proof: (1) and (2) are easy.

(3). Let $\langle A_i \mid i \in \omega \rangle$ be unbounded below a. Let $A = join_{i \in \omega} A_i$. Clearly $A \not<_\ell a$. So we need only show $A \leq_\ell a$. To see this let $B \in a$. Reduce A to B as follows:

> Suppose I plays i. You then have to reduce A_i to B, but you must move first. You are in the position of player I attempting to show $B \leq_\ell \sim A_i$. But this is true, so you have a winning strategy. Use it.

(4). Suppose $a = \check{a}$. Let $A \in a$.

Claim 1: $(\forall n)\, A_{\langle n \rangle} <_\ell A$.

That $A_{\langle n \rangle} \leq_\ell A$ is immediate. Suppose $A \leq_\ell A_n$. Then $A_n \leq_\ell \sim A_n$. Now to show $A \not\leq_\ell A_n$, let I play as follows in $G_\ell(A, A_{\langle n \rangle})$:

> Play n. Now play according to a winning strategy for II in $G_\ell(A_n, \sim A_n)$.

Claim 2: $\{A_n \mid n \in \omega\}$ is unbounded below A.

Suppose not. Pick B so that for each n, $A_n <_\ell B <_\ell A$. But in the proof of (3) we showed that $A = \text{join}_{n \in \omega} A_n \leq_\ell B$, a contradiction.

Claim 2 violates the uncountability of the cofinality of $\text{ord}_\ell(a)$. ⊣

Picture of the ℓ-degrees

```
                              cofinality > ω
          successor                ↓
     or cofinality = ω             0
              ↓
       . . .  0    0  . . .              0  . . .

                                   0
```

3. The Wadge ordering

It is not hard to see that if a is a selfdual ℓ-degree, then the next ω_1 ℓ-degrees are $\leq_W a$. Thus the following theorem provides a complete picture of the Wadge ordering.

Theorem 3.1: (Steel, Van Wesep). For $A \subseteq {}^{\omega}\omega$, $A \leq_W \sim A \Rightarrow A \leq_\ell \sim A$.

Proof: Suppose $A \leq_W \sim A$ and $A \not\leq_\ell \sim A$. Let g_1 be a strategy for Player II which witnesses the former fact and f a strategy for Player I which witnesses the latter. Let g_0 be the strategy for Player II which instructs him to copy I's moves as they are made.

Consider a sequence $\langle S_i \mid i \in \omega \rangle$, where, for each i, S_i is f, g_0 or g_1, and consider the following diagram:

$$\ldots A \overset{S_2}{\quad} A \overset{S_1}{\quad} A \overset{S_0}{\quad} A$$

We imagine filling in a column of numbers below each "A", the columns being referred to as numbered starting with zero for the rightmost column. The idea is that the entries in the $(i+1)^{th}$ and the ith columns should be a possible play of the $\begin{Bmatrix}\text{Lipschitz}\\ \text{Wadge}\end{Bmatrix}$ game in which Player $\begin{Bmatrix}\text{I}\\ \text{II}\end{Bmatrix}$ plays according to

$$S_i = \begin{Bmatrix} f \\ g_0 \quad \text{or} \quad g_1 \end{Bmatrix} \quad \text{when Player} \quad \begin{Bmatrix} \text{I's} \\ \text{II's} \end{Bmatrix} \text{ moves are represented}$$

by the entries in the ith column.

A <u>finite filling in</u> of the diagram consists of a finite set of numbers arranged in columns under the A's in such a way that for each $i \in \omega$:

(1) the ith and $(i+1)^{th}$ columns are a partial play according to S_i with the player on the left to move under the convention stated in the preceding paragraph, or

(2) for all $j \geq i$, the jth column is empty.

<u>Lemma</u>: For a given sequence $\langle S_i \mid i \in \omega\rangle$, no two finite fillings in can clash, i.e., have, for some i and n, different entries in the nth place of the ith column.

<u>Proof</u>: Since all but finitely many columns are empty we may take i_0 to be the greatest i such that the ith columns of the two finite fillings in clash. This leads easily to a contradiction. ⊣

<u>Definition</u>: The <u>filling in</u> of the diagram for $\langle S_i \mid i \leq \omega\rangle$ is the union of all finite fillings in. The filling in is well defined by virtue of the lemma.

<u>Definition</u>: The <u>filling in</u> for a finite sequence $\langle S_i \mid i \leq n\rangle$ is the union of all finite fillings in for the sequence with the nth column empty.

Since the Wadge strategy g_1 has the option of passing, occurrences of g_1 in the sequence $\langle S_i \mid i \in \omega\rangle$ must be sufficiently rare if they are not to block the complete filling in of the diagram.

Define an increasing sequence of numbers $\langle i_k \mid k \in \omega\rangle$ as follows:

> Set $i_0 = 0$. Suppose i_k is defined for $k \leq k_0$. For each sequence $0 = \langle S_i \mid i \leq i_{k_0}\rangle$ with the property that
>
> (1) $(\forall k \leq k_0)\ S_{i_k} = g_0$ or g_1
>
> and (2) $(\forall i < i_{k_0})\ [(\forall k < k_0)\ i \neq i_k \Rightarrow S_i = f]$,
>
> define $i^{\sigma}_{k_0+1}$ to be the least number $n > i_{k_0}$ such that the filling in for $\sigma f f \cdots f$ ($n - i_{k_0}$ f's) contains at least $k_0 + 1$ entries in the zeroth column. It is easy to see that $i^{\sigma}_{k_0+1}$ exists for

each σ. Let $i_{k_0+1} = \max\{i^\sigma_{k_0+1} \mid \sigma \text{ as above}\}$.

Now if $\langle S_i \mid i \in \omega\rangle$ is such that

(1) $(\forall k) S_{i_k} = g_0$ or g_1

and (2) for all other i, $S_i = f$,

then by construction there are finite fillings in for the sequence $\langle S_i \mid i \in \omega\rangle$ with arbitrarily many entries in the zeroth column. Thus the filling in for $\langle S_i \mid i \in \omega\rangle$ as above has an infinite sequence in the zeroth, a fortiori in every, column.

For any $z \in {}^\omega 2$, let $\mathfrak{g}_z : \omega \to \{f, g_0, g_1\}$ be given by

$$\mathfrak{g}_z(i) = \begin{cases} g_{z(k)} & \text{if } i = i_k \\ f & \text{otherwise.} \end{cases}$$

Define $h(z)$ to be the real produced in the zeroth column by the filling in for $\mathfrak{g}_z$. Let $T = h^{-1}A$. ⊣

Corollary: (i) $\{\emptyset\}$ and $\{{}^\omega\omega\}$ are the minimal w-degrees;

(ii) each non-self dual pair of w-degrees has a selfdual successor;

(iii) each selfdual w-degree has a non-self dual pair of successors;

(iv) each selfdual w-degree is the union of an ω_1-sequence of selfdual ℓ-degrees;

(v) a level of cofinality ω is occupied by a selfdual w-degree;

(vi) a level of uncountable cofinality is occupied by a non-self dual pair of w-degrees. Proceed as in the proof of Theorem 2. ⊣

Picture of the w-degrees

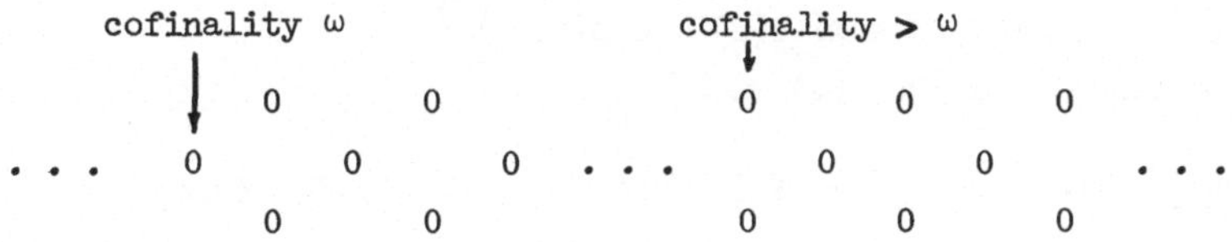

4. The order type of the $\underset{\sim}{\Delta}^1_n$ degrees

It is fairly easy to see that the order type of the Wadge degrees is $\Theta = \sup\{\xi \mid \xi \text{ is the length of a prewellordering of the reals}\}$. We seek to know the order types of the set of degrees of $\underset{\sim}{\Delta}^1_n$ sets of reals; in short, the order type of the $\underset{\sim}{\Delta}^1_n$ degrees.

It is easy to see that the set of degrees preceding a $\underset{\sim}{\Delta}^1_n$ degree has order type $< \underset{\sim}{\delta}^1_{n+1}$ (just look at the prewellordering of their codes as continuous pre-images of the given set). So the order type of the $\underset{\sim}{\Delta}^1_n$ degrees is $\le \underset{\sim}{\delta}^1_{n+1}$. In case n is odd, the inequality is strict. (Prewellorder the codes of $\underset{\sim}{\Delta}^1_n$ sets as preimages of initial segments of a $\underset{\sim}{\Pi}^1_n$-prewellordered complete $\underset{\sim}{\Pi}^1_n$ set.) We shall show that for n even, the order type of the $\underset{\sim}{\Delta}^1_n$ degrees is $\geq \underset{\sim}{\delta}^1_{n+1}$, hence $= \underset{\sim}{\delta}^1_{n+1}$. So, for n odd, $\underset{\sim}{\delta}^1_n <$ order type of $\underset{\sim}{\Delta}^1_n$ degrees $< \underset{\sim}{\delta}^1_{n+1}$.

Our proof that the order type of the $\underset{\sim}{\Delta}^1_n$ degrees is $\underset{\sim}{\delta}^1_{n+1}$ for n even, is due to Steel and proceeds via a discussion of functions from Turing degrees to ordinals which is of interest in its own right. There is also an independent direct proof of this fact due to Martin.

Consider functions from $\mathcal{D}$ (the set of Turing degrees) into ω_1, relative to the Martin measure, μ.

Definition: $f_n = \lambda d(\delta^1_n(d))$, where $\delta^1_n(d)$ is the least ordinal which is not the order type of a wellordering of ω which is δ^1_n in d.

Let $\pi : \omega_1/\mu \cong \lambda$, $\lambda \in \mathrm{Ord}$, be the canonical isomorphism.

Remark: $AD_{\mathbb{R}} \Rightarrow \lambda = \Theta$, $V = \mathbf{L}[\mathbb{R}] \Rightarrow \lambda > \Theta$.

Theorem 4.1: (Steel). For all even $n \geq 2$, $\pi([f_n]) = \underset{\sim}{\delta}^1_{n+1}$.

Remark: The full axiom of determinateness implies that $\underset{\sim}{\delta}^1_3 = \aleph_{\omega+1}$. Steel has shown that if $g_n(d) = n^{th}$ d-admissible beyond ω, then $\pi([g_n]) = \aleph_n$. His method also shows that the union of the first $\aleph_n$ Wadge degrees for $n \geq 2$ is just the nth level of the hierarchy based on the operation A.

Proof: For convenience we take $n = 2$. Let W^α be a complete $\Sigma^1_2(\alpha)$ subset of ω, uniformly in α. Let $\leq^\alpha$ be the canonical prewellordering of W^α, and let $|m|^\alpha$ be the rank of m in $\leq^\alpha$ for $m \in W^\alpha$.

Definition: We say $\langle m,s\rangle$ is a code of $f : \mathcal{D} \to \omega_1$ iff

(1) $m \in \omega$ and s is a real coding a continuous function from ${}^\omega\omega$ to ${}^\omega\omega$,

(2) for almost all α, $m \in W^{s(\alpha)}$,

(3) for almost all $d \in \mathcal{D}$, $f(d) = \sup\{|m|^{s(\alpha)} \mid \alpha \in d\}$,

(4) for almost all α, $s(\alpha) \equiv_T \alpha$.

Lemma 4.2: For any $[f] < [f_2]$, there exists a code $\langle m,s\rangle$ of f.

Proof: Consider the game,

I	II	
		I wins iff
m,α	β	$\alpha \geq_T \beta$,
		$m \in W^\alpha$,
		and $\vert m\vert^\alpha = f([\alpha])$.

Suppose II wins by the strategy σ. Let α be any real $\geq_T \sigma$ and high enough that $f([\alpha]) < \delta^1_2(\alpha)$. Let $m \in W^\alpha$ be such that $|m|^\alpha = f([\alpha])$. (The length of $\leq^\alpha$ is $\delta^1_2(\alpha)$.) The I can win by playing $\langle m,\alpha\rangle$. Contradiction.

So I wins, say be the strategy σ. Let m be I's first play by σ. Let β be any real $\geq_T \sigma$, and let II play β. Then if $\langle m,\alpha\rangle = \sigma * \beta$ we have $\alpha \equiv_T \beta$ and $|m|^\alpha = f([\alpha]) = f([\beta])$. If s codes the function $\beta \mapsto \alpha$, then $\langle m,s\rangle$ codes f. ⊣

Lemma 4.3: The set of codes is a complete Π^1_3 set of reals.

Proof: A pair $\langle m,s\rangle$ is a code iff

(i) s codes a continuous function,

(ii) $(\forall\alpha)(\exists\beta \geq_T \alpha)(\forall\beta' \equiv_T \beta)\ \beta' \in W^{s(\beta')}$,

(iii) $(\exists\alpha)(\forall\beta \geq_T \alpha)(s(\beta) \equiv_T \beta)$.

So the set of codes is Π^1_3.

To show completeness let $P = \{\beta \mid (\forall\alpha) m \in W^{\langle\alpha,\beta\rangle}\}$ be an arbitrary Π^1_3 set. Let k be such that $k \in W^{\langle\alpha,\beta\rangle} \Longleftrightarrow (\forall\alpha' \leq_T \alpha) m \in W^{\langle\alpha',\beta\rangle}$. Let s_β code the continuous function $\alpha \mapsto \langle\alpha,\beta\rangle$. Then the map $\beta \mapsto s_\beta$ is continuous, and $\beta \in P \Longleftrightarrow \langle k,s_\beta\rangle$ is a code. ⊣

Lemma 4.4: There is a Σ^1_2 relation R^+ and a Π^1_2 relation R^- such that if $n \in W^\beta$ then

$$R^+(m,\alpha,n,\beta) \Longleftrightarrow m \in W^\alpha \ \&\ |m|^\alpha \leq |n|^\beta,$$

$$R^-(m,\alpha,n,\beta) \Longleftrightarrow (m \in W^\alpha \ \&\ |m|^\alpha \leq |n|^\beta) \text{ or } \delta^1_2(\alpha) \leq |n|^\beta.$$

Proof: Define $R^+(m,\alpha,n,\beta) \Longleftrightarrow m \in W^\alpha \ \&\ (\exists f : \{m' \mid m' \leq^\alpha m\} \to \{n' \mid n' \leq^\beta n\})$ (f is order preserving). Define $R^-(m,\alpha,n,\beta) \Longleftrightarrow (\forall f)[(\mathrm{dom}(f) = \{n' \mid n' \leq^\beta n\} \ \&\ \mathrm{rng}(f) \subseteq W^\alpha \ \&\ f$ order preserving$) \Rightarrow (\exists m' \in \mathrm{rng}(f))(m \leq^\alpha m')]$. ⊣

We now define a Π^1_3-prewellordering of the codes which is very nearly the natural one.

Define $\langle m,s\rangle \leq^+ \langle n,t\rangle$ iff

(i) $\langle m,s\rangle$ and $\langle n,t\rangle$ are codes,

(ii) $\forall\alpha\, \exists\beta \geq_T \alpha\, \forall\gamma \equiv_T \beta\, \exists\gamma' \equiv_T \beta$
$R^+(m,s(\gamma),n,t(\gamma'))$.

Define $\langle m,s\rangle \leq^- \langle n,t\rangle$ iff

(i) s,t code continuous functions,

(ii) $\exists\alpha\, \forall\beta \geq_T \alpha\ \ s(\beta) \equiv_T \beta$

(iii) $\exists\alpha\, \forall\beta \geq_T \alpha\, \forall\gamma \equiv_T \beta\, \exists\gamma' \equiv \beta$
$R^-(m,s(\gamma),n,t(\gamma'))$.

Note that if $\langle n,t\rangle$ is a code and $\langle m,s\rangle \leq^- \langle n,t\rangle$, then for almost all β, for all $\gamma,\gamma' \equiv_T \beta$, we have $s(\gamma) \equiv_T \gamma \equiv_T \gamma' \equiv t(\gamma')$, so $\delta^1_2(s(\gamma)) > |n|^{t(\gamma')}$, and $R^-(m,s(\gamma),n,t(\gamma')) \Longleftrightarrow [m \in W^{s(\gamma)} \ \&\ |m|^{s(\gamma)} \leq |n|^{t(\gamma')}]$. Thus if $\langle n,t\rangle$ is a code then

(i) $\langle m,s\rangle \leq^{+} \langle n,t\rangle \Rightarrow \langle m,s\rangle$ is a code,

(ii) $\langle m,s\rangle \leq^{-} \langle n,t\rangle \Rightarrow \langle m,s\rangle$ is a code,

(iii) $\langle m,s\rangle \leq^{+} n,t\rangle \Leftrightarrow \langle m,s\rangle \leq^{-} \langle n,t\rangle$.

The relation $\leq^{o}$ on the codes defined by $\leq^{+}$ (equivalently $\leq^{-}$) is a prelinear ordering, and clearly if $\langle m,s\rangle \leq^{o} \langle n,t\rangle$ then the function coded by $\langle m,s\rangle$ is $\leq$ the function coded by $\langle n,t\rangle$. The converse is not quite true, as if $f(d)$ is almost always a limit ordinal the codes of $f(d)$ may occupy two consecutive levels of $\leq^{o}$. For suppose $\langle m,s\rangle$ codes f, and $f(d) = \eta$, a limit ordinal. Then η may be the proper supremum of $\{|m|^{s(\alpha)} \mid \alpha \in d\}$ or for some $\alpha \in d$, we may find $\eta = |m|^{s(\alpha)}$. A code for which the first case almost always obtains will be $<^{o}$ a code for which the second case almost always obtains.

Nevertheless, the order type of $\leq^{o}$ is just the order type of the functions $< f_2$. Since $\leq^{+}$ is Π^1_3, $\leq^{-}$ is Σ^1_3, and the set of codes is Π^1_3, this order type is $\underset{\sim}{\delta}^1_3$ by Moschovakis [1970]. Thus the theorem is proved. $\dashv$

<u>Theorem 4.2</u>: (Martin, Steel). The order type of the $\underset{\sim}{\Delta}^1_{2n}$ Wadge degrees is $\underset{\sim}{\delta}^1_{2n+1}$.

<u>Proof</u>: Again suppose for simplicity that $n = 1$. Let η be the order type of the $\underset{\sim}{\Delta}^1_2$ degrees. It is easy to see that $\eta \leq \underset{\sim}{\delta}^1_3$. (The prewellordering of codes of sets as preimages of a $\underset{\sim}{\Delta}^1_2$ set is $\underset{\sim}{\Delta}^1_3$).

To see that $\eta \geq \underset{\sim}{\delta}^1_3$, we exhibit a map of η cofinal in $\{g : \mathcal{D} \to \omega_1 \mid g < f_2\}$. Recall that as we are assuming full AD, $\underset{\sim}{\delta}^1_3$ is a regular cardinal. Clearly AD(L[$\mathbb{R}$]) would also suffice for this result.

<u>Definition</u>: If $\langle m,s\rangle$ is a code which works above γ, i.e. $(\forall \beta \geq_T \gamma)(s(\beta) \equiv_T \beta \;\&\; m \in W^{s(\beta)})$, let

$$C_{m,s,\gamma} = \{\langle p,\alpha\rangle \mid \alpha \geq_T \gamma \;\&\; p \leq^{s(\alpha)} m\}.$$

Let $C^{\alpha}_{m,s,\gamma} = \{\langle p,e\rangle \mid \langle p,\{e\}^{\alpha}\rangle \in C_{m,s,\gamma}\}$. Note that $\alpha \equiv_T \beta \Rightarrow C^{\alpha}_{m,s,\gamma} \equiv_T C^{\beta}_{m,s,\gamma}$.

Let $h_{m,s,\gamma} : \mathcal{D} \to \mathcal{D}$ be given by

$$h_{m,s,\gamma}(d) = d \oplus \text{degree of } C^{\alpha}_{m,s,\gamma},\ \alpha \in d.$$

<u>Claim</u>: $C_{m,s,\gamma} \leq_W C_{m',s',\gamma'} \Rightarrow h_{m,s,\gamma} \leq_T h_{m',s',\gamma'}$ a.e.

<u>Proof</u>: Let σ be a (code of a) map which reduces $C_{m,s,\gamma}$ to $C_{m',s',\gamma'}$. Let $d \geq_T \sigma$, and $\alpha \in d$. Then there are e_1 and e_2 such that for all p,e,

$$\langle p,e\rangle \in C^{\alpha}_{m,s,\gamma} \iff \langle p,\{e\}^{\alpha}\rangle \in C_{m,s,\gamma}$$

$$\iff \sigma(\langle p,\{e\}^{\alpha}\rangle) \in C_{m',s',\gamma'}$$

$$\iff \langle \{e_1\}(\alpha,p),\{\{e_2\}(p,e)\}^{\alpha}\rangle \in C_{m',s',\gamma'}$$

$$\iff \langle \{e_1\}(\alpha,p),\{e_2\}(p,e)\rangle \in C^{\alpha}_{m',s',\gamma'}.$$

So $h_{m,s,\gamma}(d) = d \oplus \deg C^{\alpha}_{m,s,\gamma} \leq_T d \oplus \deg C^{\alpha}_{m',s',\gamma'} = h_{m',s',\gamma'}(d)$. This proves the claim.

Define the map $H : \eta \to \underset{\sim}{\delta}^1_3$ as follows. Given $\xi < \eta$, look at the set of $f : \mathfrak{D} \to \omega_1$ coded by $\langle m,s\rangle$ such that for some γ, $\langle m,s\rangle$ is good above γ and $C_{m,s,\gamma}$ has ordinal $< \xi$ in the Wadge ordering of $\underset{\sim}{\Delta}^1_2$. Let $H(\xi)$ be the supremum of the ordinals of these f's.

Claim: For any $\xi < \eta$, $H(\xi) < \underset{\sim}{\delta}^1_3$.

Proof: We prove this claim by showing that for any $\langle m,s\rangle$ a code good above γ, there is a $g : \mathfrak{D} \to \omega_1$, $g < f_2$, such that for any $\langle n,t\rangle$ a code for g good above δ, $C_{n,t,\delta} \not\leq_W C_{m,s,\gamma}$.

To this end let such m, s, and γ be given, let $d \geq_T s$ be given, and let $\alpha \in d$. Uniformly in e such that $\{e\}^{\alpha} \equiv_T \alpha$, find $n_e \in W^{\{e\}^{\alpha}}$ such that $\{p \mid p \leq^{\{e\}^{\alpha}} n_e\}$ has higher Turing degree than $C^{\beta}_{m,s,\gamma}$, $\beta \in d$. Paste together all the ordinals $|n_e|^{\{e\}^{\alpha}}$ getting some $\delta^1_2(d)$ ordinal. Do this for each $\alpha \in d$. Let $g(d)$ be the least ordinal so obtained.

Clearly $g < f_2$. We now show g is as desired. So let $\langle n,t\rangle$ code g above δ. Let $d \geq_T \langle s,\delta,\gamma\rangle$. There is $\alpha \in d$ such that $\{p \mid p \leq^{t(\alpha)} n\}$ has higher Turing degree than $C^{\beta}_{m,s,\gamma}$, $\beta \in d$, i.e., higher Turing degree than $h_{m,s,\gamma}(d)$. But $\{p \mid p \leq^{t(\alpha)} n\} = \{p \mid \langle p,\alpha\rangle \in C_{n,t,\delta}\} = \{p \mid \langle p,e\rangle \in C^{\alpha}_{n,t,\delta}\}$ where $\{e\}$ is the identity. So $h_{n,t,\delta}(d) \geq_T \{p \mid p \leq^{t(\alpha)} n\} >_T h_{m,s,\gamma}(d)$. So $h_{n,t,\delta} > h_{m,s,\gamma}$ a.e., whence $C_{n,t,\delta} \not\leq_W C_{m,s,\gamma}$, and the claim is proved.

It remains only to show that $H : \lambda \to \underset{\sim}{\delta}^1_3$ is cofinal. For this we need to see that if $g : \mathfrak{D} \to \omega_1$ is less than f_2 then there is a $\underset{\sim}{\Delta}^1_2$ set A so that any $f < g$ has a code $\langle m,s\rangle$ with the property that for some γ, $\langle m,s\rangle$ is good above γ and $C_{m,s,\gamma} \leq_W A$. So let $\langle n,t\rangle$ be a code for g good above β, with the additional property that for all $d \geq_T \beta$, $f(d) = |n|^{t(\alpha)}$, $\alpha \in d$ (such codes are actually given by Lemma 4.2), and let $A = \{\langle m,s,\gamma,p,\alpha\rangle \mid \alpha \geq_T \beta \;\&\; \alpha \geq_T \gamma \;\&\; |m|^{s(\alpha)} \leq |n|^{t(\alpha)} \;\&\; p \leq^{s(\alpha)} m\}$. By virtue of Lemma 4.4, A is $\underset{\sim}{\Delta}^1_2$. Clearly if $f < g$, $\langle m,s\rangle$ is a code for f, and $\gamma \geq_T \beta$ is such that $\langle m,s\rangle$ is good above γ and $f(d) < g(d)$ for d above γ, then $\langle p,\alpha\rangle \in C_{m,s,\gamma} \iff \langle m,s,\gamma,p,\alpha\rangle \in A$, so

$C_{m,s,\gamma} \leq_w A$.

This concludes the proof of Theorem 4.2. ⊣

5. Separation, reduction, and prewellordering properties in the Wadge hierarchy

In this section we establish a number of descriptive set theoretic properties of nonselfdual classes closed under continuous preimage, assuming throughout ZF + DC + AD. In the following, then, Γ will always be a nonselfdual class of sets of reals closed under continuous preimages. The descriptive set theoretic properties we shall be concerned with are as follows.

Definition: Γ has the first separation property, $\mathrm{Sep}_I(\Gamma)$, iff

$$(\forall A,B \in \Gamma)(A \cap B = \emptyset \Rightarrow (\exists C \in \Gamma \cap \check{\Gamma})(A \subseteq C \subseteq \sim B)).$$

Γ has the second separation property, $\mathrm{Sep}_{II}(\Gamma)$, iff

$$(\forall A,B \in \Gamma)(\exists A',B' \in \check{\Gamma})(A \sim B \subseteq A' \;\&\; B \sim A \subseteq B' \;\&\; A' \cap B' = \emptyset).$$

Γ has the reduction property, $\mathrm{Red}(\Gamma)$, iff

$$(\forall A,B \in \Gamma)(\exists A',B' \in \Gamma)(A' \subseteq A, B' \subseteq B, A' \cap B' = \emptyset, \;\&\; A' \cup B' = A \cup B).$$

Γ has the prewellordering property, $\mathrm{PWO}(\Gamma)$, iff for every $A \in \Gamma$ there are relations $\leq^+$ and $\leq^-$ in Γ and $\check{\Gamma}$ respectively, and a prewellordering $\leq$ of A such that for any real $\beta \in A$ and any real α, $\alpha \leq^+ \beta \Leftrightarrow \alpha \leq^- \beta \Leftrightarrow \alpha \leq \beta$.

Before proceeding to the statement of the results of this section we take note of some simple facts. The first is that Wadge's Lemma (Lemma 2.1) has the immediate consequence that if Γ is continuously closed and nonselfdual, and $A \in \Gamma \sim \check{\Gamma}$, then A is complete for Γ, i.e., for all B in Γ, $B \leq_W A$. It is also true that Γ has a universal set. I.e., there is $A \in \Gamma$ such that for all $B \in \Gamma$ there is $\alpha \in {}^\omega\omega$ such that $\beta \in B \Leftrightarrow (\alpha,\beta) \in A$. One may take $A = \{(f,\beta) \mid f * \beta \in C\}$, where $C \in \Gamma \sim \check{\Gamma}$ and $f * \beta$ is the result of applying (the Lipschitz strategy coded by) f to β. Finally, we take note of a standard result of descriptive set theory, viz., if $\mathrm{Red}(\Gamma)$ and Γ has a universal set, then $\mathrm{Sep}_I(\check{\Gamma})$ and $\neg\,\mathrm{Sep}_I(\Gamma)$, and hence $\mathrm{Red}(\Gamma) \Rightarrow \neg\,\mathrm{Red}(\check{\Gamma})$. Since the proof of this fact is included in our proof of Theorem 5.1, we do not give it here.

We now state the results.

Theorem 5.1: (Van Wesep). For any continuously closed nonselfdual class Γ, we have $\mathrm{Sep}_I(\Gamma) \Leftrightarrow \neg\,\mathrm{Sep}_{II}(\check{\Gamma})$.

Theorem 5.2: (Van Wesep). For any continuously closed nonselfdual Γ, either $\mathrm{Sep}_{II}(\Gamma)$ or $\mathrm{Sep}_{II}(\check{\Gamma})$. Thus by Theorem 5.1, either $\neg\,\mathrm{Sep}_I(\Gamma)$ or $\neg\,\mathrm{Sep}_I(\check{\Gamma})$.

Using Wadge's characterizations of the continuously closed nonselfdual classes of Borel sets, Steel has shown that one of each nonselfdual pair of such classes has the first separation property.

Theorem 5.3: (Van Wesep). For some continuously closed nonselfdual Γ, we have $\neg\,\mathrm{Red}(\Gamma)$ and $\neg\,\mathrm{Red}(\check{\Gamma})$.

Theorem 5.4: (Van Wesep). If Γ is closed under (finite) intersections and $\mathrm{Sep}_I(\check{\Gamma})$ then $\mathrm{Red}(\Gamma)$.

Theorem 5.5: (Steel). If Γ is closed under countable unions and countable intersections, then $\mathrm{Red}(\Gamma)$ or $\mathrm{Red}(\check{\Gamma})$.

Theorem 5.6: (Kechris - Solovay). If Γ is closed under countable unions, countable intersections, projections, and coprojections, and $\Gamma \subseteq L[\mathbb{R}]$, then $\mathrm{PWO}(\Gamma)$ or $\mathrm{PWO}(\check{\Gamma})$.

We now proceed to the proofs.

Proof of Theorem 5.1: Let $A \in \Gamma \sim \check{\Gamma}$. Define

$$A_0 = \{((f_0,f_1),\alpha) : f_0 * \alpha \in A\},$$

$$A_1 = \{((f_0,f_1),\alpha) : f_1 * \alpha \in A\},$$

where $(\cdot,\cdot)$ is a reasonable pairing function for reals. Clearly A_0 and A_1 are in Γ. Moreover, the pair $\langle A_0,A_1\rangle$ is universal for pairs of sets in Γ, i.e., for any $B_0,B_1 \in \Gamma$ there is a real γ such that $B_0 = \{\alpha : (\gamma,\alpha) \in A_0\}$, and $B_1 = \{\alpha : (\gamma,\alpha) \in A_1\}$. (To see this note that by Wadge's Lemma $\Gamma = \{B : B \leq_W A\}$. Then note that this is true even if $\leq_W$ is replaced by $\leq_\ell$.) Likewise the pair $\langle \sim A_0,\sim A_1\rangle$ is universal for pairs of sets in $\check{\Gamma}$.

Now suppose $\check{\Gamma}$ has the second separation property. We shall show that Γ does not have the first separation property. Let $B_0,B_1 \in \Gamma$ be such that $A_0 \sim A_1 \subseteq B_0$, and $A_1 \sim A_0 \subseteq B_1$, with $B_0 \cap B_1 = \emptyset$. Define $C_0 = \{\alpha : (\alpha,\alpha) \in B_0\}$, $C_1 = \{\alpha : (\alpha,\alpha) \in B_1\}$. Clearly C_0 and C_1 are disjoint sets in Γ. We claim they are not separable by a set in $\Delta = \Gamma \cap \check{\Gamma}$. To see this suppose $C_0 \subseteq D$ and $D \cap C_1 = \emptyset$, with $D \in \Delta$. Let γ be such that $D = \{\alpha : (\gamma,\alpha) \in \sim A_0\}$ and $\sim D = \{\alpha : (\gamma,\alpha) \quad \sim A_1\}$. Then we have

$$\gamma \in D \Rightarrow (\gamma,\gamma) \in A_1 \sim A_0 \Rightarrow (\gamma,\gamma) \in B_1 \Rightarrow \gamma \in C_1 \Rightarrow \gamma \notin D,$$

and

$$\gamma \notin D \Rightarrow (\gamma,\gamma) \in A_0 \sim A_1 \Rightarrow (\gamma,\gamma) \in B_0 \Rightarrow \gamma \in C_0 \Rightarrow \gamma \in D,$$

which is a contradiction.

Now suppose Γ does not have the first separation property. Let C and D in Γ be disjoint and not separable by a set in Δ. Let A and B be in $\check{\Gamma}$. We shall find A' and B' in Γ such that $A \sim B \subseteq A'$, $B \sim A \subseteq B'$ and $A' \cap B' = \emptyset$.

Consider the following game:

Players I and II produce reals α and β respectively. Player I wins iff

$$\beta \in C \Rightarrow \alpha \in A \sim B,$$

$$\beta \in D \Rightarrow \alpha \in B \sim A,$$

and $$\alpha \in (A \sim B) \cup (B \sim A).$$

Equivalently, II wins iff,

$$\alpha \in A \sim B \Rightarrow \beta \in D,$$

and $$\alpha \in B \sim A \Rightarrow \beta \in C.$$

Now I cannot win this game, for if he did, say by a strategy f, which we shall view as a continuous function from ${}^{\omega}\omega$ to ${}^{\omega}\omega$, then we should have $C \subseteq E$ and $E \cap D = \emptyset$, where $E = f^{-1}(A) = \sim f^{-1}(B) \in \Gamma \cap \check{\Gamma}$, which contradicts the inseparability of C and D. Thus player II wins the game, say by the strategy f. Let $A' = f^{-1}(D)$, and $B' = f^{-1}(C)$. Then A' and B' are as desired. ⊣

<u>Proof of Theorem 5.2</u>: Let $\langle A_0, A_1 \rangle$ be a complete pair of sets in Γ (see proof of Theorem 5.1). Consider the following two games.

G_0: Players I and II play reals α and β respectively. Player II wins iff

$$\alpha \in A_0 \sim A_1 \Rightarrow \beta \in A_1 \sim A_0,$$

$$\alpha \in A_1 \sim A_0 \Rightarrow \beta \in A_0 \sim A_1,$$

and $$\beta \notin A_0 \cap A_1.$$

G_1: Players I and II play reals α and β respectively. Player II wins iff

$$\alpha \in A_0 \sim A_1 \Rightarrow \beta \in A_0 \sim A_1,$$

$$\alpha \in A_1 \sim A_0 \Rightarrow \beta \in A_1 \sim A_0,$$

and $$\beta \in A_0 \cup A_1.$$

Suppose Γ does not have the second separation property. Then II does not win G_1. For if he did, say be the strategy f, we could let $A_0^1 = f^{-1}(\sim A_1)$ and $A_1^1 = f^{-1}(\sim A_0)$. Then $A_0 \sim A_1 \subseteq A_0^1$, $A_1 \sim A_0 \subseteq A_1^1$; $A_0^1 \cap A_1^1 = \emptyset$. Since $\langle A_0, A_1 \rangle$ is complete for Γ, this contradicts our hypothesis.

Similarly, if $\check{\Gamma}$ does not have the second separation property, then II does not win G_0.

Thus, by determinateness, we shall have proved the theorem when we have shown that Player I cannot win both G_0 and G_1.

So suppose Player I wins G_0 and G_1 by the strategies f_0 and f_1 respectively. From the definition of G_0 it is apparent that for any β played by II,

$$f_0(\beta) \in A_0 \sim A_1 \ \& \ \beta \notin A_1 \sim A_0,$$

$$\text{or} \qquad f_0(\beta) \in A_1 \sim A_0 \ \& \ \beta \notin A_0 \sim A_1,$$

$$\text{or} \qquad \beta \in A_0 \cap A_1.$$

In other words,

$$\beta \in A_0 \sim A_1 \Rightarrow f_0(\beta) \in A_0 \sim A_1,$$

$$\beta \in A_1 \sim A_0 \Rightarrow f_0(\beta) \in A_1 \sim A_0,$$

$$\text{and} \qquad \beta \in (\sim A_0) \cap (\sim A_1) \Rightarrow f_0(\beta) \in (A_0 \sim A_1) \cup (A_1 \sim A_0).$$

Similarly,

$$\beta \in A_0 \sim A_1 \Rightarrow f_1(\beta) \in A_1 \sim A_0,$$

$$\beta \in A_1 \sim A_0 \Rightarrow f_1(\beta) \in A_0 \sim A_1,$$

$$\text{and} \qquad \beta \in A_0 \cap A_1 \Rightarrow f_1(\beta) \in (A_0 \sim A_1) \cup (A_1 \sim A_0).$$

Now for any $\gamma \in {}^{\omega}2$, consider the sequence $\langle f_{\gamma(n)} : n \in \omega \rangle$ of strategies for I, and consider the following diagram.

$$\begin{array}{ccccccc}
 & f_{\gamma(2)} & & f_{\gamma(1)} & & f_{\gamma(0)} & \\
\gamma^3(0) & & \gamma^2(0) & & \gamma^1(0) & & \gamma^0(0) \\
\gamma^3(1) & & \gamma^2(1) & & \gamma^1(1) & & \gamma^0(1) \\
\vdots & & \vdots & & \vdots & & \vdots
\end{array}$$

The rule of construction in this diagram is: $\gamma^n(i)$ is the response of the strategy $f_{\gamma(n)}$ to the play $\gamma^{n+1} \restriction i$, i.e., the first i numbers in the sequence γ^{n+1}. This trick was first used by Martin [1973].

We consider membership of the γ^n in A_0 and A_1 for various γ. Call $(A_0 \cap A_1) \cup ((\sim A_0) \cap (\sim A_1))$ the <u>middle</u> and $(A_1 \sim A_0) \cup (A_0 \sim A_1)$ the <u>sides</u>.

<u>Claim</u>: $\{\gamma \in {}^{\omega}2 : \gamma^0 \in \text{middle}\}$ is meager.

Suppose not. Since we are assuming the axiom of determinateness, all sets of reals have the property of Baire. So for some $s \in {}^{\omega}2$, $\{\gamma : (s^\frown\gamma)^0 \in \text{middle}\}$ is comeager. But $(s^\frown\gamma)^0 \in$ middle implies $\gamma^0 \in$ middle, because $\beta \in \text{sides} \Rightarrow f_i(\beta) \in \text{sides}$ for $i = 0,1$. So $\{\gamma : \gamma^0 \in \text{middle}\}$ is comeager. Without loss of generality assume $\{\gamma : \gamma^0 \in A_0 \cap A_1\}$ is nonmeager. For any γ in this set $(\langle 1\rangle^\frown\gamma)^0$ is in the sides. Thus $\gamma^0 \in$ sides for a nonmeager set of γ. Contradiction.

Thus $\{\gamma : \gamma^0 \in \text{sides}\}$ is nonmeager. Suppose, without loss of generality, that $\{\gamma : \gamma^0 \in A_1 \sim A_0\}$ is nonmeager. Then for some $s \in {}^{\omega}2$, $\{\gamma : (s^\frown\gamma)^0 \in A_1 \sim A_0\}$ is comeager. There are two cases.

If s contains an odd number of 1's, then $\gamma^0 \in A_0 \sim A_1 \Rightarrow (s^\frown\gamma)^0 \in A_1 \sim A_0$ and $\gamma^0 \in A_1 \sim A_0 \Rightarrow (s^\frown\gamma)^0 \in A_0 \sim A_1$. So $\{\gamma : \gamma^0 \notin A_1 \sim A_0\}$ is comeager, which contradicts our assumption.

If s contains an even number of 1's, then $\gamma^0 \in A_0 \sim A_1 \Rightarrow (s^\frown\gamma)^0 \in A_0 \sim A_1$ and $\gamma^0 \in A_1 \sim A_0 \Rightarrow (s^\frown\gamma)^0 \in A_1 \sim A_0$. So $\{\gamma : \gamma^0 \notin A_0 \sim A_1\}$ is comeager. By the claim, then, $\{\gamma : \gamma^0 \in A_1 \sim A_0\}$ is comeager. But γ in this set implies $(\langle 1\rangle^\frown\gamma)^0 \in A_0 \sim A_1$, so $\{\gamma : \gamma^0 \in A_0 \sim A_1\}$ is nonmeager. This contradiction establishes the theorem. ⊣

<u>Proof of Theorem 5.3</u>: We shall show that if Γ is a minimal continuously closed nonselfdual class including $F_\sigma \cup G_\delta$ then Reduction fails for both Γ and $\check{\Gamma}$. In Section II.4.1 of Van Wesep [1977] a more general result is proved, viz., if the Wadge order type of Γ is not of the form $\omega_1^\alpha + 1$ and $\Gamma \not\subseteq F_\sigma$, then reduction fails for Γ.

Let A be G_δ but not F_σ. Let $A_0 = \{(\alpha,\beta) \mid \alpha \in A\}$, $A_1 = \{(\alpha,\beta) \mid \beta \in A\}$. It is clear that $A_0, A_1 \in G_\delta \sim F_\sigma$ and $\langle A_0, A_1\rangle$ is complete for pairs of sets in G_δ.

Now the pair $\langle A_0, A_1\rangle$ is not reducible by sets in G_δ, for by its completeness if it were reducible by G_δ sets, then we would have $\text{Red}(G_\delta)$, which is false. Moreover, $\langle A_0, A_1\rangle$ cannot be reduced by sets in $F_\sigma \cup G_\delta$, for if $C \in F_\sigma$ were such that $C \subseteq A_0$ and $A_0 \sim A_1 \subseteq C$, then letting β be any real not in A, we have, for all reals α, $\alpha \in A \Longleftrightarrow (\alpha,\beta) \in A_0 \Longleftrightarrow (\alpha,\beta) \in C$, which is a contradiction.

We may take for A the set $\{\alpha \in {}^{\omega}\omega \mid (\forall m)(\exists n > m)\alpha(n) = 0\}$, so that for any $s \in {}^{\omega}\omega$, $A_s \equiv_W A$. Now by the above considerations, for any $s \in {}^{\omega}\omega$, $\langle A_{0_s}, A_{1_s}\rangle$ is not reducible by sets in $F_\sigma \cup G_\delta$. We shall have proved the theorem when we show that $\langle A_0, A_1\rangle$ is not reducible by sets in Γ, and therefore, by symmetry, in $\check{\Gamma}$.

Suppose $\langle B_0, B_1\rangle$ reduces $\langle A_0, A_1\rangle$, where B_0 and B_1 are in Γ. It is easy to show that for some $s \in {}^{\omega}\omega$, B_{0_s} and B_{1_s} are in $F_\sigma \cup G_\delta$. But $\langle B_{0_s}, B_{1_s}\rangle$ reduces $\langle A_{0_s}, A_{1_s}\rangle$, a contradiction. ⊣

Proof of Theorem 5.4: We use the following lemma.

Lemma: If Γ is closed under finite intersections and $\neg\operatorname{Red}(\Gamma)$, then for any $C,D \in \check{\Gamma}$, there are $C',D' \in \Gamma$ so that

$$C \sim D \subseteq C',\ C' \cap (D \sim C) = \emptyset,$$

$$D \sim C \subseteq D',\ D' \cap (C \sim D) = \emptyset,$$

$$C' \cup D' = {}^{\omega}\omega.$$

Proof: Let $\langle A,B\rangle$ be a complete pair for Γ, and let $C,D \in \check{\Gamma}$. Consider the game $G(A,B,C,D)$ defined as follows:

I plays α, II plays β. II wins iff $\beta \in A \cup B$, $\alpha \in C \sim D \Rightarrow \beta \in A \sim B$, and $\alpha \in D \sim C \Rightarrow \beta \in B \sim A$. Equivalently, I wins iff $\beta \notin A \cup B$, or $\alpha \in C \sim D$ and $\beta \notin A \sim B$, or $\alpha \in D \sim C$ and $\beta \notin B \sim A$; in other words,

$$\beta \in A \cup B \Rightarrow (\beta \in B \ \&\ \alpha \in C \sim D, \text{ or } \beta \in A \ \&\ \alpha \in D \sim C).$$

Now I cannot win $G(A,B,C,D)$, for if he did, say by the strategy f, then letting $A'' = f^{-1}(\sim C)$ and $B'' = f^{-1}(\sim D)$, we would have $A \cup B \subseteq A'' \cup B''$, $A \sim B \subseteq A''$, $B \sim A \subseteq B''$, and $A'',B'' \in \Gamma$. Thus letting $A' = A \sim A''$, $B' = B \sim B''$, we would have $A',B' \in \Gamma$, with $A \cup B = A' \cup B'$, $A \sim B \subseteq A'$, $B \sim A \subseteq B'$, but this contradicts $\neg\operatorname{Red}(\Gamma)$.

So II wins $G(A,B,C,D)$, say by f. Let $C' = f^{-1}(A)$, $D' = f^{-1}(B)$. Then C' and D' are as desired, and the lemma is proved.

To prove Theorem 5.4 suppose toward a contradiction that Γ is closed under intersection, $\operatorname{Sep}_I(\check{\Gamma})$, and $\neg\operatorname{Red}(\Gamma)$. We shall derive the absurdity, $\operatorname{Red}(\check{\Gamma})$. So let $C,D \in \check{\Gamma}$. Let C', D' be as given by the lemma. Let $E \in \Gamma \cap \check{\Gamma}$ separate $\sim C'$ and $\sim D'$. Let $C'' = C \cap \sim E$, $D'' = D \cap E$. Then $\langle C'',D''\rangle$ reduces $\langle C,D\rangle$. ⊣

Proof of Theorem 5.5: Assume the hypotheses of the theorem, and suppose that $\neg\operatorname{Red}(\Gamma)$ and $\neg\operatorname{Red}(\check{\Gamma})$. We shall show that in fact Γ has the reduction property.

Let $C,D \in \Gamma$. Define $C_{-1} = C$, $D_{-1} = D$, and for all $n \geq 0$, let C_n and D_n be C'_{n-1}, D'_{n-1} as given by the lemma in the proof of Theorem 5.4. We have C_{2n}, $D_{2n} \in \Gamma$, $C_{2n+1}, D_{2n+1} \in \check{\Gamma}$, $C_{n+1} \subseteq C_n$ and $D_{n+1} \subseteq D_n$, for each $n \geq 0$. So $C'' = \bigcap_n C_n \in \Gamma \cap \check{\Gamma}$ and C'' separates $C \sim D$ and $D \sim C$. Let $C''' = C \cap C''$, $D''' = D \cap (\sim C'')$. Then $\langle C''', D'''\rangle$ reduces $\langle C,D\rangle$. ⊣

Proof of Theorem 5.6: By Theorem 5.5 we may suppose that Γ has the reduction property. We shall show $\operatorname{PWO}(\Gamma)$. It will be useful to look at the following way of coding sets in $\Delta = \Gamma \cap \check{\Gamma}$:

Let $\langle A,B\rangle$ be a universal pair for Γ. Let $\langle A',B'\rangle$ reduce $\langle A,B\rangle$. For any real γ, if $(\forall\alpha)[(\alpha,\gamma) \in A'$ or $(\alpha,\gamma) \in B']$, then let $C_\gamma = \{\alpha \mid (\alpha,\gamma) \in A'\}$. Then $\{\gamma \mid C_\gamma$ is defined$\} \in \Gamma$, and if C_γ is defined, $C_\gamma \in \Delta$.

Lemma. (Independently also due to Steel). Let Γ satisfy the hypotheses of the theorem, and suppose Δ is not closed under wellordered unions. Then $\mathrm{Red}(\Gamma) \Rightarrow \mathrm{PWO}(\Gamma)$.

Proof: Let λ be the least ordinal so that for some sequence $\langle A_\xi \mid \xi < \lambda\rangle$ of sets $A_\xi \in \Delta$, $\bigcup_{\xi<\lambda} A_\xi \notin \Delta$. Let $\Gamma^* = \{\bigcup_{\xi<\lambda} B_\xi \mid (\forall\xi < \lambda)B_\xi \in \Delta\}$. It is apparent from the minimality of λ, then, that $\mathrm{PWO}(\Gamma^*)$. Thus we need only show that $\Gamma^* \subseteq \Gamma$ to see that $\Gamma^* = \Gamma$, and hence $\mathrm{PWO}(\Gamma)$.

Let δ be the supremum of the lengths of prewellorderings of $\mathbb{R}$ that lie in Δ. Then $\lambda \geq \delta$ by Moschovakis [1970]. But $\lambda \leq \delta$. For if not, let $\bigcup_{\xi<\lambda} A_\xi \notin \Delta$. We may assume that $\xi < \eta < \lambda \Rightarrow A_\xi \subseteq A_\eta$ by taking cumulative unions and noting the minimality of λ. Now $\langle A_\xi \mid \xi < \delta\rangle$ provides a prewellordering in Δ of length δ, a contradiction. So $\lambda = \delta$.

Now we note that the order type of the Wadge degrees in Δ is exactly δ. Showing "$\leq$" is trivial. To see the other direction note that if $\prec$ is a prewellordering in Δ, then one may define by effective transfinite induction a $\leq_w$-increasing sequence of Δ sets of length $|\prec|$. The proof of this is a routine exercise given that there are continuous functions which, acting on codes of sets in Δ, give codes for the Δ sets derived from them by the operations of union, etc., which do not lead out of Δ. Indeed, the strategies which witness the corresponding closure properties of Γ provide such functions.

Now let C be the set of α for which C_α is defined. The relation $\{\langle\alpha,\beta\rangle \mid \alpha,\beta \in C \;\&\; (C_\alpha \leq_w C_\beta$ or $C_\alpha \leq_w {\sim}C_\beta)\}$ is in Γ, as is the complement of this relation relative to $C \times C$. Thus by the Main Lemma of Moschovakis [1970], $\Gamma^* \subseteq \Gamma$. ⊣

To finish the proof of Theorem 5.6 we must show that Δ is not closed under wellordered unions. Let $\mathfrak{F}_1,\ldots,\mathfrak{F}_8$ be Gödel's operations with the property that, in the terminology of Jech [1971], any almost universal transitive class closed under $\mathfrak{F}_1,\ldots,\mathfrak{F}_8$ is a model of ZF. Set $G_i(x,y,\alpha) = \mathfrak{F}_i(x,y)$, for $i = 1,\ldots,8$, and $G_9(x,y,\alpha) = x \cap \alpha$. Let (ξ,η,θ) be the rank of $\langle\xi,\eta,\theta\rangle$ in the Gödel wellordering and let I,J, and K be such that $\xi = (I(\xi),J(\xi),K(\xi))$. Then put

$$F(\xi,\alpha) = \begin{cases} \{F(\eta,\beta) \mid \eta < \xi \;\&\; \beta \in \mathbb{R}\}, & \text{if } I(\xi) = 0 \\ G_{I(\xi)}(F(J(\xi),(\alpha)_0),F(K(\xi),(\alpha)_1),(\alpha)_2), & \text{if } 0 < I(\xi) \leq 9 \\ \{F(J(\xi),(\alpha)_0),F(K(\xi),(\alpha)_1)\}, & \text{if } I(\xi) > 9. \end{cases}$$

Then $\{F(\xi,\alpha) \mid \xi \in \mathrm{On} \,\&\, \alpha \in \mathbb{R}\} = L[\mathbb{R}]$.

Now suppose that Δ is closed under wellordered unions. By induction on the Gödel wellordering of pairs $\langle \xi,\eta \rangle$ one can show that for each ξ and η the relations

$$P^{=}_{\xi,\eta}(\alpha,\beta) \iff F(\xi,\alpha) = F(\eta,\beta)$$

and

$$P^{\in}_{\xi,\eta}(\alpha,\beta) \iff F(\xi,\alpha) \in F(\eta,\beta)$$

are in Δ. But by the preceding paragraph this means that every set of reals in $L[\mathbb{R}]$ is in Δ, whence $\Gamma \not\subseteq L[\mathbb{R}]$. This contradiction establishes Theorem 5.6. ⊣

We list some other results which partake of the flavor of those presented here:

(Steel) Any two non-Borel analytic sets are Borel isomorphic.

(Steel) Jump operators on the Turing degrees are prewellordered by the Martin measure.

(Radin, Steel, Van Wesep) Any nonselfdual continuously closed class of sets of reals may be obtained by application of a fixed ω-ary Boolean operation to sequences of open sets.

The first two of these results appear in Steel [1977] and will be published elsewhere. The last appears in the union of Steel [1977] and Van Wesep [1977].

6. Conjectures and Problems

One should like to prove one of the following two competing conjectures:

(i) If Γ is a nonselfdual continuously closed class, then $\mathrm{Sep}_I\,\Gamma$ or $\mathrm{Sep}_I\,\check{\Gamma}$. Thus the classical separation principles serve to distinguish each nonselfdual continuously closed class from its dual.

(ii) If S is a set of Wadge degrees, then for some nonselfdual degree a, we have $a \in S \iff \check{a} \in S$.

The theory of the Wadge degrees seems to sorely need results of the following sort, of which there are now essentially no examples:

> some "closure" property of the order type of $\Gamma \cap \check{\Gamma}$ (e.g., that it is a cardinal) implies some closure property for Γ (e.g., closure under intersection) or even for $\Gamma \cap \check{\Gamma}$.

References

[1976] Kechris, A. S., and R. M. Solovay, Projective algebras and the prewellordering property in $L[\mathbb{R}]$, unpublished.

[1973] Martin, D. A., The Wadge degrees are wellordered, unpublished.

[1970] Moschovakis, Y. N., Determinacy and prewellorderings of the continuum, in: Y. Bar Hillel, ed., Mathematical Logic and Foundations of Set Theory, North-Holland, Amsterdam, p. 24.

[1975] Steel, J., $^{\mathcal{D}}\aleph_1$ and Martin's measure, unpublished.

[1977] Steel, J., Ph.D. Thesis, Berkeley.

[1977] Van Wesep, R., Ph.D. Thesis, Berkeley.

__________, Separation principles and the axiom of determinateness, J. Symb. Logic, to appear.

[197?] Wadge, W., Ph.D. Thesis, Berkeley.

THE INDEPENDENCE OF DC FROM AD[1]

Robert M. Solovay

Department of Mathematics
California Institute of Technology
Pasadena, CA 91125

and

Department of Mathematics
University of California
Berkeley, CA 94720

Introduction. We essentially prove, among other things, that the axiom of dependent choice (DC) does not follow from the axiom of determinacy (AD). The assumption that $ZF + AD_{\mathbb{R}}$ is consistent is required for this theorem, where $AD_{\mathbb{R}}$ is the axiom of determinacy for reals. In fact, the actual theorem proved is $\mathrm{Con}(ZF + AD_{\mathbb{R}}) \to \mathrm{Con}(ZF + AD_{\mathbb{R}} + \neg DC)$. This is obtained by demonstrating that $ZF + AD_{\mathbb{R}} + DC$ proves $\mathrm{Con}(ZF + AD_{\mathbb{R}})$, and then quoting Gödel.

The consistency of $ZF + AD_{\mathbb{R}}$ is proved by using $AD_{\mathbb{R}}$ to get indiscernibles for certain models of set theory, and then using $AD_{\mathbb{R}} + DC$ to show that some of these models satisfy $AD_{\mathbb{R}}$. Using the indiscernibles we can define the truth set of the models, hence proving the consistency of $AD_{\mathbb{R}}$. Other theorems along this line are proved similarly.

One of the main facts needed to prove these theorems is the close connection between the amount of choice available and the cofinality of Θ, where Θ is the least ordinal which is not the surjective image of $\mathbb{R}$. Indeed, to obtain inner models of $AD_{\mathbb{R}}$ thin enough to have a definable truth set we will need that the cofinality of Θ is $> \omega$. The necessary results in this direction are proved in §2.

This line of research was suggested by work of Ramez Sami directed towards proving DC from $AD + V = L[\mathbb{R}]$. In particular, the proof of Lemma 1.6 is modeled after arguments of Sami.

§0. Preliminaries

$\mathbb{R}$ is the set of reals, which for the purposes of this paper can be thought of as elements of ${}^{\omega}\omega$ or as subsets of ω. For $A \subseteq \mathbb{R}^{\omega}$, the game G_A is defined as follows: players I and II alternate turns picking reals, producing a sequence $\langle x_0, x_1, x_2 \ldots \rangle$. I wins if $\langle x_n : n \in \omega \rangle \in A$, II wins otherwise. G_A is determined if either I or II has a winning strategy. $AD_{\mathbb{R}}$ is the axiom asserting that $\forall A \subseteq \mathbb{R}^{\omega}$ [G_A is determined].

Clearly $AD_{\mathbb{R}}$ implies AD. It is of course inconsistent with full choice, but it does imply certain forms of restricted choice. In particular let $AC_{\mathbb{R}}$ be the following:

---

(1) The results in this paper were sketched in two lectures to the Caltech-UCLA logic seminar, and then presented in more detail to Greg Ennis, whose notes formed the basis for the final form of the presentation.

$(AC_{\mathbb{R}})$ $\forall \langle S_x : x \in \mathbb{R}\rangle$ such that $\forall x \in \mathbb{R}\,(S_x \neq \emptyset$ and $S_x \subseteq \mathbb{R})$

$\exists F : \mathbb{R} \to \mathbb{R}$ such that $\forall x\; F(x) \in S_x$.

Then $AD_{\mathbb{R}}$ implies $AC_{\mathbb{R}}$. (Proof: Consider the 2 step game in which II wins iff $x_1 \in S_{x_0}$. Clearly I has no w.s., so II's w.s. yields the desired choice function.)

Let X be any nonempty set. The axiom of dependent choice on X (DC_X) is as follows:

(DC_X) Let $\mathcal{S}$ be a collection of finite sequences from X, $\emptyset \in \mathcal{S}$ and $\langle x_1 \ldots x_n\rangle \in \mathcal{S} \Rightarrow \exists x \in X\; \langle x_1 \ldots x_n\; x\rangle \in \mathcal{S}$. Then $\exists f : \omega \to X$ such that $\forall n\; f \restriction n \in \mathcal{S}$.

We have $AC_{\mathbb{R}} \to DC_{\mathbb{R}}$ and hence $AD_{\mathbb{R}} \to DC_{\mathbb{R}}$.

The full axiom of dependent choice (DC) is $\forall X[X \neq \emptyset \to DC_X]$.

For $A,B \subseteq \mathbb{R}$ we say A is Wadge reducible to B $(A \leq_w B)$ if there is a continuous $f : \mathbb{R} \to \mathbb{R}$ such that $\forall x\; (x \in A \Leftrightarrow f(x) \in B)$. $A \equiv_w B$ is $A \leq_w B$ and $B \leq_w A$, and the Wadge degree of A is its Wadge equivalence class. The Wadge hierarchy is the collection of Wadge degrees with the induced partial order, and the modified Wadge hierarchy is obtained from the Wadge hierarchy by identifying the degree of A with the degree of $\mathbb{R} - A$. Wadge showed that AD + $DC_{\mathbb{R}}$ implies the modified Wadge hierarchy is a linear ordering, and Martin showed that in fact it's well ordered. For these and other facts see Van Wesep [5].

Definition 0.1. $\Theta = \sup\{\xi \in \mathrm{ORD} : \exists f : \mathbb{R} \xrightarrow{\text{onto}} \xi\}$. Under AC + GCH, $\Theta = \omega_2$. Under AD it is much larger. A major theme of this paper is that the cofinality of Θ is related to the amount of choice available. A simple but useful fact is the following:

Lemma 0.2 (AD + $DC_{\mathbb{R}}$). Θ = order type of the modified Wadge hierarchy.

Proof. Let γ be the order type of the Wadge hierarchy. If A has Wadge degree $\eta < \gamma$, then we can define $h : \mathbb{R} \xrightarrow{\text{onto}} \eta + 1$ by $h(x)$ = Wadge ordinal of the set Wadge reducible to A via the continuous function coded by x. Hence $\gamma \leq \Theta$.

To show $\Theta \leq \gamma$, we use the following fact: there is a uniform, canonical procedure which takes a set $A \subseteq \mathbb{R}$ to a set $A' \subseteq \mathbb{R}$ with A' of higher Wadge degree than $A, \mathbb{R} - A$. (Proof: let f_x be the continuous function coded by x and put $\langle 0,x\rangle \in A' \Leftrightarrow f_x(\langle 0,x\rangle) \notin A$, $\langle 1,x\rangle \in A' \Leftrightarrow f_x(\langle 1,x\rangle) \in A$). Now if $\lambda < \Theta$, let $\varphi : \mathbb{R} \xrightarrow{\text{onto}} \lambda$. Define $\{A_\eta : \eta \leq \lambda\} \subseteq \mathcal{P}(\mathbb{R})$ by induction on $\eta : A_\eta = \{\langle y,z\rangle : \varphi(y) < \eta \wedge z \in A_{\varphi(y)}\}'$. Then $\eta < \delta \leq \lambda \Rightarrow A_\eta <_w A_\delta$, hence $\lambda < \gamma$. ⊣

§1. Choice and the cofinality of Θ

Given $A \subseteq \mathbb{R}$, let $A_{x,y} = \{z \in \mathbb{R} : \langle x,y,z \rangle \in A\}$. Collection is the following choice-like axiom:

$$\forall x \in \mathbb{R}\, \exists A \subseteq \mathbb{R}\, \langle x,A \rangle \in U \rightarrow \exists B\, \forall x \in \mathbb{R}\, \exists y \in \mathbb{R}\, \langle x,B_{x,y} \rangle \in U.$$

Lemma 1.1. Collection implies Θ is regular.

Proof. If Θ is singular then $\exists G : \mathbb{R} \xrightarrow{\text{cofinal}} \Theta$. Thus $\forall x\, \exists A$ (A is a prewellordering of $\mathbb{R}$ of length $G(x)$). By collection $\exists B \subseteq \mathbb{R}$ such that

$$\forall x\, \exists y \in \mathbb{R}\, (B_{x,y} \text{ is a prewellordering of } \mathbb{R} \text{ of length } G(x)).$$

Define $F : \mathbb{R}^3 \rightarrow \Theta$ by

$$F(x,y,z) = \begin{cases} \text{rank of } z \text{ in } B_{x,y}, & \text{if } B_{x,y} \text{ is a pwo of } \mathbb{R} \\ 0 & \text{otherwise.} \end{cases}$$

Then F is onto, a contradiction. ⊣

Theorem 1.2 ($AD + DC_{\mathbb{R}}$). Collection is equivalent to the regularity of Θ.

Proof. Suppose $\forall x \in \mathbb{R}\, \exists A \subseteq \mathbb{R}\, \langle x,A \rangle \in U$. Let $f(x) =$ least ξ such that there is an $A \subseteq \mathbb{R}$ of Wadge degree ξ with $\langle x,A \rangle \in U$. Since Θ is regular the range of f is bounded by some $\gamma < \Theta$ (using, of course, Lemma 0.2). Let B be any set of Wadge degree γ. Let $G : \mathbb{R} \rightarrow \mathcal{P}(\mathbb{R})$ be $G(x) =$ the set Wadge reducible to B via the continuous function coded by x. Let

$$\langle x,y,z \rangle \in B^* \iff z \in G(y).$$

Then $\forall x\, \exists A \in \text{range } G\, \langle x,A \rangle \in U \Rightarrow \forall x\, \exists y\, \langle x,B^*_{x,y} \rangle \in U$. ⊣

Thus under $AD_{\mathbb{R}}$, the regularity of Θ is equivalent to a choice-like axiom. We next want to show that under $AD_{\mathbb{R}} + V = L[\mathcal{P}(\mathbb{R})]$, DC is equivalent to $\text{cof } \Theta > \omega$. One direction is easy and requires no strong hypotheses:

Lemma . DC implies $\text{cof } \Theta > \omega$.

Proof. Suppose $F : \omega \rightarrow \Theta$ cofinally. Then $\forall n\, \exists S$ (S is a prewellordering of $\mathbb{R}$ of length $F(n)$). By DC (in fact, by countable choice), let h be a function with domain ω such that $\forall n$ ($h(n)$ is a pwo of $\mathbb{R}$ of length $F(n)$). Now define $G : \mathbb{R} \xrightarrow{\text{onto}} \Theta$ by $G(x) =$ rank of $\lambda m\, x(m+1)$ in the pwo $h(x(0))$ (using Church's lambda notation and thinking of $\mathbb{R}$ as ${}^{\omega}\omega$). ⊣

Theorem 1.3. $AD_{\mathbb{R}} + V = L[\mathcal{P}(\mathbb{R})] + \text{cof } \Theta > \omega \vdash DC$.

Proof. Throughout the rest of Section 1, we assume $AD_{\mathbb{R}}$, $V = L[\wp(\mathbb{R})]$, $\operatorname{cof} \Theta > \omega$ and $\neg DC$, heading for a contradiction.

Lemma 1.4. $\neg DC_{\wp(\mathbb{R})}$.

Proof. This follows from the following facts whose proofs will be suppressed:

1) If $\exists h : X \xrightarrow{\text{onto}} Y$, then $DC_X \Rightarrow DC_Y$.

2) For all ordinals λ, $DC_X \Rightarrow DC_{X\times\lambda}$.

3) $V = L[\wp(\mathbb{R})] \Rightarrow \forall$ nonempty X $\exists \lambda \in \mathrm{ORD}$ $\exists h : \wp(\mathbb{R}) \times \lambda \xrightarrow{\text{onto}} X$. ⊣

Lemma 1.5. Θ is singular.

Proof. Fix a collection of tuples from $\wp(\mathbb{R})$ $\mathcal{S}_0$ which violates $DC_{\wp(\mathbb{R})}$. Define $\mathcal{S} \subseteq \mathcal{S}_0$ as follows: $\emptyset \in \mathcal{S}$, and if $\langle X_1,\ldots,X_n\rangle \in \mathcal{S}$, then

$$\langle X_1 \ldots X_n, Y\rangle \in \mathcal{S} \iff \text{a) } \langle X_1 \ldots X_n, Y\rangle \in \mathcal{S}_0$$
$$\text{and b) } |Y|_{\text{Wadge}} \text{ is minimal among things satisfying a).}$$

Clearly $\mathcal{S}$ is a counterexample to $DC_{\wp(\mathbb{R})}$.

Now for $\langle X_1 \ldots X_n\rangle \in \mathcal{S}$ define $\gamma_0(X_1 \ldots X_n) = |X_1 \ldots X_n|_{\text{Wadge}}$, and for $m > 0$ define

$$\gamma_m(X_1 \ldots X_m) = \sup\{|X_1 \ldots X_n, Y_1 \ldots Y_m|_{\text{Wadge}} : \langle X_1 \ldots X_n, Y_1 \ldots Y_m\rangle \in \mathcal{S}\}.$$

Claim: $\exists m\ \gamma_m(\emptyset) = \Theta$.

Proof of Claim. Suppose $\forall m\ \gamma_m(\emptyset) < \Theta$. Since $\operatorname{cof} \Theta > \omega$, $\exists \eta < \Theta$ such that $\sup\{\gamma_m(\emptyset) : m \in \omega\} < \eta$. Thus $\mathcal{S}$ is a collection of tuples from $\wp_\eta(\mathbb{R})$ where

$$\wp_\eta(\mathbb{R}) =_{\text{def}} \{A \subseteq \mathbb{R} : |A|_{\text{Wadge}} < \eta\}.$$

But since there is a map $\mathbb{R} \xrightarrow{\text{onto}} \wp_\eta(\mathbb{R})$, we have $AD_{\mathbb{R}} \Rightarrow DC_{\mathbb{R}} \Rightarrow DC_{\wp_\eta(\mathbb{R})}$, hence $\mathcal{S}$ cannot be a counterexample to $DC_{\wp(\mathbb{R})}$. ⊣ for claim.

Now let m be least such that $\exists \langle X_1 \ldots X_n\rangle \in \mathcal{S}$ with $\gamma_m\langle X_1 \ldots X_n\rangle = \Theta$. Fix such a $\langle X_1 \ldots X_n\rangle$. Clearly $m \geq 1$. Let $Q = \{Y : \langle X_1 \ldots X_n, Y\rangle \in \mathcal{S}\}$. For $Y \in Q$ define $\eta(Y) = \gamma_{m-1}(X_1 \cdots X_n, Y)$. By minimality of m, $\eta(Y) < \Theta$ for all $Y \in Q$. But then $\sup\{\eta(Y) : Y \in Q\} = \gamma_m(X_1 \ldots X_n) = \Theta$. Hence η is cofinal in Θ. But Q is a fixed subset of a fixed Wadge degree, hence there is a map $\mathbb{R} \xrightarrow{\text{onto}} Q$. Hence Θ is singular. ⊣

Lemma 1.6. Let $\lambda = \operatorname{cof} \Theta$. There is a countably additive measure ν on λ such that for some ordinal β, the ultrapower β^λ/ν is not well founded.

Proof. By Lemma 1.5 $\exists g : \mathbb{R} \xrightarrow{\text{onto}} \lambda$. Let $\mathcal{D}$ be the Turing degrees and μ the Martin measure on $\mathcal{D}$. Define $h : \mathcal{D} \to \lambda$ by

$$h(d) = \sup\{g(x) : x \leq_T d\}.$$

Let ν be the measure on λ induced by h. By $DC_{\mathbb{R}}$, ν is countably additive. Let $\mathcal{S}$ be a collection of tuples from $\mathcal{P}(\mathbb{R})$ violating $DC_{\mathcal{P}(\mathbb{R})}$. For $\xi < \Theta$, let $\mathcal{S}_\xi = \{\langle X_1 \dots X_n\rangle \in \mathcal{S} : \forall i \leq n |X_i|_{\text{Wadge}} < \xi\}$. Using $DC_{\mathcal{P}_\xi(\mathbb{R})}$ we obtain a rank function on $\mathcal{S}_\xi$ for $\xi < \Theta$ as follows: define an inductive operator Φ on $\mathcal{S}_\xi$ by

$$\langle X_1 \dots X_n\rangle \in \Phi(U) \Longleftrightarrow \forall Y[\langle X_1 \dots X_n, Y\rangle \in \mathcal{S}_\xi \to \langle X_1 \dots X_n, Y\rangle \in U].$$

Let Φ^∞ be the least fixed point of the operator Φ. Suppose $\mathcal{S}_\xi - \Phi^\infty \neq \emptyset$. Then

$$\forall \langle X_1 \dots X_n\rangle \in \mathcal{S}_\xi - \Phi^\infty \;\exists Y[\langle X_1 \dots X_n, Y\rangle \in \mathcal{S}_\xi - \Phi^\infty].$$

By $DC_{\mathcal{P}_\xi(\mathbb{R})}$ $\exists f \forall n\, f \restriction n \in \mathcal{S}_\xi - \Phi^\infty$. But then $\mathcal{S}$ does not violate $DC_{\mathcal{P}(\mathbb{R})}$.

Hence $\mathcal{S}_\xi \subseteq \Phi^\infty$ and we can define for $\langle X_1 \dots X_n\rangle \in \mathcal{S}_\xi$:

> $|X_1 \dots X_n|_\xi$ = least η such that $\langle X_1 \dots X_n\rangle$ occurs in the ηth iterate of Φ. (This is of course the standard rank function on $\mathcal{S}_\xi$).

Let $\beta = \sup\{|X_1 \dots X_n|_\xi : \xi < \Theta, \langle X_1 \dots X_n\rangle \in \mathcal{S}_\xi\}$. Let $k : \lambda \xrightarrow[\text{o.p.}]{\text{cofinal}} \Theta$, and for $\langle X_1 \dots X_n\rangle \in \mathcal{S}$ define

$$|X_1 \dots X_n|_\nu = [\lambda\xi |X_1 \dots X_n|_{k(\xi)}]_\nu \in \beta^\lambda/\nu .$$

Note that $|X_1 \dots X_n|_{k(\xi)}$ is actually only defined for ξ such that $k(\xi) > \sup\{|X_i|_{\text{Wadge}} : i \leq n\}$, but these ξ's form a set of ν-measure 1.

Clearly if $\langle X_1 \dots X_n, Y\rangle, \langle X_1 \dots X_n\rangle \in \mathcal{S}$ then $|X_1 \dots X_n, Y|_\nu < |X_1 \dots X_n|_\nu$. Hence $\{|X_1 \dots X_n|_\nu : \langle X_1 \dots X_n\rangle \in \mathcal{S}\}$ is a subset of β^λ/ν with no least element. ⊣

We can now complete the proof of Theorem 1.3. Fix $S \subseteq \beta^\lambda/\nu$, $S \neq \emptyset$ such that S has no least element. For $\xi < \Theta$, let S_ξ be those members of S which have representatives ordinal definable from a set of Wadge degree ξ and a real.

Claim. For all $\xi < \Theta$, every nonempty $A \subseteq S_\xi$ has a least element.

Proof of Claim. Fix ξ, $A \subseteq S_\xi$. Let $B = \{h \in \beta^\lambda : h$ is ordinal definable from C and a real, and $[h]_\nu \in A\}$, where C is a set of Wadge degree ξ. There is a fixed function F such that every element of B is $F(\eta, C, x)$ for some ordinal η and some real x. As B is a set we need only a set of such η's, hence B is the surjective image of $\mathbb{R} \times \alpha$ for some ordinal α. Hence DC_B. If A has no least element, then $\forall h \in B \exists g \in B([h]_\nu > [g]_\nu)$, so by DC_B there is a function G with domain ω such that

$$\forall n([G(n)]_\nu > [G(n+1)]_\nu)$$

which violates the countable additivity of ν. ⊣ claim

Let b_ξ = least element of S_ξ. Suppose $\exists \xi_0 < \Theta \forall \eta < \Theta\, b_{\xi_0} \leq b_\eta$. Since $V = L[\mathcal{P}(\mathbb{R})]$, $S = \bigcup_{\eta<\Theta} S_\eta$. Hence b_{ξ_0} is the least element of S, contrary to assumption on S.

Hence $\forall \xi < \Theta \exists \eta < \Theta[\eta > \xi$ and $b_\xi > b_\eta]$. Let $\xi_0 = 0$, ξ_{n+1} = least ordinal $> \xi_n$ such that $b_{\xi_{n+1}} < b_{\xi_n}$. Since $\operatorname{cof} \Theta > \omega$, $\gamma = \sup\{\xi_n : n < \omega\} < \Theta$. But then $\{b_{\xi_n} : n < \omega\}$ is a subset of S_γ with no least element. This contradiction completes the proof of Theorem 1.3. ⊣

§2. Inner models for $AD_{\mathbb{R}}$

For $\xi < \Theta$, let $\mathcal{P}_\xi(\mathbb{R}) = \{A \subseteq \mathbb{R} : |A|_{\text{Wadge}} < \xi\}$. The main result of this section is the construction of a function $g : \Theta \to \Theta$ such that if ξ is closed under G then $L[\mathcal{P}_\xi(\mathbb{R})] \models AD_{\mathbb{R}}$ (assuming $AD_{\mathbb{R}}$). $L[\mathcal{P}_\xi(\mathbb{R})]$ is the smallest model of ZF containing $\mathbb{R}$ and $\mathcal{P}_\xi(\mathbb{R})$.

Lemma 2.1. ($AD_{\mathbb{R}}$). Let $A \subseteq \mathbb{R}$. Then $\exists$ $B \subseteq \mathbb{R}$ such that for every game Wadge reducible to A there is a winning strategy for some player which is Wadge reducible to B.

Proof. Consider the following game: I plays x_0 which codes a game G_{x_0} (i.e. x_0 codes the continuous function which reduces the payoff set of G_{x_0} to A). II then chooses to be player I or II in G_{x_0}. Then, I, II play G_{x_0}. Clearly I has no winning strategy, since if he did, some G_{x_0} would have winning strategies for both players. So II has a winning strategy, which can be coded as a $B \subseteq \mathbb{R}$. Clearly this B works. ⊣

Lemma 2.2. ($AD_{\mathbb{R}}$). $\forall A \subseteq \mathbb{R}$ $\exists B \subseteq \mathbb{R}$ (B is not ordinal definable from A and a real).

Proof. Suppose A is such that $\forall B \subseteq \mathbb{R}$ (B is ordinal definable from A and a real). For each $x \in \mathbb{R}$, let $S_x = \{y \in \mathbb{R} : y$ is ordinal definable from $x, A\}$. By Myhill-Scott, S_x is wellorderable for each x. So by AD, each S_x is countable. By $AC_{\mathbb{R}}$ there is a function F such that $\forall x \in \mathbb{R}\, F(x) \in \mathbb{R} - S_x$. By assumption, F is ordinal definable from A and a real x_0, hence $F(x_0)$ is ordinal definable from A and x_0, hence $F(x_0) \in S_{x_0}$, contradiction. ⊣

Definition 2.3. For $A \subseteq B \subseteq \mathbb{R}$, let $L[A,B]$ be the smallest model of ZF containing all ordinals, containing B as a subset and A as an element.

Corollary 2.4. ($AD_{\mathbb{R}}$). $\forall A \subseteq \mathbb{R}$ $V \neq L[A,\mathbb{R}]$.

Proof. $V = L[A,\mathbb{R}]$ implies that everything is ordinal definable from A and a real. ⊣

We shall see later that in fact $AD_{\mathbb{R}}$ implies "$A^{\#}$ exists" for every $A \subseteq \mathbb{R}$.

Theorem 2.5. ($AD_{\mathbb{R}}$). There is a monotone $G : \Theta \to \Theta$, such that if $\eta < \Theta$ is closed under G, then $L[\mathcal{P}_\eta(\mathbb{R})] \models AD_{\mathbb{R}}$ and $\mathcal{P}(\mathbb{R}) \cap L[\mathcal{P}_\eta(\mathbb{R})] = \mathcal{P}_\eta(\mathbb{R})$.

Proof. Let $G_0(\eta)$ = least ξ such that no set of reals of Wadge degree ξ is ordinal definable from a real and a set of Wadge degree η.

Let $G_1(\eta)$ = least ξ such that every game of Wadge complexity $\leq \eta$ has a winning strategy of complexity $\leq \xi$.

Let $G(\eta) = \max(G_0(\eta), G_1(\eta))$. Suppose $G : \eta \to \eta$. If $B \subseteq \mathbb{R}$ is in $L[\mathcal{P}_\eta(\mathbb{R})]$, then $\exists A$ with $|A|_{\text{Wadge}} = \delta < \eta$ such that in $L[\mathcal{P}_\eta(\mathbb{R})]$, B is ordinal definable from A and a real. Hence B is ordinal definable from A and a real. Hence $|B|_{\text{Wadge}} \leq G_0(\delta) < \eta$, i.e., $B \in \mathcal{P}_\eta(\mathbb{R})$.

Now given any game in $L[\mathcal{P}_\eta(\mathbb{R})]$, by the above its Wadge complexity is $< \eta$. Since η is closed under G_1, there is a winning strategy for the game in $\mathcal{P}_\eta(\mathbb{R})$. Hence the game is determined in $L[\mathcal{P}_\eta(\mathbb{R})]$, i.e. $L[\mathcal{P}_\eta(\mathbb{R})] \models AD_{\mathbb{R}}$. ⊣

§3. A measure on the set of countable subsets of $\mathbb{R}$

Let $\Omega = \{S \subseteq \mathbb{R} : S \text{ countable}\}$. For $A \subseteq \Omega$, consider the following game G_A: I and II alternately play (possibly empty) finite subsets of $\mathbb{R}$, and II wins iff the countable set produced by both players is in A. Since finite subsets of $\mathbb{R}$ can be coded as individual reals, $AD_{\mathbb{R}}$ implies $\forall A \subseteq \Omega$ (G_A is determined).

Define for $A \subseteq \Omega$,

$$A \in U \iff \text{II has a winning strategy in } G_A.$$

Lemma 3.1. ($AD_{\mathbb{R}}$). U is a countably complete ultrafilter on Ω such that

1. $\forall x \in \mathbb{R}\{S : x \in S\} \in U$

2. (Normality) Let $T : \Omega \to \Omega$ satisfy $\forall S \in \Omega$, $T(S) \subseteq S$, and $\{S : T(S) \neq \emptyset\} \in U$. Then $\exists x \in \mathbb{R}$ such that $\{S : x \in T(S)\} \in U$.

Proof. Clearly $B \supseteq A \in U \Rightarrow B \in U$. Suppose $A \notin U$. Let τ be a w.s. for I in G_A. II wins $G_{\Omega-A}$ as follows: I plays s_0. II ignores this play and plays s_1 according to τ. I plays s_2. II plays $s_3 = \tau(s_0 \cup s_2)$. From this point on II plays τ against I's moves. Thus $A \notin U \iff \Omega - A \in U$.

To show U is countably complete: suppose $A_n \in U$ for $n \in \omega$, and then I and II play G_{A_n}, we see that there is a sequence of strategies τ_n so that $\forall n$ (τ_n is a w.s. for II in G_{A_n}). Let $\langle\ ,\ \rangle : \omega \times \omega \xrightarrow[\text{onto}]{1-1} \omega$ be a recursive pairing function such that $m < k \Rightarrow \langle n,m\rangle < \langle n,k\rangle$. Let $B = \bigcap_n A_n$, and II can win G_B as follows: for all $n,m < \omega$, on the $\langle n,m\rangle^{th}$ move of II, II plays according to τ_n, treating all reals played by either player since II's $\langle n,m-1\rangle$th move as a single play of I. Thus on II's plays at steps $\langle n,0\rangle, \langle n,1\rangle, \ldots$ II is ensuring that the set produces is in A_n.

Thus U is a countably complete ultrafilter. Property 1 is obvious. To show 2, it is easy to check that 2 is equivalent to the following diagonal intersection property.

2'. Let $\langle M_x : x \in \mathbb{R}\rangle$ be a collection of elements of U. Then $\{S \in \Omega : \forall x \in S(S \in M_x)\} \in U$.

To verify 2', first note that by considering the game in which I passes, II plays $x \in \mathbb{R}$, and then I and II play G_{M_x}, we get a collection of strategies indexed by reals $\{\tau_x : x \in \mathbb{R}\}$ such that for all x, τ_x is a w.s. for II in G_{M_x}. The proof of 2' is now similar to the proof of countable completeness - again, II must ensure that the set produced is in each of countably many sets, i.e. S must be in M_x for each $x \in S$. The difference between this and the countable completeness situation is that the countably many sets II is trying to get into are not given before the game is played but are determined as play progresses. But this is no real problem - each time an x appears in a play of either player, II resolves to play via τ_x infinitely often (i.e. a dovetailing procedure). When II plays τ_x for some x, all moves of both players since the last time II used τ_x are treated as a single move of player I. By playing in this way II ensures that $\forall x \in S\ (S \in M_x)$. ⊣

§4. Sharps for sets of reals

Let $A \subseteq S \subseteq \mathbb{R}$. By $L[A,S]$ we mean the constructible universe built above S as a set of urelements with A as an additional predicate. U will always denote a normal ultrafilter on $\Omega = \{S \subseteq \mathbb{R} : S$ countable$\}$ as in §3.

<u>Lemma</u> 4.1 ($AD_{\mathbb{R}}$). For any $A \subseteq \mathbb{R}$, $\{S \in \Omega : \mathbb{R} \cap L[A \cap S,S] = S\} \in U$.

<u>Proof</u>. Suppose not. Then for almost all S, $\exists y \in \mathbb{R} \cap L[A \cap S,S] - S$. Fixing any standard coding of pairs of reals by reals, it is easy to see by normality that S is closed under the coding for almost all S. For such S we have a canonical map τ enumerating $L[A \cap S,S]$, i.e.

$$L[A \cap S,S] = \{\tau(\xi,S,x) : \xi \in \mathrm{ORD},\ x \in S\}.$$

Thus for almost all $S\ \exists x \in S\ \exists \xi\ \tau(\xi,A,x) \notin S$. Let $T(S) = \{x \in S : \exists \xi\ \tau(\xi,S,x) \notin S\}$. By normality $\exists x_0 \in \mathbb{R}$ such that $\{s : x_0 \in T(S)\} \in U$. Let ξ_S = least ξ such that $\tau(\xi,S,x_0) \notin S$. Let $y_S = \tau(\xi_S,S,x_0)$. Thinking of $\mathbb{R}$ as $\mathcal{P}(\omega)$ put $n \in y \iff \{S : y_S = y\} \in U$. By countable additivity, $\{S : y_S = y\} \in U$. But then

$$\{S : y \notin S\} \in U, \text{ a contradiction.} \qquad \dashv$$

Henceforth we will concentrate on those S,A such that $\mathbb{R} \cap L[A \cap S,S] = S$. Assume that $A \subseteq S$. For such A,S we want to define a set of reals $\langle A,S\rangle^{\#}$ so that $\langle A,S\rangle^{\#}$ exists iff $L[A,S]$ contains a class of ordinal indiscernibles with certain properties. $\langle A,S\rangle^{\#}$ will consist of the codes of a certain set of formulas of a language $\mathcal{L}_S$ (we need reals to code formulas since we need constants for the reals in S). The main difference between the present situation and the situation involving L and $0^{\#}$ is the following: $ZF + V = L$ has definable Skolem functions, whereas $ZF + V = L[\mathbb{R}]$ does not. However, $ZF + V = L[\mathbb{R}]$ does imply that everything is definable from $ORD \cup \mathbb{R}$, and this is how we will generate submodels from classes of ordinals.

The language $\mathcal{L}_S$ is formally defined as follows: For $S \subseteq \mathbb{R}$ we can uniformly (i.e. not depending on S) code the language of set theory as reals, with the following additional distinguished terms inductively defined:

1. $\langle 0,x\rangle$ is a term for $x \in S$ (in the models $L[A,S]$, $\langle 0,x\rangle$ will denote x).
2. $\langle 1,0\rangle$ is a term ($\langle 1,0\rangle$ will denote A).
3. For $j < \omega$, $\langle 2,j\rangle$ is a term ($\langle 2,j\rangle$ denotes the jth indiscernible).
4. Let e be the code of a formula φ having at most the variables $v_1 \ldots v_m$ free (containing no terms). Then for all terms $t_1 \ldots t_m$, $\langle 3,e,t_1 \ldots t_m\rangle$ is a term. (If t_i denotes x_i in $L[A,s]$, and if $L[A,s] \models \exists x\ \varphi(x,x_1 \ldots x_m)$, then $\langle 3,e,t_1 \ldots t_m\rangle$ denotes this x. It denotes 0 otherwise.)

Recall the construction of the models $\Gamma(0^{\#},\xi)$: $0^{\#}$ is a set of sentences in a language with constants $\langle c_n : n < \omega\rangle$. To get $\Gamma(0^{\#},\xi)$ we form a new language with constants $\langle d_\eta : \eta < \xi\rangle$, and then we define an equivalence relation on terms of the form

$$\tau_\varphi(d_{\eta_1} \ldots d_{\eta_k})$$

(where τ_φ is a Skolem function) by referring to $0^{\#}$ on similar terms involving the c_n's. In our context, if Σ is a set of sentences in $\mathcal{L}_S$ including the sentences asserting that the terms $\langle 2,j\rangle, j < \omega$ are indiscernibles, we form $\Gamma(\Sigma,\xi)$ by forming a new language with terms $\langle 2,\eta\rangle$, $\eta < \xi$. An equivalence relation is then defined on the terms in the new language by referring to Σ on similar terms involving $\langle 2,j\rangle$, $j < \omega$. It is left to the reader to fill in the details of this construction.

A crucial fact about the models $\Gamma(0^{\#},\xi)$ is that $\Gamma(0^{\#},\xi) \models \varphi(d_{\eta_1} \cdots d_{\eta_n}) \Longleftrightarrow \varphi(c_{j_1} \cdots c_{j_n}) \in 0^{\#}$, where $j_1 \cdots j_n < \omega$ are similarly ordered as $\eta_1 \cdots \eta_n$. The proof is by induction on φ, with the definable Skolem functions used for the existential quantifier case. To get the analogous result for $\langle A,S\rangle^{\#}$ we must stipulate as one of our axioms on $\langle A,S\rangle^{\#}$ that it satisfy the <u>witness condition</u> (*):

<u>Definition</u> 4.2. If $\Sigma \subseteq \mathcal{L}_S$, we say Σ satisfies the <u>witness condition</u> if:

(*) Whenever $\exists x\varphi(x) \in \Sigma$, then for some term t involving no indiscernibles not appearing in $\varphi(x)$, $\varphi(t) \in \Sigma$.

We can now make the following definition:

<u>Definition</u> 4.3. Let $A \subseteq S \subseteq \mathbb{R}$. A set of sentences $\Sigma \subseteq \mathcal{L}_S$ is a <u>remarkable character</u> for A,S if

1. Σ is a complete, consistent extension of $ZF + V = L[A,S]$.
2. Σ satisfied the obvious analogues of the syntactical conditions required of $0^{\#}$ (including the "remarkable" condition).
3. For some uncountable $\xi, \Gamma(\Sigma,\xi)$ is wellfounded.
4. Σ satisfies the witness condition (*).

As expected, the remarkable character for A,S is unique (if it exists), and is denoted $\langle A,S\rangle^{\#}$. The existence of $\langle A,S\rangle^{\#}$ is equivalent to the existence of a canonical closed, unbounded class I of indiscernibles for $L[A,S]$, which generates $L[A,S]$ in the sense that everything in $L[A,S]$ is definable from $I \cup S \cup \{A\}$.

Now suppose $S \subseteq \mathbb{R}$ is countable. Then for any $A \subseteq \mathbb{R}$, $\langle A \cap S,S\rangle$ can be coded into a single real x. Indiscernibles for $L[x]$ are thus also indiscernibles for $L[A \cap S,S]$. Thus since AD yields indiscernibles for $L[x]$ for all real x, we get $AD \rightarrow \forall S, A \subseteq \mathbb{R}$ [S countable $\Rightarrow \langle A \cap S,S\rangle^{\#}$ exists].

<u>Theorem</u> 4.4 ($AD_{\mathbb{R}}$). $\forall A \subseteq \mathbb{R}\ \langle A,\mathbb{R}\rangle^{\#}$ exists.

<u>Proof</u>. For a sentence φ of the language $\mathcal{L}_{\mathbb{R}}$, put $\varphi \in B \Longleftrightarrow \{S : S$ countable, $L[A \cap S,S] \cap \mathbb{R} = S$ and $\varphi \in \langle A \cap S,S\rangle^{\#}\} \in U$. We claim that B is $\langle A,\mathbb{R}\rangle^{\#}$. We must show that B is a complete, consistent theory, certain sentences are in it, B satisfies the witness condition (*), and the model $\Gamma(B,\omega_1)$ is wellfounded. Most of this routine - for example, to check that B is complete, suppose $\varphi \notin B$. Then $\{S : \varphi \in \langle A \cap S,S\rangle^{\#}\} \notin U$, so $\{S : \varphi \notin \langle A \cap S,S\rangle^{\#}\} \in U$, hence $\{S : \neg\varphi \in \langle A \cap S,S\rangle^{\#}\} \in U$.

To check the witness condition: suppose $\exists y\ \varphi(y) \in B$. Thus $\{S : \exists y\ \varphi(y) \in \langle A \cap S,S\rangle^{\#}\} \in U$. Since each $\langle A \cap S,S\rangle^{\#}$ satisfies the witness condition, $\{S : \exists x \in S\ \exists$ term t_S (t_S "contains" only reals recursive in x and $\varphi(t_S) \in \langle A \cap S,S\rangle^{\#})\} \in U$. Let

$T(S) = \{x \in S : \exists$ term t_S containing only reals recursive in x such that $\varphi(t_S) \in \langle A \cap S, S\rangle^{\#}\}$.

Thus $\{S : T(S) \neq \emptyset\} \in U$, so by normality there is a fixed $x_0 \in \mathbb{R}$ such that $x_0 \in T(S)$ a.e. (U). But there are only countably many terms containing only reals recursive in x_0, so by countable additivity there is a fixed term t_0 so that $\varphi(t_0) \in \langle A \cap S, S\rangle^{\#}$ a.e. (U). Hence $\varphi(t_0) \in B$.

To check that $\Gamma(B,\omega_1)$ is well-founded: Any subset of the ordinals of $\Gamma(B,\omega_1)$ with no least element is the surjective image of $\mathbb{R} \times \xi$ for some (real) ordinal ξ. Hence $DC_{\mathbb{R}}$ implies there is an infinite descending chain. But then countable additivity of U yields an infinite descending chain in $\Gamma(\langle A \cap S, S\rangle^{\#}, \omega_1)$ for some S, a contradiction. $\dashv$

§5. The main theorems

Lemma 5.1. Let $A \subseteq \mathbb{R}$, $\lambda_1 < \lambda_2$ limit ordinals, with $i : \lambda_1 \to \lambda_2$ the inclusion map. Let $M_1 = \Gamma(A^{\#},\lambda_1)$, $M_2 = \Gamma(A^{\#},\lambda_2)$, and let $i_* : M_1 \to M_2$ the induced elementary embedding. Let $a \in OR^{M_1}$, $b \in OR^{M_2}$, $b <_{M_2} i_*(a)$. Then for some $b' \in OR^{M_1}$, $b = i_*(b')$.

Proof. WLOG a is an indiscernible in M_1, i.e. $a = [c_\alpha]$ for some ordinal $\alpha < \lambda_1$, where $\{c_\xi : \xi < \lambda_1\}$ are the constants denoting the indiscernibles in the language used to construct $\Gamma(A^{\#},\lambda_1)$, and [ ] denotes equivalence class. Suppose $b = [\tau(y,c_{\alpha_1} \dots c_{\alpha_m}, c_{\gamma_1} \dots c_{\gamma_n})]$, where τ is a term, $y \in \mathbb{R}$, $\alpha_1 < \dots < \alpha_m \leq \gamma_1 < \dots < \gamma_n$. Since λ_1 is limit, we can find $\gamma_1' < \dots < \gamma_n' < \lambda_1$ with $\alpha \leq \gamma_1'$. Since $A^{\#}$ is remarkable,

$$b = [\tau(y,c_{\alpha_1} \dots c_{\alpha_m}, c_{\gamma_1'} \dots c_{\gamma_n'})]$$

so b is in the range of i_*. $\dashv$

Lemma 5.2. Let $\lambda_1 < \lambda_2$ be limit ordinals, and let $\langle i_\xi : \xi \in ORD\rangle$ be the canonical class of indiscernibles for $L[A,\mathbb{R}]$. Then $L_{i_{\lambda_1}}[A,\mathbb{R}] \prec L_{i_{\lambda_2}}[A,\mathbb{R}]$.

Proof. Let M_1, M_2, i_* be as in 5.1, and let $M_1 \simeq L_{\gamma_1}[A,\mathbb{R}]$, $M_2 \simeq L_{\gamma_2}[A,\mathbb{R}]$ via transitive collapse. Let $j_* : L_{\gamma_1}[A,\mathbb{R}] \to L_{\gamma_2}[A,\mathbb{R}]$ be the elementary embedding induced by i_*. By 5.1, j_* is the identity on reals and ordinals, hence j_* is the inclusion map. By a proof similar to that of 5.1, if $b \in OR^{M_2}$ is less than $[c_{\lambda_1}]$, then $b \in$ range i_*. Hence $\gamma_1 =$ image of $[c_{\lambda_1}] = i_{\lambda_1}$, and similarly $\gamma_2 = i_{\lambda_2}$. Hence $L_{i_{\lambda_1}}[A,\mathbb{R}] \prec L_{i_{\lambda_2}}[A,\mathbb{R}]$. $\dashv$

Corollary 5.3. If $\langle A,\mathbb{R}\rangle^{\#}$ exists then the truth set of $L[A,\mathbb{R}]$ is definable.

Proof. Using 5.2, as in the case of indiscernibles for L.

Theorem 5.4. ZF + $AD_{\mathbb{R}} \vdash Con(ZF + AD)$.

Proof. $AD_{\mathbb{R}}$ implies $L[\mathbb{R}] = L[\emptyset,\mathbb{R}] \models AD$, and since $\langle\emptyset,\mathbb{R}\rangle^{\#}$ exists, the truth set of $L[\mathbb{R}]$ is definable. $\dashv$

Lemma 5.5. Let $A \subseteq \mathbb{R}$ have Wadge degree η, and let I be the canonical class of indiscernibles for $L[A,\mathbb{R}]$. Then for $\gamma < \delta$ in I, $L_\gamma[\mathcal{P}_\eta(\mathbb{R})] \prec L_\delta[\mathcal{P}_\eta(\mathbb{R})]$.

Proof. $L[\mathcal{P}_\eta(\mathbb{R})]$ is a definable subclass of $L[A,\mathbb{R}]$, and η is definable in $L[A,\mathbb{R}]$, since every set of lower Wadge degree is in $L[A,\mathbb{R}]$, and the Wadge ordering is absolute. Hence η is less than every indiscernible.

Suppose $x_1 \dots x_n \in L_\gamma[\mathcal{P}_\eta(\mathbb{R})]$, and let φ be a formula. Let $\psi(\lambda,\xi,x_1 \dots x_n) \Leftrightarrow L_\lambda[\mathcal{P}_\xi(\mathbb{R})] \models \varphi(x_1 \dots x_n)$. Since η is less than every indiscernible, we can find $\gamma_1 < \dots < \gamma_m \in I$ so that γ,δ, and every indiscernible necessary to define $x_1 \dots x_n$ are less than γ_1 and $\eta = \tau(A,y,\gamma_1 \dots \gamma_m)$ for some $y \in \mathbb{R}$. Hence in $L[A,\mathbb{R}]$,

$$\psi(\gamma,\tau(A,y,\gamma_1 \dots \gamma_m),x_1 \dots x_n) \Leftrightarrow$$

$$\psi(\delta,\tau(A,y,\gamma_1 \dots \gamma_m),x_1 \dots x_n),$$

hence $L_\gamma[\mathcal{P}_\eta(\mathbb{R})] \models \varphi(x_1 \dots x_n) \Leftrightarrow L_\delta[\mathcal{P}_\eta(\mathbb{R})] \models \varphi(x_1 \dots x_n)$. $\dashv$

Corollary 5.6. $AD_{\mathbb{R}}$ implies the truth set of $L[\mathcal{P}_\eta(\mathbb{R})]$ is definable, for all $\eta < \Theta$.

Proof. As usual, using 5.5. $\dashv$

Theorem 5.7. ZF + $AD_{\mathbb{R}}$ + $cof(\Theta) > \omega \vdash Con(AD_{\mathbb{R}})$.

Proof. Work in $L[\mathcal{P}(\mathbb{R})]$. Let G be the function given by Theorem 2.5. Then $cof(\Theta) > \omega$ implies $\exists\, \eta < \Theta$ (η is closed under G). Hence $L[\mathcal{P}_\eta(\mathbb{R})] \models AD_{\mathbb{R}}$. Since the truth set of $L[\mathcal{P}_\eta(\mathbb{R})]$ is definable, the theorem follows.

Corollary 5.8. ZF + $AD_{\mathbb{R}}$ + DC $\vdash Con(ZF + AD_{\mathbb{R}})$.

Proof. DC implies $cof\ \Theta > \omega$. $\dashv$

Theorem 5.9. ZF + $AD_{\mathbb{R}}$ + Θ regular $\vdash Con(ZF + AD_{\mathbb{R}} + DC)$.

Proof. Work in $L[\mathcal{P}(\mathbb{R})]$, and let G be as before. As Θ is regular, $\exists\, \eta < \Theta$ of cofinality ω_1 which is closed under G. Then $L[\mathcal{P}_\eta(\mathbb{R})] \models AD_{\mathbb{R}}$. But by Theorem 2.5, $L[\mathcal{P}_\eta(\Theta)] \models \eta = \Theta$, hence $L[\mathcal{P}_\eta(\mathbb{R})] \models cof\ \Theta > \omega$. So by Theorem 1.3,

$L[\wp_\eta(\mathbb{R})] \models DC$. By Corollary 5.6, the truth set of $L[\wp_\eta(\mathbb{R})]$ is definable, hence the theorem follows. ⊣

By Gödel we thus get the following:

Theorem 5.10.

1. If ZF + AD is consistent, then ZF + AD does not prove $AD_{\mathbb{R}}$.
2. If ZF + $AD_{\mathbb{R}}$ is consistent, then ZF + $AD_{\mathbb{R}}$ does not prove $\operatorname{cof} \Theta > \omega$, hence ZF + $AD_{\mathbb{R}}$ does not prove DC.
3. If ZF + $AD_{\mathbb{R}}$ + DC is consistent, then ZF + $AD_{\mathbb{R}}$ + DC does not prove Θ is regular.

§6. Open questions

1. Does AD imply $DC_{\mathbb{R}}$? We conjecture that it does not.

2. Recall the functions G_0 and G_1 used to define the function G of Section 4. Is $G_0(\eta) < G_1(\eta)$?

It is known (Mycielski-Blass) that $AD_{\mathbb{R}}$ is equivalent to $AD(\omega^2)$, which asserts the determinacy of all games in which players play integers in a sequence of ω^2 moves. (See Blass [1].) It is known that the analogous axiom $AD(\omega_1)$ is false.

3. Conjecture: $AD(\omega^3) \vdash \mathrm{Con}(AD(\omega^2) + \Theta \text{ regular})$.

4. $AD(\omega^3)$ should give stronger measures.

5. Does $AD_{\mathbb{R}} + \Theta$ regular imply every set of reals has a scale? (For the notion of a scale see Kechris [2].) We know that AD does not imply every set has a scale, in fact $V = L[\mathbb{R}]$ yields a set without a scale.

REFERENCES

1. A. Blass, Equivalence of two strong forms of determinacy, Proc. A.M.S. 52, 1975, 373-376.
2. A. Kechris, AD and projective ordinals, this volume.
3. Y. Moschovakis, Determinacy and prewellorderings of the continuum, in Mathematical logic and foundations of set theory, edited by Y. Bar Hillel, North Holland, Amsterdam, 1970, 24-62.
4. J. Mycielski, On the axiom of determinateness I, II; Fund. Math. 53, 1963-64, 205-224, Fund. Math. 59, 1966, 203-212.
5. R. Van Wesep, Wadge degrees and descriptive set theory, this volume.

INDUCTIVE SCALES ON INDUCTIVE SETS

Yiannis N. Moschovakis
Department of Mathematics
University of California
Los Angeles, California 90024

Let $\mathfrak{X} = X_1 \times \dots \times X_n$ be any product of copies of ω and $\mathcal{R} = {}^{\omega}\omega$. A pointset $P \subseteq \mathfrak{X}$ is inductive if there is a projective set $Q \subseteq \mathfrak{X} \times \mathcal{R}$ such that

(1) $$P(x) \iff \{(\forall\alpha_1)(\exists\alpha_2)(\forall\alpha_3)(\exists\alpha_4)\dots\}(\exists n)Q(x,\langle\alpha_1,\dots,\alpha_n\rangle),$$

where $\langle\alpha_1,\dots,\alpha_n\rangle$ is the usual (recursive) coding of tuples and the (open) game quantifier is interpreted in the obvious way. If Q is analytical in some $\beta_0 \in \mathcal{R}$, we call P inductive in β_0 and if we can choose β_0 recursive, we call P absolutely inductive. Sets which are both inductive and coinductive are hyperprojective; "hyperprojective in β_0" and "absolutely hyperprojective" sets are defined in the obvious way.

The theory of inductive sets (on arbitrary structures) is developed in some detail in [4] which we will cite EIAS. Our purpose here is to outline a proof of the following.

Main Theorem. If every hyperprojective game is determined, then every absolutely inductive pointset admits an absolutely inductive scale; it follows that inductive sets admit inductive scales and hyperprojective sets admit hyperprojective scales.

Part of the interest in this result lies in the fact that the collection of inductive sets is the largest collection of pointsets for which we can presently establish the scale property, from any hypotheses.

1. Proof of the Main Theorem

We assume that the tupling function $\langle\alpha_1,\dots,\alpha_n\rangle$ is defined for the empty tuple $(n = 0)$,

$$\langle\ \rangle = \lambda t 1$$

and that concatenation is given by a recursive function $*$ on the codes,

$$\langle\alpha_1,\dots,\alpha_n\rangle * \langle\beta_1,\dots,\beta_m\rangle = \langle\alpha_1,\dots,\alpha_n,\beta_1,\dots,\beta_m\rangle.$$

We will prove the main theorem in a sequence of simple lemmas.

Generalizing slightly the definition (1) above, for any given $Q \subseteq \mathfrak{X} \times \mathcal{R}$, put

(2) $$R(x,\alpha) \iff \{(\forall\alpha_1)(\exists\alpha_2)(\forall\alpha_3)(\exists\alpha_4)\ldots\}(\exists n)Q(x,\alpha * \langle\alpha_1,\ldots,\alpha_n\rangle)$$

and for each ordinal ξ define by induction

(3) $$R_\xi(x,\alpha) \iff Q(x,\alpha) \vee (\forall\beta)(\exists\gamma)(\exists\eta < \xi)R_\eta(x,\alpha*\langle\beta,\gamma\rangle).$$

Lemma 1. $R(x,\alpha) \iff (\exists\xi)R_\xi(x,\alpha)$.

Proof. First check by a simple induction on ξ that

$$R_\xi(x,\alpha) \Rightarrow R(x,\alpha).$$

For the converse, assume $(\forall\xi)\neg R_\xi(x,\alpha)$ and show by applying (3) repeatedly that in that case

$$\{(\exists\alpha_1)(\forall\alpha_2)(\exists\alpha_3)(\forall\alpha_4)\ldots\}(\forall n)\neg Q(x,\alpha * \langle\alpha_1,\ldots,\alpha_n\rangle)$$

which is equivalent to $\neg R(x,\alpha)$. ⊣

It follows from results of EIAS that if Q is analytical (or projective), then

$$R(x,\alpha) \iff (\exists\xi < \varkappa)R_\xi(x,\alpha),$$

where $\varkappa$ is the closure ordinal for positive elementary inductive definitions on $\mathcal{R}$, or alternatively,

$$\varkappa = \text{supremum}\{\text{rank}(\leq) : \leq \text{ is a hyperprojective prewellordering of } \mathcal{R}\}.$$

Suppose now that R and R_ξ are defined by (2) and (3) and suppose we are given a scale $\bar{\varphi}^0 = \{\varphi_n^0\}_{n\in\omega}$ on $Q = R_0$. We will define by induction on ξ a scale $\bar{\varphi}^\xi$ on each R_ξ.

Assuming that $\bar{\varphi}^\eta$ has been defined for each $\eta < \xi$, define first on

$$R^{<\xi} = \bigcup_\eta R_\eta$$

the sequence of norms $\bar{\psi}^\xi = \{\psi_n^\xi\}_{n\in\omega}$ by

(4) $$\psi_0^\xi(x,\alpha) = \text{least } \eta \text{ such that } R_\eta(x,\alpha),$$

(5) $$\psi_{n+1}^\xi(x,\alpha) = \langle\zeta,\varphi_n^\zeta(x,\alpha)\rangle, \quad \text{where } \zeta = \psi_0^\xi(x,\alpha).$$

Lemma 2. If each $\bar{\varphi}^\eta$ is a scale on R_η $(\eta < \xi)$, then (4) and (5) define a scale $\bar{\psi}^\xi$ on $R^{<\xi}$.

Proof. If $(x_i,\alpha_i) \to (x,\alpha)$ and all norms $\psi_n^\xi(x_i,\alpha_i)$ are ultimately fixed, then in particular

$$\zeta = \psi_0^\xi(x_i,\alpha_i) = \text{least } \eta \text{ such that } R_\eta(x_i,\alpha_i)$$

is ultimately fixed; hence $R_\zeta(x,\alpha)$ holds and for all n we have

(6) $$\varphi_n^\zeta(x,\alpha) \leq \varphi_n^\zeta(x_i,\alpha_i) \qquad \text{(all large } i).$$

⊣

Now if $\psi_0^\xi(x,\alpha) = \zeta$, then (6) implies directly that for all n, $\psi_{n+1}^\xi(x,\alpha) = \langle \zeta, \varphi_n^\zeta(x,\alpha)\rangle \le \langle \zeta, \varphi_n^\zeta(x_i,\alpha_i)\rangle = \psi_{n+1}^\xi(x_i,\alpha_i)$ for all large i. If on the other hand $\psi_0^\xi(x,\alpha) = \zeta' < \zeta$, then the lexicographic ordering used in (5) implies that

$$\psi_{n+1}^\xi(x,\alpha) = \langle \zeta', \varphi_n^{\zeta'}(x,\alpha)\rangle < \langle \zeta, \varphi_n^\zeta(x_i,\alpha_i)\rangle = \psi_{n+1}^\xi(x_i,\alpha_i),$$

so that in either case, $\overline{\psi}^\xi$ has the critical lower semicontinuity property of scales.
⊣

To extend the construction and define a scale on R_ξ from given scales on all the R_η, $\eta < \xi$, we need the determinacy hypothesis.

Lemma 3. Assume that every hyperprojective game is determined and that both $R^{<\xi}$ and the scale $\overline{\psi}^\xi$ on $R^{<\xi}$ constructed above is hyperprojective; then we can construct a hyperprojective scale on R_ξ.

Proof. This is immediate from (3B-2) and (3C-1) of [1], taking Γ = all hyperprojective sets and using the rather obvious fact that Γ is adequate and closed under $\exists^{\mathcal{R}}, \forall^{\mathcal{R}}$.
⊣

To complete the construction, we now put a scale $\overline{\psi}^\varkappa$ on

$$R = R^{<\varkappa}$$

by taking $\xi = \varkappa$ in Lemma 2.

To prove the main theorem, it will be enough to show that if P is defined from an analytical pointset Q by (1) and R is defined by (2), then this construction actually gives an inductive scale $\overline{\psi}^\varkappa$ on R, since we can then obviously get an inductive scale on P by setting

$$\varphi_n(x) = \psi_n^\varkappa(x, \langle\ \rangle).$$

The proof naturally rests on some elementary properties of inductive and hyperprojective sets, all proved in EIAS.

To begin with, the collection of absolutely inductive sets is adequate in the sense of [1] and it is closed under $\exists^\omega, \forall^\omega, \exists^{\mathcal{R}}, \forall^{\mathcal{R}}$ - this is almost obvious. The hyperprojective sets are additionally closed under $\neg$ and continuous substitutions.

It is a bit less obvious that the inductive sets can be parameterized in $\mathcal{R}$ by absolutely inductive universal sets; i.e. for each $\mathfrak{X}$, there is an absolutely inductive $G^{\mathfrak{X}} = G \subseteq \mathcal{R} \times \mathfrak{X}$ such that for $P \subseteq \mathfrak{X}$,

$$P \text{ is inductive} \iff \text{for some } \alpha \in \mathcal{R},$$
$$P = G_\alpha = \{x : G(\alpha, x)\}.$$

This can be proved by checking that we can always take Q to be $\underset{\sim}{\Sigma}_2^1$ - see 5D.2 of EIAS.

The most important fact about inductive sets that we need is the <u>Stage Comparison Theorem</u>, 2A.2 of EIAS: <u>if Q is analytical in</u> (2), <u>then the obvious norm on R given by</u>

$$\varphi(x,\alpha) = \underline{\text{least}}\ \xi\ \underline{\text{such that}}\ R_\xi(x,\alpha)$$

<u>is an absolutely inductive norm.</u> In particular, the pointclasses of aboslutely inductive and inductive sets have <u>the prewellordering property</u>; see 2A of [1].

For each $\mathfrak{X}$, fix an absolutely inductive $G^{\mathfrak{X}} \subseteq \mathcal{R} \times \mathfrak{X}$ which parametrizes the inductive subsets of $\mathfrak{X}$ as above and put

$$\langle\alpha,\beta\rangle \in C \iff (\forall x \in \mathfrak{X})[G_\alpha(x) \iff \neg G_\beta(x)].$$

Each $\gamma = \langle\alpha,\beta\rangle \in C$ <u>codes</u> a hyperprojective subset of $\mathfrak{X}$,

$$H_\gamma = G_\alpha = \mathfrak{X} - G_\beta$$

and every hypreprojective subset of $\mathfrak{X}$ receives at least one code in C. It is not very hard to check that <u>if the system</u> $\{G^{\mathfrak{X}}\}$ <u>of universal sets is chosen carefully, then the hyperprojective sets are effectively closed in this coding under every operation which preserves both absolute inductiveness and absolute co-inductiveness;</u> for example, there is a recursive function

$$f : \mathcal{R} \times \mathcal{R} \to \mathcal{R}$$

such that if $\gamma,\delta \in C$, then

$$\varepsilon = f(\gamma,\delta) \in C$$

and

$$H_\varepsilon = H_\gamma \cap H_\delta .$$

The key to this argument is to choose universal sets so that for suitable recursive functions $S^{\mathfrak{X},\mathfrak{Y}} = S$,

$$G(\alpha,x,y) \iff G(S(\alpha,x),y)$$

and such that in addition,

$$P \text{ is absolutely inductive} \Rightarrow P = G_\alpha \text{ with a recursive } \alpha.$$

We get effective closure under & (as an example) using such universal sets by setting

$$P_0(\gamma,\delta,x) \iff G((\gamma)_0,x)\ \&\ G((\delta)_0,x),$$
$$P_1(\gamma,\delta,x) \iff G((\gamma)_1,x) \vee G(\delta)_1,x),$$

choosing recursive $\varepsilon_0,\varepsilon_1$ so that

$$P_0(\gamma,\delta,x) \iff G(\varepsilon_0,\gamma,\delta,x) \iff G(S(\varepsilon_0,\gamma,\delta),x)$$

$$P_1(\gamma,\delta,x) \iff G(\varepsilon_1,\gamma,\delta,x) \iff G(S(\varepsilon_1,\gamma,\delta),x)$$

and taking

$$f(\gamma,\delta) = \langle S(\varepsilon_0,\gamma,\delta),S(\varepsilon_1,\gamma,\delta)\rangle.$$

It is simple to compute that if $\gamma,\delta \in C$ and $\varepsilon = f(\gamma,\delta)$, then $\varepsilon \in C$ and $H_\varepsilon = H_\gamma \cap H_\delta$.

One can always find "good" universal sets with these properties - see 9C.7 of EIAS.

After these preliminaries we can now describe the main construction for the proof.

If $\overline{\varphi} = \{\varphi_n\}_{n\in\omega}$ is a scale on a set $A \subseteq \mathfrak{X}$, then a (hyperprojective) <u>code for</u> $\overline{\varphi}$ is any $\alpha \in C$ which codes a set $H_\alpha \subseteq \omega \times \mathfrak{X} \times \mathfrak{X}$ such that

$$y \in A \Rightarrow (\forall x)\{[x \in A \;\&\; \varphi_n(x) \leq \varphi_n(y)] \Longleftrightarrow (n,x,y) \in H_\alpha\}.$$

<u>Lemma 4</u>. Assume that every hyperprojective game is determined and that

$$R(x,\alpha) \Longleftrightarrow \{(\forall\alpha_1)(\exists\alpha_2)(\forall\alpha_3)(\exists\alpha_4)\ldots\}(\exists n)Q(x,\alpha*\langle\alpha_1\ldots\alpha_n\rangle)$$

with Q analytical, let

$$R_\xi(x,\alpha) \Longleftrightarrow (\forall\beta)(\exists\gamma)[Q(x,\alpha) \vee (\exists\eta < \xi)R_\eta(x,\alpha*\langle\beta,\gamma\rangle)]$$

as above and for $(x,\alpha) \in R$, put

$$|x,\alpha| = \text{least } \xi \text{ such that } R_\xi(x,\alpha).$$

For each ξ, there is a scale $\overline{\varphi}^\xi = \{\varphi_n^\xi\}_{n\in\omega}$ on R_ξ such that for some recursive function

$$\pi : \mathfrak{X} \times \mathcal{R} \to \mathcal{R},$$

if $R(x,\alpha)$ and $|x,\alpha| = \xi$, then $\pi(x,\alpha)$ is a code of $\overline{\varphi}^\xi$.

In particular, each $\overline{\varphi}^\xi$ is a hyperprojective scale on R_ξ.

<u>Proof</u>. The scales $\overline{\varphi}^\xi$ will be exactly those defined in Lemmas 2 and 3 - we will have to make sure that the hypothesis of Lemma 3 is satisfied.

Since the definition of $\overline{\varphi}^\xi$ is by transfinite induction and we want π to be recursive, we must apply Kleene's well-known method of definition of <u>effective transfinite induction</u>, using the <u>recursion theorem</u>; i.e. we will set

$$\pi(x,\alpha) = \{\varepsilon_0\}(x,\alpha)$$

where ε_0 is a fixed recursive real satisfying

$$\{\varepsilon_0\}(x,\alpha) = f(\varepsilon_0,x,\alpha)$$

with a recursive (possibly partial) $f : \mathcal{R} \times \mathfrak{X} \times \mathcal{R} \to \mathcal{R}$. (The necessary coding machinery for recursive partial functions on the reals can be found in [3] or [5].)

Now the key to the construction is the fact that the transfer theorems for scales (3B-2) and (3C-1) of [1] have effective proofs. Combined with the remarks about effective closure of the hyperprojective sets above, this implies easily that there are recursive functions $f_1(\alpha,\gamma)$, $f_2(\alpha,\gamma)$ <u>such that if</u> α <u>happens to code a</u>

scale on H_γ $(\gamma \in C)$, then $f_1(\alpha,\gamma)$ codes a scale on $\exists^{\mathcal{R}} H_\gamma$ and $f_2(\alpha,\gamma)$ codes a scale on $\forall^{\mathcal{R}} H_\gamma$. Considering the positive analytical definition of R_ξ from $R^{<\xi}$, this in turn implies that there is a recursive function $f_3(\alpha,\gamma)$ such that if $R^{<\xi} = H_\alpha$ and γ happens to code a scale on $R^{<\xi}$, then $f_3(\alpha,\gamma)$ codes a scale on R_ξ. (Notice that the hypothesis here implies that both $R^{<\xi}$ and the scale on it are hyperprojective, so the appropriate games in Lemma 3 are determined.)

At the same time, the Stage Comparison Theorem implies easily that for a suitable recursive $f_4(x,\alpha)$, if $R(x,\alpha)$ holds and $|x,\alpha| = \xi$, then $f_4(x,\alpha)$ is a hyperprojective code of $R^{<\xi}$. Finally, looking at (4) and (5), it is not hard to believe that there is a further recursive function $f_5(\varepsilon,x,\alpha)$ with the following property: if there are scales $\overline{\varphi}^\eta$ on R_η for each $\eta < \xi$ and if for each $(x',\alpha') \in R$ such that $|x',\alpha'| = \eta < \xi$, $\{\varepsilon\}(x',\alpha')$ is defined and codes $\overline{\varphi}^\eta$, then $f_5(\varepsilon,x,\alpha)$ codes the scale $\overline{\psi}^\xi$ defined from the $\overline{\varphi}^\eta$ by (4), (5).

Now choose a recursive $\varepsilon^* \in R$ such that

$$\begin{aligned} f_5(\varepsilon,x,\alpha) &= \{\varepsilon^*\}(\varepsilon,x,\alpha) \\ &= \{S(\varepsilon^*,\varepsilon)\}(x,\alpha) \end{aligned}$$

using the appropriate recursive S for our coding of recursive partial functions and set

$$f(\varepsilon,x,\alpha) = f_3(f_4(x,\alpha),S(\varepsilon^*,\varepsilon)).$$

If we pick ε_0 by the recursion theorem and set

$$\pi(x,\alpha) = \{\varepsilon_0\}(x,\alpha) = f(\varepsilon_0,x,\alpha),$$

it is then trivial to verify by induction on ξ that π has the required property. ⊣

This lemma completes the proof of the main theorem, since in the scale $\overline{\psi}^\kappa$ we put on $R = R^{<\kappa}$ by (4) and (5) the first norm ψ_0^κ is absolutely inductive by the Stage Comparison Theorem and the later norms can be computed easily using the recursive function π.

2. Corollaries and remarks.

Of course the main corollary concerns uniformization.

Corollary 1. If every hyperprojective game is determined, then the collection of absolutely inductive pointsets has the uniformization property.

Proof. See (3A-1) of [1]. ⊣

To look a bit closer at one of the more interesting consequences of this uniformization result, let us go back and reconsider the open games on $\mathcal{R}$ that we used to define inductive sets. In a definition of the form

$$P(x) \iff \{(\forall\alpha_1)(\exists\alpha_2)(\forall\alpha_3)(\exists\alpha_4)\ldots\}(\exists n)Q(x,\langle\alpha_1,\ldots,\alpha_n\rangle)$$

we normally interpret the expression on the right as asserting the existence of a winning strategy for player II in the obvious game. This understanding of the definitions will force us to use the axiom of choice in the proof of Lemma 2 which is best avoided in the context of determinacy; instead we can reinterpret the infinite alternating string using multiple-valued strategies (quasistrategies in EIAS) and if we do that, then the proof of the main theorem is easily given in ZF + DC.

On the other hand, once we have the theorem, we should point out that if $P(x)$ holds, then II can win the game on the right using a (single-valued) strategy which is hyperprojective in x.

Corollary 2. Assume that every hypreprojective game is determined and suppose that

$$\{(\forall\alpha_1)(\exists\alpha_2)(\forall\alpha_3)(\exists\alpha_4)\ldots\}(\exists n)Q(\langle\alpha_1,\ldots,\alpha_n\rangle),$$

where Q is hyperprojective in some β_0. Then there exists a function

$$\sigma:\mathcal{R}\to\mathcal{R}$$

which is hyperprojective in β_0 such that

$$(*)\qquad \{(\forall\alpha_1)(\forall\alpha_3)(\forall\alpha_5)\ldots\}(\exists n)Q(\langle\alpha_1,\alpha_2,\alpha_3,\alpha_4,\ldots,\alpha_n\rangle),$$

where for even n,

$$\alpha_n = \sigma(\langle\alpha_1,\ldots,\alpha_{n-1}\rangle).$$

Proof. Put

$$P(\alpha,\beta) \iff \{(\forall\alpha_{n+1})(\exists\alpha_{n+2})\ldots\}(\exists k)Q(\alpha*\langle\beta,\alpha_{n+1},\alpha_{n+2},\ldots,\alpha_k\rangle),$$

so that P is inductive in β_0. Let P^* uniformize P and set for even n

$$\sigma(\langle\alpha_1,\ldots,\alpha_{n-1}\rangle) = \beta \iff P^*(\langle\alpha_1,\ldots,\alpha_{n-1}\rangle,\beta).$$

⊣

A further consequence of this remark concerns the absoluteness properties of the partially playful universe associated with the inductive sets.

Corollary 3. Assume that every hyperprojective game is determined and let T be the tree on $\omega\times\kappa$ associated with some absolutely inductive scale on some universal inductive set, let $L[T]$ be the associated model as in [1]. If Q is analytical and for $x\in L[T]$

$$\{(\forall\alpha_1)(\exists\alpha_2)(\forall\alpha_3)(\exists\alpha_4)\ldots\}(\exists n)Q(x,\langle\alpha_1,\ldots,\alpha_n\rangle)$$

holds (in the world), then the relativization of this open game assertion to $L[T]$ also holds.

Proof. $L[T]$ is easily closed under functions hyperprojective in x, with $x\in L[T]$.

⊣

The corollary is quite strong, as there are a lot of things that one can say with open game assertions - particularly if one assumes full determinacy and uses the coding Lemma 3 of [3].

The proof of the main theorem makes it clear that the result goes through for various pointclasses (included in the inductive sets) which enjoy some of the structural and closure properties of the inductive sets. In particular, the careful reader will notice that closure of the inductive sets under $\exists^{\mathcal{R}}$ is not used in the proof - although we did use (effective) closure of the hyperprojective sets under $\exists^{\mathcal{R}}$. Using this observation, any one who is somewhat familiar with recursion in higher types (as presented in [2], for example) will verify easily

Corollary 4 (to the proof). If every game which is Kleene recursive in ^3E and some $\beta_0 \in \mathcal{R}$ is determined, then every pointset which is semirecursive in ^3E admits a scale which is semirecursive in ^3E.

In particular, the collection of pointsets semirecursive in ^3E has the uniformization property (under this determinacy hypotheses). ⊣

In the other direction, it is hard to see how one could extend the class of sets now known to possess definable scales without using some entirely new principle of constructing scales. It appears that the question whether coinductive sets admit definable scales is one of the critical open problems in this theory.

References

[1] A. S. Kechris and Y. N. Moschovakis, Notes on the theory of scales, this volume.

[2] A. S. Kechris and Y. N. Moschovakis, Recursion in higher types, Handbook of Mathematical Logic, ed. K. J. Barwise, North Holland, 1977, p. 681-737.

[3] Y. N. Moschovakis, Determinacy and prewellorderings of the continuum, Mathematical Logic and Foundations of Set Theory, ed. Y. Bar Hillel, North Holland, 1970, p. 24-62.

[4] Y. N. Moschovakis, Elementary induction on abstract structures, North Holland, 1974.

[5] Y. N. Moschovakis, Descriptive set theory, in preparation.

ON VAUGHT'S CONJECTURE

John R. Steel
Department of Mathematics
University of California
Los Angeles, CA 90024

0. Introduction

We have two purposes in the present paper. The first is to survey some of the descriptive set theory and model theory directly relevant to Vaught's conjecture. This is done in §1. The survey is a bit distorted by our second purpose, which is to prove a special case of the conjecture. That is done in §2, where we show that if φ is an $\mathcal{L}_{\omega_1\omega}$ sentence all of whose models are trees (i.e. partially ordered sets so that the set of predecessors of any element is linearly ordered) then φ has either countably many or $2^{\aleph_0}$ countable models. (This extends earlier work of L. Marcus [11], A. Miller [13] and M. Rubin [16], [17].)

It is the author's pleasure to acknowledge his debt to Arnold Miller, both for many conversations on this topic of a general nature, and for some of the key ideas involved in the proof in case 2 of Theorem 2.1.1.

1. Background

1.1. The conjecture.

All languages in this paper will be countable. If $\mathcal{L}$ is a countable list of symbols, then $\mathcal{L}_{\omega_1\omega}$ is the set of formulae built up from symbols of $\mathcal{L}$ using countable conjections and disjunctions as well as the usual logical operations.

Vaught's conjecture is the statement: for any sentence φ in $\mathcal{L}_{\omega_1\omega}$, φ has either countably many or $2^{\aleph_0}$ countable models (up to isomorphism of course.)

Note that countable structures can be coded by reals. (For us, a real is an element of ${}^{\omega}2$, the Cantor space.) The strong Vaught conjecture states: For any φ in $\mathcal{L}_{\omega_1\omega}$, either φ has countably many countable models, or there is a perfect $P \subseteq {}^{\omega}2$ so that any two distinct elements of P code non-isomorphic models of φ.

One reason to concentrate on the strong conjecture is to avoid trivialities involving the Continuum Hypothesis. Vaught's conjecture is provable in ZF + CH, and provably equivalent in ZF + $\neg$CH to the strong conjecture (cf. 1.2.1). The strong conjecture, restricted to φ in $L_{\omega\omega}$, is provably equivalent in ZF to a Σ^1_2 sentence. So the strong conjecture is probably decidable in ZF.

Closely related to Vaught's conjecture is the ordinal spectrum conjecture, of J. Barwise. Recall that if $x \in {}^{\omega}2$, then ω_1^x is the least ordinal not recursive in x; equivalently, ω_1^x is the least $\mu > \omega$ so that $L_\mu[x]$ is admissible. Let $\omega_1^{\emptyset} = \omega_1^{ck}$. If $\mathfrak{A}$ is a countable structure, then

$$\omega_1^{\mathfrak{A}} = \min\{\omega_1^x \mid x \in {}^{\omega}2 \wedge x \text{ codes } \mathfrak{A}\}.$$

(In the notation of [1], $\omega_1^{\mathfrak{A}} = \max(\omega_1^{ck}, o(\mathfrak{A}^+))$). The ordinal spectrum conjecture states: Let φ be a sentence in $\mathcal{L}_{\omega_1\omega} \cap L_{\omega_1^{ck}}$. Then either $\{\omega_1^{\mathfrak{A}} \mid \mathfrak{A} \models \varphi\} = \{\omega_1^{ck}\}$ or $\{\omega_1^{\mathfrak{A}} \mid \mathfrak{A} \models \varphi\} = \{\mu < \aleph_1 \mid \mu$ is admissible$\}$. We prove this conjecture for trees in §2.

1.2 Effective descriptive set theory.

Some information about Vaught's conjecture can be obtained simply from the fact that the relation of coding isomorphic structures is a Σ_1^1 equivalence relation on ${}^{\omega}2$. The first result in this direction is due to J. Silver. We say that an equivalence relation E on ${}^{\omega}2$ has perfectly many classes if $\exists P \subseteq {}^{\omega}2$ (P is perfect $\wedge \forall x,y \in P$ $(x \neq y \Rightarrow \neg\, xEy)$).

<u>Theorem 1.2.1</u>: (Silver; see [7] and [18]). Let E be a $\underset{\sim}{\Pi}_1^1$ equivalence relation on ${}^{\omega}2$. Then there are either $\leq \aleph_0$ or perfectly many E-equivalence classes.

<u>Corollary 1.2.2</u>: (J. Burgess [2]). Let E be a $\underset{\sim}{\Sigma}_1^1$ equivalence relation on ${}^{\omega}2$. Then there are either $\leq \aleph_1$ or perfectly many E-equivalence classes.

<u>Proof</u>: For notational simplicity let E be Σ_1^1 lightface. Thus xEy iff T_{xy} is well-founded, where T_{xy} is a tree on ω depending recursively on x and y. For $\mu < \aleph_1$, let

$$xE_\mu y \text{ iff } \neg(|T_{xy}| < \mu).$$

Here $|T_{xy}|$ is the ordinal rank of T_{xy}, or $\aleph_1$ if T_{xy} is not well-founded. Note that each E_μ is Borel, $E_\mu \subseteq E_\nu$ if $\nu < \mu$, and $E = \bigcap_{\mu<\aleph_1} E_\mu$.

If μ is admissible, then E_μ is an equivalence relation. For example, to see that E_μ is transitive, suppose not and let $\nu < \mu$ be such that $S \neq \emptyset$, where

$$S = \{\langle x,y,z\rangle \mid xE_\mu y \wedge yE_\mu z \wedge \neg\, xE_\nu z\}.$$

Pick an $r \in {}^{\omega}2$ so that $\omega_1^r = \mu$. Then S is $\Sigma_1^1(r)$, so by Gandy's Basis Theorem we have $\langle x,y,z\rangle \in S$ so that $\omega_1^{\langle x,y,z\rangle} \leq \mu$. But notice $\forall x,y,\nu(\nu = \omega_1^{x,y} \wedge xE_\nu y \Rightarrow xEy)$. Thus $xEy \wedge yEz \wedge \neg\, xE_\nu z$, a contradiction.

We can now show that Corollary 1.2.2 is true in any countable model $\mathfrak{m}$ of a sufficiently large fragment of ZF, thereby proving the corollary. For let $\mathfrak{m}$ be such. Let $\mathfrak{n}$ be a $\mathbb{P}$-generic extension of $\mathfrak{m}$, where $\mathbb{P}$ is the Levy algebra for collapsing $\aleph_1^{\mathfrak{m}}$ to ω. If $\mathfrak{m} \models$ "E has at least $\aleph_2$ classes", then $\mathfrak{n} \models$ "$E_{\aleph_1^{\mathfrak{m}}}$ has at least $\aleph_1$ classes". (Since $\aleph_1^{\mathfrak{n}} = \aleph_2^{\mathfrak{m}}$, and inequivalent reals in $\mathfrak{m}$ remain so in $\mathfrak{n}$.) By 1.2.1 in $\mathfrak{n}$, $\mathfrak{n} \models$ "$E_{\aleph_1^{\mathfrak{m}}}$ has perfectly many classes." Thus $\mathfrak{n} \models$ "E has perfectly many classes". By Shoenfield's Absoluteness Theorem, $\mathfrak{m} \models$ "E has perfectly many classes". ⊠

<u>Remark</u>: In [2] Burgess has given a purely combinatorial (forcing-free) proof of an improvement of 1.2.2. A similar improvement is proved model-theoretically in [6]. In [3], Burgess has shown that the classes of a Σ_1^1 equivalence relation with $\aleph_1$ classes are the "levels" of some $\Delta_2^1(0^{\#})$ prewellordering of ${}^{\omega}2$ (provided $0^{\#}$ exists.)

<u>Corollary 1.2.3</u>: Let E be a $\underset{\sim}{\Sigma}_1^1$ equivalence relation, and $S \subseteq {}^{\omega}2$ be $\underset{\sim}{\Sigma}_1^1$. Then there are either $\leq \aleph_1$ or perfectly many classes of E represented by elements of S.

<u>Proof</u>: Let $S = \mathrm{ran}(f)$, where $f : {}^{\omega}2 \to {}^{\omega}2$ is Borel. Let xE^*y iff $f(x)Ef(y)$. Apply 1.2.2 to E^*. ⊠

We say a class $\mathcal{C}$ of $\mathcal{L}$ structures is $PC_{\mathcal{L}_{\omega_1\omega}}$ iff there is a (countable) language $\mathcal{L}' \supseteq \mathcal{L}$ and $\sigma \in \mathcal{L}'_{\omega_1\omega}$ and list $\vec{R}$ of relation symbols in $\mathcal{L}' - \mathcal{L}$ so that

$$\mathcal{C} = \{\mathfrak{A} \mid \mathfrak{A} \models \exists\vec{R}\sigma\}.$$

($\mathcal{C}$ is $EC_{\mathcal{L}_{\omega_1\omega}}$ iff $\mathcal{C} = \{\mathfrak{A} \mid \mathfrak{A} \models \varphi\}$ for some $\varphi \in \mathcal{L}_{\omega_1\omega}$.)

<u>Corollary 1.2.4</u>: (Morley [14]). Let $\mathcal{C}$ be $PC_{\mathcal{L}_{\omega_1\omega}}$. Then $\mathcal{C}$ has either $\leq \aleph_1$ or perfectly many countable members, up to isomorphism.

1.3 Invariant descriptive set theory.

For simplicity, let $\mathcal{L}$ have one binary relation symbol only. Consider the space $2^{\omega\times\omega}$ with the product topology induced by the discrete topology on $\omega \times \omega$. A set $\mathcal{C} \subseteq 2^{\omega\times\omega}$ is invariant iff $\forall R,S((R \in \mathcal{C} \wedge (\omega,R) \cong (\omega,S)) \Rightarrow S \in \mathcal{C})$.

<u>Theorem 1.3.1</u>: Let $\mathcal{C} \subseteq 2^{\omega\times\omega}$. Then

(a) $\mathcal{C}$ is $\underset{\sim}{\Sigma}_1^1$ and invariant iff $\exists\mathcal{C}^*$ ($\mathcal{C}^*$ is $PC_{\mathcal{L}_{\omega_1\omega}} \wedge \mathcal{C} = \{R \mid (\omega,R) \in \mathcal{C}^*\}$);

(b) $\mathcal{C}$ is $\underset{\sim}{\Delta}_1^1$ and invariant iff $\exists\mathcal{C}^*$ ($\mathcal{C}^*$ is $EC_{\mathcal{L}_{\omega_1\omega}} \wedge \mathcal{C} = \{R \mid (\omega,R) \in \mathcal{C}^*\}$).

<u>Proof</u>: (a) is immediate. (b) follows directly from (a) and the Interpolation Theorem ([9], p. 19) for $\mathcal{L}_{\omega_1\omega}$. ⊠

<u>Remark</u>: A lightface version of 1.3.1 can be proved. Also, there is a "level by level" version of 1.3.1 (b).

In view of 1.3.1, it is natural to ask which theorems of classical descriptive set theory hold in the more general setting of invariant descriptive set theory. Vaught's conjecture states that the classical theorem on the cardinality of Borel sets generalizes to an invariant version.

Invariant descriptive set theory has been investigated in [19].

1.4 Counterexamples.

Life is sometimes difficult.

1.4.1. Let xEy iff $\omega_1^x = \omega_1^y$. Then E is a Σ_1^1 equivalence relation, and each E-equivalence class is Borel. (By 1.5.2, isomorphism has the latter property as well.) However, E has $\aleph_1$ but not perfectly many equivalence classes. There appear to be no definability-theoretic grounds on which to distinguish between E and isomorphism.

1.4.2. (H. Friedman). There is a $PC_{\mathcal{L}_{\omega_1\omega}}$ class with exactly $\aleph_1$ countable members. For example, let

$$\mathbf{C} = \{\langle A,<\rangle \mid \exists \mathfrak{m}\ (\mathfrak{m} \text{ is an } \omega\text{-model of } KP \wedge \langle A,<\rangle \cong \langle OR^{\mathfrak{m}},<\rangle)\} .$$

H. Friedman has shown in [5] that the countable members of C are precisely the orderings α or $\alpha + \alpha \cdot \eta$, where $\alpha < \aleph_1$ is admissible. (Here η is the order type of the rationals.) So C has exactly $\aleph_1$ models. C doesn't have perfectly many models (cf. 1.5.5).

1.4.3. (K. Kunen). There is a $PC_{\mathcal{L}_{\omega\omega}}$ class with exactly $\aleph_1$ countable models. For let

$$C = \{\langle A,<\rangle \mid \langle A,<\rangle \text{ is a linear order } \wedge \forall a,b \in A\ \exists F\ (F \text{ is an automorphism of } \langle A,<\rangle \wedge F(a) = b)\} .$$

Let Z be the ordered set of integers, and let

$$Z^0 = \{0\} ;$$

$$Z^{\alpha+1} = Z^\alpha \times Z, \text{ ordered lexicographically} ;$$

$$Z^\lambda = \text{direct limit of } \langle Z^\beta \mid \beta < \lambda\rangle,\ \lambda \text{ limit} .$$

Then one can show that $\langle A,<\rangle \in C$ iff $\exists \alpha\ (\langle A,<\rangle \cong Z^\alpha \vee \langle A,<\rangle = Z^\alpha \cdot \eta)$. (As a first step, notice every order in C is either dense or discrete.) Thus C has $\aleph_1$ countable models. Again, C doesn't have perfectly many models (cf. 1.5.5).

Examples 1.4.2 and 1.4.3 indicate a failure of the analogy between classical and invariant descriptive set theory. They also indicate an important restriction: a proposed proof of Vaught's conjecture must differentiate between $EC_{\mathcal{L}_{\omega_1\omega}}$ and $PC_{\mathcal{L}_{\omega_1\omega}}$ classes. The work of §2 may be of interest in this regard.

1.5 Some model theory for $\mathcal{L}_{\omega_1\omega}$.

[1] and [9] are references for this material.

A. Scott sentences.

For $\varphi \in \mathcal{L}_{\omega_1\omega}$ we define $qr(\varphi)$, the quantifier rank of φ, by induction on φ:

(i) $qr(\varphi) = 0$ for φ atomic;

(ii) $qr(\neg\varphi) = qr(\varphi)$;

$qr(\bigwedge\Phi) = qr(\bigvee\Phi) = \sup\{qr(\varphi) \mid \varphi \in \Phi\}$;

(iii) $qr(\forall v\, \varphi) = qr(\exists v\, \varphi) = qr(\varphi) + 1$

Let $\mathfrak{A} \equiv_\nu \mathfrak{B}$ iff $\mathfrak{A}$ and $\mathfrak{B}$ satisfy the same sentences of quantifier rank $\leq \nu$.

We use "$|\mathfrak{A}|$" to denote the universe of $\mathfrak{A}$.

For any $\mathfrak{A}$ there is a least ordinal ν so that $\forall n \forall \vec{a} \in |\mathfrak{A}|^n \forall \vec{b} \in |\mathfrak{A}|^n$

$$((\mathfrak{A},\vec{a}) \equiv_\nu (\mathfrak{A},\vec{b}) \Rightarrow (\mathfrak{A},\vec{a}) \equiv_{\nu+1} (\mathfrak{A},\vec{b})) .$$

We call ν the Scott rank of $\mathfrak{A}$, and write "$\nu = sr(\mathfrak{A})$". Note that if M is admissible, then $\{\langle\mathfrak{A},\nu\rangle \in M \mid sr(\mathfrak{A}) = \nu\}$ is Δ over M. (However, $\mathfrak{A} \in M$ does not imply $sr(\mathfrak{A}) \in M$.)

Let $\mathfrak{A}$ be given. For $\vec{a} \in |\mathfrak{A}|^n$ and $\vec{v} = \langle v_1 \cdots v_n\rangle$ we define a formula $\varphi^{\alpha}_{\vec{a}}(\vec{v})$ by induction on α:

$$\varphi^0_{\vec{a}}(\vec{v}) = \bigwedge \{\theta(\vec{v}) \mid \theta \text{ is atomic or negation of atomic} \wedge \mathfrak{A} \models \theta[\vec{a}]\} ;$$

$$\varphi^{\alpha+1}_{\vec{a}} = \bigwedge_{b\in|\mathfrak{A}|} \exists v_{n+1}\, \varphi^{\alpha}_{\vec{a},b}(\vec{v},v_{n+1}) \wedge \forall v_{n+1} \bigvee_{b\in|\mathfrak{A}|} \varphi^{\alpha}_{\vec{a},b}(\vec{v},v_{n+1}) ;$$

$$\varphi^{\lambda}_{\vec{a}}(\vec{v}) = \bigwedge_{\beta<\lambda} \varphi^{\beta}_{\vec{a}}(\vec{v}), \quad \text{for } \lambda \text{ limit} .$$

Let ν be least so that

$$\mathfrak{A} \models \varphi^{\nu}_{\emptyset} \wedge \bigwedge_n \bigwedge_{\vec{a}\in|\mathfrak{A}|^n} \forall \vec{v}\, (\varphi^{\nu}_{\vec{a}}(\vec{v}) \Rightarrow \varphi^{\nu+1}_{\vec{a}}(\vec{v})) .$$

Then $\nu = sr(\mathfrak{A})$, and the displayed sentence is called the canonical Scott sentence of $\mathfrak{A}$ and denoted by "$CSS(\mathfrak{A})$".

Theorem 1.5.1: Let $\mathfrak{A}$ and $\mathfrak{B}$ be countable. Then $\mathfrak{A} \cong \mathfrak{B}$ iff $\mathfrak{L} \models CSS(\mathfrak{A})$.

Remark: Let C be a $PC_{\mathcal{L}_{\omega_1\omega}}$ class with at least $\aleph_1$ non-isomorphic countable models of Scott rank $< \nu$, where $\nu < \aleph_1$. By 1.5.1 there are $\aleph_1$ $\equiv_{\nu+\omega}$-classes of countable elements of C. Since $\equiv_{\nu+\omega}$ is a Borel equivalence, 1.2.1 and the proof of 1.2.3 yield perfectly many models in C.

Theorem 1.5.2: (Nadel [15]). For $\mathfrak{A}$ countable, $sr(\mathfrak{A}) \leq \omega_1^{\mathfrak{A}}$.

Proof: Notice that $\{\langle\vec{a},\vec{b}\rangle \mid \exists \nu\, ((\mathfrak{A},\vec{a}) \not\equiv_\nu (\mathfrak{A},\vec{b}))\}$ is definable by a Σ_1 positive inductive definition over $L_{\omega_1^x}[x]$, whenever $x \in {}^{\omega}2$ codes $\mathfrak{A}$. Such a definition terminates in $\leq \omega_1^x$ steps.

Theorem 1.5.3: (Sacks). Suppose C is $PC_{\mathcal{L}_{\omega_1\omega}}$ and $\forall \mathfrak{A} \in C$ $(sr(\mathfrak{A}) < \omega_1^{\mathfrak{A}})$. Then C has either $\leq \aleph_0$ or perfectly many countable models.

Proof: View countable elements of C as reals. By hypothesis then

$$\forall x \in {}^{\omega}2 \quad (x \in C \Rightarrow \exists e\ (\{e\}^x \text{ is a well-order} \wedge |\{e\}^x| = sr(x)))\ .$$

This is of the form $\forall x \exists e\ P(x,e)$ where P is $\underset{\sim}{\Pi}_1^1$. Let $f : {}^{\omega}2 \to \omega$ be $\underset{\sim}{\Delta}_1^1$ so that $\forall x\ P(x,f(x))$. Then $\{\{f(x)\}^x \mid x \in {}^{\omega}2\}$ is a $\underset{\sim}{\Sigma}_1^1$ set of well-orders, hence $\exists \nu < \aleph_1\ \forall x\ (|\{f(x)\}^x| < \nu)$. But then $sr(x) < \nu$, whenever $x \in C$. The remark after 1.5.1 completes the proof.

There are $\mathfrak{A}$ so that $sr(\mathfrak{A}) = \omega_1^{\mathfrak{A}}$. E.g., let $\mathfrak{A} = \langle \omega_1^{ck} + \omega_1^{ck} \cdot \eta, < \rangle$ Then $sr(\mathfrak{A}) = \omega_1^{\mathfrak{A}} = \omega_1^{ck}$.

There is a game theoretic characterization of $\equiv_\nu$, due to M. Benda and C. Karp. Namely, $\mathfrak{A} \equiv_\nu \mathfrak{B}$ iff II has a winning strategy in the following game: On the ith move I plays an ordinal $\nu_i < \nu_{i-1}$ (where $\nu_0 = \nu$) and an element of one of the structures; II then plays an element of the other structure. So eventually I plays $\nu_n = 0$. Let $a_1 \cdots a_n$ be the first n elements of $|\mathfrak{A}|$ played, and $b_1 \cdots b_n$ the corresponding elements of $|\mathfrak{B}|$ played. Then II wins iff $\langle \mathfrak{A}, a_1 \cdots a_n \rangle \equiv_0 \langle \mathfrak{B}, b_1 \cdots b_n \rangle$.

Remark 1.5.4. This characterization provides a trivial proof of the following. Suppose $\mathfrak{A}$ and $\mathfrak{B}$ are $\mathcal{L}$ structures, where for simplicity $\mathcal{L}$ has one binary relation symbol R. Let $\mathcal{P}$, $\mathcal{Q}$ be partitions of $|\mathfrak{A}|$ and $|\mathfrak{B}|$, and $\pi : \mathcal{P} \xrightarrow[\text{onto}]{1\text{-}1} \mathcal{Q}$ so that $\forall A \in \mathcal{P}\ (\langle A, R^{\mathfrak{A}} \rangle \equiv_\nu \langle \pi(A), R^{\mathfrak{B}} \rangle)$. Suppose that whenever $\langle A, R^{\mathfrak{A}}, a \rangle \equiv_0 \langle \pi(A), R^{\mathfrak{B}}, c \rangle$ and $\langle B, R^{\mathfrak{A}}, b \rangle \equiv_0 \langle \pi(B), R^{\mathfrak{B}}, d \rangle$ and $A \neq B$, then $R^{\mathfrak{A}}(a,b)$ iff $R^{\mathfrak{B}}(c,d)$. Then $\mathfrak{A} \equiv_\nu \mathfrak{B}$. This remark is used repeatedly in §2.

Remark 1.5.5. Let C be $PC_{\mathcal{L}_{\omega_1\omega}}$. Then C has perfectly many models iff $\exists \nu < \aleph_1$ (C has uncountably many models of Scott rank $< \nu$). The "if" part follows from 1.5.1 and 1.2.1. For the "only if" direction, let $P \subseteq {}^{\omega}2$ be a perfect set of codes for nonisomorphic models in C. Regard P itself as a real. For uncountably many $x \in P$ then $\omega_1^x \leq \omega_1^P$. For such x, $sr(x) \leq \omega_1^P$ by 1.5.2.

B. Prime models.

A fragment of $\mathcal{L}_{\omega_1\omega}$ is a set of formulae closed under the finitary logical operations, substitution of terms, and subformulae. For 1.5.6 through 1.5.10, fix a countable fragment F and a complete theory T in F.

Definition 1.5.6. Let $\varphi(\vec{v}) \in F$. Then $\varphi(\vec{v})$ is complete over T iff $\forall \psi(\vec{v}) \in F$ ("$\forall \vec{v}\ (\varphi(\vec{v}) \to \psi(\vec{v}))$" $\in T$ or "$\forall \vec{v}\ (\varphi(\vec{v}) \to \neg \psi(\vec{v})) \in T$).

Definition 1.5.7: Let $\Sigma \subseteq F$. Then Σ is an n-type of T iff $\exists \mathfrak{A}\, (\mathfrak{A} \models T \wedge \exists \vec{a} \in |\mathfrak{A}|^n\, (\Sigma = \{\theta(\vec{v}) \mid \mathfrak{A} \models \theta[\vec{a}]\}))$. We say that $\vec{a}$ realizes Σ in $\mathfrak{A}$. Σ is principal over T iff $\exists \theta \in \Sigma$ (θ is complete over T).

Let $\mathrm{Th}(\mathfrak{A}) = \{\varphi \in \mathcal{L}_{\omega_1\omega} \mid \mathfrak{A} \models \varphi\}$.

Definition 1.5.8: $\mathfrak{A}$ is prime over F iff every n-type realized in $\mathfrak{A}$ is principal over $\mathrm{Th}(\mathfrak{A}) \cap F$.

Let "$\mathfrak{A} \preccurlyeq_F \mathfrak{B}$" mean that there is an embedding of $\mathfrak{A}$ into $\mathfrak{B}$ which preserves all formulae of F. "$\mathfrak{A} \prec_F \mathfrak{B}$" means the inclusion map is such an embedding.

Theorem 1.5.9: Suppose $\forall n$ (there are $\leq \aleph_0$ n-types over T). Then T has a countable model $\mathfrak{A}$, prime over F, which is unique up to isomorphism. Moreover, $\forall \mathfrak{B}\, (\mathfrak{B} \models T \Rightarrow \mathfrak{A} \preccurlyeq_F \mathfrak{B})$.

In §2, the hypothesis of 1.5.9 will usually be available, due to 1.5.10.

Theorem 1.5.10: Let C be $PC_{\mathcal{L}_{\omega_1\omega}}$ without perfectly many models. Then $\{\Sigma \subseteq F \mid \exists \mathfrak{A} \in C\, \exists n$ (Σ is an n-type realized in $\mathfrak{A}$$\}$ is countable.

Proof: Otherwise there are at least $\aleph_1$ $(\nu+1)$-equivalence classes in C, where $\nu = \sup\{qr(\varphi) \mid \varphi \in F\}$. This contradicts the remark after 1.5.1.

Our use for prime models in §2 stems from the observation of M. Nadel that if $\mathfrak{A}$ is prime over F, then $sr(\mathfrak{A}) \leq \sup\{qr(\varphi) \mid \varphi \in F\}$.

These concepts yield a slightly different proof of Theorem 1 of [6]. Say that $\varphi \in \mathcal{L}_{\omega_1\omega}$ is a counterexample if φ has at least $\aleph_1$ but not perfectly many countable models. A counterexample φ is minimal if $\forall \theta \in \mathcal{L}_{\omega_1\omega}$ (either $\varphi \wedge \theta$ or $\varphi \wedge -\theta$ has $\leq \aleph_0$ countable models.

Theorem 1.5.11: Let φ be a counterexample. Then there is a minimal counterexample ψ so that $\models \psi \rightarrow \varphi$.

Proof: Let φ be given, and suppose no such ψ exists. It is easy to construct by induction on the length of s for $s \in 2^{<\omega}$, structures $\mathfrak{A}_s$ and fragments F_s so that $\varphi \in F_\emptyset$ and

(i) $F_s \subseteq F_t$ if $s \subseteq t$;

(ii) $\mathfrak{A}_{s^\frown i}$ is prime over F_s, $i \in \{0,1\}$, and $\mathfrak{A}_{s^\frown 0} \not\equiv_{F_s} \mathfrak{A}_{s^\frown 1}$;

(iii) $\mathrm{Th}(\mathfrak{A}_{s^\frown i}) \cap F_s$ has $\aleph_1$ models, and $\mathrm{Th}(\mathfrak{A}_{s^\frown i}) \cap F_s = \mathrm{Th}(\mathfrak{A}_{s^\frown i^\frown j}) \cap F_s$, for $i,j \in \{0,1\}$;

(iv) $\mathfrak{A}_s \models \varphi$.

Then by (ii), (iii), and 1.5.9, $\mathfrak{A}_{s^\frown i} \preccurlyeq \mathfrak{A}_{s^\frown i^\frown j}$ for $i,j \in \{0,1\}$. For $x \in {}^{\omega}2$ let $\mathfrak{A}_x$ be the direct limit of $\langle \mathfrak{A}_{x \restriction n} \mid n \in \omega \rangle$. Then $x \neq y \Rightarrow \mathfrak{A}_x \not\equiv \mathfrak{A}_y$, a contradiction.

2. Trees

2.1 Some lemmas.

We now descend to a much more concrete level. A structure $\mathfrak{A} = \langle A,<\rangle$ is a tree iff $<$ partially orders A so that $\forall a \in A$ ($\{b \mid b < a\}$ is linearly ordered by $<$). The main goal of §2 is to prove the following theorem.

Theorem 2.1.1: Let $\varphi \in \mathcal{L}_{\omega_1\omega}$. If every model of φ is a tree, then φ has either $\leq \aleph_0$ or perfectly many countable models.

Throughout the proof of 2.1.1 we shall be computing the Scott ranks of various structures. In general the computation is done simply by writing down sentences which characterize the structure and observing their quantifier rank. The next lemma contains a typical example of such a computation.

If $\mathfrak{A}$ is a tree, then for $a \in |\mathfrak{A}|$ let

$$C_a = \{b \in |\mathfrak{A}| \mid \exists c\ (c \leq^{\mathfrak{A}} b \wedge c \leq^{\mathfrak{A}} a)\}$$

and

$$\mathfrak{C}_a = \langle C_a, <^{\mathfrak{A}} \restriction C_a \times C_a\rangle .$$

The $\mathfrak{C}_a$'s partition $\mathfrak{A}$ into its maximal connected subtrees. $\mathfrak{A}$ is connected iff $\exists a \in |\mathfrak{A}|$ $(|\mathfrak{A}| = C_a)$.

Lemma 2.1.2: Suppose $\mathfrak{A}$ is a tree and $sr(\mathfrak{A}) = \omega_1^{\mathfrak{A}}$. Then $\exists a \in |\mathfrak{A}|$ $(sr(\mathfrak{C}_a) = \omega_1^{\mathfrak{A}})$.

Proof: Assume not. Let $x \in {}^{\omega}2$ code $\mathfrak{A}$, $\omega_1^x = \omega_1^{\mathfrak{A}}$. Notice that the map $a \mapsto \mathfrak{C}_a$ is Δ over $L_{\omega_1^x}[X]$. So then is the map $a \mapsto sr(\mathfrak{C}_a)$. By assumption this map is total, so the boundedness principle for admissible sets yields a $\nu < \omega_1^{\mathfrak{A}}$ so that $\forall a \in |\mathfrak{A}|$ $(sr(\mathfrak{C}_a) < \nu)$. We show $sr(\mathfrak{A}) < \nu + \omega 2$, a contradiction.

The direct method for doing this is to produce for each $\vec{a} \in |\mathfrak{A}|^n$ a sentence ψ with $qr(\psi) < \nu + \omega 2$ so that $\forall \mathfrak{B}$ countable $\forall \vec{b} \in |\mathfrak{B}|^n$

$$(\mathfrak{B},\vec{b}) \cong (\mathfrak{A},\vec{a}) \quad \text{iff} \quad (\mathfrak{B},\vec{b}) \models \psi .$$

So let $\vec{a} \in |\mathfrak{A}|^n$ be given. For simplicity, let $n = 1$, $\vec{a} = \langle a\rangle$. Let $S = \{CSS(\mathfrak{C}_b) \mid b \in |\mathfrak{A}|\}$, and for $\varphi \in S$ let K_φ be the number of distinct $\mathfrak{C}_b$'s so that $\mathfrak{C}_b \models \varphi$. Let θ_φ say that exactly K_φ maximal connected subtrees satisfy φ; e.g., if $K_\varphi = \omega$

$$\theta_\varphi = \bigwedge_{i \leq K_\varphi} \exists v_1 \cdots v_i \left(\bigwedge_{k,j \leq i} \neg \exists u\ (u \leq v_k \wedge u \leq v_j) \wedge \bigwedge_{j \leq i} \varphi^{\exists u(u \leq v_0 \wedge u \leq v_j)} \right) .$$

(Here $\varphi^{\theta(v_0)}$ is φ with all quantifiers relativised to $\theta(v_0)$.)

Let $\theta = \mathrm{CSS}(\langle \mathfrak{C}_a, a\rangle)$. The desired sentence ψ is given by:

$$\psi = \theta^{\exists u(u \leq v_0 \wedge u \leq a)} \wedge \bigwedge_{\varphi \in S} \theta_\varphi \wedge \forall x \left(\bigvee_{\varphi \in S} \varphi^{\exists u(u \leq v_0 \wedge u \leq x)} \right).$$

In the future we'll indicate proofs like that of 2.1.2 in an elliptical fashion. The next lemma will be our other main tool for evaluating Scott ranks. Its proof is more subtle than that of 2.1.2.

Let $\mathfrak{A}$ be a tree. For $a <^{\mathfrak{A}} b$, let

$$[a,b]^{\mathfrak{A}} = \{c \mid a \leq^{\mathfrak{A}} c \wedge \neg(b <^{\mathfrak{A}} c)\}.$$

Thus when $\mathfrak{A}$ is a linear order, $[a,b]^{\mathfrak{A}}$ is just a closed interval. We also use "$[a,b]^{\mathfrak{A}}$" to denote the structure $\langle [a,b]^{\mathfrak{A}}, <^{\mathfrak{A}}, a, b\rangle$.

Lemma 2.1.3: Let $\mathfrak{A}$ be a connected tree, and L a maximal linearly ordered subset of $|\mathfrak{A}|$. Let λ be a limit ordinal so that $\forall a,b \in L$ $(\mathrm{sr}([a,b]^{\mathfrak{A}}) < \lambda)$ and there are fewer than perfectly many structures λ-equivalent to $\langle \mathfrak{A}, L\rangle$. Then $\mathrm{sr}(\langle \mathfrak{A}, L\rangle) \leq \lambda \cdot 3$.

Proof: Let λ, $\mathfrak{A}$, and $L \subseteq |\mathfrak{A}|$ be as in the hypotheses. Suppose $\mathrm{sr}(\langle \mathfrak{A}, L\rangle) > \lambda \cdot 3$. Let F be the smallest fragment of $\mathfrak{L}_{\omega_1\omega}$ so that whenever $\vec{a} \in |\mathfrak{A}|^n$ and $\alpha < \lambda \cdot 2$ then $\varphi^{\alpha}_{\vec{a}} \in F$. ($\mathfrak{L}$ is the language of $\langle \mathfrak{A}, L\rangle$, and $\varphi^{\alpha}_{\vec{a}}$ is computed in the sense of $\langle \mathfrak{A}, L\rangle$.) By our hypotheses and 1.5.10, $\mathrm{Th}(\langle \mathfrak{A}, L\rangle) \cap F$ has a model $\langle \mathfrak{B}, M\rangle$ prime over F. Since $\sup\{\mathrm{qr}(\psi) \mid \psi \in F\} = \lambda \cdot 2$, $\mathrm{sr}(\langle \mathfrak{B}, M\rangle) \leq \lambda \cdot 2$. But by the choice of F, $\langle \mathfrak{B}, M\rangle \equiv_{\lambda \cdot 2} \langle \mathfrak{A}, L\rangle$. Thus $\mathrm{sr}(\langle \mathfrak{B}, M\rangle) = \lambda \cdot 2$. We obtain a contradiction by showing $\langle \mathfrak{A}, L\rangle \cong \langle \mathfrak{B}, M\rangle$.

It is enough to get an F-elementary embedding j of $\langle \mathfrak{B}, M\rangle$ into $\langle \mathfrak{A}, L\rangle$ so that $j''M$ is cofinal in L in both directions. For suppose we had such a j. Let $i \mapsto b_i$ map I order-preservingly and cofinally into M, where I is one of the ordered sets ω, ω^*, or $\omega^* + \omega$. Say for definiteness that $I = \omega$. For $i \in \omega$ let

$$\pi_i : [b_i, b_{i+1}]^{\mathfrak{B}} \cong [j(b_i), j(b_{i+1})]^{\mathfrak{A}}.$$

Such a π_i exists because j is F-elementary and $\mathrm{sr}([j(b_i), j(b_{i+1})]^{\mathfrak{A}}) < \lambda$. Now for $b \in |\mathfrak{B}|$ let $\pi(b) = \pi_i(b)$ where $b \in [b_i, b_{i+1}]^{\mathfrak{B}}$. Then $\pi : \langle \mathfrak{B}, M\rangle \cong \langle \mathfrak{A}, L\rangle$, as desired. (That $\mathrm{dom}(\pi) = |\mathfrak{B}|$ and $\mathrm{ran}(\pi) = |\mathfrak{A}|$ follows from the maximality of L in $\mathfrak{A}$, the connectedness of $\mathfrak{A}$, and the corresponding properties of M and $\mathfrak{B}$.)

Now by 1.5.9, there is an F-elementary map j of $\langle \mathfrak{B}, M\rangle$ into $\langle \mathfrak{A}, L\rangle$. If we weave into the usual construction of j (cf. [9], p. 61) steps designed to make $\mathrm{ran}(j)$ as large as possible, we can arrange that $\forall a \in (|\mathfrak{A}| - \mathrm{ran}(j))\ \exists \vec{b} \in \mathrm{ran}(j)^n$ ($\langle a, \vec{b}\rangle$ satisfies $\langle \mathfrak{A}, L\rangle$ no complete formula of $\mathrm{Th}(\langle \mathfrak{A}, L\rangle) \cap F$). We show now that if j has this property, then $j''M$ is cofinal in L.

Let $R = \{a \in |\mathfrak{A}| \mid \exists\, b,c \in j''M\ (a \in [b,c]^{\mathfrak{A}})\}$. Let

$$\mathfrak{C} = \langle (|\mathfrak{A}| - R) \cup \operatorname{ran}(j), <^{\mathfrak{A}}, L\rangle .$$

Then $\mathfrak{C} \prec_F \langle \mathfrak{A}, L\rangle$. In fact, if $\vec{c} \in |\mathfrak{C}|^n$, then $\langle \mathfrak{A}, L, \vec{c}\rangle \cong \langle \mathfrak{C}, \vec{c}\rangle$ (the proof being like the proof that a cofinal embedding yields an isomorphism).

Let F' be the smallest fragment of $\mathcal{L}_{\omega_1\omega}$ containing each $\varphi^{\alpha}_{\vec{a}}$ for $\alpha < \lambda$ and $\vec{a} \in |\mathfrak{A}|^n$, and let $\langle \mathfrak{B}', M'\rangle \prec_{F'} \langle \mathfrak{B}, M\rangle$ with $sr(\langle \mathfrak{B}', M'\rangle) = \lambda$. Define

$$\mathfrak{C}' = \langle (|\mathfrak{A}| - R) \cup \operatorname{ran}(j \upharpoonright |\mathfrak{B}'|), L, <^{\mathfrak{A}}\rangle .$$

It is convenient to have an endpoint for L, so pick $a \in L \cap \operatorname{ran}(j \upharpoonright |\mathfrak{B}'|)$. Then either $sr(\langle \{b \in |\mathfrak{A}| \mid a \leq^{\mathfrak{A}} b\}, L, <^{\mathfrak{A}}\rangle) \geq \lambda \cdot 3$ or $sr(\langle \{b \in |\mathfrak{A}| \mid a \not\leq^{\mathfrak{A}} b\}, L, <^{\mathfrak{A}}\rangle) \geq \lambda \cdot 3$. (Otherwise, by direct computation, as in 2.1.2, $sr(\langle \mathfrak{A}, L\rangle) \leq \lambda \cdot 3$.) We assume the former alternative; the proof in the other case is the same. It follows that there is a $b \in L$ so that $\forall c \in j''M \ (c <^{\mathfrak{A}} b)$. For otherwise $j''M$ is cofinal upward in L, and j would yield an isomorphism of $\langle \{b \in |\mathfrak{A}| \mid a \leq^{\mathfrak{A}} b\}, L, <^{\mathfrak{A}}\rangle$ with $\langle \{b \in |\mathfrak{B}| \mid j^{-1}(a) \leq^{\mathfrak{B}} b\}, M, <^{\mathfrak{B}}\rangle$, which is impossible since the latter structure has Scott rank $\leq \lambda \cdot 2$.

Now by Remark 1.5.4 we have $\langle \mathfrak{C}, a, b\rangle \equiv_{\lambda} \langle \mathfrak{C}', a, b\rangle$. Since $\mathfrak{C} \prec_F \langle \mathfrak{A}, L\rangle$, $sr([a,b]^{\mathfrak{C}}) < \lambda$, whence we have an isomorphism π from $[a,b]^{\mathfrak{C}}$ onto $[a,b]^{\mathfrak{C}'}$. Clearly π preserves L. π cannot map $\operatorname{ran}(j)$ onto $\operatorname{ran}(j \upharpoonright |\mathfrak{B}'|)$, for then we have an isomorphism of $\langle \{b \in |\mathfrak{B}| \mid j^{-1}(a) \leq b\}, <^{\mathfrak{B}}, M\rangle$ onto $\langle \{b \in |\mathfrak{B}'| \mid j^{-1}(a) \leq b\}, <^{\mathfrak{B}'}, M\rangle$, which is impossible since the first structure has Scott rank $\lambda \cdot 2$ and the second has Scott rank λ. There are now two cases.

<u>Case 1</u>: $\exists c\ (c \in \operatorname{ran}(j) \cap [a,b]^{\mathfrak{C}} \wedge \pi(c) \notin \operatorname{ran}(j \upharpoonright |\mathfrak{B}'|))$.
Using 1.5.4 we have

$$\langle [a,b]^{\mathfrak{C}'}, \pi(c)\rangle \equiv_{\lambda} \langle [a,b]^{\mathfrak{C}}, \pi(c)\rangle .$$

Since the latter structure has Scott rank $< \lambda$ (recall $\mathfrak{C} \prec_F \langle \mathfrak{A}, L\rangle$), we have an isomorphism σ from the former to the latter. Clearly σ preserves L. Then $\sigma \circ \pi$ is an automorphism of $\langle [a,b]^{\mathfrak{C}}, <^{\mathfrak{C}}, L\rangle$ taking c to $\pi(c)$, where $c \in \operatorname{ran}(j)$ and $\pi(c) \notin \operatorname{ran}(j)$.

Since $\pi(c) \notin \operatorname{ran}(j)$, the construction of j yields $\vec{d} \in \operatorname{ran}(j)^n$ so that $\langle \pi(c), \vec{d}\rangle$ satisfies in $\langle \mathfrak{A}, L\rangle$, hence in $\mathfrak{C}$, no complete formula of $Th(\langle \mathfrak{A}, L\rangle) \cap F$. But $\sigma \circ \pi$ extends trivially to an automorphism of $\mathfrak{C}$ taking some $\vec{e} \in \operatorname{ran}(j)^n$ to $\vec{d}$. Thus $\langle c, \vec{e}\rangle$ satisfies in $\mathfrak{C}$, hence in $\langle \mathfrak{A}, L\rangle$, no complete formula of $Th(\langle \mathfrak{A}, L\rangle) \cap F$. Since $c \in \operatorname{ran}(j)$, this is a contradiction.

<u>Case 2</u>: $\exists c\ (c \in [a,b]^{\mathfrak{C}} - \operatorname{ran}(j) \wedge \pi(c) \in \operatorname{ran}(j \upharpoonright |\mathfrak{B}'|))$.

The proof in this case is similar to that in case 1. The diagram below illustrates the proof in the case $\mathfrak{A}$ is a linear order (so that $L = |\mathfrak{A}|$ and $M = |\mathfrak{B}|$).

In the case that $\mathfrak{A}$ is a linear order, 2.1.3 has a simple formulation, which would become still simpler if the hypothesis on the number of models $\equiv_{\lambda} \mathfrak{A}$ could be dropped. We conjecture that this can be done.

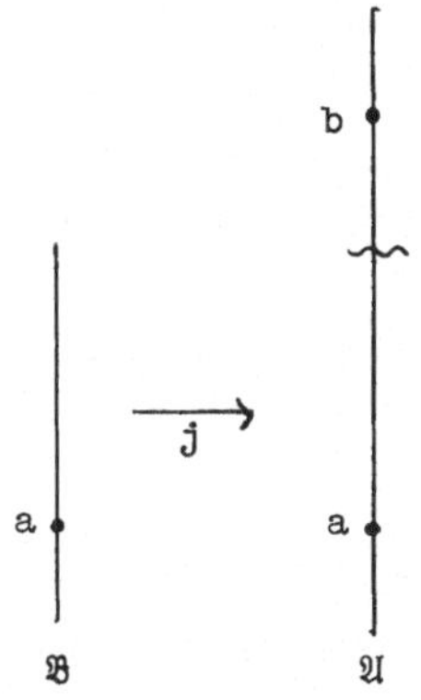

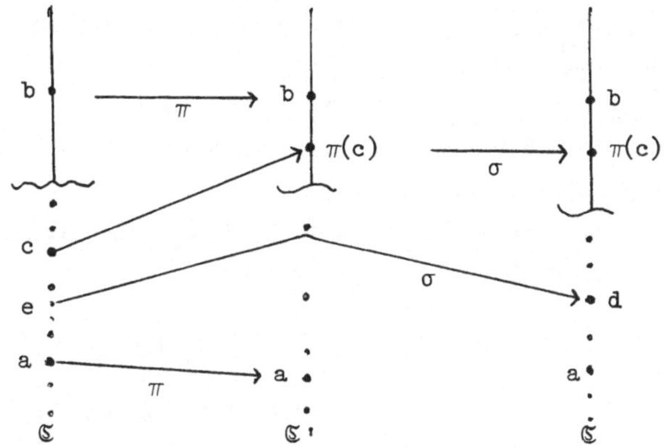

2.2 Proof of Theorem 2.1.1.

Let $\varphi \in \mathcal{L}_{\omega_1\omega}$ be a counterexample to 2.1.1. Then by 1.5.3 (essentially) we have a structure $\mathfrak{A} \models \varphi$ so that $qr(\varphi) < \omega_1^{\mathfrak{A}}$ and $sr(\mathfrak{A}) = \omega_1^{\mathfrak{A}}$. Let $\lambda = qr(\varphi) + \omega$. Our plan is to construct for each $x \subseteq \omega$ a structure $\mathfrak{A}_x \equiv_\lambda \mathfrak{A}$ so that $x \neq y \Rightarrow \mathfrak{A}_x \not\cong \mathfrak{A}_y$. The real x is coded into the isomorphism type of $\mathfrak{A}_x$ by manipulating the Scott ranks of "pieces" of $\mathfrak{A}_x$.

We may assume $\mathfrak{A}$ is connected. For by 2.1.2, $sr(\mathfrak{C}_a) = \omega_1^{\mathfrak{C}_a}$ for some $a \in |\mathfrak{A}|$. But then the proof to follow would produce perfectly many structures $\equiv_\lambda \mathfrak{C}_a$, and thus by 1.5.4 perfectly many structures $\equiv_\lambda \mathfrak{A}$.

For $a \in |\mathfrak{A}|$ let $\mathfrak{A}_a = \langle \{b \mid a \leq^{\mathfrak{A}} b\}, <^{\mathfrak{A}} \rangle$. Then for some $a \in |\mathfrak{A}|$, $sr(\mathfrak{A}_a) = \omega_1^{\mathfrak{A}}$. For pick any $a \in |\mathfrak{A}|$. If $sr(\mathfrak{A}_a) < \omega_1^{\mathfrak{A}}$, then consider $\mathfrak{B} = \langle \{c \in |\mathfrak{A}| \mid \neg(a <^{\mathfrak{A}} c)\}, <^{\mathfrak{A}} \rangle$. By direct computation, $sr(\mathfrak{B}) = \omega_1^{\mathfrak{A}}$. Let $L = \{c \mid c \leq^{\mathfrak{A}} a\}$, so that L is a maximal linearly ordered subset of $|\mathfrak{B}|$ definable in $\mathfrak{B}$ (from a). We may assume there are fewer than perfectly many models $\equiv_\lambda \mathfrak{B}$, as otherwise 1.5.4 would bring our proof to an early end. Thus there are fewer than perfectly many models $\equiv_\lambda \langle \mathfrak{B}, L \rangle$. If $\forall b,c \in L$ $(sr([b,c]^{\mathfrak{B}} < \omega_1^{\mathfrak{A}})$, then by boundedness and 2.1.3 $sr(\langle \mathfrak{B}, L\rangle) < \omega_1^{\mathfrak{A}}$, so $sr(\mathfrak{B}) < \omega_1^{\mathfrak{A}}$. So we have $sr([b,c]^{\mathfrak{B}}) = \omega_1^{\mathfrak{A}}$ for some $b,c \in L$. But then $sr(\mathfrak{A}_b) = \omega_1^{\mathfrak{A}}$, as desired.

The proof now splits into two cases. In case 1, $\mathfrak{A}$ behaves like the graph of a unary function; in case 2 $\mathfrak{A}$ behaves like a linear order.

<u>Case 1</u>: $\forall a \in |\mathfrak{A}|\ (sr(\mathfrak{A}_a) = \omega_1^{\mathfrak{A}} \Rightarrow \exists b,c \in |\mathfrak{A}|\ (sr(\mathfrak{A}_b) = sr(\mathfrak{A}_c) = \omega_1^{\mathfrak{A}} \wedge a <^{\mathfrak{A}} b \wedge a <^{\mathfrak{A}} c \wedge b,c$ are incomparable$))$.

<u>Proof</u>: Find sequences $\langle b_i \mid i \in \omega \rangle$ and $\langle a_i \mid i \in \omega \rangle$ so that for each i:

$$b_i < b_{i+1} \ \wedge\ b_i < a_i \ ;$$

$$a_i,\ b_{i+1} \text{ are incomparable } ;$$

$$sr(\mathfrak{A}_{b_i}) = sr(\mathfrak{A}_{a_i}) = \omega_1^{\mathfrak{A}} .$$

If $c \in |\mathfrak{A}|$, let $|\mathfrak{B}_c| = \{d \mid c \leq d$ or $\exists e < c\ (\forall i\ (e \not\leq b_i) \wedge d \in [e,c])\}$, and let $\mathfrak{B}_c = \langle |\mathfrak{B}_c|, <^{\mathfrak{A}} \rangle$.

Let $\langle \lambda_i \mid i \in \omega \rangle$ be a strictly increasing sequence of limit ordinals with limit $\omega_1^{\mathfrak{A}}$, and $\lambda < \lambda_0$. Let $x \subseteq \omega$ be infinite. For each i, choose $\mathfrak{C}_i \prec_{\lambda_k} \mathfrak{B}_{a_i}$ so that $sr(\mathfrak{C}_i) = \lambda_k$, where k is the ith element of x. (Such a $\mathfrak{C}_i$ can be found by 1.5.9 and 1.5.10; by 1.5.4 we can assume $\mathfrak{B}_{a_i}$ has fewer than perfectly many λ-equivalent models.)

For $a \notin \bigcup_i |\mathfrak{B}_{a_i}|$, let $\mathfrak{C}_a \prec_\lambda \mathfrak{B}_a$ and $sr(\mathfrak{C}_a) = \lambda$. Choose the $\mathfrak{C}_a$'s so that $\mathfrak{B}_a = \mathfrak{B}_b \Rightarrow \mathfrak{C}_a = \mathfrak{C}_b$. Finally, let

$$\mathfrak{A}_x = \langle \{c \mid \exists\, i\ (c \overset{\mathfrak{A}}{\leq} b_i)\} \cup \bigcup_i |\mathfrak{C}_i| \cup \bigcup_a |\mathfrak{C}_a|, \overset{\mathfrak{A}}{<} \rangle .$$

By 1.5.4, $\mathfrak{A}_x \equiv_\lambda \mathfrak{A}$. To see that x can be recovered from the isomorphism type of $\mathfrak{A}_x$, notice that $\exists\, i\ (c \leq b_i)$ iff $sr((\mathfrak{A}_x)_c) = \omega_1^{\mathfrak{A}}$. Thus $\{c \mid \exists\, i\ (c \leq b_i)\}$ can be recovered. But then $\langle \gamma_k \mid k \in x \rangle$, and thus x, can be recovered. Since $\exists\, \nu < \aleph_1\ \forall x\ (sr(\mathfrak{A}_x) < \nu)$, 1.2.1 yields perfectly many models of φ.

<u>Case 2</u>: Otherwise.

<u>Proof</u>: We have an $a \in |\mathfrak{A}|$ so that $sr(\mathfrak{A}_a) = \omega_1^{\mathfrak{A}}$ and $M = \{b \in |\mathfrak{A}_a| \mid sr(\mathfrak{A}_b) = \omega_1^{\mathfrak{A}}\}$ is linearly ordered by $\overset{\mathfrak{A}}{<}$. We claim $\exists\, b\ (sr([a,b]^{\mathfrak{A}}) = \omega_1^{\mathfrak{A}})$. For suppose first that $\forall c \in M\ (c < b)$. Then $sr(\mathfrak{A}_b) < \omega_1^{\mathfrak{A}}$, and so by direct computation $sr([a,b]^{\mathfrak{A}}) = \omega_1^{\mathfrak{A}}$. If no upper bound for M exists, then M is a maximal linearly ordered subset of $|\mathfrak{A}_a|$. A boundedness argument yields a $\nu < \omega_1^{\mathfrak{A}}$ so that $sr(\mathfrak{A}_b) < \nu$ whenever $b \in |\mathfrak{A}_a| - M$. Thus M is definable in $\mathfrak{A}_a$ by a formula of $qr < \omega_1^{\mathfrak{A}}$. We may assume there are fewer than perfectly many models $\equiv_\lambda \mathfrak{A}_a$. If $\forall b \in M\ (sr([a,b]^{\mathfrak{A}}) < \omega_1^{\mathfrak{A}})$ then boundedness and 2.1.3 imply $sr(\mathfrak{A}_a) < \omega_1^{\mathfrak{A}}$, a contradiction. Thus we have a b so that $sr([a,b]^{\mathfrak{A}}) = \omega_1^{\mathfrak{A}}$.

Our efforts will now be bent toward finding perfectly many models $\equiv_\lambda [a,b]^{\mathfrak{A}}$; by 1.5.4 this will complete the proof.

Let $L = \{c \mid a \leq c \leq b\}$. For $c,d \in L$ let $c \sim d$ iff $sr([c,d]^{\mathfrak{A}} < \omega_1^{\mathfrak{A}}$. It is easy to check that $\sim$ is an equivalence relation partitioning L into segments. For $d \in L$, let $C_d = \{c \mid \exists\, e\ (e \sim d \wedge (c \in [e,d]^{\mathfrak{A}} \vee c \in [d,e]^{\mathfrak{A}}))\}$. We call C_d a component. Clearly $e \not\sim d$ iff $C_e \cap C_d = \emptyset$.

We claim the components are densely ordered (where $C_d < C_e$ iff $d < e \wedge d \not\sim e$). For suppose $d < e$ and C_d, C_e are immediately adjacent components. Then $\forall c \in [d,e] \cap L\ (sr([c,e]^{\mathfrak{A}}) < \omega_1^{\mathfrak{A}} \vee sr([d,c]^{\mathfrak{A}}) < \omega_1^{\mathfrak{A}})$. By boundedness this is true with "ν" in place of "$\omega_1^{\mathfrak{A}}$" for some $\nu < \omega_1^{\mathfrak{A}}$. But then by 2.1.3 each of

$$\langle \bigcup_{\substack{c\sim d\\ d<c}} [d,c]^{\mathfrak{A}}, \overset{\mathfrak{A}}{<}, L \rangle$$

and

$$\langle \bigcup_{\substack{c\sim e\\ c<e}} [c,e]^{\mathfrak{A}}, \overset{\mathfrak{A}}{<}, L \rangle$$

has Scott rank $< \omega_1^{\mathfrak{A}}$. Let P be the set of points in $[d,e]$ appearing in neither of the above structures. Then by 2.1.2 essentially, $sr(P, \overset{\mathfrak{A}}{<}) < \omega_1^{\mathfrak{A}}$. By direct computation it follows that $sr([d,e]^{\mathfrak{A}}) < \omega_1^{\mathfrak{A}}$, a contradiction.

Our plan now is to juggle the Scott ranks of the components, roughly as we did in case 1. There are two subcases according to the distribution of these Scott ranks in $[a,b]^{\mathfrak{A}}$.

<u>Subcase A</u>: $\forall \upsilon < \omega_1^{\mathfrak{A}}$ $(\{C_d \mid sr(\langle C_d, <^{\mathfrak{A}}, L\rangle) > \upsilon\})$ is a dense set of components.

<u>Proof</u>: Recall λ is given; we are to find perfectly many models $\equiv_\lambda [a,b]^{\mathfrak{A}}$. Let $\langle C_i \mid i < \omega\rangle$ be a strictly increasing sequence of components so that $sr(\langle C_i, <^{\mathfrak{A}}, L\rangle) \geq \lambda + \omega \cdot 2$. Let $x \in {}^{\omega}2$ be given. For each i, choose $C_i' \subseteq C_i$ so that $\langle C_i', <^{\mathfrak{A}}, L\rangle <_\lambda \langle C_i, <^{\mathfrak{A}}, L\rangle$ and $sr(\langle C_i', <^{\mathfrak{A}}, L) = \lambda + \omega + \omega \cdot x(i)$. If C is a component so that $sr(\langle C, <^{\mathfrak{A}}, L\rangle) \leq \lambda + \omega \cdot 2$ and $C \notin \{C_i \mid i < \omega\}$ let $C' \subseteq C$ be so that $\langle C', <^{\mathfrak{A}}, L\rangle <_\lambda \langle C, <^{\mathfrak{A}}, L\rangle$ and $sr(\langle C', <^{\mathfrak{A}}, L\rangle) \leq \lambda$. Otherwise for C a component, let $C' = C$. (The existence of the C's follows from 1.5.4, 1.5.9, 1.5.10, and the assumption that there are fewer than perfectly many models $\equiv_\lambda [a,b]^{\mathfrak{A}}$.)

Let $A_x = \bigcup \{C' \mid C$ a component$\} \cup \{c \in [a,b]^{\mathfrak{A}}] \mid c$ is a member of no components$\}$, and let $\mathfrak{A}_x = \langle A_x, <^{\mathfrak{A}}\rangle$.

By 1.5.4, $\mathfrak{A}_x \equiv_\lambda \mathfrak{A}$. To see that $x \neq y \Rightarrow \mathfrak{A}_x \not\equiv \mathfrak{A}_y$, it suffices to show that $\forall c,d \in L$ $(sr([c,d]^{\mathfrak{A}_x}) < \omega_1^{\mathfrak{A}}$ iff $sr([c,d]^{\mathfrak{A}}) < \omega_1^{\mathfrak{A}})$. The "if" direction is clear. Now suppose $sr([c,d]^{\mathfrak{A}_x}) = \upsilon < \omega_1^{\mathfrak{A}}$ and $sr([c,d]^{\mathfrak{A}}) = \omega_1^{\mathfrak{A}}$. By the subcase hypothesis and 2.1.3 we can find $i,j \in [c,d]^{\mathfrak{A}} \cap L$ so that $\omega_1^{\mathfrak{A}} > sr([i,j]^{\mathfrak{A}}) > \nu + \lambda + \omega \cdot 2$. But then by construction of $\mathfrak{A}_x$, $[i,j]^{\mathfrak{A}} = [i,j]^{\mathfrak{A}_x}$. Then $sr([c,d]^{\mathfrak{A}_x}) < sr([i,j]^{\mathfrak{A}_x})$, which is impossible.

To obtain perfectly many $\mathfrak{A}_x$'s, note that there is a fixed $z \in {}^{\omega}2$ so that $\forall x\ (\mathfrak{A}_x \in L_{\omega_1^{\langle x,z\rangle}}[x,z])$. But then 1.5.2 and the existence of perfectly many x so that $\omega_1^{\langle x,z\rangle} = \omega_1^z$ give the desired conclusion (cf. 1.5.5).

<u>Subcase B</u>: Otherwise.

<u>Proof</u>: Let $\upsilon < \omega_1^{\mathfrak{A}}$ be so that the components of Scott rank $\geq \upsilon$ are not dense. By moving a and b we can assume there are no such components.

Pick an increasing sequence $\langle c_i \mid i < \omega\rangle$ so that $\forall i$ $(a < c_i < b \wedge c_i \not\sim c_{i+1})$. Thus $sr([c_i, c_{i+1}]^{\mathfrak{A}}) = \omega_1^{\mathfrak{A}}$. Let $x \in {}^{\omega}2$. Choose $\mathfrak{C}_i <_\lambda [c_{2i}, c_{2i+1}]^{\mathfrak{A}}$ so that $sr(\mathfrak{C}_i) = \upsilon + \lambda + \omega + \omega \cdot x(i)$. Let $A_x = \bigcup_i |\mathfrak{C}_i| \cup ([a,b]^{\mathfrak{A}} - \bigcup_i [c_{2i}, c_{2i+1}]^{\mathfrak{A}})$, and let $\mathfrak{A}_x = \langle A_x, <^{\mathfrak{A}}\rangle$.

Again, $\mathfrak{A}_x \equiv_\lambda [a,b]^{\mathfrak{A}}$ by 1.5.4. To show that there are perfectly many $\mathfrak{A}_x$'s, one proceeds along the lines of Subcase A.

The proof of Theorem 2.1.1 is now complete.

2.3 Remarks and corollaries.

Two corollaries of 2.1.1 had been proved earlier and by different methods.

Corollary 2.3.1: (A. Miller [13] and L. Marcus [11] for $\mathcal{L}_{\omega\omega}$.) Let φ be an $\mathcal{L}_{\omega_1\omega}$ sentence all of whose models are graphs of unary functions. Then φ has $\leq \aleph_0$ or perfectly many models.

Corollary 2.3.2: (M. Rubin [16], [17].) Let φ be an $\mathcal{L}_{\omega_1\omega}$ sentence all of whose models are linear orders. Then φ has $\leq \aleph_0$ or perfectly many models.

The proof of 2.1.1 is somewhat simpler than the earlier proofs of these corollaries. In particular, in the special case of trees anyway, there is no advantage in restricting one's attention to $\mathcal{L}_{\omega\omega}$. The machinery of Scott sentences and admissible sets leads naturally into $\mathcal{L}_{\omega_1\omega}$.

It would be natural to attempt to extend 2.1.1 from trees to arbitrary partial orders. However, Arnold Miller has shown Vaught's conjecture for partial orders is equivalent to the full conjecture, and his proof shows that partial orders are really no simpler in this context than arbitrary binary relations. What classes lie between trees and partial orders we leave to the reader's imagination.

There are effective refinemenets of 2.1.1, for which we need the following lemma.

Lemma 2.3.3: Let $\varphi \in \mathcal{L}_{\omega_1\omega} \cap L_{\omega_1^{ck}}$ and suppose $\forall \mathfrak{A}\ (\mathfrak{A} \models \varphi \Rightarrow \mathrm{sr}(\mathfrak{A}) < \omega_1^{ck})$. Then: (a) If φ has $\leq \aleph_0$ countable models, then every countable model of φ is isomorphic to an element of $L_{\omega_1^{ck}}$; (b) If φ has $> \aleph_0$ countable models, then $\{\omega_1^{\mathfrak{A}} \mid \mathfrak{A} \models \varphi\} = \{\mu \mid \omega < \mu < \aleph_1 \wedge \mu$ is admissible$\}$.

Proof: By boundedness let $\nu < \omega_1^{ck}$ be so that $\mathfrak{A} \models \varphi \Rightarrow \mathrm{sr}(\mathfrak{A}) < \nu$. Adopt a reasonable coding of formulae of $\mathcal{L}_{\omega_1\omega}$ with q.r. $< \nu$ by reals, so that each formula gets exactly one code. The set of codes will be a lightface Δ_1^1 set of reals. Let $D = \{x \mid \exists \mathfrak{A}\ (\mathfrak{A} \models \varphi \wedge x$ codes $\mathrm{CSS}(\mathfrak{A}))\}$. By 2.14 of [10], D is Δ_1^1. (D is clearly Σ_1^1. But $D = \{x \mid \exists \mathfrak{A} \in L_{\omega_1^x}[x]\ (\mathfrak{A} \models \varphi \wedge x$ codes $\mathrm{CSS}(\mathfrak{A})\}$, so D is Π_1^1.)

(a). If φ has $\leq \aleph_0$ models, D is countable, hence $D \subseteq L_{\omega_1^{ck}}$. Since $\mathfrak{A}$ has an isomorphic copy inside any admissible set which thinks $\mathrm{CSS}(\mathfrak{A})$ is countable, (a) is proved.

(b). If D is uncountable, then D has members of every hyperdegree above Kleene's $\mathcal{O}$ (cf. [12]). Since our coding can be taken so that any admissible set is closed under the map $\mathfrak{A} \mapsto$ code for $\mathrm{CSS}(\mathfrak{A})$ (for $\mathfrak{A} \models \varphi$!), this implies $\{\omega_1^{\mathfrak{A}} \mid \mathfrak{A} \models \varphi\}$ is cofinal in $\aleph_1$. By Gandy's Basis theorem relative to any x so that $\omega_1^x = \mu$, we have: $\exists \mathfrak{A}\ (\mathfrak{A} \models \varphi \wedge \omega_1^{\mathfrak{A}} = \mu)$.

Corollary 2.3.4: Let $\varphi \in \mathcal{L}_{\omega_1\omega} \cap L_{\omega_1^{ck}}$. Suppose φ has $\leq \aleph_0$ countable models, all of which are trees. Then every model of φ is isomorphic to an element of $L_{\omega_1^{ck}}$.

Proof. If the hypothesis of 2.3.3 is not satisfied, then by Gandy's Basis theorem

$\exists \mathfrak{A}\ (\mathfrak{A} \models \varphi \wedge sr(\mathfrak{A}) = \omega_1^{\mathfrak{A}} = \omega_1^{ck})$. The proof of 2.1.1 then yields perfectly many models of φ.

Corollary 2.3.5: Let $\varphi \in \mathcal{L}_{\omega_1\omega} \cap L_{\omega_1^{ck}}$, and suppose every model of φ is a tree. Then either $\{\omega_1^{\mathfrak{A}} \mid \mathfrak{A} \models \varphi\} \subseteq \{\omega_1^{ck}\}$ or $\{\omega_1^{\mathfrak{A}} \mid \mathfrak{A} \models \varphi\} = \{\mu \mid \omega < \mu < \aleph_1 \wedge \mu$ is admissible$\}$.

Proof: If the hypothesis of 2.3.3 is not satisfied, then $\exists \mathfrak{A}\ (\mathfrak{A} \models \varphi \wedge sr(\mathfrak{A}) = \omega_1^{\mathfrak{A}} = \omega_1^{ck})$. But then the proof of 2.1.1 implies $\{\omega_1^{\mathfrak{A}} \mid \mathfrak{A} \models \varphi\}$ is cofinal in $\aleph_1$. Thus $\{\omega_1^{\mathfrak{A}} \mid \mathfrak{A} \models \varphi\} = \{\mu \mid \omega < \mu < \aleph_1 \wedge \mu$ is admissible$\}$.

3. Determinateness

We wish to conclude by mentioning some consequences of the hypothesis that all $\underset{\sim}{\Sigma}_1^1$ games are determined which shed some light on 1.5.3 and the counterexamples of 1.4. We omit proofs.

If we rephrase 1.5.3 in the language of invariant descriptive set theory, we obtain the following statement: Let $C \subseteq {}^{\omega}2$ be $\underset{\sim}{\Sigma}_1^1$ and invariant, and suppose $\forall x \in C\ \ \exists \alpha < \omega_1^x\ (\{y \mid y \cong x\}$ is $\underset{\sim}{\Sigma}_{\alpha}^0)$; then C has $\leq \aleph_0$ or perfectly many classes.

Theorem 3.1: Assume all $\underset{\sim}{\Sigma}_1^1$ games are determined. Let $C \subseteq {}^{\omega}2$ be $\underset{\sim}{\Sigma}_1^1$ and invariant, and suppose $\forall x \in C\ (\{y \mid y \cong x\}$ is $\underset{\sim}{\Sigma}^0_{\omega_1^x+2})$; then C has $\leq \aleph_0$ or perfectly many classes.

By 1.5.1 and 1.5.2, $\{y \mid y \cong x\}$ is always $\underset{\sim}{\Pi}^0_{\omega_1^x+2}$, and so the examples of 1.4 show that 3.1 cannot be improved.

Theorem 3.2: Assume all $\underset{\sim}{\Sigma}_1^1$ games are determined. Let C be a $PC_{\mathcal{L}_{\omega_1\omega}}$ class with uncountably many but not perfectly many countable models. Then $\exists \langle \mathfrak{A}_\nu \mid \nu < \aleph_1\rangle$ so that

(i) $\forall \nu\ (\mathfrak{A}_\nu \in C)$;

(ii) $\{\mathfrak{B} \mid \exists \nu\ (\mathfrak{B} \cong \mathfrak{A}_\nu)\}$ is $PC_{\mathcal{L}_{\omega_1\omega}}$

(iii) let $\alpha_\nu = \omega_1^{\mathfrak{A}_\nu}$; then $\nu \mapsto \alpha_\nu$ is strictly increasing and continuous;

(iv) $sr(\mathfrak{A}_\nu) = \alpha_\nu$;

(v) $\mathfrak{A}_\nu \prec_{\alpha_\nu} \mathfrak{A}_\mu$ if $\nu < \mu$;

(vi) (Saturation) Suppose $\langle \mathfrak{A}_\mu, b_0 \cdots b_n\rangle \equiv_{\alpha_\nu} \langle \mathfrak{A}_\nu, j(b_0) \cdots j(b_n)\rangle$, where $\nu \leq \mu$. Then j can be extended to an α_ν-elementary embedding of $\mathfrak{A}_\mu$ into $\mathfrak{A}_\nu$.

We shall describe only the game which enters into the proofs of 3.1 and 3.2. Let $WO = \{x \in {}^{\omega}2 \mid x$ codes a well-order of $\omega\}$, and for $x \in WO$ let $|x|$ = length of well-order coded by x. Let C be $PC_{\mathcal{L}_{\omega_1\omega}}$ with at least $\aleph_1$ but not perfectly many models. Consider the following Solovay-style game: I plays x, II plays y.

Player I loses unless $x \in WO$. If $x \in WO$, then II loses unless y codes a model $\mathfrak{A} \in C$ so that $|x| \leq sr(\mathfrak{A})$. By a standard argument, I can have no winning strategy in this game, so by determinateness II has a winning strategy.

The existence of such a strategy, together with forcing arguments like those of [8], implies 3.1 and 3.2.

References

[1] Barwise, J., Admissible sets and structures, Springer Verlag.

[2] Burgess, J., Infinitary languages and descriptive set theory, Ph.D. thesis, Berkeley, 1974.

[3] Burgess, J., Equivalences generated by families of Borel sets, to appear in Proc. AMS.

[4] Burgess, J., Effective enumeration of classes in a Σ_1^1 equivalence relation, to appear.

[5] Friedman, H., Countable models of set theories, in: Cambridge Summer School in Mathematical Logic, ed. by A.R.D. Mathias and H. Rogers, Springer Lecture Notes No. 337.

[6] Harnik, V., and M. Makkai, A tree argument in infinitary model theory, to appear in Proc. AMS.

[7] Harrington, L., A powerless proof of a theorem of Silver, unpublished notes.

[8] Harrington, L., Analytic determinacy and $0^{\#}$, to appear.

[9] Keisler, H.J., Model theory for infinitary logic, North Holland, 1971.

[10] Makkai, M., An "admissible" generalization of a theorem on countable Σ_1^1 sets of reals with applications, Annals of Math. Logic, June 1977.

[11] Marcus, L., The number of countable models of a theory of one unary function, to appear.

[12] Martin, D., Proof of a conjecture of Friedman, Proc. AMS.

[13] Miller, A., Ph.D. thesis, Berkeley, 1977.

[14] Morley, M., The number of countable models, JSL v. 35 (1970), pp. 14-18.

[15] Nadel, M., Model theory in admissible sets, Ph.D. thesis, University of Wisconsin, 1971.

[16] Rubin, M., Vaught's conjecture for linear orderings, Notices of AMS, vol. 24, p. A 390.

[17] Rubin, M., Theories of linear order, Israel J. Math., v. 17, 1974, pp. 392-443.

[18] Silver, J., $\underset{\sim}{\Pi}_1^1$ equivalence relations, unpublished notes.

[19] Vaught, R., Invariant sets in topology and logic, Fund. Math., v. 82, 1974, pp. 269-293.

ON RECURSION IN $\underset{\sim}{E}$ AND SEMI-SPECTOR CLASSES[1]

Phokion G. Kolaitis

Department of Mathematics
University of California
Los Angeles, CA 90024

In this paper we survey the theory of recursion in $\underset{\sim}{E}$ on an arbitrary acceptable structure and compare it to the theory of (positive elementary) induction. Since the class of the semirecursive relations in $\underset{\sim}{E}$ on an acceptable structure is the main example of a semi-Spector class, we will develop in effect some parts of the general theory of semi-Spector classes and will compare it to the theory of Spector classes.

We approach recursion in $\underset{\sim}{E}$ here as a branch of the general theory of inductive definability according to the program introduced by Moschovakis [1976] and developed in Kechris-Moschovakis [1977]. We should point out that all the notions which will be introduced and studied here are "lightface", i.e., only parameters from ω are allowed. This is necessary, because in general semi-Spector classes are not closed under $\exists$, so that one has to introduce "lightface" notions and pursue "lightface" versions of results. It turns out that in most cases a theorem about the "boldface" inductive relations or about "boldface" Spector classes has a natural analog about the ("lightface") semirecursive relations in $\underset{\sim}{E}$ or about ("lightface") semi-Spector classes.

The basic definitions and the necessary background material are included in the first section of the paper. In Section 2 we present the Stage Comparison Theorems in the context of functional induction and in Section 3 we apply the Second Stage Comparison Theorem in order to obtain a general version of the Spector-Gandy Theorem for recursion in normal, definable functionals. Both the semirecursive relations in $\underset{\sim}{E}$ and the inductive relations on an acceptable structure can be characterized in terms of nonmonotone inductive definability. These results are due to Grilliot [1971] and in Section 4 we prove them in the spirit of the theory of semi-Spector classes. Finally, in Section 5 we discuss certain generalizations and state some open problems.

1. Positive Inductive Definability and Functional Induction.

1.1. Let $\mathfrak{A} = \langle A, R_1, \ldots, R_n, f_1, \ldots f_m, c_1, \ldots, c_k \rangle$ be a structure such that $\omega \subseteq A$, $c_1, \ldots, c_k \in A$, each R_i, $1 \leq i \leq n$, is a relation on A and each f_j, $1 \leq j \leq m$, is a total function on A. We will be working with the "lightface" first order language $\mathcal{L}^{\mathfrak{A}}$ of the structure $\mathfrak{A}$ which has as constants the natural numbers and

[1]The author is grateful to Professors A. S. Kechris and Y. N. Moschovakis for numerous helpful discussions during the preparation of this paper.

the distinguished elements $c_1, c_2, \ldots, c_k$. More precisely, the language $\mathfrak{L}^{\mathfrak{A}}$ of the structure $\mathfrak{A}$ has an infinite list $x, y, z, \ldots$ of individual variables, constant symbols $\underset{\sim}{c}_\lambda$, where $1 \leq \lambda \leq k$, a constant symbol $\underset{\sim}{n}$ for each $n \in \omega$, an infinite list $S, T, V, \ldots$ of m-ary relation variables for each $m \geq 1$, relation symbols $\underset{\sim}{R}_i$, where $1 \leq i \leq n$, function symbols $\underset{\sim}{f}_j$, where $1 \leq j \leq n$, the equality symbol $=$ and the logical symbols $\neg, \vee, \wedge, \rightarrow, \exists, \forall$. The formulas of the language $\mathfrak{L}^{\mathfrak{A}}$ are defined in the usual way with the quantifiers ranging over the individual variables only. A relation $R \subseteq A^m$ is elementary on $\mathfrak{A}$ if it can be defined by a formula of the language $\mathfrak{L}^{\mathfrak{A}}$. We are interested in structures $\mathfrak{A}$ such that the relations ω and $\leq_\omega = \{(m,n) \in \omega \times \omega : m \leq n\}$ are elementary on $\mathfrak{A}$.

A quantifier on A is a set Q such that $\emptyset \subsetneq Q \subsetneq P(A)$ and such that if $X \subseteq Y \subseteq A$ and $X \in Q$, then $Y \in Q$ (monotonicity property). With each quantifier Q on A we associate the dual quantifier $\breve{Q}$, where $(X \in \breve{Q} \Leftrightarrow A - X \notin Q)$. Given a quantifier Q on A we expand the language $\mathfrak{L}^{\mathfrak{A}}$ of the structure $\mathfrak{A}$ to the language $\mathfrak{L}^{\mathfrak{A}}(Q)$ which has in addition the symbols Q and $\breve{Q}$, so that if φ is a formula of $\mathfrak{L}^{\mathfrak{A}}(Q)$, then so are $Qx\varphi(x)$ and $\breve{Q}x\varphi(x)$. For the interpretation we add the obvious clause

$$Qx\varphi(x) \text{ is true} \Leftrightarrow \{x : \varphi(x)\} \in Q.$$

1.2. Let $\mathfrak{A}$ be a structure such that $\omega, \leq_\omega$ are elementary on $\mathfrak{A}$ and let Q be a quantifier on A. If $\varphi(x_1, \ldots, x_n, S)$ is a formula of the language $\mathfrak{L}^{\mathfrak{A}}(Q)$ in which S is an n-ary relation variable occurring positively, then $\varphi(x_1, \ldots, x_n, S)$ defines a transfinite sequence $\{I_\varphi^\xi\}$ of subsets of A^n, where

$$I_\varphi^\xi = \{\vec{x} : \varphi(\vec{x}, I_\varphi^{<\xi})\} \quad \text{and} \quad I_\varphi^{<\xi} = \bigcup_{\eta<\xi} I_\varphi^\eta.$$

We put $I_\varphi = \bigcup_\xi I_\varphi^\xi$ and call I_φ the set inductively defined by φ.

A relation $R \subseteq A^m$ is Q-inductive on $\mathfrak{A}$ if there is a formula $\varphi(\vec{u}, \vec{v}, S)$ of the language $\mathfrak{L}^{\mathfrak{A}}(Q)$ with S occurring positively and constants $\vec{n} = (n_1, \ldots, n_k)$ from ω such that

$$R(\vec{y}) \Leftrightarrow (\vec{n}, \vec{y}) \in I_\varphi.$$

A relation $R \subseteq A^m$ is Q-hyperelementary on $\mathfrak{A}$ if both R and $A^m - R$ are Q-inductive on $\mathfrak{A}$. These definitions relativize to any finite sequence $\vec{x} = (x_1, \ldots, x_k)$ from A, i.e., $R \subseteq A^m$ is Q-inductive on $\mathfrak{A}$ from $\vec{x}$ if there is some Q-inductive relation P on $\mathfrak{A}$ such that $R(\vec{y}) \Leftrightarrow P(\vec{x}, \vec{y})$. Similarly, $R \subseteq A^m$ is Q-hyperelementary on $\mathfrak{A}$ from $\vec{x}$ if both R and $A^m - R$ are Q-inductive on $\mathfrak{A}$ from $\vec{x}$. We introduce the following notation to denote the various classes of relations defined above:

$\mathrm{IND}(\mathfrak{A}, Q) = \mathrm{IND}(Q) =$ all Q-inductive relations on $\mathfrak{A}$,

$\mathrm{HYP}(\mathfrak{A}, Q) = \mathrm{HYP}(Q) =$ all Q-hyperelementary relations on $\mathfrak{A}$,

$\mathrm{IND}(\mathfrak{A}, Q, \vec{x}) = \mathrm{IND}(Q, \vec{x}) =$ all Q-inductive relations on $\mathfrak{A}$ from $\vec{x}$,

$\mathrm{HYP}(\mathfrak{A}, Q, \vec{x}) = \mathrm{HYP}(Q, \vec{x}) =$ all Q-hyperelementary relations on $\mathfrak{A}$ from $\vec{x}$.

The theory of positive inductive definability has been developed in Moschovakis [EIAS]. It should be pointed out that the language used there is the "boldface" language $\underset{\sim}{\mathfrak{L}}^{\mathfrak{A}}(Q)$, which has in addition a constant symbol $\underset{\sim}{a}$ for each $a \in A$. Also the classes of relations studied by Moschovakis [EIAS] are the "boldface" Q-inductive and Q-hyperelementary relations, i.e. the class $\underset{\sim}{\mathrm{IND}}(\mathfrak{A},Q) = \underset{\sim}{\mathrm{IND}}(Q) = \bigcup_{\vec{x}\in A^{<\omega}} \mathrm{IND}(\mathfrak{A},Q,\vec{x})$ and the class

$$\underset{\sim}{\mathrm{HYP}}(\mathfrak{A},Q) = \underset{\sim}{\mathrm{HYP}}(Q) = \bigcup_{\vec{x}\in A^{<\omega}} \mathrm{HYP}(\mathfrak{A},Q,\vec{x}), \quad \text{where } A^{<\omega} = \bigcup_{n\in\omega} A^n.$$

Let $\mathfrak{A} = \langle A, R_1, \dots, R_n, f_1, \dots, f_m, c_1, \dots, c_k \rangle$ be a structure such that ω, $\leq_\omega$ are elementary on $\mathfrak{A}$; if we do not consider an additional quantifier Q on $\mathfrak{A}$ and we iterate formulas of the language $\mathcal{L}^{\mathfrak{A}}$, then we have the notions of the <u>inductive</u> and the <u>hyperelementary relations on</u> $\mathfrak{A}$. We write $\mathrm{IND}(\mathfrak{A})$ and $\mathrm{HYP}(\mathfrak{A})$ for these classes of relations; we also write $\mathrm{IND}(\mathfrak{A},\vec{x})$, $\mathrm{HYP}(\mathfrak{A},\vec{x})$ for the <u>inductive relations from</u> $\vec{x}$ and the <u>hyperelementary relations from</u> $\vec{x}$ <u>on</u> $\mathfrak{A}$. Finally, we put $\underset{\sim}{\mathrm{IND}}(\mathfrak{A}) = \bigcup_{\vec{x}\in A^{<\omega}} \mathrm{IND}(\mathfrak{A},\vec{x})$ and $\underset{\sim}{\mathrm{HYP}}(\mathfrak{A}) = \bigcup_{\vec{x}\in A^{<\omega}} \mathrm{HYP}(\mathfrak{A},\vec{x})$.

1.3. In order to prove some of the more interesting theorems of this paper, we must restrict ourselves to structures which possess a definable coding apparatus. More precisely, a structure $\mathfrak{A} = \langle A, R_1, \dots, R_m, f_1, \dots, f_n, c_1, \dots, c_k \rangle$ with $\omega \subseteq A$ is <u>acceptable</u> if ω and $\leq_\omega$ are elementary on $\mathfrak{A}$ and there is a total one-to-one function $\langle\ \rangle : A^{<\omega} \to A$ such that the relation seq and the functions lh and q are elementary on $\mathfrak{A}$, where

$$\mathrm{seq}(x) \Longleftrightarrow \text{there are } x_1, \dots, x_n \text{ such that } x = \langle x_1, \dots, x_n \rangle,$$

$$\mathrm{lh}(x) = \begin{cases} 0, & \text{if } \neg\mathrm{seq}(x), \\ n, & \text{if } \mathrm{seq}(x) \text{ and } x = \langle x_1, \dots, x_n \rangle, \end{cases}$$

$$q(x,i) = (x)_i = \begin{cases} x_i, & \text{if } x = \langle x_1, \dots, x_n \rangle \text{ and } 1 \leq i \leq n, \\ 0, & \text{otherwise.} \end{cases}$$

If $\mathfrak{A}$ is an acceptable structure and $\langle\ \rangle : A^{<\omega} \to A$ is a coding function, then the functions s and $*$ are also elementary on $\mathfrak{A}$, where

$$s(x) = \langle x \rangle,$$

$$x * y = \begin{cases} \langle x_1, \dots, x_n, y_1, \dots, y_m \rangle, & \text{if } x = \langle x_1, \dots, x_n \rangle \text{ and } y = \langle y_1, \dots, y_m \rangle, \\ 0 & \text{, otherwise.} \end{cases}$$

We can assume without loss of generality that the function $\langle\ \rangle : A^{<\omega} \to A$ agrees with the standard coding on ω.

1.4. If A is a set such that $\omega \subseteq A$, then a <u>functional</u> (on A with values in ω) is a partial mapping

$$\Phi : A^n \times \mathcal{P}\mathfrak{F}_{k_1} \times \dots \times \mathcal{P}\mathfrak{F}_{k_m} \to \omega$$

which is monotone, i.e. if $f_1 \subseteq g_1, \ldots, f_m \subseteq g_m$ and $\Phi(\vec{x}, f_1, \ldots, f_m) = w$, then $\Phi(\vec{x}, g_1, \ldots, g_m) = w$ (here $\mathcal{P}\mathfrak{F}_k = \{f : f$ is a k-ary partial function from A into $\omega\}$). The signature of such a functional Φ is the tuple $(n, k_1, \ldots, k_m)$; a functional is operative if its signature is of the form $(n, n, k_1, \ldots, k_m)$. An operative functional $\Phi(\vec{x}, f, g_1, \ldots, g_m)$ defines a transfinite sequence $\{\Phi^\xi\}$ of functionals by the equations

$$\Phi^\xi(\vec{x}, \vec{g}) = \Phi(\vec{x}, \lambda\vec{y}\Phi^{<\xi}(\vec{y}, \vec{g}), \vec{g}),$$

where

$$\Phi^{<\xi}(\vec{y}, \vec{g}) = w \iff \text{there is } \eta < \xi \text{ such that } \Phi^\eta(\vec{y}, \vec{g}) = w.$$

We put

$$\Phi^\infty(\vec{y}, \vec{g}) = w \iff \text{there is an ordinal } \xi \text{ such that } \Phi^\xi(\vec{y}, \vec{g}) = w.$$

Note that if Φ is of signature (n,n), then Φ^∞ is actually an n-ary partial function; the least ordinal ξ such that $\Phi^\xi = \Phi^{<\xi}$ is called the closure ordinal of Φ and is denoted by $\|\Phi\|$.

Let $\mathfrak{F}$ be a class of functionals on A. A functional $\Psi(\vec{x}, \vec{g})$ is $\mathfrak{F}$-recursive if there is a functional $\Phi(\vec{u}, \vec{x}, f, \vec{g})$ in $\mathfrak{F}$ and a sequence $\vec{n} = (n_1, \ldots, n_k)$ from ω such that

$$\Psi(\vec{x}, \vec{g}) = \Phi^\infty(\vec{n}, \vec{x}, \vec{g}).$$

A relation $R \subseteq A^n$ is $\mathfrak{F}$-semirecursive if it is the domain of an n-ary $\mathfrak{F}$-recursive partial function $\varphi : A^n \to \omega$; R is $\mathfrak{F}$-recursive if its characteristic function χ_R is $\mathfrak{F}$-recursive, where

$$\chi_R(\vec{x}) = \begin{cases} 0, & \text{if } R(\vec{x}), \\ 1, & \text{if } \neg R(\vec{x}). \end{cases}$$

These notions relativize directly to any finite sequence $\vec{x} = (x_1, \ldots, x_n)$ from A. We say, for example, that a partial function f is $\mathfrak{F}$-recursive from $\vec{x}$ if there is an $\mathfrak{F}$-recursive partial function g such that

$$f(\vec{y}) = g(\vec{x}, \vec{y}).$$

1.5. A class $\mathfrak{F}$ of functionals is suitable if it contains the functions and functionals (i)-(v) and is closed under the rules (vi)-(x) below:

(i) $$\chi_\omega(a) = \begin{cases} 0, & \text{if } a \in \omega, \\ 1, & \text{if } a \notin \omega. \end{cases}$$

(ii) $$\varphi_1(a) = \begin{cases} a, & \text{if } a \in \omega, \\ 0, & \text{if } a \notin \omega. \end{cases}$$

(iii) $\varphi_2(a) = \begin{cases} a+1, & \text{if } a \in \omega, \\ 0, & \text{if } a \notin \omega. \end{cases}$

(iv) $\varphi_3(a,b) = \begin{cases} 0, & \text{if } a,b \in \omega \text{ and } a = b, \\ 1, & \text{if } a,b \in \omega \text{ and } a \neq b, \\ 0, & \text{otherwise.} \end{cases}$

(v) Evaluation Functional: $\Phi(\vec{x},f) = f(\vec{x})$

(vi) Addition of variables: If $\Phi(\vec{x},\vec{f})$ is in $\mathfrak{F}$, so is

$$\Psi(\vec{z},\vec{x},\vec{y},\vec{h},\vec{f},\vec{g}) = \Phi(\vec{x},\vec{f})$$

(vii) Composition: If $\Phi(a,\vec{x},\vec{f})$ and $X(\vec{x},\vec{f})$ are in $\mathfrak{F}$, so is

$$\Psi(\vec{x},\vec{f}) = \Phi(X(\vec{x},\vec{f}),\vec{x},\vec{f})$$

(viii) Definition by cases: If $\Phi(\vec{x},\vec{f})$ and $X(\vec{x},\vec{f})$ are in $\mathfrak{F}$, so is

$$\Psi(a,\vec{x},\vec{f}) = \begin{cases} \Phi(\vec{x},\vec{f}), & \text{if } a = 0, \\ \Psi(\vec{x},\vec{f}), & \text{if } a \in \omega \text{ and } a = 0, \\ 0, & \text{if } a \notin \omega. \end{cases}$$

(ix) Substitution by projections: A projection is mapping π such that $\pi(x_1,\ldots,x_n) = x_i$, where $1 \leq i \leq n$. If $\Phi(y_1,\ldots,y_m,\vec{f})$ is in $\mathfrak{F}$ and $\pi_1,\ldots,\pi_m$ are projections, then $\Psi(\vec{x},\vec{f})$ is in $\mathfrak{F}$, where

$$\Psi(\vec{x},\vec{f}) = \Phi(\pi_1(\vec{x}),\ldots,\pi_m(\vec{x}),\vec{f}).$$

(x) Functional Substitution: If $\Phi(\vec{x},g_1,\ldots,g_m)$ and $X_1,\ldots,X_m$ are in $\mathfrak{F}$, so is

$$\Psi(\vec{x},\vec{f}) = \Phi(\vec{x},\lambda\vec{y}_1\, X_1(\vec{y}_1,\vec{x},\vec{f}),\ldots,\lambda\vec{y}_m X_m(\vec{y}_m,\vec{x},\vec{f})).$$

We introduce the following abbreviations:

$$\Phi(\vec{x},\vec{f})\downarrow \iff \Phi(\vec{x},\vec{f}) \text{ is defined,}$$

$$\Phi(\vec{x},\vec{f})\uparrow \iff \Phi(\vec{x},\vec{f}) \text{ is undefined,}$$

$$f(\vec{x}) \neq n \iff f(\vec{x}) \text{ is defined and has value } \neq n.$$

1.6. Each quantifier Q on a set A with $\omega \subseteq A$ gives rise to two functionals $\underset{\sim}{F}_Q$ and $\underset{\sim}{F}^{\#}_Q$ of signature $(0,1)$, where

$$\underset{\sim}{F}_Q(f) = \begin{cases} 0, & \text{if } f \text{ is total and } Qx(f(x) = 0), \\ 1, & \text{if } f \text{ is total and } \breve{Q}x(f(x) \neq 0), \\ \text{undefined}, & \text{if } f \text{ is not total;} \end{cases}$$

$$F^{\#}_Q(f) = \begin{cases} 0, & \text{if } Qx(f(x) = 0), \\ 1, & \text{if } \breve{Q}x(f(x) \neq 0), \\ \text{undefined}, & \text{otherwise.} \end{cases}$$

If $\exists = \{Y \subseteq A : Y \neq \emptyset\}$ is the existential quantifier on A, then we write $\underset{\sim}{E}$ for the functional $\underset{\sim}{F}^{\#}_{\exists}$ and $\underset{\sim}{E}^{\#}$ for the functional $\underset{\sim}{F}^{\#}_{\exists}$, i.e.

$$\underset{\sim}{E}(f) = \begin{cases} 0 & \text{, if } f \text{ is total and } (\exists x)(f(x) = 0), \\ 1 & \text{, if } f \text{ is total and } (\forall x)\ (f(x) \neq 0), \\ \text{undefined,} & \text{otherwise;} \end{cases}$$

$$\underset{\sim}{E}^{\#}(f) = \begin{cases} 0 & \text{, if } (\exists x)(f(x) = 0), \\ 1 & \text{, if } (\forall x)(f(x) \neq 0), \\ \text{undefined,} & \text{otherwise.} \end{cases}$$

If $Q = \{Y \subseteq A : Y \cap \omega \neq \emptyset\}$, then we write $\underset{\sim}{E}^{\#}_{\omega}$ for the functional $\underset{\sim}{F}^{\#}_{Q}$, i.e.

$$\underset{\sim}{E}^{\#}_{\omega}(f) = \begin{cases} 0 & \text{, if } (\exists\, x \in \omega)(f(x) = 0), \\ 1 & \text{, if } (\forall\, x \in \omega)(f(x) \neq 0), \\ \text{undefined,} & \text{otherwise.} \end{cases}$$

Let $\mathfrak{A} = \langle A, R_1, \ldots, R_n, f_1, \ldots, f_m, c_1, \ldots, c_k \rangle$ be a structure such that ω, $\leq_\omega$ are elementary on $\mathfrak{A}$ and let Q be a quantifier on A; if $\vec{\Phi} = (\Phi_1, \ldots, \Phi_s)$ is a finite sequence of functionals on A, then

$$\mathfrak{F}[\mathfrak{A}, \vec{\Phi}] = \mathfrak{F}[\vec{\Phi}]$$

is the smallest suitable class of functionals on A containing the functionals $\Phi_1, \ldots, \Phi_s$, the characteristic functions of $R_1, \ldots, R_n$, $=$ and closed under the functions $f_1, \ldots, f_m$ and the constant functions $x \to c_\ell$, $1 \leq \ell \leq k$. (In general, a class $\mathfrak{F}$ of functionals on A is <u>closed under a function</u> $f : A^n \to A$ if whenever $\Phi(x, \vec{y}, \vec{g})$ is in $\mathfrak{F}$, so is the functional $\Psi(\vec{z}, \vec{y}, \vec{g}) = \Phi(f(\vec{z}), \vec{y}, \vec{g})$).

We put (for $\vec{x} \in A^{<\omega}$)

$$\mathrm{ENV}[\mathfrak{A}, \vec{\Phi}] = \mathrm{ENV}[\vec{\Phi}] = \text{all } \mathfrak{F}[\mathfrak{A}\ \vec{\Phi}]\text{-semirecursive relations,}$$

$$\mathrm{ENV}[\mathfrak{A}, \vec{\Phi}, \vec{x}] = \mathrm{ENV}[\vec{\Phi}, \vec{x}] = \text{all } \mathfrak{F}[\mathfrak{A}, \vec{\Phi}]\text{-semirecursive relations from } \vec{x},$$

$$\mathrm{SEC}[\mathfrak{A}, \vec{\Phi}] = \mathrm{SEC}[\vec{\Phi}] = \text{all } \mathfrak{F}[\mathfrak{A}, \vec{\Phi}]\text{-recursive relations,}$$

$$\mathrm{SEC}[\mathfrak{A}, \vec{\Phi}, \vec{x}] = \mathrm{SEC}[\vec{\Phi}, \vec{x}] = \text{all } \mathfrak{F}[\mathfrak{A}, \vec{\Phi}]\text{-recursive relations from } \vec{x}.$$

The class $\mathrm{ENV}[\mathfrak{A}, \vec{\Phi}]$ is the <u>envelope of</u> $\Phi_1, \ldots, \Phi_s$ and its elements are the <u>semi-recursive relations in</u> $\Phi_1, \ldots, \Phi_s$; the class $\mathrm{SEC}[\mathfrak{A}, \vec{\Phi}]$ is the <u>section of</u> $\Phi_1, \ldots, \Phi_s$ and its elements are the <u>recursive relations in</u> $\Phi_1, \ldots, \Phi_s$.

In this paper we will study the classes $\mathrm{ENV}[\underset{\sim}{E}, \underset{\sim}{F}_Q]$ and $\mathrm{SEC}[\underset{\sim}{E}, \underset{\sim}{F}_Q]$ and will compare them to the classes $\mathrm{ENV}[\underset{\sim}{E}^{\#}, \underset{\sim}{F}^{\#}_Q]$ and $\mathrm{SEC}[\underset{\sim}{E}^{\#}, \underset{\sim}{F}^{\#}_Q]$. It is a well known fact that the last two classes are nothing else but the classes of the Q-inductive and the Q-hyperelementary relations on $\mathfrak{A}$, so that we have the following

1.7. <u>Theorem</u>. Let $\mathfrak{A}$ be a structure such that $\omega \subseteq A$ and let Q be a quantifier on A. A relation R is Q-inductive on $\mathfrak{A}$ if and only if R is semirecursive in $\underset{\sim}{E}^{\#},\underset{\sim}{F}^{\#}_{Q}$, i.e.

$$\mathrm{IND}(\mathfrak{A},Q) = \mathrm{ENV}[\underset{\sim}{E}^{\#},\underset{\sim}{F}^{\#}_{Q}].$$

Proof. The theorem follows easily from the following two facts:

Fact 1: For each functional $\Phi(\vec{x},\vec{f})$ of signature $(n,k_1,\ldots,k_m)$ in $\mathfrak{I}[\underset{\sim}{E}^{\#},\underset{\sim}{F}^{\#}_{Q}]$ there is a formula $\varphi(\vec{x},u,S_1,\ldots,S_m)$ of the language $\mathcal{L}^{\mathfrak{A}}(Q)$ in which each S_i is a (k_i+1)-ary relation variable occurring positively $(1 \le i \le k)$ and such that

$$\Phi(\vec{x},f_1,\ldots,f_m) = u \Longleftrightarrow \varphi(\vec{x},u,\text{graph } f_1,\ldots,\text{graph } f_m).$$

This fact is proved by induction on the construction of the suitable class $\mathfrak{I}[\underset{\sim}{E}^{\#},\underset{\sim}{F}^{\#}_{Q}]$.

Fact 2: For each formula $\varphi(x_1,\ldots,x_n,S)$ of the language $\mathcal{L}^{\mathfrak{A}}(Q)$ in which S is an m-ary relation variable occurring positively there is a functional $\Phi(\vec{x},f)$ of signature (n,m) in $\mathfrak{I}[\underset{\sim}{E}^{\#},\underset{\sim}{F}^{\#}_{Q}]$ such that for any relation $R \subseteq A^m$

(i) $\varphi(\vec{x},R) \Longleftrightarrow \Phi(\vec{x},\psi_R) = 0$,

(ii) $\neg\varphi(\vec{x},R) \Longleftrightarrow \Phi(\vec{x},\psi_R)$ is undefined,

where

$$\psi_R(\vec{y}) = \begin{cases} 0, & \text{if } \vec{y} \in R, \\ \text{undefined}, & \text{if } \vec{y} \notin R. \end{cases}$$

This fact is proved by induction on the construction of the formula $\varphi(\vec{x},S)$. ⊣

1.8. If Γ is a class of relations on a set A with $\omega \subseteq A$, then the dual of Γ is the class $\check{\Gamma} = \{R : R \text{ is an n-ary relation } A\ (n \in \omega) \text{ and } A^n - R \in \Gamma\}$; we also put $\Delta = \Gamma \cap \check{\Gamma}$. Let Q be a quantifier on A; we say that Γ is closed under the deterministic Q-rule if whenever $P(y,\vec{x})$ and $S(y,\vec{x})$ are disjoint relations in Γ, then the relation R is also in Γ, where

$$R(\vec{x}) \Longleftrightarrow (\forall y)\,(P(y,\vec{x}) \vee S(y,\vec{x}))\ \&\ QyP(y,\vec{x}).$$

We say that Γ has the prewellordering property if for any relation $R \in \Gamma$ there is a map $\sigma : R \to$ Ordinals such that the relations $\le^*_\sigma$ and $<^*_\sigma$ are in Γ, where

(i) $\vec{x} \le^*_\sigma \vec{y} \Longleftrightarrow R(\vec{x})\ \&\ (\neg R(\vec{y}) \vee \sigma(\vec{x}) \le \sigma(\vec{y}))$,

(ii) $\vec{x} <^*_\sigma \vec{y} \Longleftrightarrow R(\vec{x})\ \&\ (\neg R(\vec{y}) \vee \sigma(\vec{x}) < \sigma(\vec{y}))$.

A class Γ of relations on A is ω-parametrized if for any $n \in \omega$ there is an $(n+1)$-ary relation $U \in \Gamma$ such that for any $P \subseteq A^n$

$$P \in \Gamma \Longleftrightarrow \text{there is } m \in \omega \text{ such that } P = U_m = \{\vec{x} : U(m,\vec{x})\}.$$

The next theorem summarizes the closure and structural properties of the Q-inductive relations on a structure $\mathfrak{A}$ and the semirecursive relations in $\underset{\sim}{E},\underset{\sim}{F}_{Q}$; it follows from results in Moschovakis [EIAS] and Kechris-Moschovakis [1977].

1.9. Theorem. Let $\mathfrak{A} = \langle A, R_1, \ldots, R_n, f_1, \ldots, f_m, c_1, \ldots, c_k \rangle$ be a structure such that ω, $\leq_\omega$ are elementary on A and let Q be a quantifier on A.

(i) The class $\mathrm{ENV}[\underset{\sim}{E}, \underset{\sim}{F}_Q]$ is closed under &, $\vee$, $\forall$ and the deterministic $\exists$, Q and $\breve{Q}$-rules; a relation $R \subseteq A^n$ is in $\mathrm{SEC}[\underset{\sim}{E}, \underset{\sim}{F}_Q]$ if and only if R and $A^n - R$ are in $\mathrm{EVN}[\underset{\sim}{E}, \underset{\sim}{F}_Q]$.

(ii) The class $\mathrm{IND}(\mathfrak{A}, Q) = \mathrm{ENV}[\underset{\sim}{E}^{\#}, \underset{\sim}{F}_Q^{\#}]$ is closed under &, $\vee$, $\forall$, $\exists$, Q, $\breve{Q}$.

(iii) The classes $\mathrm{ENV}[\underset{\sim}{E}, \underset{\sim}{F}_Q]$ and $\mathrm{IND}(\mathfrak{A}, Q)$ have the prewellordering property.

(iv) If the structure $\mathfrak{A}$ is acceptable, then the classes $\mathrm{ENV}[\underset{\sim}{E}, \underset{\sim}{F}_Q]$ and $\mathrm{IND}(\mathfrak{A}, Q)$ are ω-parametrized. Moreover in this case the class $\mathrm{ENV}[\underset{\sim}{E}, \underset{\sim}{F}_Q]$ is closed under $\exists^\omega$, i.e., if $R(\vec{x}, y)$ is in $\mathrm{ENV}[\underset{\sim}{E}, \underset{\sim}{F}_Q]$, so is the relation $\exists^\omega R$, where

$$\exists^\omega R(\vec{x}) \iff (\exists n \in \omega) R(\vec{x}, n). \qquad \dashv$$

1.10. By abstracting the properties of the semirecursive relations in $\underset{\sim}{E}$ and the inductive relations on an acceptable structure $\mathfrak{A}$ we arrive at the notions of a ("lightface") semi-Spector class and a ("lightface") Spector class.

Let $\mathfrak{A} = \langle A, R_1, \ldots, R_n, f_1, \ldots, f_m, c_1, \ldots, c_k \rangle$ be a structure such that ω, $\leq_\omega$ are elementary on $\mathfrak{A}$. A collection Γ of relations on A is a semi-Spector class on $\mathfrak{A}$ if

(i) Γ contains all the elementary relations on $\mathfrak{A}$ and is closed under &, $\vee$, $\forall$ and the deterministic $\exists$-rule.

(ii) Γ contains a coding scheme on A, i.e. there is a coding function $\langle\ \rangle : A^{<\omega} \to A$ such that the corresponding relation seq and the functions ℓh and q are in Δ.

(iii) Γ has the prewellordering property and is ω-parameterized.

A Spector class on $\mathfrak{A}$ is a semi-Spector class Γ which is also closed under $\exists$.

The notions of a semi-Spector class and a Spector class were introduced by Moschovakis [EIAS] and [1974b]. In order to avoid confusion, however, we should point out that the notions introduced by Moschovakis were the "boldface" versions of the ones defined here, i.e. arbitrary parameters from A were allowed in the elementary relations and A-parametrization was required in the definitions. The Q-inductive relations on $\mathfrak{A}$ and the semirecursive relations in $\underset{\sim}{E}, \underset{\sim}{F}_Q$ have the following important minimality characterizations, which follow from Theorem 9A.2 in Moschovakis [EIAS] and the Main Lemma in Moschovakis [1974a]:

1.11. Theorem. Let $\mathfrak{A} = \langle A, R_1, \ldots, R_n, f_1, \ldots, f_m, c_1, \ldots, c_k \rangle$ be an acceptable structure and let Q be a quantifier on A.

(i) The class $\mathrm{IND}(\mathfrak{A},Q)$ of the Q-inductive relations on $\mathfrak{A}$ is the smallest Spector class on $\mathfrak{A}$ closed under Q and $\breve{Q}$.

(ii) The class $\mathrm{ENV}[\underset{\sim}{E},\underset{\sim}{F}_Q]$ of the semirecursive relations in $\underset{\sim}{E},\underset{\sim}{F}_Q$ is the smallest semi-Spector class on $\mathfrak{A}$ closed under the deterministic Q and $\breve{Q}$ rules. Moreover,

$$\mathrm{ENV}[\underset{\sim}{E},\underset{\sim}{F}_Q] = \mathrm{ENV}[\underset{\sim}{E},\underset{\sim}{F}_Q,\underset{\sim}{E}^{\#}_{\omega}].$$ ⊣

1.12. Let $\mathfrak{A} = \langle A,R_1,\ldots,R_n,f_1,\ldots,f_m,c_1,\ldots,c_k\rangle$ be an acceptable structure and Q a quantifier on $\mathfrak{A}$. The next table provides a comparison between the theories of positive elementary induction and recursion in $\underset{\sim}{E},\underset{\sim}{F}_Q$. In Sections 2 and 3 of this paper we will establish some of the results about recursion in $\underset{\sim}{E},\underset{\sim}{F}_Q$ which are the natural analogs of theorems about positive elementary induction.

Theory	Positive Elementary Induction	Recursion in $\underset{\sim}{E},\underset{\sim}{F}_Q$
Class	$\mathrm{IND}(\mathfrak{A},Q)$	$\mathrm{ENV}[\underset{\sim}{E},\underset{\sim}{F}_Q]$
closed under	$\&,\vee,\forall,\exists,Q,\breve{Q}$	$\&,\vee,\forall,\exists^{\omega}$ deterministic $\exists,Q,\breve{Q}$-rules
structural properties	prewellordering property ω-parametrization	prewellordering property ω-parametrization
minimality characterization	smallest Spector class on $\mathfrak{A}$	smallest semi-Spector class on $\mathfrak{A}$
other recursive theoretic characterizations	$\mathrm{IND}(\mathfrak{A},Q) = \mathrm{ENV}[\underset{\sim}{E}^{\#},\underset{\sim}{F}^{\#}_Q]$	$\mathrm{ENV}[\underset{\sim}{E},\underset{\sim}{F}_Q] = \mathrm{ENV}[\underset{\sim}{E},\underset{\sim}{F}_Q,\underset{\sim}{E}^{\#}_{\omega}]$
second order property	second stage comparison theorem	second stage comparison theorem
normal form theorem	Spector-Gandy Theorem $R(\vec{x}) \Leftrightarrow (\exists Y \in \mathrm{HYP}(Q,\vec{x}))\varphi(Y,\vec{x})$	Spector-Gandy Theorem $R(\vec{x}) \Leftrightarrow (\exists Y \in \mathrm{SEC}[\underset{\sim}{E},\underset{\sim}{F}_Q,\vec{x}])\varphi(Y,\vec{x})$
model theoretic characterizations	i) $\underset{\sim}{\mathrm{HYP}}(\mathfrak{A})$ is the smallest model of the schema of Δ^1_1-comprehension ii) $\underset{\sim}{\mathrm{HYP}}(\mathfrak{A})$ is the smallest model of the schemata of Σ^1_1-collection and Δ^0_∞-comprehension	i) ? ii) $\mathfrak{R}=\{\mathrm{SEC}[\underset{\sim}{E},\underset{\sim}{F}_Q,\vec{x}]:\vec{x}\in A^{<\omega}\}$ is the smallest indexed family on A satisfying the schemata of Σ^1_1-collection and Δ^0_∞-comprehension
relation to non-monotone inductive definability	$\mathrm{IND}(\mathfrak{A}) = \Sigma^0_2(\mathfrak{A})$ - IND	$\mathrm{ENV}[\underset{\sim}{E}] = \Pi^0_1(\mathfrak{A})$ - IND
syntactic characterization	$\mathrm{IND}(\mathfrak{A}) = \Pi^1_1(\mathfrak{A})$ (for $\mathfrak{A}$ a countable acceptable structure)	?

2. Normality, the Stage Comparison Theorem and the Second Stage Comparison Theorem

2.1. Many interesting structural results about positive elementary induction and recursion in $\underset{\sim}{E},\underset{\sim}{F}_Q$ (for example the prewellordering property and its consequences) follow from the Stage Comparison Theorems. Moschovakis [1969b] and [EIAS] established the Second Stage Comparison Theorem for positive elementary induction on an acceptable structure and Aczel [1972] extended it to any structure. This result is an important second order property of the inductive relations and it is the key technical tool in proving the Spector-Gandy Theorem and model theoretic characterizations in the theory of positive elementary reduction. Moschovakis [1976] introduced a notion of normality which makes it possible to obtain the Stage Comparison Theorem in the context of functional induction. A Second Stage Comparison Theorem for normal classes of functionals was established in Kolaitis [1977] and a corollary of it was the corresponding theorem for recursion in $\underset{\sim}{E},\underset{\sim}{F}_Q$. In what follows in this part of the paper we define the notion of normality and outline the proofs of the Stage and the Second Stage Comparison Theorem for normal classes of functionals.

2.2. If $\Phi(\vec{x},f)$ is an operative functional of signature (n,n) on a set A, then with each $\vec{x} \in A^n$ we associate the ordinal

$$|\vec{x}|_\Phi = \begin{cases} \text{least } \xi \text{ such that } \Phi^\xi(\vec{x})\downarrow, & \text{if } \Phi^\infty(\vec{x})\downarrow, \\ (\text{card}(A))^+ = \infty, & \text{if } \Phi^\infty(\vec{x})\uparrow. \end{cases}$$

If $\Phi(\vec{x},f)$ and $\Psi(\vec{y},g)$ are two operative functionals on A, we put

$$\vec{x} \leq^*_{\Phi,\Psi} \vec{y} \iff \Phi^\infty(\vec{x})\downarrow \text{ and } |\vec{x}|_\Phi \leq |\vec{y}|_\Psi,$$

$$\vec{x} <^*_{\Phi,\Psi} \vec{y} \iff \Phi^\infty(\vec{x})\downarrow \text{ and } |\vec{x}|_\Phi < |\vec{y}|_\Psi.$$

We write $\leq^*_\Phi$, $<^*_\Phi$ for $\leq^*_{\Phi,\Phi}$, $<^*_{\Phi,\Phi}$ respectively.

The <u>stage comparison partial function</u> χ <u>for the functionals</u> Φ,Ψ is

$$\chi(\vec{x},\vec{y}) = \begin{cases} 0, & \text{if } \vec{x} \leq^*_{\Phi,\Psi} \vec{y}, \\ 1, & \text{if } \vec{y} <^*_{\Psi,\Phi} \vec{x}, \\ \text{undefined, otherwise.} \end{cases}$$

2.3. <u>Definition</u>. Let $\mathfrak{F}$ be a class of functionals on A and let $\Phi(\vec{x},\vec{f})$ be a functional in $\mathfrak{F}$; Φ is <u>normal in</u> $\mathfrak{F}$ if there is an $\mathfrak{F}$-recursive functional $\Delta_\Phi(\vec{x},\vec{f},\vec{\delta})$ taking values 0,1 only such that

(i) if $\Phi(\vec{x},\vec{f}\restriction\{\vec{y} : \vec{\delta}(\vec{y}) = 0\})\downarrow$, then $\Delta_\Phi(\vec{x},\vec{f},\vec{\delta}) = 0$,

(ii) if $\vec{\delta}$ is total, $\{\vec{y} : \vec{\delta}(\vec{y}) = 0\} \subseteq \text{dom } \vec{f}$ and $\Phi(\vec{x},\vec{f}\restriction\{\vec{y} : \delta(\vec{y}) = 0\})\uparrow$, then

$\Delta_\Phi(\vec{x},\vec{f},\vec{\delta}) = 1$.

We say that Φ is <u>strongly normal in</u> $\mathfrak{F}$ if Δ_Φ is actually in $\mathfrak{F}$. In either case, we call Δ_Φ a <u>normalizing functional for</u> Φ. A class $\mathfrak{F}$ of functionals is <u>normal</u> if every functional in $\mathfrak{F}$ is normal in $\mathfrak{F}$; $\mathfrak{F}$ is <u>strongly normal</u> if every functional in $\mathfrak{F}$ is strongly normal in $\mathfrak{F}$.

The following theorem in Kechris-Moschovakis [1977] provides a simple criterion for normality and establishes that the classes of functionals we are interested in are strongly normal.

2.4. <u>Theorem</u>. Let $\mathfrak{A} = \langle A,R_1,\ldots,R_n,f_1,\ldots,f_m,c_1,\ldots,c_k\rangle$ be a structure such that ω, $\leq_\omega$ are elementary on $\mathfrak{A}$ and let Q be a quantifier on A.

(i) If $\vec{\Phi} = (\Phi_1,\ldots,\Phi_s)$ is a finite sequence of functionals on A such that each Φ_i $(1 \leq i \leq s)$ is normal in $\mathfrak{F}[\vec{\Phi}]$, then $\mathfrak{F}[\vec{\Phi}]$ is normal. Similarly with "strongly normal" in place of "normal".

(ii) The functionals $\underset{\sim}{E}$ and $\underset{\sim}{F}_Q$ are strongly normal in $\mathfrak{F}[\underset{\sim}{E},\underset{\sim}{F}_Q]$ and hence the class $\mathfrak{F}[\underset{\sim}{E},\underset{\sim}{F}_Q]$ is strongly normal.

(iii) The functional $\underset{\sim}{F}^\#_Q$ is strongly normal in $\mathfrak{F}[\underset{\sim}{F}^\#_Q]$ and hence the classes $\mathfrak{F}[\underset{\sim}{F}^\#_Q]$, $\mathfrak{F}[\underset{\sim}{E}^\#,\underset{\sim}{F}^\#_Q]$, $\mathfrak{F}[\underset{\sim}{E},\underset{\sim}{F}_Q,\underset{\sim}{E}^\#_\omega]$ are strongly normal.

We will give now a proof of the Stage Comparison Theorem which is somewhat simpler than the one in Kechris-Moschovakis [1977].

2.5. <u>The Stage Comparison Theorem</u>. Let $\mathfrak{F}$ be a suitable class of functionals on A and let $\Phi(\vec{x},f)$, $\Psi(\vec{y},g)$ be two functionals in $\mathfrak{F}$ which are normal in $\mathfrak{F}$; then the stage comparison function χ of Φ,Ψ is $\mathfrak{F}$-recursive.

<u>Proof</u>. We will prove the theorem in the case $\Phi = \Psi$, so that for a given functional $\Phi \in \mathfrak{F}$ which is normal in $\mathfrak{F}$ we want to show that the stage comparison function

$$\chi(\vec{x},\vec{y}) = \begin{cases} 0, & \text{if } \vec{x} \leq^*_\Phi \vec{y}, \\ 1, & \text{if } \vec{y} <^*_\Phi \vec{x}, \\ \text{undefined}, & \text{otherwise}, \end{cases}$$

is $\mathfrak{F}$-recursive. It is enough to find an $\mathfrak{F}$-recursive functional $X(\vec{x},\vec{y},h)$ taking values $0,1$ such that $\chi \subseteq X^\infty$, because then

$$\chi(\vec{x},\vec{y}) = \begin{cases} \Phi^\infty(\vec{x}) \cdot 0, & \text{if } X^\infty(\vec{x},\vec{y}) = 0, \\ \varphi(\Psi^\infty(\vec{y})), & \text{if } X^\infty(\vec{x},\vec{y}) = 1, \end{cases}$$

where $\varphi(a) = 1$, for all $a \in A$.

Let $\Delta_\Phi(\vec{x},f,\delta)$ be a normalizing functional for Φ and assume that $\chi(\vec{x},\vec{y}) = 0$, i.e. $\vec{x} \leq^*_\Phi \vec{y}$. Then

$$\vec{x} \leq^*_\Phi \vec{y} \Rightarrow \Phi^{|\vec{x}|}(\vec{x})\downarrow \Rightarrow \Phi(\vec{x},\Phi^\infty \restriction \{\vec{x}\,' : \vec{x}\,' <^*_\Phi \vec{y}\})\downarrow \Rightarrow$$

$$\Phi(\vec{x},\Phi^\infty \restriction \{\vec{x}\,' : \chi(\vec{y},\vec{x}\,') = 1\})\downarrow \Rightarrow \Delta_\Phi(\vec{x},\Phi^\infty,\lambda\, \vec{x}\,' \dot{\neg}\, \chi(\vec{y},\vec{x}\,')) = 0,$$

where in general

$$\dot{\neg}\, h(\vec{z}) = \begin{cases} 1, & \text{if } h(\vec{z}) = 0, \\ 0, & \text{if } h(\vec{z}) \neq 0. \end{cases}$$

This argument shows that

$$\chi(\vec{x},\vec{y}) = 0 \Rightarrow \Delta_\Phi(\vec{x},\Phi^\infty, \lambda\vec{x}\,' \dot{\neg}\, \chi(\vec{y},\vec{x}\,')) = 0.$$

Using a similar argument we can show that

$$\chi(\vec{x},\vec{y}) = 1 \Rightarrow \Delta_\Phi(\vec{x},\Phi^\infty, \lambda\vec{x}\,' \dot{\neg}\, \chi(\vec{y},\vec{x}\,')) = 1.$$

With this as motivation, we put

$$X(\vec{x},\vec{y},h) = \Delta_\Phi(\vec{x},\Phi^\infty, \lambda\vec{x}\,' \dot{\neg}\, h(\vec{y},\vec{x}\,')).$$

The functional X is $\mathfrak{F}$-recursive and therefore (by the Induction Completeness Theorem of Kechris-Moschovakis [1976]) the function $X^\infty(\vec{x},\vec{y})$ is also $\mathfrak{F}$-recursive. Finally, we prove by simultaneous induction on ξ that

(i) if $\vec{x} \leq^*_\Phi \vec{y}$ and $|\vec{x}|_\Phi = \xi$, then $X^\infty(\vec{x},\vec{y}) = 0$,

(ii) if $\vec{y} <^*_\Phi \vec{x}$ and $|\vec{y}|_\Phi = \xi$, then $X^\infty(\vec{x},\vec{y}) = 1$.

Assume that (i) and (ii) hold for all $\eta < \xi$. If $\vec{x} \leq^*_\Phi \vec{y}$ and $|\vec{x}|_\Phi = \xi$, then $\Phi(\vec{x},\Phi^\infty \restriction \{\vec{x}\,' : |\vec{x}\,'|_\Phi < \xi$ and $\vec{x}\,' <^*_\Phi \vec{y}\})\downarrow$, so that

$$X^\infty(\vec{x},\vec{y}) = \Delta_\Phi(\vec{x},\Phi^\infty, \lambda\vec{x}\,' \dot{\neg}\, X^\infty(\vec{y},\vec{x}\,')) = 0.$$

If $\vec{y} <^*_\Phi \vec{x}$ and $|\vec{y}|_\Phi = \xi$, then $\Phi(\vec{x},\Phi^\infty \restriction \{\vec{x}\,' : \vec{x}\,' <^*_\Phi \vec{y}\})\uparrow$. Using the induction hypothesis and part (i) for ξ we can easily see that the function $\lambda\vec{x}\,' X^\infty(\vec{y},\vec{x}\,')$ is total and $(X^\infty(\vec{y},\vec{x}\,') = 1 \Longleftrightarrow \vec{x}\,' <^*_\Phi \vec{y})$. We have therefore that $\Phi(\vec{x},\Phi^\infty \restriction \{\vec{x}\,' : \dot{\neg} X^\infty(\vec{y},\vec{x}\,') = 0\})\uparrow$ and hence

$$X^\infty(\vec{x},\vec{y}) = \Delta_\Phi(\vec{x},\vec{y}, \lambda\, \vec{x}\,' \dot{\neg}\, X^\infty(\vec{y},\vec{x}\,')) = 1. \qquad \dashv$$

The general case of the stage comparison theorem follows from what we proved above and the Simultaneous Induction Lemma of Kechris-Moschovakis [1976]. Another way to establish the general case directly is to consider the simultaneous induction

$$X_1(\vec{x},\vec{y},h_1,h_2) = \Delta_\Phi(\vec{x},\Phi^\infty, \lambda\, \vec{x}\,' \dot{\neg}\, h_2(\vec{y},\vec{x}\,'))$$

$$X_2(\vec{x},\vec{y},h_1,h_2) = \Delta_\Psi(\vec{y},\Psi^\infty, \lambda\, \vec{y}\,' \dot{\neg}\, h_1(\vec{x},\vec{y}\,'))\ .$$

Here X_1^∞ generates the stage comparison function for Φ,Ψ and X_2^∞ the stage comparison function for Ψ,Φ.

2.6. Let $\mathfrak{F}$ be a normal class of functionals on A and let $\Phi(\vec{x},g)$ be an operative functional in $\mathfrak{F}$. For each ordinal $\xi < \|\Phi\|$ let Φ^ξ be the ξ-th stage of Φ, i.e.

$$\Phi^\xi(\vec{x}) = \Phi(\vec{x},\Phi^{<\xi})$$

and let Ψ^ξ be the stage comparison function for Φ up to ξ, i.e.

$$\Psi^\xi(\vec{x},\vec{y}) = \begin{cases} 0, & \text{if } \vec{x} \leq_\Phi^* \vec{y} \text{ and } |\vec{x}|_\Phi \leq \xi, \\ 1, & \text{if } \vec{y} <_\Phi^* \vec{x} \text{ and } |\vec{y}|_\Phi \leq \xi. \end{cases}$$

We fix a normalizing functional $\Delta_\Phi(\vec{x},g,\delta)$ for Φ and we put

$$X(\vec{x},\vec{y},f,g) = \Delta_\Phi(\vec{x},g,\lambda\vec{x}\,'\dot{-}f(\vec{y},\vec{x}\,')).$$

We also put $P(f,g)$ for the conjunction of the following four conditions:

(i) If $g(\vec{x}) = w$, then $\Phi(\vec{x},g\restriction\{\vec{x}\,' : f(\vec{x},\vec{x}\,') = 1\}) = w$.

(ii) If $f(\vec{x},\vec{y}) = 0$, then $g(\vec{x})$ is defined and the functions $\lambda\vec{x}\,' f(\vec{x}\,',\vec{x})$, $\lambda\vec{y}\,' f(\vec{x},\vec{y}\,')$ are total.

(iii) If $f(\vec{x},\vec{y}) = 1$, then $g(\vec{y})$ is defined and the functions $\lambda\vec{x}\,' f(\vec{x}\,',\vec{y})$, $\lambda\vec{y}\,' f(\vec{y},\vec{y}\,')$ are total.

(iv) Range $f \subseteq \{0,1\}$ and if $f(\vec{x},\vec{y}) = w$, then $X(\vec{x},\vec{y},f,g) = w$.

It is now very easy to show the following

2.7. <u>Lemma</u>. Let $\Phi(\vec{x},g)$ be an operative functional in the normal class $\mathfrak{F}$ and let ξ be an ordinal such that $\xi < \|\Phi\|$.

a) The functions Ψ^ξ and Φ^ξ satisfy the conditions (i)-(iv) introduced above, i.e. $P(\Psi^\xi,\Phi^\xi)$ holds.

b) If $|\vec{x}|_\Phi = \xi$, then the functions Ψ^ξ and Φ^ξ are $\mathfrak{F}$-recursive from $\vec{x}$ and their domains are $\mathfrak{F}$-recursive sets from $\vec{x}$.

⊣

From this lemma we see that the statement $P(f,g)$ introduced above describes some "first order" properties of the functions Ψ^ξ and Φ^ξ. It turns out that these properties are also sufficient to describe "approximations" of the functions Ψ^ξ and Φ^ξ. This is the content of the Basic Lemma in Kolaitis [1977] which shows that if f and g are partial functions such that $P(f,g)$ holds, then f and g agree with Ψ^ξ and Φ^ξ in the intersections of their domains. We will now outline a proof of this result.

2.8. The Basic Lemma. Let $\mathfrak{F}$ be a normal class of functionals, $\Phi(\vec{x},g)$ an operative functional in $\mathfrak{F}$ and f,g partial functions such that $P(f,g)$ holds.

(i) If $\vec{x} \leq^*_\Phi \vec{y}$ and $f(\vec{x},\vec{y})$ is defined, then $f(\vec{x},\vec{y}) = 0$ and $g(\vec{x}) = \Phi^{|\vec{x}|_\Phi}(\vec{x})$.

(ii) If $\vec{y} <^*_\Phi \vec{x}$ and $f(\vec{x},\vec{y})$ is defined, then $f(\vec{x},\vec{y}) = 1$ and $g(\vec{y}) = \Phi^{|\vec{y}|_\Phi}(\vec{y})$.

Proof. We will prove by simultaneous induction ξ that

(i) If $\vec{x} \leq^*_\Phi \vec{y}$, $|\vec{x}|_\Phi = \xi$ and $f(\vec{x},\vec{y})$ is defined, then $f(\vec{x},\vec{y}) = 0$ and $g(\vec{x}) = \Phi^\xi(\vec{x})$.

(ii) If $\vec{y} <^*_\Phi \vec{x}$, $|\vec{y}|_\Phi = \xi$ and $f(\vec{x},\vec{y})$ is defined, then $f(\vec{x},\vec{y}) = 1$ and $g(\vec{y}) = \Phi^\xi(\vec{y})$.

If (i) and (ii) hold for all $\eta < \xi$, then notice that $(f \restriction \mathrm{dom}\, \Psi^{<\xi}) \subseteq \Psi^{<\xi}$, where

$$\Psi^{<\xi}(\vec{x}\,', \vec{y}\,') = \begin{cases} 0, & \text{if } \vec{x}\,' \leq^*_\Phi \vec{y}\,' \text{ and } |\vec{x}\,'|_\Phi < \xi, \\ 1, & \text{if } \vec{y}\,' <^*_\Phi \vec{x}\,' \text{ and } |\vec{y}\,'|_\Phi < \xi. \end{cases}$$

(i) Assume that $\vec{x} \leq^*_\Phi \vec{y}$, $|\vec{x}|_\Phi = \xi$ and $f(\vec{x},\vec{y})$ is defined. We claim that $f(\vec{x},\vec{y}) = 0$; if not, then $f(\vec{x},\vec{y}) = 1$ and hence $X(\vec{x},\vec{y},f,g) = f(\vec{x},\vec{y}) = 1$. We will show that $X(\vec{x},\vec{y},f \restriction \mathrm{dom}\, \Psi^{<\xi},g) = 0$ and hence $X(\vec{x},\vec{y},f,g) = 0$, contradicting the hypothesis $f(\vec{x},\vec{y}) \neq 0$. Since $X(\vec{x},\vec{y},f \restriction \mathrm{dom}\, \Psi^{<\xi},g) = \Delta_\Phi(\vec{x},g,\lambda \vec{x}\,' \dot{\neg} (f \restriction \Psi^{<\xi})(\vec{y},\vec{x}\,'))$, suffices to show that $\Phi(\vec{x},g \restriction \{\vec{x}\,' : \dot{\neg} (f \restriction \Psi^{<\xi})(\vec{y},\vec{x}\,') = 0\})\downarrow$. But $\Phi(\vec{x},\Phi^{<\xi})\downarrow$ and so it is enough to show that

$$\Phi^{<\xi}(\vec{x}\,') = w \Rightarrow g(\vec{x}\,') = w \quad \text{and} \quad (f \restriction \mathrm{dom}\, \Psi^{<\xi})(\vec{y},\vec{x}\,') = 1.$$

Let $\vec{x}\,'$ be such that $|\vec{x}\,'|_\Phi = \eta < \xi$ and $\Phi^{<\xi}(\vec{x}\,') = w$. Since $f(\vec{x},\vec{y}) = 1$, we have that $f(\vec{y},\vec{x}\,')$ is defined and therefore the induction hypothesis implies immediately that $f(\vec{y},\vec{x}\,') = 1$ and $g(\vec{x}\,') = \Phi^\xi(\vec{x}\,') = \Phi^{<\xi}(\vec{x}\,')$.

This completes the proof that $f(\vec{x},\vec{y}) = 0$; we have to show also that $g(\vec{x}) = \Phi^\xi(\vec{x})$. Since $f(\vec{x},\vec{y}) = 0$ we have that $g(\vec{x})$ is defined and hence

$$g(\vec{x}) = \Phi(\vec{x},g \restriction \{\vec{x}\,' : f(\vec{x},\vec{x}\,') = 1\}).$$

Using the induction hypothesis and what we just proved we can easily show that $g \restriction \{\vec{x}\,' : f(\vec{x},\vec{x}\,') = 1\} = \Phi^{<\xi}$ and hence $g(\vec{x}) = \Phi(\vec{x},\Phi^{<\xi}) = \Phi^\xi(\vec{x})$.

Finally, we can establish (ii) with a similar argument using the induction hypothesis and part (i) for ξ. $\dashv$

If $\mathfrak{F}$ is a class of functionals on A and $\vec{x} = (x_1,\ldots,x_n) \in A^n$, then we put

$$\mathfrak{F}\text{-REC}_{\vec{x}} = \{f : f \text{ is an } \mathfrak{F}\text{-recursive partial function from } \vec{x} \text{ such that its domain is an } \mathfrak{F}\text{-recursive set from } \vec{x}\}.$$

2.9. <u>The Second Stage Comparison Theorem</u>. Let $\mathfrak{F}$ be a suitable normal class of functionals on A, $\Phi(\vec{x},g)$ an operative functional in $\mathfrak{F}$ and let $P(f,g)$ be the condition introduced in 2.6.

i) If $\vec{x} \leq^*_\Phi \vec{y}$, then $(\exists f \in \mathfrak{F}\text{-REC}_{\vec{x}})(\exists g \in \mathfrak{F}\text{-REC}_{\vec{x}})(P(f,g) \text{ and } f(\vec{x},\vec{y}) = 0)$.

ii) If $\vec{y} <^*_\Phi \vec{x}$, then $\neg (\exists f)(\exists g)(P(f,g) \text{ and } f(\vec{x},\vec{y}) = 0)$.

iii) If $\vec{y} <^*_\Phi \vec{x}$, then $(\exists f \in \mathfrak{F}\text{-REC}_{\vec{y}})(\exists g \in \mathfrak{F}\text{-REC}_{\vec{y}})(P(f,g) \text{ and } f(\vec{x},\vec{y}) = 1)$.

iv) If $\vec{x} \leq^*_\Phi \vec{y}$, then $\neg (\exists f)(\exists g)(P(f,g) \text{ and } f(\vec{x},\vec{y}) = 1)$.

v) If $\Phi^\infty(\vec{x})\downarrow$ or $\Phi^\infty(\vec{y})\downarrow$, then

$$\vec{x} \leq^*_\Phi \vec{y} \iff (\exists f \in \mathfrak{F}\text{-REC}_{\vec{x}})(\exists g \in \mathfrak{F}\text{-REC}_{\vec{x}})(P(f,g) \text{ and } f(\vec{x},\vec{y}) = 0)$$
$$\iff (\exists f)(\exists g)(P(f,g) \text{ and } f(\vec{x},\vec{y}) = 0);$$

$$\vec{y} <^*_\Phi \vec{x} \iff (\exists f \in \mathfrak{F}\text{-REC}_{\vec{y}})(\exists g \in \mathfrak{F}\text{-REC}_{\vec{y}})(P(f,g) \text{ and } f(\vec{x},\vec{y}) = 1)$$
$$\iff (\exists f)(\exists g)(P(f,g) \text{ and } f(\vec{x},\vec{y}) = 1).$$

⊣

This result follows now easily from the preceding Lemma 2.7 and the Basic Lemma 2.8. The Second Stage Comparison Theorem has important corollaries for the specific classes of functionals we are interested in.

Let $\mathfrak{A} = \langle A,R_1,\ldots,R_n,f_1,\ldots,f_m,c_1,\ldots,c_k\rangle$ be a structure such that ω, $\leq_\omega$ are elementary on $\mathfrak{A}$, let Q be a quantifier on A and let $\Phi(\vec{x},\vec{f})$ be a functional on A; we say that Φ is Q-<u>definable on</u> A if there is a formula $\varphi(\vec{x},u,\vec{S})$ of the language $\mathcal{L}^{\mathfrak{A}}(Q)$ such that

$$\Phi(\vec{x},\vec{f}) = u \iff \varphi(\vec{x},u,\overrightarrow{\text{graph } f}).$$

A class $\mathfrak{F}$ of functionals is Q-<u>definable on</u> $\mathfrak{A}$ if every functional in $\mathfrak{F}$ is Q-definable on $\mathfrak{A}$. We can easily show that if $\vec{\Phi} = (\Phi_1,\ldots,\Phi_s)$ is a finite sequence of Q-definable functionals on $\mathfrak{A}$, then the suitable class $\mathfrak{F}[\vec{\Phi}]$ is Q-definable. Note that if $\mathfrak{F}$ is a Q-definable class of functionals on $\mathfrak{A}$ which is strongly normal and Φ is an operative functional in $\mathfrak{F}$, then the condition $P(f,g)$ corresponding to Φ is actually elementary, i.e. there is a formula $\varphi(Y_1,Y_2)$ of the language $\mathcal{L}^{\mathfrak{A}}(Q)$ such that for any partial functions f,g

$$P(f,g) \iff \varphi(\text{graph } f, \text{ graph } g).$$

The classes $\mathfrak{F}[\underset{\sim}{E},\underset{\sim}{F}_Q]$, $\mathfrak{F}[\underset{\sim}{E},\underset{\sim}{F}_Q,\underset{\sim}{E}^\#_\omega]$, $\mathfrak{F}[\underset{\sim}{E}^\#,\underset{\sim}{F}^\#_Q]$ are typical examples of Q-definable, strongly normal classes of functionals. These comments and the Second Stage Comparison Theorem establish the following.

2.10. <u>Theorem</u>. Let $\mathfrak{A} = \langle A,R_1,\ldots,R_n,f_1,\ldots,f_m,c_1,\ldots,c_k\rangle$ be a structure such that $\omega, \leq_\omega$ are elementary on $\mathfrak{A}$, let Q be a quantifier on $\mathfrak{A}$ and let $\vec{\Phi} = (\Phi_1,\ldots,\Phi_s)$ be a finite sequence of Q-definable functionals on $\mathfrak{A}$ which are strongly normal in $\mathfrak{F}[\vec{\Phi}]$. If $\Phi(\vec{x},g)$ is an operative functional in $\mathfrak{F}[\vec{\Phi}]$, then there is a formula $\varphi(Y_1,Y_2)$ of the language $\mathcal{L}^{\mathfrak{A}}(Q)$ such that:

i) If $\vec{x} \leq^*_\Phi \vec{y}$, then $(\exists Y_1,Y_2 \in SEC[\vec{\Phi},\vec{x}])(\varphi(Y_1,Y_2)$ and $(\vec{x},\vec{y},0) \in Y_1)$.

ii) If $\vec{y} <^*_\Phi \vec{x}$, then $\neg(\exists Y_1,Y_2)(\varphi(Y_1,Y_2)$ and $(\vec{x},\vec{y},0) \in Y_1)$.

iii) If $\vec{y} <^*_\Phi \vec{x}$, then $(\exists Y_1,Y_2 \in SEC[\vec{\Phi},\vec{y}])(\varphi(Y_1,Y_2)$ and $(\vec{x},\vec{y},1) \in Y_1)$.

iv) If $\vec{x} <^*_\Phi \vec{y}$, then $\neg(\exists Y_1,Y_2)(\varphi(Y_1,Y_2)$ and $(\vec{x},\vec{y},1) \in Y_1)$.

v) If $\Phi^\infty(\vec{x})\downarrow$ or $\Phi^\infty(\vec{y})\downarrow$, then

$$\vec{x} \leq^*_\Phi \vec{y} \Longleftrightarrow (\exists Y_1,Y_2 \in SEC[\vec{\Phi},\vec{x}])(\varphi(Y_1,Y_2) \text{ and } (\vec{x},\vec{y},0) \in Y_1)$$
$$\Longleftrightarrow (\exists Y_1,Y_2)(\varphi(Y_1,Y_2) \text{ and } (\vec{x},\vec{y},0) \in Y_1);$$
$$\vec{y} <^*_\Phi \vec{x} \Longleftrightarrow (\exists Y_1,Y_2 \in SEC[\vec{\Phi},\vec{y}])(\varphi(Y_1,Y_2) \text{ and } (\vec{x},\vec{y},1) \in Y_1)$$
$$\Longleftrightarrow (\exists Y_1,Y_2)(\varphi(Y_1,Y_2) \text{ and } (\vec{x},\vec{y},1) \in Y_1).$$

⊣

3. <u>The Spector-Gandy Theorem and Model Theoretic Characterizations</u>

3.1. Moschovakis [1969a] and [EIAS] established the following generalization of a classical theorem of Spector [1960] and Gandy [1960] about the Π^1_1 relations on ω: a relation R is inductive on an acceptable structure $\mathfrak{A}$ if and only if there is a formula $\varphi(Y,\vec{x})$ of the language $\mathcal{L}^{\mathfrak{A}}$ such that

$$R(\vec{x}) \Longleftrightarrow (\exists Y \in HYP(\mathfrak{A},\vec{x}))\varphi(Y,\vec{x}).$$

Aczel [1972] obtained a new proof of this result and extended it to the class of the Q-inductive relations on $\mathfrak{A}$, where Q is a quantifier on A. The Spector-Gandy Theorem for recursion in $\underset{\sim}{E},\underset{\sim}{F}_Q$ and recursion in $\underset{\sim}{E},\underset{\sim}{F}^{\#}_Q$ was established in Kolaitis [1977]; the proof used the Second Stage Comparison Theorem for normal classes of functionals and was patterned after the proof of the Spector-Gandy Theorem for positive elementary induction in Aczel [1972]. In what follows here we will outline a proof of these results in a slightly more general context.

3.2. Let A be a set, Γ a collection of relations on A and $\vec{x} = (x_1,\ldots,x_n)$ a finite sequence from A; the <u>relativization of</u> Γ <u>to</u> $\vec{x}$ is the class

$$\Gamma_{\vec{x}} = \{R : \text{there is } P \in \Gamma \text{ such that } (\forall\vec{y})(R(\vec{y})) \Longleftrightarrow P(\vec{x},\vec{y})\}.$$

We put $\breve{\Gamma}_{\vec{x}} = \{R \subseteq A^n : (A^n - R) \in \Gamma_{\vec{x}}\}$ and $\Delta_{\vec{x}} = \Gamma_{\vec{x}} \cap \breve{\Gamma}_{\vec{x}}$.

If Γ is a semi-Spector class on an acceptable structure $\mathfrak{A}$, then there is a relation $U \in \Gamma$ which is <u>universal for</u> Γ, i.e. a relation $R \subseteq A^n$ is in Γ if and only if there is $n \in \omega$ such that

$$R(\vec{y}) \Longleftrightarrow U(\langle m,\langle\vec{y}\rangle\rangle),$$

where $\langle\ \rangle : A^{<\omega} \to A$ is the coding function of the structure $\mathfrak{A}$.

In general, if $P \subseteq A$ and $a \in A$, we put $P_a = \{y : P(\langle a,y\rangle)\}$. The next result is a well known fact about semi-Spector classes and it is the analog of Theorem 9C.8 in Moschovakis [EIAS].

3.3. Parametrization Theorem for semi-Spector classes. Let $\mathfrak{A}$ be an acceptable structure and Γ a semi-Spector class on $\mathfrak{A}$. Then there are subsets of A $I,H,\check{H}$ such that $I,H,A-\check{H}$ are in Γ and such that for each $\vec{x} \in A^{<\omega}$:

(i) If $R \subseteq A$ is in $\Delta_{\vec{x}}$, then there is some $m \in \omega$ such that $\langle \vec{x},m\rangle \in I$ and $R = H_{\langle \vec{x},m\rangle} = \{y : H(\langle\langle \vec{x},m\rangle,y\rangle)\}$

(ii) If $m \in \omega$ is such that $\langle \vec{x},m\rangle \in I$, then $H_{\langle \vec{x},m\rangle} = \check{H}_{\langle \vec{x},m\rangle}$, i.e. $\langle\langle \vec{x},m\rangle,y\rangle \in H \Leftrightarrow \langle\langle \vec{x},m\rangle,y\rangle \in \check{H}$.

Proof. Let $U \subseteq A$ be a relation in Γ, which is universal for Γ and let $\sigma : U \to$ Ordinals be a Γ-norm on U (i.e. the associated relations $\leq^*_\sigma$ and $<^*_\sigma$ are in Γ). We put

$$I = \{\langle \vec{x},n\rangle : (\forall y)[(\langle (n)_1,\langle \vec{x},y\rangle\rangle \leq^*_\sigma \langle (n)_2,\langle \vec{x},y\rangle\rangle) \vee (\langle (n)_2,\langle \vec{x},y\rangle\rangle <^*_\sigma \langle (n)_1,\langle \vec{x},y\rangle\rangle)]\},$$

$$H = \{\langle\langle \vec{x},n\rangle,y\rangle : \langle \vec{x},n\rangle \in I \text{ and } (\langle (n)_1,\langle \vec{x},y\rangle\rangle \leq^*_\sigma \langle (n)_2,\langle \vec{x},y\rangle\rangle)\},$$

$$\check{H} = \{\langle\langle \vec{x},n\rangle,y\rangle : \neg (\langle (n)_2,\langle \vec{x},y\rangle\rangle <^*_\sigma \langle (n)_1,\langle \vec{x},y\rangle\rangle)\}.$$

⊣

3.4. Let $\mathfrak{A} = \langle A,R_1,\ldots,R_n,f_1,\ldots,f_m,c_1,\ldots,c_k\rangle$ be an acceptable structure and Q a quantifier on A. Let $\vec{\Phi} = (\Phi_1,\ldots,\Phi_s)$ be a finite sequence of Q-definable functionals on $\mathfrak{A}$ which are strongly normal in $\mathfrak{F}[\vec{\Phi}]$ and such that the functionals $\underset{\sim}{E},\underset{\sim}{F}_Q,\underset{\sim}{E}^\#_\omega$ are in $\mathfrak{F}[\vec{\Phi}]$. It follows from general results in Kechris-Moschovakis [1977] that the class $\mathrm{ENV}[\vec{\Phi}]$ of the semirecursive relations in $\Phi_1,\ldots,\Phi_s$ is a semi-Spector class on $\mathfrak{A}$ closed under the deterministic Q and $\breve{Q}$-rules. The examples we have in mind of course are the classes $\mathrm{ENV}[\underset{\sim}{E}^\#,\underset{\sim}{F}^\#_Q]$, $\mathrm{ENV}[\underset{\sim}{E}^\#,\underset{\sim}{F}_Q]$, $\mathrm{ENV}[\underset{\sim}{E},\underset{\sim}{F}_Q] = \mathrm{ENV}[\underset{\sim}{E},\underset{\sim}{F}_Q,\underset{\sim}{E}^\#_\omega]$, $\mathrm{ENV}[\underset{\sim}{E},\underset{\sim}{F}^\#_Q] = \mathrm{ENV}[\underset{\sim}{E},\underset{\sim}{F}^\#_Q,\underset{\sim}{E}^\#_\omega]$.

Let $\vec{\Phi} = (\Phi_1,\ldots,\Phi_s)$ be a sequence of Q-definable functionals on $\mathfrak{A}$, strongly normal in $\mathfrak{F}[\vec{\Phi}]$ and such that $\underset{\sim}{E},\underset{\sim}{F}_Q,\underset{\sim}{E}^\#_\omega$ are in $\mathfrak{F}[\vec{\Phi}]$; let X be an operative functional in $\mathfrak{F}[\vec{\Phi}]$ and such that $U = \mathrm{dom}\, X^\infty$ is a universal set for $\mathrm{ENV}[\vec{\Phi}]$ and for each $\lambda < \|X\|$ put $\Lambda^\lambda = \mathrm{dom}\, X^\lambda$ and $\Lambda^{<\lambda} = \mathrm{dom}\, X^{<\lambda} = \bigcup_{\mu<\lambda} \Lambda^\mu$. Since U is a universal relation for the semi-Spector class $\mathrm{ENV}[\vec{\Phi}]$ we can find $i,h,\check{h} \in \omega$ such that $I = U_i, \check{H} = U_{\check{h}}, A - H = U_n$, where $I,H,\check{H}$ are the sets in Theorem 3.3. If $n \in \omega$ and $\lambda < \|X\|$ we put

$$I^\lambda_n = \{\langle x_1,\ldots,x_n,m\rangle \in (\Lambda^{<\lambda})_i : ((\Lambda^{<\lambda})_h)_{\langle \vec{x},m\rangle} = A - ((\Lambda^{<\lambda})_{\check{h}})_{\langle \vec{x},m\rangle}\}.$$

For each $\vec{x} \in A^n$ and each $\lambda < \|X\|$ we put

$$\mathrm{REC}^\lambda_{\vec{x}}[\vec{\Phi}] = \{H_{\langle \vec{x},m\rangle} : \langle \vec{x},m\rangle \in I^\lambda_n\}$$

Using the parametrization Theorem 3.3 and standard diagonal arguments we can prove the following

3.5. Hierarchy Lemma.

i) If $\lambda < \mu < \|X\|$, then $\mathrm{REC}^\lambda_{\vec{x}}[\vec{\Phi}] \subseteq \mathrm{REC}^\mu_{\vec{x}}[\vec{\Phi}]$.

ii) $\bigcup_{\lambda<\|X\|} \mathrm{REC}^\lambda_{\vec{x}}[\vec{\Phi}] = \{R \subseteq A : R \in \mathrm{SEC}[\vec{\Phi},\vec{x}]\}$

iii) If $m \in \omega$ is such that $\langle m,\langle \vec{x}\rangle\rangle \in U = \mathrm{dom}\, X^\infty$ and $|\langle m,\langle \vec{x}\rangle\rangle|_X = \lambda$, then

$$\mathrm{REC}^{\lambda}_{\vec{x}}[\vec{\Phi}] \subsetneq \{R \subseteq A : R \in \mathrm{SEC}[\vec{\Phi},\vec{x}]\}.$$

We may assume, without loss of generality, that there is some $n_0 \in \omega$ such that $n_0 \notin U = \mathrm{dom}\, X^{\infty}$. It is now easy to construct a functional $\Psi(a,x_1,\dots,x_n,f,g)$ of signature $(n+1,1,1)$ in $\mathfrak{F}[\vec{\Phi}]$ such that if Y is a subset of A and χ_Y is its characteristic function, then

$$\Psi(a,\vec{x},f,\chi_Y) = \begin{cases} X(a,f \upharpoonright \{y : y \neq n_0\}), & \text{if } a \neq n_0, \\ 0, & \text{if } a = n_0 \text{ and } (\exists n \in \omega)P(Y,f \upharpoonright \{y : y \neq n_0\},\vec{x},m), \\ X(n_0,f \upharpoonright \{y : y \neq n_0\}), & \text{otherwise}, \end{cases}$$

where

$$P(Y,f,\vec{x},m) \iff (\langle\vec{x},m\rangle \in (\mathrm{dom}\, f)_i) \;\&\; (Y \subseteq ((\mathrm{dom}\, f)_h)_{\langle\vec{x},m\rangle}) \;\&\; (A - Y \subseteq ((\mathrm{dom}\, f)_h^{\vee})_{\langle\vec{x},m\rangle}).$$

If for each set $Y \subseteq A$ and each $\vec{x} \in A^n$ we put

$$\Psi_{Y,\vec{x}}(a,f) = \Psi(a,\vec{x},f,\chi_Y),$$

then $\Psi_{Y,\vec{x}}$ is an operative functional of signature $(1,1)$ and defines a sequence $\{\Psi^{\mu}_{Y,\vec{x}}\}$, where

$$\Psi^{\mu}_{Y,\vec{x}}(a) = \Psi(a,\vec{x},\lambda b \Psi^{\lhd\mu}_{Y,\vec{x}}(b),\chi_Y)$$

It is now easy to check that for each ordinal $\lambda < \|X\|$

$$n_0 \in \mathrm{dom}\, \Psi^{\lambda}_{Y,\vec{x}}(n_0) \iff Y \in \mathrm{REC}^{\lambda}[\vec{\Phi},\vec{x}]$$

and hence using the hierarchy lemma 3.5 we conclude that

$$\Psi^{\infty}_{Y,\vec{x}}(n_0)\downarrow \iff Y \in \mathrm{SEC}[\vec{\Phi},\vec{x}].$$

From the last two facts and part (iii) of the hierarchy lemma if follows that

$$\langle m,\langle\vec{x}\rangle\rangle \in U = \mathrm{dom}\, X^{\infty} \iff (\exists Y \in \mathrm{SEC}[\vec{\Phi},\vec{x}])(\langle m,\langle\vec{x}\rangle\rangle <^{*}_{\Psi_{Y,\vec{x}}} n_0).$$

Using the Second Stage Comparison 2.10 and the fact that the functionals $\Psi_{Y,\vec{x}}$ are uniform in Y and $\vec{x}$ we can find a formula $\psi(Y,Y_1,\vec{x})$ of the language $\mathfrak{L}^{\mathfrak{A}}(Q)$ such that

$$\begin{aligned} \langle m,\langle\vec{x}\rangle\rangle \in U &\iff (\exists Y \in \mathrm{SEC}[\vec{\Phi},\vec{x}])(\langle m,\langle\vec{x}\rangle\rangle <^{*}_{\Psi_{Y,\vec{x}}} n_0) \\ &\iff (\exists Y \in \mathrm{SEC}[\vec{\Phi},\vec{x}])(\exists Y_1 \in \mathrm{SEC}[\vec{\Phi},\vec{x}])\psi(Y,Y_1,\vec{x}). \end{aligned}$$

This completes the proof of the nontrivial direction in the following theorem (the other direction of which follows immediately from the parametrization Theorem 3.3 and the closure properties of the class $\mathrm{ENV}[\vec{\Phi}]$).

3.6. <u>The Spector-Gandy Theorem</u>. Let $\mathfrak{A} = \langle A, R_1, \dots, R_n, f_1, \dots, f_m, c_1, \dots, c_k \rangle$ be an acceptable structure, let Q be a quantifier on A and let $\vec{\Phi} = (\Phi_1, \dots, \Phi_s)$ be a finite sequence of Q-definable functionals on $\mathfrak{A}$ which are strongly normal in $\mathfrak{J}[\vec{\Phi}]$ and such that the functionals $\underset{\sim}{E}, \underset{\sim}{F}_Q, \underset{\sim}{E}^{\#}_{\omega}$ are in $\mathfrak{J}[\vec{\Phi}]$. Then a relation R is semirecursive in $\Phi_1, \dots, \Phi_s$ if and only if there is some formula $\varphi(Y, \vec{x})$ of the language $\mathfrak{L}^{\mathfrak{A}}(Q)$ such that

$$R(\vec{x}) \Longleftrightarrow (\exists Y \in \mathrm{SEC}[\vec{\Phi}, \vec{x}])\varphi(Y, \vec{x}).$$

⊣

Among the special cases of this result is the Spector-Gandy Theorem for positive elementary induction (i.e. recursion in $\underset{\sim}{E}^{\#}, \underset{\sim}{F}^{\#}_Q$) and recursion in $\underset{\sim}{E}, \underset{\sim}{F}_Q$ ($\mathrm{ENV}[\underset{\sim}{E}, \underset{\sim}{F}_Q] = \mathrm{ENV}[\underset{\sim}{E}, \underset{\sim}{F}_Q, \underset{\sim}{E}^{\#}_{\omega}]$). Also the theorem relativizes directly to a finite sequence $\vec{x} \in A^{<\omega}$, i.e. (under the above hypotheses) a relation R is semirecursive in $\Phi_1, \dots, \Phi_s$ from $\vec{x}$ if and only if there is a formula of the language $\mathfrak{L}^{\mathfrak{A}}(Q)$ such that

$$R(\vec{y}) \Longleftrightarrow (\exists Y \in \mathrm{SEC}[\vec{\Phi}, \vec{x}, \vec{y}])\varphi(Y, \vec{x}, \vec{y}).$$

3.7. The "boldface" language $\underset{\sim}{\mathfrak{L}}^{\mathfrak{A}}$ of a structure $\mathfrak{A}$ is obtained from the language $\mathfrak{L}^{\mathfrak{A}}$ by adding a constant symbol $\underset{\sim}{a}$ for each $a \in A$. A collection $\mathcal{G}$ of relations on A satisfies the <u>schema of Δ^1_1-Comprehension on</u> $\mathfrak{A}$ if for any formulas $\varphi(\vec{x}, Z_1, \vec{V}), \psi(\vec{x}, Z_2, \vec{V})$ of the language $\underset{\sim}{\mathfrak{L}}^{\mathfrak{A}}$ and any relations $\vec{Y} \in \mathcal{G}$ if

$$(\forall \vec{x})\{(\exists Z_1 \in \mathcal{G})\varphi(\vec{x}, \vec{Z}_1, \vec{Y}) \Longleftrightarrow (\forall Z_2 \in \mathcal{G})\psi(\vec{x}, \vec{Z}_2, \vec{Y})\},$$

then

$$(\exists W \in \mathcal{G})(\forall \vec{x})[\vec{x} \in W \Longleftrightarrow (\exists Z_1 \in \mathcal{G})\varphi(\vec{x}, Z_1, \vec{Y})].$$

We say that $\mathcal{G}$ satisfies the <u>schema of Δ^0_∞-Comprehension on</u> $\mathfrak{A}$ if for any formula $\varphi(\vec{x}, \vec{W})$ of $\underset{\sim}{\mathfrak{L}}^{\mathfrak{A}}$ and any relations $\vec{Y} \in \mathcal{G}$ there is some $Z \in \mathcal{G}$ such that

$$(\forall \vec{x})(\vec{x} \in Z \Longleftrightarrow \varphi(\vec{x}, \vec{Y})).$$

A collection $\mathcal{G}$ of relations on A satisfies the <u>schema of Σ^1_1-Collection on</u> $\mathfrak{A}$ if for any formula $\varphi(\vec{x}, \vec{Z}, \vec{V})$ of $\underset{\sim}{\mathfrak{L}}^{\mathfrak{A}}$ and any relations $\vec{Y} \in \mathcal{G}$ we have that

$$(\forall \vec{x})(\exists Z \in \mathcal{G})\varphi(\vec{x}, \vec{Z}, \vec{Y}) \Longleftrightarrow (\exists W \in \mathcal{G})(\forall \vec{x})(\exists a)\varphi(\vec{x}, W_a, \vec{Y}),$$

where $W_a = \{\vec{y} : (a, \vec{y}) \in W\}$.

Moschovakis [1969b] and [EIAS] obtained the following characterization of the "boldface" hyperelementary relations on an acceptable structure, generalizing results of Kleene [1959a] and Kreisel [1961] about the hyperarithmetic relations on $\langle \omega, +, \cdot \rangle$.

3.8. <u>Theorem</u>. Let $\mathfrak{A}$ be an acceptable structure and let $\underset{\sim}{\mathrm{HYP}}(\mathfrak{A}) = \bigcup_{\vec{x} \in A^{<\omega}} \mathrm{HYP}(\mathfrak{A}, \vec{x})$ be the collection of the "boldface" hyperelementary relations on $\mathfrak{A}$.

i) The class $\underset{\sim}{\mathrm{HYP}}(\mathfrak{A})$ is the smallest collection of relations on A which is a model of the schema of Δ^1_1-Comprehension.

ii) The class $\underset{\sim}{\mathrm{HYP}}(\mathfrak{A})$ is the smallest collection of relations on A which is a model of the schemata of Δ^0_∞-Comprehension and Σ^1_1-Collection.

⊣

Harrington-Kirousis-Schlipf [1977] obtained a generalization of these results for the class of the "boldface" Q-hyperelementary relations on $\mathfrak{A}$, where Q is a quantifier on A. In Kolaitis [1977] "lightface" versions of the schemata of Δ^0_∞-Comprehension and Σ^1_1-Collection were introduced and in terms of these a model theoretic characterization of the recursive in $\underset{\sim}{E},\underset{\sim}{F}_Q$ relations was obtained. In what follows here we will give the definitions of these schemata and state this characterization.

3.9. Let $\mathfrak{A} = \langle A,R_1,\dots,R_n,f_1,\dots,f_m,c_1,\dots,c_k\rangle$ be a structure such that $\omega,\leq_\omega$ are elementary on $\mathfrak{A}$ and let Q be a quantifier on A. An <u>indexed family on</u> A is a set $\Lambda = \{\Lambda_{\vec{x}} : \vec{x} \in A^{<\omega}\}$ having the following properties

i) If $\vec{x} \in A^{<\omega}$, then $\Lambda_{\vec{x}}$ is a nonempty collection of relations on A.

ii) If $\vec{x} \in A^{<\omega}$ and $\vec{m} \in \omega^{<\omega}$, then $\Lambda_{\vec{x}} = \Lambda_{\vec{x},\vec{m}}$.

iii) If $\vec{x} = (x_1,\dots,x_n) \in A^{<\omega}$, then $\Lambda_{\vec{x}}$ depends only on the set $\{x_1,\dots,x_n\}$, i.e. if $\vec{x} = (x_1,\dots,x_n), \vec{y} = (y_1,\dots,y_m)$ are finite sequences from A and $\{y_j : j=1,\dots,m\} \subseteq \{x_i : i = 1,\dots,n\}$, then $\Lambda_{\vec{y}} \subseteq \Lambda_{\vec{x}}$.

iv) If $\vec{x} \in A^n$, $\vec{y} \in A^m$ and $R \subseteq A^{m+k}$ is an element of $\Lambda_{\vec{x}}$, then $\{\vec{z} : R(\vec{y},\vec{z})\} \in \Lambda_{\vec{x},\vec{y}}$.

If Γ is a semi-Spector class on $\mathfrak{A}$, then the collection $\mathfrak{D} = \{\Delta_{\vec{x}} : \vec{x} \in A^{<\omega}\}$ is an indexed family on A. In particular, the collections $\mathfrak{R} = \{SEC[\underset{\sim}{E},\underset{\sim}{F}_Q,\vec{x}] : \vec{x} \in A^{<\omega}\}$ and $\mathfrak{H} = \{HYP(Q,\vec{x}) : \vec{x} \in A^{<\omega}\}$ are examples of indexed families.

We say that the indexed family $\Lambda = \{\Lambda_{\vec{x}} : \vec{x} \in A^{<\omega}\}$ on A satisfies the <u>schema of Σ^1_1-Collection on</u> $\mathfrak{A}$ if for any $\vec{x} \in A^{<\omega}$, any formula $\varphi(\vec{u},\vec{v},\vec{Z},\vec{V})$ of $\mathfrak{L}^{\mathfrak{A}}(Q)$ and any relations $\vec{Y} \in \Lambda_{\vec{x}}$ we have that:

$$(\forall\vec{y})(\exists Z \in \Lambda_{\vec{x},\vec{y}})\varphi(\vec{x},\vec{y},\vec{Z},\vec{Y}) \Longleftrightarrow (\exists W \in \Lambda_{\vec{x}})(\forall\vec{y})(\exists n \in \omega)\varphi(\vec{x},\vec{y},W_{\vec{y},n},\vec{Y}),$$

where $W_{\vec{y},n} = \{\vec{z} : W(\vec{y},n,\vec{z})\}$.

The indexed family $\Lambda = \{\Lambda_{\vec{x}} : \vec{x} \in A^{<\omega}\}$ on A satisfies the <u>schema of Δ^0_∞-Comprehension on</u> $\mathfrak{A}$ if for any $\vec{x} \in A^{<\omega}$, any formula $\varphi(\vec{u},\vec{v},\vec{W})$ of the language $\mathfrak{L}^{\mathfrak{A}}(Q)$ and any relations $\vec{Y} \in \Lambda_{\vec{x}}$ there is some $Z \in \Lambda_{\vec{x}}$ such that

$$(\forall\vec{y})(\vec{y} \in Z \Longleftrightarrow \varphi(\vec{x},\vec{y},\vec{Y})).$$

3.10. <u>Theorem</u>. Let $\mathfrak{A} = \langle A,R_1,\dots,R_n,f_1,\dots,f_m,c_1,\dots,c_k\rangle$ be an acceptable structure and Q a quantifier on A; then the indexed family $\mathfrak{R} = \{SEC[\underset{\sim}{E},\underset{\sim}{F}_Q,\vec{x}] : \vec{x} \in A^{<\omega}\}$ is the smallest indexed family on A satisfying the schemata of Σ^1_1-Collection and Δ^0_∞-Comprehension on $\mathfrak{A}$, i.e. if $\Lambda = \{\Lambda_{\vec{x}} : \vec{x} \in A^{<\omega}\}$ is any other such family then $SEC[\underset{\sim}{E},\underset{\sim}{F}_Q,\vec{x}] \subseteq \Lambda_{\vec{x}}$ for each $\vec{x} \in A^{<\omega}$.

⊣

4. Nonmonotone Inductive Definability and Semi-Spector Classes

4.1. Grilliot [1971] established that the semi-recursive relations in $\underset{\sim}{E}$ on an acceptable structure $\mathfrak{A}$ are the $\Pi_1^0(\mathfrak{A})$-nonmonotone inductive relations; he also proved in the same paper that the inductive relations on $\mathfrak{A}$ are the $\Sigma_2^0(\mathfrak{A})$-nonmonotone inductive relations. The theory of nonmonotone induction was developed by Richter [1971], Aczel [1972] and Richter-Aczel [1974], who characterized nonmonotone inductive definitions in terms of reflecting properties of ordinals. Nonmonotone induction in an abstract form was studied by Moschovakis [1974b] who introduced the notion of compactness and applied the theory of "boldface" Spector classes in order to characterize various classes of nonmonotone inductive relations. In what follows here we introduce a natural "lightface" version of compactness and prove a nonmonotone first recursion theorem for semi-Spector classes, due to Moschovakis. Finally, using this theorem we give a proof of both of Grilliot's results in the spirit of the theory of semi-Spector classes.

If Γ is a class of relations on a set A, then recall that the dual of Γ is the class $\breve{\Gamma} = \{R : R$ is an n-ary relation on A and $A^n - R \in \Gamma\}$ and the self-dual of Γ is the class $\Delta = \Gamma \cap \breve{\Gamma}$. If $\vec{x} \in A^{<\omega}$, the relativization of Γ to $\vec{x}$ is the class $\Gamma_{\vec{x}} = \{R :$ there is some $P \in \Gamma$ such that $(\forall\vec{y})(R(\vec{y}) \Longleftrightarrow P(\vec{x},\vec{y}))\}$. In a similar way we define the class $\breve{\Gamma}_{\vec{x}}$ and we put $\Delta_{\vec{x}} = \Gamma_{\vec{x}} \cap \breve{\Gamma}_{\vec{x}}$; we also consider the "boldface" classes $\underset{\sim}{\Gamma} = \bigcup_{\vec{x}\in A^{<\omega}}\Gamma_{\vec{x}}$ and $\underset{\sim}{\Delta} = \bigcup_{\vec{x}\in A^{<\omega}}\Delta_{\vec{x}}$. Note that if Γ is a semi-Spector class on an acceptable structure $\mathfrak{A}$, then the class $\underset{\sim}{\Gamma}$ is a "boldface" Spector class on $\mathfrak{A}$. We say that a relation $R \subseteq A^{n+1}$ parametrizes the n-ary relations in the semi-Spector class Γ in case a relation $R \subseteq A^n$ is in Γ if and only if there is some $e \in \omega$ such that $(R(\vec{y}) \Longleftrightarrow U(e,\vec{y}))$.

We now state and prove some well known results in the theory of semi-Spector classes which we will use heavily in the sequel.

4.2. The s-m-n Theorem. Let $\mathfrak{A} = \langle A,R_1,\ldots,R_n,f_1,\ldots,f_m,c_1,\ldots,c_k\rangle$ be an acceptable structure and let Γ be a semi-Spector class on $\mathfrak{A}$. Then for each $s \in \omega$ there is a sequence $\{U^n\}_{n\in\omega}$ of relations in Γ such that $U^n \subseteq A^{s+n+1}$, each U^n parametrizes the (s+n)-ary relations in Γ and for each m,n there is an elementary total function $s_n^m : \omega^{m+1} \to \omega$ such that for all $\vec{x} = (x_1,\ldots,x_s) \in A^s$, $\vec{y} = (y_1,\ldots,y_n) \in A^n$, $(e,k_1,\ldots,k_m) \in \omega^{m+1}$

$$U^{m+n}(e,\vec{x},k_1,\ldots,k_m,\vec{y}) \Longleftrightarrow U^n(s_n^m(e,k_1,\ldots,k_m),\vec{x},\vec{y}).$$

Proof. Let $V \subseteq A^{s+3}$ be a relation in Γ which parametrizes the (s+2)-ary relations in Γ and for each $n \in \omega$ put

$$U^n(e,\vec{x},\vec{y}) \Longleftrightarrow V((e)_1,\vec{x},(e)_2,\langle\vec{y}\rangle).$$

Clearly, $U^n \subseteq A^{s+n+1}$ is in Γ and parametrizes the (s+n)-ary relations in Γ. Now fix $m,n \in \omega$ and put

$$R(\vec{x},z,y) \Longleftrightarrow U^{m+n}((z)_1,\vec{x},(z)_2,\ldots,(z)_{m+1},(y)_1,\ldots,(y)_n).$$

Since $R \subseteq A^{s+2}$ is in Γ, there is some $e_0 \in \omega$ such that

$$R(\vec{x},z,y) \Longleftrightarrow V(e_0,\vec{x},z,y).$$

Then the required s-m-n function is

$$s^m_n(e,k_1,\ldots,k_m) = \langle e_0,\langle e,k_1,\ldots,k_m\rangle\rangle. \qquad \dashv$$

If $\{U^n\}_{n\in\omega}$ is a sequence of relations in Γ satisfying the conclusion of the s-m-n Theorem, then we say that the sequence $\{U^n\}_{n\in\omega}$ is a <u>good universal system for</u> Γ.

4.3. <u>The (Second) Recursion Theorem</u>. Let $\mathfrak{A} = \langle A,R_1,\ldots,R_n,f_1,\ldots,f_m,c_1,\ldots,c_k\rangle$ be an acceptable structure, let Γ be a semi-Spector class on $\mathfrak{A}$ and let $\vec{x} = (x_1,\ldots,x_s) \in A^s$. If the sequence $\{U^n\}_{n\in\omega}$ with $U^n \subseteq A^{s+n+1}$ is a good universal system for Γ, then for each relation $R \subseteq A^{n+1}$ in $\Gamma_{\vec{x}}$ there is some $e \in \omega$ such that

$$R(e,\vec{y}) \Longleftrightarrow U^n(e,\vec{x},\vec{y}).$$

<u>Proof</u>. Let $P(k,\vec{y}) \Longleftrightarrow R(s^1_n(k,k),\vec{y})$; then $P \in \Gamma_{\vec{x}}$ and hence there is some $e_0 \in \omega$ such that $(P(k,\vec{y}) \Longleftrightarrow U^{n+1}(e_0,\vec{x},k,\vec{y}))$. But then

$$R(s^1_n(k,k),\vec{y}) \Longleftrightarrow U^{n+1}(e_0,\vec{x},k,\vec{y}) \Longleftrightarrow U^n(s^1_n(e_0,k),\vec{x},\vec{y}),$$

so that $e = s^1_n(e_0,e_0)$ is the required index. $\dashv$

4.4. Let $\mathfrak{A} = \langle A,R_1,\ldots,R_n,f_1,\ldots,f_m,c_1,\ldots,c_k\rangle$ be an acceptable structure and let Γ be a semi-Spector class on $\mathfrak{A}$. With each $\vec{x} \in A^{<\omega}$ there is naturally associated the ordinal

$$\delta_{\vec{x}} = O(\Delta_{\vec{x}}) = \text{supremum}\{\text{rank}(<) : < \text{ is a prewellordering in } \Delta_{\vec{x}}\}.$$

Assume that $\vec{x} \in A^s$, $\{U^n\}_{n\in\omega}$ with $U^n \subseteq A^{s+n+1}$ is a good universal system for Γ and $\sigma : U^1 \to$ Ordinals is a Γ-norm on U^1; then we associate with $\vec{x}$ the ordinal

$$\lambda_{\vec{x}} = \text{supremum}\{\sigma((e,\vec{x},n)) : (e,\vec{x},n) \in U^1 \text{ and } e,n \in \omega\},$$

i.e. $\lambda_{\vec{x}}$ is the supremum of the "notations of ordinals recursive in $\vec{x}$". It is a well known fact that

$$\lambda_{\vec{x}} = \delta_{\vec{x}}.$$

To see this, note first of all that if $(e,\vec{x},n) \in U^1$, then the relation

$$< \; = \{(\vec{y},\vec{z}) \in A^{s+2} \times A^{s+2} : \vec{y} <^*_\sigma \vec{z} \text{ and } \vec{z} \leq^*_\sigma (e,\vec{x},n)\}$$

is a prewellordering in $\Delta_{\vec{x}}$ of rank $\sigma(e,\vec{x},n)+1$, so that $\lambda_{\vec{x}} \leq \delta_{\vec{x}}$. On the other hand, if $< \subseteq A \times A$ is a prewellordering in $\Delta_{\vec{x}}$, then put

$$y <' z \iff [(y)_1 = 0 \;\&\; (z)_1 = 0 \;\&\; ((y)_2 < (z)_2)] \vee [(y)_1 = 0 \;\&\; (z)_1 = 1 \;\&$$
$$\&\; ((y)_2 \in \mathrm{Field}(<)) \;\&\; (z)_2 = 1].$$

It is clear that $<'$ is a prewellordering in $\Delta_{\vec{x}}$ such that $\mathrm{rank}(<') = \mathrm{rank}(<)+1$. We consider now the relation

$$P(m,z) \iff (\forall y)(y <' z \rightarrow (m,\vec{x},y) <^*_\sigma (m,\vec{x},z)).$$

Using the Second Recursion Theorem 4.3 we can find some $e \in \omega$ such that

$$U^1(m,\vec{x},z) \iff (\forall y)(y <' z \rightarrow (m,\vec{x},y) <^*_\sigma (m,\vec{x},z)).$$

It follows now immediately that if $z \in \mathrm{Field}(<')$, then $(m,\vec{x},z) \in U^1$ and $\mathrm{rank}_{<'}(z) \leq \sigma(m,\vec{x},z)$. Therefore we have that

$$\mathrm{rank}(<) = \mathrm{rank}_{<'}(\langle 1,1\rangle) \leq \sigma(m,\vec{x},\langle 1,1\rangle)$$

and hence $\delta_{\vec{x}} \leq \lambda_{\vec{x}}$.

4.5. A <u>second order relation on</u> A is a relation with arguments elements and relations on A, i.e. a relation of the form

$$\varphi(\vec{x},\vec{Y}) \iff \varphi(x_1,\ldots,x_n,Y_1,\ldots,Y_m),$$

where $\vec{x} = (x_1,\ldots,x_n) \in A^n$ and each Y_i $(1 \leq i \leq m)$ is a k_i-ary relation on A. The <u>signature</u> of such a second order relation φ is the sequence $(n,k_1,\ldots,k_m)$. If the signature of φ is of the form (n,n), then we say that φ is an <u>operative</u> second order relation. In this case φ defines a transfinite sequence $\{\varphi^\xi\}$ of n-ary relations on A by the equations

$$\varphi^\xi = \varphi^{<\xi} \cup \{\vec{x} : \varphi(\vec{x},\varphi^{<\xi})\}, \quad \text{where} \quad \varphi^{<\xi} = \bigcup_{\eta<\xi} \varphi^\eta.$$

The relation φ^ξ is the ξ-th <u>iterate</u> or the ξ-th <u>stage</u> of φ, while the <u>fixed point</u> of φ is

$$\varphi^\infty = \bigcup_\xi \varphi^\xi$$

and the <u>closure ordinal</u> of φ is the least stage at which no new elements are thrown in, i.e.

$$\|\varphi\| = \text{least } \xi \text{ such that } \varphi^\xi = \varphi^{<\xi}.$$

Note that if φ is an operative second order relation which is monotone (i.e. $S \subseteq T$ & $\varphi(\vec{x},S) \Rightarrow \varphi(\vec{x},T)$), then $\varphi^{\xi} = \{\vec{x} : \varphi(\vec{x},\varphi^{<\xi})\}$.

Assume now that Γ is a semi-Spector class on a structure $\mathfrak{A}$ and that φ is a second order operative relation on A. One of the main problems in inductive definability is to find natural conditions under which the fixed point φ^{∞} of φ is in the class Γ. Such results are the key tools in obtaining minimality characterizations of certain classes of relations or in comparing and identifying various inductively definable classes of relations. If φ is a monotone operative relation on A, then a simple condition for φ^{∞} to be in Γ is that Γ is closed under φ. More specifically, let $\mathfrak{A} = \langle A,R_1,\ldots,R_n,f_1,\ldots,f_m,c_1,\ldots,c_k\rangle$ be an acceptable structure, let Γ be a semi-Spector class on $\mathfrak{A}$ and let $\varphi(\vec{x},S)$ be a second order relation on A; we say that Γ is closed under φ if for each $P(\vec{y},\vec{z}) \in \Gamma$ the relation

$$R(\vec{x},\vec{z}) \Longleftrightarrow \varphi(\vec{x},\{\vec{y} : P(\vec{y},\vec{z})\})$$

is also in Γ. We have then the following theorem of Moschovakis [1974a]:

4.6. The (Monotone) First Recursion Theorem. Let $\mathfrak{A}$ be an acceptable structure, let Γ be a semi-Spector class on $\mathfrak{A}$ and let $\varphi(\vec{x},S)$ be a monotone operative relation on A. If Γ is closed under φ, then the fixed point φ^{∞} of φ is in Γ.

$\dashv$

4.7. Moschovakis [1974b] introduced the notions of "compactness" and "Δ on Δ" which are sufficient conditions to guarantee that $\varphi^{\infty} \in \Gamma$ in the case that Γ is a "boldface" Spector class on $\mathfrak{A}$ and φ is a nonmonotone operative relation on A. It turns out that a similar result can be established for semi-Spector classes by introducing the proper "lightface" version of these notions.

Let Γ be a semi-Spector class on $\mathfrak{A}$ and let $\varphi(\vec{x},S_1,\ldots,S_k)$ be a second order relation on A of signature $(n,m_1,\ldots,m_k)$; we say that Γ is φ-compact if for every $\vec{x} \in A^n$, every $R_1,\ldots,R_k \in \Gamma_{\vec{x}}$ and every $R_1^0,\ldots,R_k^0$ in $\Delta_{\vec{x}}$, $R_1^0 \subseteq R_1,\ldots$ $\ldots,R_k^0 \subseteq R_k$,

$$\varphi(\vec{x},R_1,\ldots,R_k) \Rightarrow \text{there exist } R_1^*,\ldots,R_k^* \text{ in } \Delta_{\vec{x}}, R_1^0 \subseteq R_1^* \subseteq R_1,\ldots,R_k^0 \subseteq R_k^* \subseteq R_k,$$

such that $\varphi(\vec{x},R_1^*,\ldots,R_k^*)$.

If $\mathcal{G}$ is a collection of second order relations on A, then we say that Γ is $\mathcal{G}$-compact in case Γ is φ-compact for every φ in $\mathcal{G}$. The following consequence of compactness follows easily by analyzing the proof of Theorem 12 in Moschovakis [1974b].

4.8. Lemma. Let Γ be a semi-Spector class on $\mathfrak{A}$ and let $\mathcal{G}$ be a collection of second order relations on A containing the second relations on A definable by universal formulas of the language $\mathcal{L}^{\mathfrak{A}}$, closed under & and such that Γ is $\mathcal{G}$-compact. For every $\vec{x} \in A^{<\omega}$, every $\varphi \in \mathcal{G}$, every $R \in \Gamma_{\vec{x}}$ and every $\Gamma_{\vec{x}}$-norm σ on R with $|\sigma| = \sup\{\sigma(\vec{y}) : \vec{y} \in R\} \geq \delta_{\vec{x}}$,

$$\varphi(\vec{x},R) \Rightarrow \text{there exists an ordinal } \xi < |\sigma| \text{ such that } \varphi(\vec{x},\{\vec{y} \in R : \sigma(\vec{y}) < \xi\}).$$

⊣

4.9. Let Γ be a semi-Spector class on $\mathfrak{A}$ and let $\varphi(\vec{x},Y_1,\ldots,Y_k)$ be a second order relation on A; we say that φ is <u>Γ on Δ</u> if for any relations R_i, S_i $(1 \leq i \leq k)$ in Γ there is a relation P in Γ with the property that:

for every $\vec{x}_1,\ldots,\vec{x}_k$ such that $(\forall i \leq k)(\forall \vec{z}_i)[R_i(\vec{x}_i,\vec{z}_i) \Longleftrightarrow \neg S_i(\vec{x}_i,\vec{z}_i)]$ we have

$$P(\vec{x},\vec{x}_1,\ldots,\vec{x}_k) \Longleftrightarrow \varphi(\vec{x},\{\vec{z}_1 : R_1(\vec{x}_1,\vec{z}_1)\},\ldots,\{\vec{z}_k : R_k(\vec{x}_k,\vec{z}_k)\}).$$

We see that φ being Γ on Δ means that the semi-Spector class Γ is "uniformly closed under Δ substitutions in φ". We also say that φ is <u>Δ on Δ</u> if both φ and $\neg\varphi$ are Γ on Δ.

4.10. Theorem. Let Γ be a Spector class on $\mathfrak{A}$ and let $\mathcal{G}$ be a collection of second order relations on A such that $\mathcal{G}$ contains all relations definable by universal formulas of the language $\mathcal{L}^{\mathfrak{A}}$, $\mathcal{G}$ is closed under & and such that every φ in $\mathcal{G}$ is Δ on Δ. Then the Spector class Γ is $\mathcal{G}$-compact if and only if for every relation $\varphi(\vec{x},Y_1,\ldots,Y_k)$ in $\mathcal{G}$, every $R_1,\ldots,R_k$ in $\underset{\sim}{\Gamma}$ and every $R_1^0,\ldots,R_k^0$ in $\underset{\sim}{\Delta}$, $R_1^0 \subseteq R_1,\ldots,R_k^0 \subseteq R_k$,

$\varphi(\vec{x},R_1,\ldots,R_k) \Rightarrow$ there exist $R_1^*,\ldots,R_k^*$ in $\underset{\sim}{\Delta}$, $R_1^0 \subseteq R_1^* \subseteq R_1,\ldots,R_k^0 \subseteq R_k^* \subseteq R_k$, such that $\varphi(\vec{x},R_1^*,\ldots,R_k^*)$.

Proof. Towards proving the nontrivial direction, let us assume that $\vec{x} \in A^{<\omega}$, φ is in $\mathcal{G}$, $R \in \Gamma_{\vec{x}}$, $R_0 \in \Delta_{\vec{x}}$, $R_0 \subseteq R$ and $\varphi(\vec{x},R)$ holds. We want to find a relation $R^* \in \Delta_{\vec{x}}$, $R_0 \subseteq R^* \subseteq R$ such that $\varphi(\vec{x},R^*)$ holds. Let U be a universal set for Γ, let $\sigma : U \to$ Ordinals be a Γ-norm on U and let $e \in \omega$ be such that $(\forall\vec{y})(R(\vec{y}) \Longleftrightarrow U(e,\vec{x},\vec{y}))$. Using the hypothesis and the argument in the proof of Theorem 12 of Moschovakis [1974b] we can find an ordinal $\xi < |\sigma|$ such that $R_0 \subseteq \{\vec{y} \in R : \sigma((e,\vec{x},\vec{y})) < \xi\}$ and $\varphi(\vec{x},\{\vec{y} \in R : \sigma((e,\vec{x},\vec{y})) < \xi\})$. Let ξ_0 be the least ordinal having these properties and let $R^* = \{\vec{y} \in R : \sigma((e,\vec{x},\vec{y})) < \xi_0\}$. Clearly $R_0 \subseteq R^* \subseteq R$ and $\varphi(\vec{x},R^*)$, so in order to complete the proof we have to show that $R^* \in \Delta_{\vec{x}}$. Put $P(\vec{z}) \Longleftrightarrow (\vec{z} \in U)$ & $(\sigma(\vec{z}) = \xi_0)$. Then

$$\vec{y} \in R^* \Longleftrightarrow (\exists\vec{z})(\vec{z} \in P \;\&\; ((e,\vec{x},\vec{y}) <_\sigma^* \vec{z}))$$

and
$$\vec{y} \notin R^* \Longleftrightarrow (\forall\vec{z})(\vec{z} \in P \to \vec{z} \leq_\sigma^* (e,\vec{x},\vec{y})),$$

hence it is enough to show that $P \in \Delta_{\vec{x}}$. Since φ is Δ on Δ, we can find relations S and T in $\Gamma_{\vec{x}}$, such that

$$S(\vec{z}) \Longleftrightarrow (\vec{z} \in U) \;\&\; (\forall\vec{y})(\vec{y} \in R_0 \to (e,\vec{x},\vec{y}) <_\sigma^* \vec{z}) \;\&\; \varphi(\vec{x},\{\vec{y} \in R : (e,\vec{x},\vec{y}) <_\sigma^* \vec{z}\})$$

and

$$T(\vec{z}) \Longleftrightarrow (\vec{z} \in U) \,\&[(\exists \vec{y})(\vec{y} \in R_0 \,\&\, (\vec{z} <_\sigma^* (e,\vec{x},\vec{y}))) \vee \neg\varphi(\vec{x}, \{\vec{y} \in R : (e,\vec{x},\vec{y}) <_\sigma^* \vec{z}\})].$$

But then

$$\vec{z} \in P \Longleftrightarrow S(\vec{z}) \,\&\, (\forall \vec{z}\,')(\vec{z}\,' <_\sigma^* \vec{z} \to T(\vec{z}\,'))$$

and

$$\vec{z} \notin P \Longleftrightarrow (\exists \vec{z}\,')[(\vec{z}\,' \in P) \,\&\, (\vec{z}\,' <_\sigma^* \vec{z} \vee \vec{z} <_\sigma^* \vec{z}\,')].$$ ⊣

4.11. <u>Corollary</u>. Let Γ be a Spector class on $\mathfrak{A}$ and let $\mathcal{G}$ be a collection of second order relations on A such that $\mathcal{G}$ contains all relations definable by universal formulas of the language $\mathcal{L}^{\mathfrak{A}}$, $\mathcal{G}$ is closed under $\&$ and such that every relation in $\mathcal{G}$ is Δ on Δ. If Γ is $\mathcal{G}$-compact, then Γ is also $\exists^A\mathcal{G}$-compact, where

$$\exists^A\mathcal{G} = \{\varphi(\vec{x},\vec{Y}) : \text{there is some } \psi(y,\vec{x},\vec{Y}) \text{ in } \mathcal{G} \text{ such that } \varphi(\vec{x},\vec{Y}) \Longleftrightarrow (\exists y)\psi(y,\vec{x},\vec{Y})\}$$

⊣

The previous Theorem 4.10 is due to Kechris [1973]; note that it states in effect that for Spector classes and relations which are Δ on Δ the notion of compactness coincides with the notion of "boldface" compactness introduced by Moschovakis [1974b]. It should be pointed out that this is not the case for semi-Spector classes and as we shall see later on both Theorem 4.10 and Corollary 4.11 do not hold if Γ is only a semi-Spector class.

The next result, due to Moschovakis, is the key closure property of semi-Spector classes under nonmonotone inductions.

4.12. <u>The Nonmonotone First Recursion Theorem</u>. Let Γ be a semi-Spector class on a structure $\mathfrak{A}$ and let $\mathcal{G}$ be a collection of second order relations on A containing the relations definable by universal formulas of the language $\mathcal{L}^{\mathfrak{A}}$ and closed under $\&$. If Γ is $\mathcal{G}$-compact and every relation in $\mathcal{G}$ is Δ on Δ, then for every operative second order relation φ in $\mathcal{G}$ the fixed point φ^∞ of φ is in Γ.

<u>Proof</u>. Let Γ be a semi-Spector class on $\mathfrak{A}$ and let $\mathcal{G}$ be a collection of second order relations on A satisfying the hypotheses of the theorem. If $\varphi(\vec{x},S)$ is an operative relation of signature (n,n) in $\mathcal{G}$, then for each $\vec{x} \in \varphi^\infty$ we put $|\vec{x}|_\varphi$ = the least ordinal ξ such that $\vec{x} \in \varphi^\infty$. Let $U^1 \subseteq A^{n+2}$ be a relation on Γ which belongs in a good universal system $\{U^m\}_{m\in\omega}$ for Γ and let $\sigma : U^1 \to$ Ordinals be a Γ-norm on U^1; recall that for each $\vec{x} \in A^n$, $\lambda_{\vec{x}} = \sup\{\sigma(e,\vec{x},m) : (e,\vec{x},m) \in U^1 \,\&\, e,m \in \omega\}$. In order to show that $\varphi^\infty \in \Gamma$ we will establish first that:

i) There are relations $R \subseteq A^{2n+2}, S \subseteq A^{2n+2}$ which are in Γ and such that

$$R(\vec{x},\vec{y}) \Longleftrightarrow (\vec{y} \in U^1) \,\&\, (\vec{x} \in \varphi^{\sigma(\vec{y})}),$$
$$S(\vec{x},\vec{y}) \Longleftrightarrow (\vec{y} \in U^1) \,\&\, (\vec{x} \notin \varphi^{\sigma(\vec{y})}).$$

ii) If $\vec{x} \in \varphi^\infty$, then $|\vec{x}|_\varphi < \lambda_{\vec{x}}$.

From i) and ii) and the fact that Γ is closed under existential quantification over ω it will follow immediately that φ^{∞} is in Γ, since then

$$\vec{x} \in \varphi^{\infty} \Leftrightarrow \vec{x} \in \varphi^{|\vec{x}|\varphi} \Leftrightarrow \vec{x} \in \varphi^{<\lambda_{\vec{x}}} \Leftrightarrow (\exists e \in \omega)R(\vec{x},(e)_1,\vec{x},(e)_2).$$

The relations R and S can be defined in the following way using the Second Recursion Theorem 4.3, the assumption that φ is Δ on Δ and the fact that Γ is closed under the deterministic $\exists$-rule:

$$R(\vec{x},\vec{y}) \Leftrightarrow (\vec{y} \in U^1) \,\&\, \{[\sigma(\vec{y}) = 0 \,\&\, \varphi(\vec{x},\emptyset)] \vee$$
$$\vee [\sigma(\vec{y}) > 0 \,\&\, [(\exists \vec{y}\,')(\vec{y}\,' <^*_\sigma \vec{y} \,\&\, R(\vec{x},\vec{y}\,')) \vee$$
$$\vee \varphi(\vec{x},\{\vec{x}\,' : (\exists \vec{y}\,')(\vec{y}\,' <^*_\sigma \vec{y} \,\&\, R(\vec{x}\,',\vec{y}\,'))\})]]\},$$

$$S(\vec{x},\vec{y}) \Leftrightarrow (\vec{y} \in U^1) \,\&\, \{[\sigma(\vec{y}) = 0 \,\&\, \neg\varphi(\vec{x},\emptyset)] \vee$$
$$\vee [\sigma(\vec{y}) > 0 \,\&\, (\forall \vec{y}\,')(\vec{y}\,' <^*_\sigma \vec{y} \rightarrow S(\vec{x},\vec{y}\,')) \,\&$$
$$\& \neg\varphi(\vec{x},\{\vec{x}\,' : (\exists \vec{y}\,')(\vec{y}\,' <^*_\sigma \vec{y} \,\&\, R(\vec{x}\,',\vec{y}\,'))\})]\}.$$

Finally we show that $(\vec{x} \in \varphi^{\infty} \Rightarrow |\vec{x}|_\varphi < \lambda_{\vec{x}})$ by induction on ξ, for $\xi < \|\varphi\|$. Assume that the induction hypothesis holds for all ordinals $\eta < \xi$ and let $\vec{x} \in \varphi^{\infty}$ be such that $|\vec{x}|_\varphi = \xi$. Then $\varphi^{<\xi} = \{\vec{z} : \vec{z} <^*_\varphi \vec{x}\}$ and $\varphi(\vec{x},\varphi^{<\xi})$. We claim that:

(a) the set $\varphi^{<\xi}$ is in $\Gamma_{\vec{x}}$,

(b) the mapping $\tau : \varphi^{<\xi} \rightarrow \xi$ such that $\tau(\vec{z}) = |\vec{z}|_\varphi$ is a $\Gamma_{\vec{x}}$-norm on $\varphi^{<\xi}$.

We first show (a) as follows using the induction hypothesis and the fact that Γ is closed under the deterministic $\exists$-rule:

$$\vec{z} \in \varphi^{<\xi} \Leftrightarrow \vec{z} <^*_\varphi \vec{x} \Leftrightarrow (\exists e \in \omega)[((e)_1,\vec{z},(e)_2) \in U^1) \,\&$$
$$(\exists \vec{y})(\vec{y} \leq^*_\sigma ((e)_1,\vec{z},(e)_2) \,\&\, R(\vec{z},\vec{y}) \,\&\, S(\vec{x},\vec{y}))]$$
$$\Leftrightarrow (\exists e \in \omega)\{(((e)_1,\vec{z},(e)_2) \in U^1) \,\&$$
$$(\forall \vec{y})[(((e)_1,\vec{z},(e)_2) <^*_\sigma \vec{y}) \vee ((\vec{y} \leq^*_\sigma ((e)_1,\vec{z},(e)_2)) \,\&$$
$$(S(\vec{z},\vec{y}) \vee R(\vec{x},\vec{y}))) \vee ((\vec{y} \leq^*_\sigma ((e)_1,\vec{z},(e)_2)) \,\&\, R(\vec{z},\vec{y}) \,\&\, S(\vec{x},\vec{y}))] \,\&$$
$$(\exists \vec{y})[(\vec{y} \leq^*_\sigma ((e)_1,\vec{z},(e)_2)) \,\&\, R(\vec{z},\vec{y}) \,\&\, S(\vec{x},\vec{y})]\}.$$

We can show now that the map τ is a $\Gamma_{\vec{x}}$-norm on $\varphi^{<\xi}$ using a similar argument. We have for example that

$$\vec{z} \leq^*_\tau \vec{w} \Leftrightarrow (\exists e \in \omega)\{(((e)_1,\vec{z},(e)_2) \in U^1) \,\&\, (\vec{z} \in \varphi^{<\xi}) \,\&$$
$$(\exists \vec{y})[(\vec{y} \leq^*_\sigma ((e)_1,\vec{z},(e)_2)) \,\&\, R(\vec{z},\vec{y}) \,\&\, (\forall \vec{y}\,')(\vec{y}\,' <^*_\sigma \vec{y} \rightarrow S(\vec{y}\,',\vec{w}))]\}.$$

We have thus established that the set $\varphi^{<\xi}$ is in $\Gamma_{\vec{x}}$ and the map $\tau : \vec{z} \to |\vec{z}|_\varphi$ is a $\Gamma_{\vec{x}}$-norm on $\varphi^{<\xi}$. We also have that $\varphi(\vec{x},\varphi^{<\xi})$ holds and $|\tau| = |\vec{x}|_\varphi = \xi$. We claim that these facts imply immediately that $|\vec{x}|_\varphi < \lambda_{\vec{x}}$. In fact, recall from 4.4 that $\lambda_{\vec{x}} = \delta_{\vec{x}}$, so that if $|\vec{x}|_\varphi \geq \lambda_{\vec{x}}$, then using Lemma 4.8 we can find an ordinal $\eta < |\tau| = |\vec{x}|_\varphi$ such that $\varphi(\vec{x},\{\vec{z} \in \varphi^{<\xi} : \tau(\vec{z}) < \eta\})$. But then $\varphi(\vec{x},\varphi^{<\eta})$ holds and so $|\vec{x}|_\varphi \leq \eta$, contradiction.

$\dashv$

4.13. Let $\mathfrak{A} = \langle A,R_1,\ldots,R_n,f_1,\ldots,f_m,c_1,\ldots,c_k\rangle$ be an acceptable structure and let $C = \langle \omega,\leq_\omega,seq,lh,q\rangle$ be the elementary coding scheme on A associated with a coding function $\langle\ \rangle : A^{<\omega} \to A$. We consider the expanded structure

$$\mathfrak{A}(C) = \langle A,R_1,\ldots,R_n,\omega, \leq_\omega,seq,lh,q,f_1,\ldots,f_m,c_1,\ldots,c_k\rangle$$

and we say that a formula φ of the language $\mathcal{L}^{\mathfrak{A}(C)}$ <u>is restricted on</u> $\mathfrak{A}$ <u>relative to</u> C if all quantifiers in φ occur in one of the following two forms:

$$(\exists x)[x \leq_\omega t(\vec{y})\ \&\ \ldots], \qquad (\forall x)[x \leq_\omega t(\vec{y}) \to \ldots],$$

where t is a term of the language $\mathcal{L}^{\mathfrak{A}(C)}$. We say that a formula of the language $\mathcal{L}^{\mathfrak{A}(C)}$ is Σ^0_k <u>on</u> $\mathfrak{A}$ <u>relative to</u> C or $\Sigma^0_k(C)$ on $\mathfrak{A}$ if it is of the form

$$(\exists \vec{z}_1)(\forall \vec{z}_2)\ \ldots\ (Q\vec{z}_k)\ \varphi\ (\vec{y},\vec{z}_1,\ldots,\vec{z}_k,\vec{Y}),$$

where φ is a restricted formula on $\mathfrak{A}$ relative to C. In a similar way we define the notion of a Π^0_k <u>formula on</u> $\mathfrak{A}$ <u>relative to</u> C or simply $\Pi^0_k(C)$ on $\mathfrak{A}$; we say that a second order relation P on A is $\Sigma^0_k(C)$ <u>on</u> $\mathfrak{A}$ (or $\Pi^0_k(C)$ on $\mathfrak{A}$), if P is definable by a $\Sigma^0_k(C)$ formula on $\mathfrak{A}$ (or by a $\Pi^0_k(C)$ formula on $\mathfrak{A}$).

In general, if $\mathcal{G}$ is a collection of second order relations on A, then a relation P on A is $\mathcal{G}$-<u>inductive</u> if there is an operative relation φ in $\mathcal{G}$ and constants $\vec{n} = (n_1,\ldots,n_k)$ from ω such that $P(\vec{y}) \Leftrightarrow \varphi^\infty(\vec{n},\vec{y})$. If $\vec{x} \in A^{<\omega}$, then P is $\mathcal{G}$-<u>inductive from</u> $\vec{x}$ if there is a $\mathcal{G}$-inductive relation R such that $P(\vec{y}) \Leftrightarrow R(\vec{x},\vec{y})$. We put

$$\mathcal{G}\text{-IND} = \{P : P \text{ is a } \mathcal{G}\text{-inductive relation on } A\},$$

$$\mathcal{G}\text{-IND}(\vec{x}) = \{P : P \text{ is a } \mathcal{G}\text{-inductive relation on } A \text{ from } \vec{x}\}.$$

A relation P on A is $\mathcal{G}$-<u>hyperelementary</u> if both P and $\neg P$ are $\mathcal{G}$-inductive; we write $\mathcal{G}$-HYP for the collection of the $\mathcal{G}$-hyperelementary relations on A. In a similar way we define the collection $\mathcal{G}\text{-HYP}(\vec{x})$ of the $\mathcal{G}$-hyperelementary relations from $\vec{x}$.

We will prove here two results of Grilliot [1971] about the classes of the $\Pi^0_1(C)$-IND and $\Sigma^0_2(C)$-IND relations on an acceptable structure $\mathfrak{A}$; it is well known (see for example Theorem 11 in Moschovakis [1974b]) that if C and C' are two different coding schemes on an acceptable structure $\mathfrak{A}$, then for all $k \geq 1$ and $\ell \geq 2$

$$\Pi_k^0(C)\text{-IND} = \Pi_k^0(C')\text{-IND} \quad \text{and} \quad \Sigma_\ell^0(C)\text{-IND} = \Sigma_\ell^0(C')\text{-IND}.$$

In what follows we consider an acceptable structure $\mathfrak{A} = \langle A, R_1, \ldots, R_n, f_1, \ldots, f_m, c_1, \ldots, c_k \rangle$ and a fixed coding scheme $C = \langle \omega, \leq_\omega, \text{seq}, \ell h, q \rangle$ on A and for $k \geq 1$, $\ell \geq 2$ we put

$$\Pi_k^0(\mathfrak{A})\text{-IND} = \{P : P \text{ is a } \Pi_k^0(C)\text{-inductive relation on } A\},$$
$$\Sigma_\ell^0(\mathfrak{A})\text{-IND} = \{P : P \text{ is a } \Sigma_\ell^0(C)\text{-inductive relation on } A\}.$$

The closure and structural properties of these classes are summarized in the following well known result whose proof can be found in Moschovakis [1974b]:

4.14. Theorem. $\mathfrak{A} = \langle A, R_1, \ldots, R_n, f_1, \ldots, f_m, c_1, \ldots, c_k \rangle$ be an acceptable structure. Then

i) The class $\Pi_1^0(\mathfrak{A})$-IND is a semi-Spector class on $\mathfrak{A}$.

ii) For each $k \geq 2$ the classes $\Pi_k^0(\mathfrak{A})$-IND and $\Sigma_k^0(\mathfrak{A})$-IND are Spector classes on $\mathfrak{A}$. ⊣

We will prove now that the $\Pi_1^0(\mathfrak{A})$-inductive relations are just the semirecursive relations in $\underset{\sim}{E}$ and that the $\Sigma_2^0(\mathfrak{A})$-inductive relations are the (positive elementary) inductive relations. We start by stating the following lemma which can be easily proved by induction on the construction of the restricted formulas on $\mathfrak{A}$ relative to C.

4.15. Lemma. Let $\mathfrak{A}$ be an acceptable structure, let C be a coding scheme on A and let $\varphi(v_1, \ldots, v_n, Y)$ be a formula of the language $\mathcal{L}^{\mathfrak{A}(C)}$ which is restricted on $\mathfrak{A}$ relative to C and such that Y is a unary relation variable and all negation symbols occurring in φ apply only to atomic subformulas of φ.

i) Let $t_1(\vec{v}, \vec{w}), \ldots, t_m(\vec{v}, \vec{w})$ be all the terms which occur in φ in the form $t_i \in Y$. For each $\vec{x} \in A^n$ there are finitely many elements $a_1, a_2, \ldots, a_s$ of A of the form $t_i(\vec{x}, \vec{j})$ where $\vec{j}$ is a sequence from ω, such that for any unary relations R, S on A if $S \subseteq R$ and for all $\ell \leq s$ $(a_\ell \in R \Rightarrow a_\ell \in S)$, then $\varphi(\vec{x}, R) \Rightarrow \varphi(\vec{x}, S)$.

ii) If $\vec{x} \in A^n$ and $\{S_k\}_{k \in \omega}$ is a sequence of unary relations on A such that $\varphi(\vec{x}, S_k)$ and $S_k \subseteq S_{k+1}$ for all $k \in \omega$, then $\varphi(\vec{x}, \bigcup_{k \in \omega} S_k)$. ⊣

This simple lemma is the key property of the restricted formulas and we will use it now in order to establish the following result which is essentially due to Grilliot [1971].

4.16. Theorem. Let $\mathfrak{A}$ be an acceptable structure, let C be a coding scheme on A and let Γ be a semi-Spector class on $\mathfrak{A}$. If φ is a second order relation on A which is $\Pi_1^0(C)$ on $\mathfrak{A}$, then Γ is φ-compact.

Proof. Let Γ be a semi-Spector class on $\mathfrak{A}$ and let φ be a second order relation on A which is $\Pi_1^0(C)$ on $\mathfrak{A}$. Let us also assume for simplicity that φ is of the form $\varphi(x,Y)$, where Y ranges over the unary relations on A and that there is a formula $\psi(x,z,Y)$ restricted on $\mathfrak{A}$ relative to C such that all negation symbols occurring in ψ apply only to atomic subformulas of ψ and such that

$$\varphi(x,Y) \Longleftrightarrow (\forall z)\psi(x,z,Y).$$

Assume now that $x \in A$, $R \in \Gamma_x$, $R_0 \in \Delta_x$, $R_0 \subseteq R \subseteq A$ and $\varphi(x,R)$. We have to show that there is a relation $R^* \in \Delta_x$ such that $R_0 \subseteq R^* \subseteq R$ and $\varphi(x,R^*)$. Let $U^1 \subseteq A^2$ be a relation in Γ which belongs in a good universal system $\{U^n\}_{n\in\omega}$ for Γ and let $\sigma : U^1 \to$ Ordinals be a Γ-norm on U^1. Since U^1 parametrizes the relations in Γ, we can find some $e \in \omega$ such that

$$(\forall y)(R(y) \Longleftrightarrow U^1(e,y)).$$

As usually, for $(a,b) \in U^1$ we put $R^{<\sigma((a,b))} = \{y \in R : (e,y) <_\sigma^* (a,b)\}$. A standard diagonal argument shows that the set $W_x = \{k \in \omega : (k,x) \in U^1\}$ is in the class $\Gamma_x - \Delta_x$; notice also that $\delta_x = \lambda_x = \sup\{\sigma((k,x)) : k \in W_x\}$. We consider then the relation

$$\begin{aligned} P(k,\ell) \Longleftrightarrow{}& (k \in W_x) \,\&\, (\ell \in W_x) \,\&\, ((k,x) <_\sigma^* (\ell,x)) \,\&\\ &(\forall z)(\exists n \in \omega)[((k,x) <_\sigma^*(n,\langle x,z\rangle) <_\sigma^*(\ell,x)) \,\&\, \psi(x,z,R^{<\sigma((n,\langle x,z\rangle))})].\end{aligned}$$

It is clear that the relation P is in the class Γ_x. Moreover, we claim that:

$$(\forall k)[k \in W_x \Rightarrow (\exists \ell)(\ell \in W_x \,\&\, P(k,\ell))].$$

To prove this claim notice first of all that if $k \in W_x$, then

$$(\forall z)(\exists n \in \omega)[((n,\langle x,z\rangle) \in U^1) \,\&\, ((k,x) <_\sigma^*(n,\langle x,z\rangle)) \,\&\, \psi(x,z,R^{<\sigma((n,\langle x,z\rangle))})].$$

In fact, if $k \in W_x$, then for any $z \in A$ we can find $n \in \omega$ such that $(n,\langle x,z\rangle) \in U^1$, $\sigma((k,x)) < \sigma((n,\langle x,z\rangle))$ and $\max_{1\leq i\leq s}\{\sigma((e,a_i)) : a_i \in R\} < \sigma((n,\langle x,z\rangle))$, where $a_1,\ldots,a_s$ are the elements of A corresponding to (x,z) and the restricted formula ψ as in Lemma 4.15. But also $\psi(x,z,R)$ holds, hence by part i) of Lemma 4.15 we have that $\psi(x,z,R^{<\sigma((n,\langle x,z\rangle))})$. We can now complete the proof of the claim by noticing that if there is some $k_0 \in W_x$ such that $\neg(\exists \ell)(\ell \in W_x \,\&\, P(k,\ell))$, then

$$\ell \notin W_x \Longleftrightarrow (\forall z)(\exists n \in \omega)[((k,x) <_\sigma^* (n,\langle x,z\rangle) <_\sigma^*(\ell,x)) \,\&\, \psi(x,z,R^{<\sigma((n,\langle x,z\rangle))})]$$

and hence $W_x \in \Delta_x$, contradiction. Thus we have established that:

$$(\forall k)[k \in W_x \Rightarrow (\exists \ell)(\ell \in W_x \ \& \ P(k,\ell))].$$

Since $P \in \Gamma_x$ and P is a relation on ω, we can find a relation $P^* \subseteq P$, $P^* \in \Gamma_x$ and such that for each $k \in W_x$ there is exactly one integer $\ell \in W_x$ such that $P^*(k,\ell)$. Also, since $R_0 \in \Delta_x$ and $W_x \in \Gamma_x - \Delta_x$, there is a $k_0 \in W_x$ such that $R_0 \subseteq R^{<\sigma((k_0,x))}$. Therefore we can find a sequence $\{k_m\}_{m\in\omega}$ with the property that:

$$R_0 \subseteq R^{<\sigma((k_0,x))} \quad \text{and} \quad P^*(k_m,k_{m+1}) \text{ for all } m \in \omega.$$

We put now $R^* = \bigcup_{m\in\omega} R^{<\sigma((k_m,x))}$ and we claim that $R^* \in \Delta_x$ and $\varphi(x,R^*)$. It is clear that $R^* \in \Gamma_x$, but also

$$y \notin R^* \Longleftrightarrow (\forall n \geq 2)(\exists u \in \omega)[\mathrm{seq}(u) \ \& \ (\ell h(u) = n) \ \& \ ((u)_1 = k_0) \ \& \\ (\forall i)(i < n \Rightarrow P^*((u)_i,(u)_{i+1})) \ \& \ (((u)_n,x) <^*_\sigma (e,y))]$$

and hence $\neg R^* \in \Gamma_x$. We conclude the proof by showing that $(\forall z)\psi(x,z,R^*)$; for any fixed $z \in A$ and any $m \in \omega$ there is an integer n_m such that

$$(k_m,x) <^*_\sigma (n_m,\langle x,z\rangle) <^*_\sigma (k_{m+1},x) \quad \text{and} \quad \psi(x,z,R^{<\sigma((n_m,\langle x,z\rangle))}).$$

Then $R^* = \bigcup_{m\in\omega} R^{<\sigma((k_m,x))} = \bigcup_{m\in\omega} R^{<\sigma((n_m,\langle x,z\rangle))}$ and by part ii) of Lemma 4.15 $\psi(x,z,\bigcup_{m\in\omega} R^{<\sigma((n_m,\langle x,z\rangle))})$, hence $\psi(x,z,R^*)$. ⊣

In general, if φ is a $\Pi^0_k(C)$ or a $\Sigma^0_k(C)$ relation on $\mathfrak{A}$ and Γ is a semi-Spector class on $\mathfrak{A}$, then clearly φ is Δ on Δ. From Theorem 4.16 and Corollary 4.11 we derive now the following result which is also essentially due to Grilliot [1971].

4.17. <u>Corollary</u>. Let $\mathfrak{A}$ be an acceptable structure, let C be a coding scheme on A and let Γ be a Spector class on $\mathfrak{A}$. If φ is a second order relation on A which is $\Sigma^0_2(C)$ on $\mathfrak{A}$, then Γ is φ-compact. ⊣

Finally, by putting together Theorems 1.11, 4.12, 4.16 and Corollary 4.17, we derive immediately the main result of this section, which is due to Grilliot [1971].

4.18. <u>Theorem</u>. If $\mathfrak{A} = \langle A,R_1,\ldots,R_n,f_1,\ldots,f_m,c_1,\ldots,c_k\rangle$ is an acceptable structure, then

i) $\mathrm{ENV}[\underset{\sim}{E}] = \Pi^0_1(\mathfrak{A})\text{-IND}$ = the smallest semi-Spector class on $\mathfrak{A}$,

ii) $\mathrm{IND}(\mathfrak{A}) = \Sigma^0_2(\mathfrak{A})\text{-IND}$ = the smallest Spector class on $\mathfrak{A}$. ⊣

As a concluding remark, notice that if $\mathfrak{A}$ is an acceptable structure such that $\mathrm{ENV}[\underset{\sim}{E}] \subsetneq \mathrm{IND}(\mathfrak{A})$ (consider for example the structure of analysis), then $\mathrm{ENV}[\underset{\sim}{E}]$ is not $\Sigma^0_2(C)$-compact, although it is $\Pi^0_1(C)$-compact. Of course in this case $\mathrm{ENV}[\underset{\sim}{E}]$

is not a Spector class and this shows that both Theorem 4.10 and Corollary 4.11 are false, if we assume that Γ is only a semi-Spector class. This comment shows also that Theorem 14 of Moschovakis [1974b] (which is the "boldface" analog of the Non-monotone First Recursion Theorem 4.12) is not true for semi-Spector classes.

5. Generalizations and Open Problems.

5.1. Let $\mathfrak{A} = \langle A, R_1, \ldots, R_n, f_1, \ldots, f_m, c_1, \ldots, c_k \rangle$ be an acceptable structure and let Q be a quantifier on A. It is natural to ask if there is a generalization of the two Theorems of Grilliot [1971] to inductions involving the quantifier Q. The functional

$$\hat{F}_Q(f) = \begin{cases} 0, & \text{if } Qx(f(x) = 0), \\ 1, & \text{if } f \text{ is total and } \breve{Q}x(f(x) \neq 0), \\ \text{undefined, otherwise}, \end{cases}$$

was introduced and studied in Kolaitis [1978]. In general, recursion in $\hat{F}_Q$ does not coincide with recursion in $\hat{F}_{\breve{Q}}$, in contrast to the situation with the functional F_Q or the functional $F^{\#}_Q$. It turns out that the classes $\mathfrak{F}[E, \hat{F}_Q]$ and $\mathfrak{F}[E^{\#}, \hat{F}_Q]$ are strongly normal classes of functionals on A. This fact implies that the class $ENV[E, \hat{F}_Q]$ of the semirecursive relations in $E, \hat{F}_Q$ is the smallest semi-Spector class on $\mathfrak{A}$ closed under Q and the deterministic $\breve{Q}$-rule. Also, the class $ENV[E^{\#}, \hat{F}_Q]$ of the semirecursive relations in $E^{\#}, \hat{F}_Q$ is the smallest Spector class on $\mathfrak{A}$ closed under Q and the deterministic $\breve{Q}$-rule. In Kolaitis [1978] it is proved that for certain quantifiers Q the classes $ENV[E, \hat{F}_Q]$ and $ENV[E^{\#}, \hat{F}_Q]$ can be characterized in terms of nonmonotone inductions involving Q. One result of this type is the following generalization of the Theorems of Grilliot [1971]:

5.2. Theorem. Let $\mathfrak{A} = \langle A, R_1, \ldots, R_n, f_1, \ldots, f_m, c_1, \ldots, c_k \rangle$ be an acceptable structure, let Q be a quantifier on A and let $\mathcal{G}$ be the collection of all second order relations on A definable by a formula of the language $\mathcal{L}^{\mathfrak{A}}(Q)$ of the form $Q_1 x_1 \ldots Q_n x_n \psi(\vec{x}, \vec{y}, \vec{Y})$, where $n \in \omega$, $Q_i = \forall$ or $Q_i = Q$ for $1 \leq i \leq n$ and ψ is restricted on $\mathfrak{A}$ relative to a coding scheme $\mathbf{C}$ on A. If Q is an ω-complete filter, then

i) $ENV[E, \hat{F}_Q] = \mathcal{G}\text{-IND}$,

ii) $ENV[E^{\#}, \hat{F}_Q] = (\exists^A \mathcal{G})\text{-IND}$. ⊣

We conclude this paper by listing some open problems which are related to the results we presented here.

5.3. There are several specific problems on the comparison of recursion in E and recursion in $E^{\#}$; we state one of them, which was raised by Moschovakis.

Characterize the least ordinal $\lambda > \omega$ such that the semirecursive relations in $\underset{\sim}{E}$ on the structure $\underset{\sim}{\lambda} = \langle \lambda, \leq \rangle$ are not closed under $\exists$, i.e. characterize the least $\lambda > \omega$ such that $\mathrm{ENV}[\underset{\sim}{E}] \subsetneq \mathrm{IND}(\underset{\sim}{\lambda})$.

5.4. If $\mathfrak{A}$ is an acceptable structure, then the class $\underset{\sim}{\mathrm{HYP}}(\mathfrak{A})$ of the "boldface" hyperelementary relations on $\mathfrak{A}$ is the smallest model of the schema of Δ^1_1-Comprehension on $\mathfrak{A}$. Recall that in 3.7 we introduced "lightface" versions of the schemata of Σ^1_1-Collection and Δ^0_∞-Comprehension on $\mathfrak{A}$ and that by Theorem 3.10 the collection $\mathfrak{R} = \{\mathrm{SEC}[\underset{\sim}{E}, \underset{\sim}{F}_Q, \vec{x}] : \vec{x} \in A^{<\omega}\}$ (where Q is a quantifier on A) is the smallest indexed family on A satisfying these schemata. We can also introduce a "lightface" version of the schema of Δ^1_1-Comprehension on $\mathfrak{A}$ and then a natural question to ask is if the collection $\mathfrak{R} = \{\mathrm{SEC}[\underset{\sim}{E}, \underset{\sim}{F}_Q, \vec{x}] : \vec{x} \in A^{<\omega}\}$ is the smallest model of the schema of "lightface" Δ^1_1-Comprehension on $\mathfrak{A}$.

Moschovakis [1969b] and [EIAS] obtained a "construction of the "boldface" hyperelementary relations from below" by establishing that each relation in $\underset{\sim}{\mathrm{HYP}}(\mathfrak{A})$ is Δ^1_1 definable with basis the class of previously constructed relations in $\underset{\sim}{\mathrm{HYP}}(\mathfrak{A})$. We consider a "lightface" version of this construction by defining a sequence $\{\mathfrak{D}^\xi\}_\xi$ of classes of relations on A as follows:

$\mathfrak{D}^0$ = all relations on A definable by a formula of the language $\mathcal{L}^{\mathfrak{A}}(Q)$,

$\mathfrak{D}^\xi$ = all relations R on A for which there are formulas $\varphi(\vec{x},\vec{Y},\vec{Z}), \psi(\vec{x},\vec{Y},\vec{W})$ of the language $\mathcal{L}^{\mathfrak{A}}(Q)$ and relations $\vec{S},\vec{T}$ in $\mathfrak{D}^{<\xi} = \bigcup_{\eta<\xi} \mathfrak{D}^\eta$ such that

$$R(\vec{x}) \Longleftrightarrow (\exists Y \in \mathfrak{D}^{<\xi}_{\vec{x}})\, \varphi(\vec{x},\vec{Y},\vec{S}) \Longleftrightarrow (\exists Y)\, \varphi(\vec{x},\vec{Y},\vec{S}),$$

$$\neg R(\vec{x}) \Longleftrightarrow (\exists Y \in \mathfrak{D}^{<\xi}_{\vec{x}})\, \psi(\vec{x},\vec{Y},\vec{T}) \Longleftrightarrow (\exists Y)\, \psi(\vec{x},\vec{Y},\vec{T}).$$

We put $\mathfrak{D} = \bigcup_\xi \mathfrak{D}^\xi$ and then the question is if $\mathfrak{D} = \mathrm{SEC}[\underset{\sim}{E}, \underset{\sim}{F}_Q]$. A positive answer will also imply that the collection $\mathfrak{R} = \{\mathrm{SEC}[\underset{\sim}{E}, \underset{\sim}{F}_Q, \vec{x}] : \vec{x} \in A^{<\omega}\}$ is in fact the smallest indexed family on A satisfying the schema of Δ^1_1-Comprehension on $\mathfrak{A}$.

5.5. Spector [1961] proved that the inductive relations on the structure $\langle \omega, +, \cdot \rangle$ coincide with the Π^1_1 relations. Barwise-Gandy-Moschovakis [1971] and Moschovakis [1970] extended this theorem by showing that a relation R is Π^1_1 on a countable acceptable structure $\mathfrak{A}$ if and only if it is inductive on $\mathfrak{A}$. It is not known, however, if there is a natural analog of this result for the Q-inductive relations on $\mathfrak{A}$, where Q is a quantifier on A.

Kleene [1959b] showed that a relation R on ω is Π^1_1 if and only if it is semirecursive in $\underset{\sim}{E}$ on the structure $\langle \omega, +, \cdot \rangle$; it is well known, however, that there are countable acceptable structures $\mathfrak{A}$ such that $\mathrm{ENV}[\underset{\sim}{E}] \subsetneq \Pi^1_1(\mathfrak{A}) = \mathrm{IND}(\mathfrak{A})$. This raises the problem if there is a "syntactic" characterization of the semirecursive relations in $\underset{\sim}{E}$ that will extend the classical theorem of Kleene to abstract structures.

REFERENCES

P. Aczel [1972], Stage Comparison theorems and game playing with inductive definitions, unpublished notes.

K. J. Barwise, R. O. Gandy and Y. N. Moschovakis [1971], The next admissible set, J. Symb. Logic 36 (1971), 108-120.

R. O. Gandy [1960], Proof of Mostowski's conjecture. Bull. Acad. Polon. Sci. Ser. Sci. Math. Astron. Phys. 8 (1960), 571-575.

T. J. Grilliot [1971], Inductive Definitions and Computability, Trans. Amer. Math. Soc. 158 (1971), 309-317.

L. A. Harrington, L. M. Kirousis, T. S. Schlipf [1977], A generalized Kleene-Moschovakis theorem, to appear in Proc. of Amer. Math. Soc.

A. S. Kechris [1973], The structure of envelopes: a survey of recursion theory in higher types, M.I.T. Logic Seminar notes.

A. S. Kechris and Y. N. Moschovakis [1977], Recursion in Higher Types, J. Barwise (ed.), Handbook of Mathematical Logic, North Holland, 681-737.

S. C. Kleene [1959a], Quantification of number theoretic functions, Compositio Math. 14 (1959), 23-40.

S. C. Kleene [1959b], Recursive functionals and quantifiers of finite types I, Trans. Amer. Math. Soc. 91 (1959), 1-52.

Ph. G. Kolaitis [1977], Recursion in a quantifier vs. elementary induction, to appear in J. Symb. Logic.

Ph. G. Kolaitis [1978], Doctoral Dissertation, University of California, Los Angeles.

G. Kreisel [1961], Set theoretic problems suggested by the notion of potential totality, Infinitistic Methods, Pergamon (1961), 103-140.

Y. N. Moschovakis [1969a], Abstract first order computability II, Trans. Amer. Math. Soc. 138 (1969), 464-504.

Y. N. Moschovakis [1969b], Abstract computability and invariant definability, J. Symb. Logic 34 (1969), 605-633.

Y. N. Moschovakis [1970], The Suslin-Kleene theorem for countable structures, Duke Math. J. 37 (1970), 341-352.

Y. N. Moschovakis [EIAS], Elementary Induction on Abstract Structures, North Holland (1974).

Y. N. Moschovakis [1974a], Structural characterizations of classes of relations, J. E. Fenstad - P. G. Hinman (eds.), Generalized Recursion Theory, North Holland (1974), 53-79.

Y. N. Moschovakis [1974b], On nonmonotone inductive definability, Fundamenta Mathematicae 82 (1974), 39-83.

Y. N. Moschovakis [1976], On the basic notions in the theory of induction, to appear in Proc. Fifth International Congress of Logic, Methodology and Philosophy of Science.

W. Richter [1971], Recursively Mahlo ordinals and inductive definitions, R. O. Gandy - C. E. M. Yates (eds.), Logic Colloquium 69, North Holland (1971), 273-288.

W. Richter and P. Aczel [1974], Inductive definitions and reflecting properties of admissible ordinals, Generalized Recursion Theory, J. E. Fenstad - P. G. Hinman (eds.), North Holland (1974), 301-381.

C. Spector [1960], Hyperarithmetical quantifiers, Fundamenta Mathematicae 48 (1960), 313-320.

C. Spector [1961], Inductively defined sets of natural numbers, Infinitistic Methods, Pergamon (1961), 97-102.

ON SPECTOR CLASSES

Alexander S. Kechris[1]
Department of Mathematics
California Institute of Technology
Pasadena, California 91125

This paper contains an exposition of certain aspects of the theory of Spector classes. The general plan is as follows: In Section 1 we present a review of the structure theory, mainly developed in Chapter 9 of Moschovakis [1]. In Section 2 we give a comprehensive list of examples from various areas of definability theory. Finally, in Section 3, we deal with characterization and representation theorems.

1. INTRODUCTION TO SPECTOR CLASSES

1.1. The definition (Moschovakis [1]). Let $\mathcal{A} = \langle A, R_1, \ldots, R_\ell \rangle$ be an infinite structure. A collection Γ of relations on A is called a Spector class on $\mathcal{A}$ if

(i) Γ is closed under $\wedge, \vee, \exists, \forall$.

(ii) Γ contains all the relations on A, which are first order definable with parameters in $\mathcal{A}$.

(iii) Γ contains a coding scheme on A.

(iv) Γ is A-parametrized.

(v) Γ is normed (or has the prewellordering property).

Explanations. For (iii): A coding scheme on A is a triple $C = \langle N, \leq, \langle\,\rangle \rangle$, where $\langle N, \leq \rangle \cong \langle \omega, \leq \rangle$ and $\langle\,\rangle : \bigcup_{n\in\omega} A^n \xrightarrow{1-1} A$. We usually identify $\langle N, \leq \rangle$ with $\langle \omega, \leq \rangle$. Associated with the coding scheme C are the following relations and functions:

$$\mathrm{Seq}(x) \Leftrightarrow \exists x_1 \ldots x_n \, (x = \langle x_1 \ldots x_n \rangle)$$

$$\ell h(x) = \begin{cases} n, & \text{if } x = \langle x_1 \ldots x_n \rangle \\ 0, & \text{if } \neg \mathrm{Seq}(x) \end{cases}$$

$$(x)_i = \begin{cases} x_i, & \text{if } x = \langle x_0 \ldots x_n \rangle \ \&\ i \leq n \\ 0, & \text{otherwise.} \end{cases}$$

To say that Γ contains C means that $N, \leq, \mathrm{Seq}, \ell h, (x)_i$ are in $\Delta = \Gamma \cap \breve{\Gamma}$, where $\breve{\Gamma} = \{A^n - R : R \in \Gamma\}$ = the dual of Γ. (A function is in Δ iff its graph is in Δ). It should be noted here that an equivalent version of (iii) (and one which is easier to verify when checking that a given collection of relations is a Spector

[1]Preparation of this paper was partially supported by NSF Grant MCS 76-17254. We would like to thank L. Kirousis and P. Kolaitis for many valuable comments and suggestions on an earlier draft of this paper.

class is the following

(iii)* There is N, $\leq$ and $<>: A^2 \xrightarrow{1-1} A$ such that $\langle N,\leq\rangle \cong \langle\omega,\leq\rangle$ and N, $\leq$, <> are all in Δ.

(The proof of the equivalence follows from the fact that there is a coding scheme hyperelementary in $\langle\mathfrak{A}, N, \leq, <>\rangle$ and Theorem 1.4 below.)

For (iv): Γ is A-parametrized means that for each $n \geq 1$ there is $W^n \subseteq A^{n+1}$ such that $W^n \in \Gamma$ and for any $R \subseteq A^n$ we have

$$R \in \Gamma \Leftrightarrow \exists a \in A(R = W^n_a =^{\text{def}} \{\bar{x} : W^n(a,\bar{x})\}).$$

We call such a W^n universal (for the n-ary relations in Γ). If $R = W^n_a$ we call a a code of R.

For (v): If R is a relation on A, a Γ-norm on R is a map $\sigma : R \to$ Ordinals such that the following two relations $\leq^*_\sigma$, $<^*_\sigma$ are in Γ:

$$\bar{x} \leq^*_\sigma \bar{y} \Leftrightarrow \bar{x} \in R \;\&\; [\bar{y} \notin R \vee \sigma(\bar{x}) \leq \sigma(\bar{y})],$$

$$\bar{x} <^*_\sigma y \Leftrightarrow \bar{x} \in R \;\&\; [\bar{y} \notin R \vee \sigma(\bar{x}) < \sigma(\bar{y})].$$

If we put $\sigma(\bar{y}) = \infty$ if $\bar{y} \notin R$ then the above relations become

$$\bar{x} \leq^*_\sigma \bar{y} \Leftrightarrow \bar{x} \in R \;\&\; \sigma(\bar{x}) \leq \sigma(\bar{y})$$

$$\bar{x} <^*_\sigma \bar{y} \Leftrightarrow \bar{x} \in R \;\&\; \sigma(\bar{x}) < \sigma(\bar{y}) \Leftrightarrow \sigma(\bar{x}) < \sigma(\bar{y}).$$

Equivalently, we can easily see that σ is a Γ-norm iff there are relations S,T in $\Gamma,\check{\Gamma}$ resp. such that

$$\bar{y} \in R \Rightarrow [\bar{x} \in R \;\&\; \sigma(\bar{x}) \leq \sigma(\bar{y}) \Leftrightarrow S(\bar{x},\bar{y}) \Leftrightarrow T(\bar{x},\bar{y})]$$

(this means that the initial segments of the prewellordering associated with σ are uniformly in Δ).

We call now Γ normed if every $R \in \Gamma$ admits a Γ-norm.

Remark. Note that every Spector class Γ on $\mathfrak{A}$ is closed under substitution by constants from A i.e. if $P(x,\bar{y}) \in \Gamma$ and $a_0 \in A$ then $Q(y) \Leftrightarrow P(a_0,\bar{y})$ is also in Γ, since $Q(\bar{y}) \Leftrightarrow (\exists x)(x = a_0 \wedge P(x,\bar{y}))$. If we restrict (ii) so that only first-order definable without parameters over $\mathfrak{A}$ relations are admitted in Γ and modify (iv) to state that Γ is ω-parametrized (i.e. $R \in \Gamma \Leftrightarrow \exists \ell \in \omega\ (R = W^n_\ell)$), then we get the notion of a lightface Spector class.

1.2. The Second Recursion Theorem

An important consequence of the parametrization property is the Second Recursion Theorem, a basic tool in the study of Spector classes. As usual we need to prove first

1.2.1. The s-m-n- Theorem. Given a Spector class Γ on $\mathfrak{A}$ we can always find a system of relations $\{W^n\}_{n=1,2,\ldots}$ all in Γ such that

(i) W^n is universal for the n-ary relations in Γ.

(ii) For each $m,n \geq 1$ there is a function S^m_n (whose graph is) in Δ and

$$W^{m+n}(a,x_1 \ldots x_m,y_1 \ldots y_n) \Leftrightarrow W^n(S^m_n(a,x_1 \ldots x_m),y_1 \ldots y_n).$$

Proof. Let $V \subseteq A^3$ be universal for the binary relations in Γ. Put

$$W^k(a,x_1 \ldots x_k) \Leftrightarrow V((a)_0, (a)_1, \langle x_1 \ldots x_k\rangle).$$

Then W^k is universal for the k-ary relations in Γ. Put for $m \geq 1$, $n \geq 1$

$$Q(s,t) \Leftrightarrow W^{m+n}((s)_0 \ldots (s)_m, (t)_0 \ldots (t)_{n-1}).$$

Then $Q \in \Gamma$, so for some $b_0 \in A$

$$Q(s,t) \Leftrightarrow V(b_0,s,t).$$

Put

$$S^m_n(a,x_1 \ldots x_m) = \langle b_0,\langle a,x_1 \ldots x_m\rangle\rangle. \qquad \dashv$$

From now on and without explicit mentioning we shall assume that universal sets belong to systems for which the s-m-n property holds.

1.2.2 The Second Recursion Theorem

Let Γ be a Spector class on $\mathfrak{A}$. Let W^n be universal in Γ. Then for each $R(a,\bar{x})$ in Γ there is $a_0 \in A$ such that

$$W^n_{a_0}(\bar{x}) \Leftrightarrow R(a_0,\bar{x}).$$

Proof. We are trying to find an a so that

$$R(a,\bar{x}) \Leftrightarrow W^n(a,\bar{x}).$$

Change the "unknown" a to the "unknown" b by the transformation $a = S(b,b) = S^1_n(b,b)$. Then we must have

$$W(S(b,b),\bar{x}) \Leftrightarrow R(S(b,b),\bar{x})$$

or by the basic property of S

$$W(b,b,\bar{x}) \Leftrightarrow R(S(b,b),\bar{x}).$$

But such a b is easy to find. Indeed find b_0 such that

$$W_{b_0}(b,\bar{x}) \Leftrightarrow R(S(b,b),\bar{x}).$$

Then $W(b_0,b_0,\bar{x}) \Leftrightarrow R(S(b_0,b_0),\bar{x})$ and we are done. $\dashv$

1.3. A basic example

For any structure $\mathfrak{A}$, let $\mathrm{IND}(\mathfrak{A}) = \{R \subseteq A^n : R \text{ is inductive in } \mathfrak{A}\}$. We call a structure $\mathfrak{A}$ almost acceptable if it admits a hyperelementary coding scheme.

1.3.1 Theorem (Moschovakis [1]). For each almost acceptable structure $\mathfrak{A}$, $\mathrm{IND}(\mathfrak{A})$ is a Spector class.

Proof. Closure under $\wedge, \vee, \exists, \forall$ follows from the substitution theorem for positive first order induction (see Moschovakis [1]). Conditions (ii), (iii) are immediate, while (v) follows from the Stage Comparison Theorem (see Moschovakis [1]). We outline now a proof of (iv):

First Gödel number all first-order formulas $\varphi(\bar{x},R)$ which are positive in R. Denote by $\ulcorner\varphi\urcorner$ the Gödel number of φ. Then (by the usual inductive analysis of the definition of truth) find a relation $\Phi(u,x,S)$ which is naturally positive inductive (i.e. for some first order $\Psi(\bar{v},u,x,T,S)$ positive in T,S we have $\Phi(u,x,S) \Leftrightarrow \Psi^{\infty}(\bar{a},u,x,S)$ for some fixed $\bar{a}$) and the following holds:

$$\varphi(x_1 \ldots x_n,R) \Leftrightarrow \Phi(\ulcorner\varphi\urcorner,\langle x_1 \ldots x_n\rangle,\langle R\rangle)$$

where if $R \subseteq A^k$, $\langle R\rangle \subseteq A$ and

$$y \in \langle R\rangle \Leftrightarrow \exists x_1 \ldots \exists x_k\ (y = \langle x_1 \ldots x_k\rangle \ \&\ R(x_1 \ldots x_k)).$$

Put now

$$\chi(e,x,T) \Leftrightarrow e \in \omega \ \&\ \Phi(e,x,\{\langle (x)_0 \ \ldots\ (x)_{n(e)\dot{-}1}\rangle : (e,x) \in T\}),$$

where if $e = \ulcorner\varphi(x_1 \ldots x_n,R)\urcorner$ then $n(e) = n$. Then it is easy to check that if $\varphi(x_1 \ldots x_n,R)$ is first order positive in the n-ary R then

$$\varphi^{\xi}(x_1 \ldots x_n) \Leftrightarrow \chi^{\xi}(\ulcorner\varphi\urcorner,\langle x_1 \ldots x_n\rangle)$$

so $\varphi(\bar{x}) \Leftrightarrow \chi^{\infty}(\ulcorner\varphi\urcorner,\langle\bar{x}\rangle)$. By the Completeness Theorem for positive induction (see Moschovakis [1]) χ^{∞} is inductive on $\mathfrak{A}$. Finally put

$$W^n(a,x_1 \ldots x_n) \Leftrightarrow \chi^{\infty}((a)_0,\langle (a)_1 \ldots (a)_{\ell h(a)\dot{-}1},x_1 \ldots x_n\rangle).$$

Then W^n is universal. ⊣

Remark. For a countable almost acceptable structure $\mathfrak{A}$ we also have

$$\mathrm{IND}(\mathfrak{A}) = \Pi^1_1(\mathfrak{A}) = \{\forall X\,\varphi(\bar{x},X) : \varphi \text{ is first-order over } \mathfrak{A}\}.$$

The Spector class $\mathrm{IND}(\mathfrak{A})$ has also the following basic minimality property.

1.3.2. Theorem (Moschovakis [1]). Let $\mathfrak{A}$ be an almost acceptable structure. Then $\mathrm{IND}(\mathfrak{A})$ is the smallest Spector class on $\mathfrak{A}$.

This result is an immediate consequence of the First Recursion Theorem for Spector classes given next.

1.4 The First Recursion Theorem

Definition. Let Γ be a Spector class on $\mathfrak{A}$. Let $\varphi(\bar{x},S)$ be a second order relation on A. We say that Γ is closed under φ if for each $P(\bar{y},\bar{z}) \in \Gamma$ the relation

$$P_\varphi(\bar{x},\bar{z}) \Longleftrightarrow \varphi(\bar{x},\{\bar{y} : P(\bar{y},\bar{z})\})$$

is also in Γ. We now have

The First Recursion Theorem (Moschovakis [2]). Let Γ be a Spector class on $\mathfrak{A}$. Let $\varphi(\bar{x},S)$ be a monotone operator on A (i.e. S is n-ary, where $\bar{x} = x_1 \ldots x_n$ and $\varphi(\bar{x},S)$ & $S \subseteq T \Rightarrow \varphi(\bar{x},T)$). If Γ is closed under φ, then $\varphi^\infty \in \Gamma$.

Proof. Since φ is monotone, φ^∞ is the smallest relation R with the property

$$(*) \qquad \varphi(\bar{x},R) \Longleftrightarrow R(\bar{x}).$$

Although it is easy to find a relation R in Γ which satisfies (*) using the second recursion theorem (let $R = W_a$; then (*) becomes $\varphi(\bar{x},\{\bar{y} : W(a,\bar{y})\}) \Longleftrightarrow W_a(\bar{x})$), there is no guarantee that we will have the smallest solution of (*). What is needed is the following key characterization of φ^∞:

φ^∞ is the unique relation R which admits a norm $\tau : R \to$ Ordinals such that

$$R(\bar{x}) \Longleftrightarrow \varphi(\bar{x},\{\bar{y} : \bar{y} <^*_\tau \bar{x}\}).$$

(Note that no definability restrictions are imposed on τ.)

Using this characterization we can now apply the second recursion theorem as follows: We are looking for a code a for φ^∞ i.e. an a such that

$$\varphi^\infty = W_a.$$

Now on any W_a we have a natural norm, namely

$$\tau(\bar{x}) = \sigma(a,\bar{x}),$$

where σ is a Γ-norm on W. By the key characterization we only have to choose a so that

$$W_a(\bar{x}) \Longleftrightarrow \varphi(\bar{x},\{\bar{y} : (a,\bar{y}) <^*_\sigma (a,\bar{x})\}).$$

The existence of such an a is guaranteed to course by the second recursion theorem.

To prove now the characterization of φ^∞ we note first that φ^∞ satisfies it by taking the relevant norm to be $|\bar{x}|_\varphi$ = least ξ such that $\bar{x} \in \varphi^\xi$. For the converse assume that R,τ satisfy it and prove first that $R \subseteq \varphi^\infty$ by induction on $\tau(\bar{x})$, for $\bar{x} \in R$. Then prove that $\varphi^\infty \subseteq R$ by induction on $|\bar{x}|_\varphi$ as follows: Assume $\varphi^{<\xi} \subseteq R$. Let $\bar{x} \in \varphi^\xi$. Then $\varphi(\bar{x},\varphi^{<\xi})$ so $\varphi(\bar{x},R)$. If $\bar{x} \notin R$ then $\{\bar{y} : \bar{y} <^*_\tau \bar{x}\} = R$, so $\bar{x} \in R$, a contradiction. $\dashv$

1.5. Simple structural consequences of the prewellordering property.

1.5.1 Theorem. Let Γ be a Spector class on $\mathfrak{A}$. Then we have

(i) Reduction (Γ): For any $P,Q \subseteq A^n$ in Γ there are P^*,Q^* in Γ such that $P^* \subseteq P$, $Q^* \subseteq Q$, $P^* \cap Q^* = \emptyset$, $P^* \cup Q^* = P \cup Q$.

(ii) Separation ($\breve{\Gamma}$): For any disjoint $P,Q \subseteq A^n$ in $\breve{\Gamma}$, there is $S \in \Delta$ such that $P \subseteq S$, $S \cap Q = \emptyset$.

(iii) $\neg$ Separation (Γ): There are disjoint $P,Q \subseteq A^n$ in Γ which are not separated by any $S \in \Delta$.

Proof. (i) If $P,Q \in \Gamma$ let $R = (P\times\{0\}) \cup (Q\times\{1\})$. Then $R \in \Gamma$. Let σ be a Γ-norm on R and put

$$\bar{x} \in P^* \iff (0,\bar{x}) \leq^*_\sigma (1,\bar{x})$$

$$\bar{x} \in Q^* \iff (1,\bar{x}) <^*_\sigma (0,\bar{x}).$$

(ii) To separate P,Q reduce their complements.

(iii) Let W be universal for the unary relations in Γ. Put

$$W_0(a,b) \iff W((a)_0,b)$$

$$W_1(a,b) \iff W((a)_1,b).$$

Then W_0,W_1 are doubly universal sets i.e. for any $P,Q \subseteq A$ in Γ there is a such that

$$P(b) \iff W_0(a,b),\ Q(b) \iff W_1(a,b).$$

Reduce W_0,W_1 to W_0^*,W_1^* in Γ. If $S \in \Delta$ separates W_0^*,W_1^* then S is universal for the unary relations in Δ, a contradiction. ⊣

Corollary. If Γ be a Spector class on $\mathfrak{A}$. Given a relation $R(\bar{x},\bar{y})$ in Γ, there are $R_1(\bar{x},\bar{y})$, $R_2(\bar{x},\bar{y})$ in Γ, $\breve{\Gamma}$ respectively such that for any $\bar{x} \in \mathrm{dom}(R) = \{\bar{x} : \exists\bar{y}R(\bar{x},\bar{y})\}$ we have

(i) $R_1(x,y) \iff R_2(x,y)$ and $R_1(x,y) \Rightarrow R(x,y)$

(ii) $\exists\bar{y}R_1(\bar{x},\bar{y})$.

Picture

For $\bar{x} \in \mathrm{dom}(R)$,
$R^*(\bar{x},\bar{y}) \Leftrightarrow R_1(\bar{x},\bar{y})$
$\Leftrightarrow R_2(\bar{x},\bar{y})$

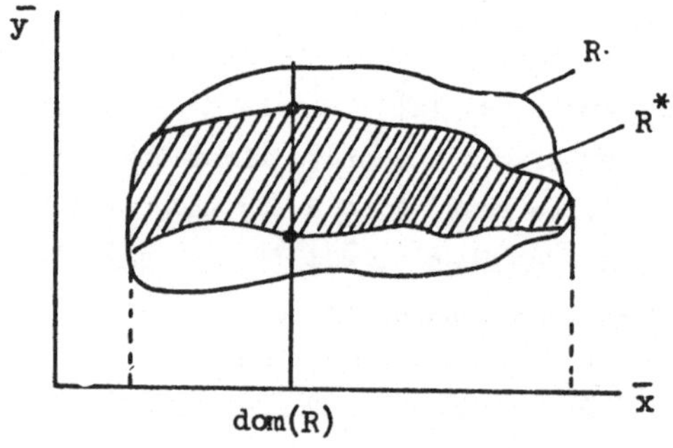

$R^* \subseteq R$

$\exists\bar{y}R(\bar{x},\bar{y}) \Rightarrow \exists\bar{y}R^*(\bar{x},\bar{y})$

Proof. Let σ be a Γ-norm on R. Put $R^*(\bar{x},\bar{y}) \Leftrightarrow R(\bar{x},\bar{y})$ & $\sigma(\bar{x},\bar{y})$ is least. Then let

$$R_1(\bar{x},\bar{y}) \Leftrightarrow R(\bar{x},\bar{y}) \wedge \forall \bar{y}'((\bar{x},\bar{y}) \leq^*_\sigma (\bar{x},\bar{y}'))$$

$$R_2(\bar{x},\bar{y}) \Leftrightarrow \neg \exists \bar{y}'((\bar{x},\bar{y}') <^*_\sigma (\bar{x},\bar{y})). \qquad \dashv$$

Corollary. (i) In the notation of the previous theorem if $B \in \Delta$ and $\forall \bar{x} \in B$ $\exists \bar{y} R(\bar{x},\bar{y})$ there is $R^* \in \Delta$, $R^* \subseteq R$ such that $\forall \bar{x} \in B \; \exists \bar{y} R^*(\bar{x},\bar{y})$.

(ii) In the notation of the previous theorem, if $B \in \Delta$, $C \in \Delta$ and C carries a Δ-wellordering, then if $\forall \bar{x} \in B \exists \bar{y} \in C\, R(\bar{x},\bar{y})$ there is $R^* \subseteq R$, $R^* \in \Delta$ such that $\forall \bar{x} \in B \exists ! \bar{y} \in C\, R^*(\bar{x},\bar{y})$.

1.6 The ordinal of a Spector class

Let Γ be a Spector class on $\mathfrak{a}$. We put

κ^Γ = least ordinal not the length of a prewellordering in Δ.

Let R be a relation on A. A norm $\sigma : R \to$ Ordinals is called regular if σ maps R onto an initial segment of the ordinals. We put $|\sigma|$ = length of σ = least ordinal not in the range of σ. Clearly for every norm σ' on R there is a unique regular norm σ on R which induces the same prewellordering on R.

The following is a basic result on Spector classes.

1.6.1. The Boundedness Theorem (Moschovakis [1]). Let Γ be a Spector class on $\mathfrak{a}$. Let $P \in \Gamma$ and let σ be a regular Γ-norm on P. Then

(i) $|\sigma| \leq \kappa^\Gamma$

(ii) $|\sigma| < \kappa^\Gamma \Leftrightarrow P \in \Delta$.

Proof. (i) is obvious. Also if $P \in \Delta$ clearly $|\sigma| < \kappa^\Gamma$.

Assume now $|\sigma| < \kappa^\Gamma$. Find a prewellordering $\precsim$ in Δ so that

$$|\sigma| = |\precsim|.$$

Let $\prec$ be the strict part of $\precsim$ i.e.

$$\bar{x} \prec \bar{y} \Leftrightarrow \bar{x} \precsim \bar{y} \;\&\; \bar{y} \not\precsim \bar{x}.$$

Then $\prec$ is a wellfounded relation in Δ of length $|\sigma|$. Notice now the following

Length Comparison Lemma. If $\prec_1$, $\prec_2$ are wellfounded relations in $\breve{\Gamma}$ and Γ respectively, there is $R \in \Gamma$ such that

$$\bar{x} \in \mathrm{Field}(\prec_1) \Rightarrow [R(\bar{x},\bar{y}) \Leftrightarrow \bar{y} \in \mathrm{Field}(\prec_2) \;\&\; |\bar{x}|_{\prec_1} \leq |\bar{y}|_{\prec_2}].$$

Proof of Lemma. Put

$$\varphi(\bar{x},\bar{y},S) \Leftrightarrow \bar{y} \in \mathrm{Field}(\prec_2) \;\& \forall \bar{x}'[\bar{x}' \prec_1 \bar{x} \Rightarrow \exists \bar{y}' (\bar{y}' \prec_2 \bar{y} \;\&\; S(\bar{x}',\bar{y}'))]$$

and take $R(\bar{x},\bar{y}) \Leftrightarrow \varphi^\infty(x,y)$. By the first recursion theorem $R \in \Gamma$. $\dashv$

Apply now the lemma to $\prec_1 = \prec$ and $\prec_2 = \{(\bar{y},\bar{y}') : \bar{y},\bar{y}' \in P \ \&\ \sigma(\bar{y}) < \sigma(\bar{y}')\}$. Then

$$\forall \bar{x} \in \mathrm{Field}(\prec) \exists \bar{y}\, R(\bar{x},\bar{y})$$

$\therefore$ there is $R^* \subseteq R$, $R^* \in \Delta$ such that

$$\forall \bar{x} \in \mathrm{Field}(\prec) \exists \bar{y}\, R^*(\bar{x},\bar{y}).$$

Then

$$\bar{y} \in P \Longleftrightarrow \exists \bar{x} \in \mathrm{Field}(\prec) \exists \bar{y}'[R^*(\bar{x},\bar{y}') \ \&\ \sigma(\bar{y}) \leq \sigma(\bar{y}')]$$

so $P \in \Delta$. ⊣

Corollary. If Γ is a Spector class on $\mathfrak{A}$, then

$$\varkappa^\Gamma = \sup\{\xi : \xi \text{ is the length of a } \breve{\Gamma} \text{ wellfounded relation}\}.$$

Proof. Let $\prec$ be a wellfounded relation in $\breve{\Gamma}$. Put

$$\varphi(\bar{x},S) \Longleftrightarrow \forall \bar{y}(\bar{y} \prec \bar{x} \Rightarrow \bar{y} \in S).$$

Then Γ is closed under φ and so by the proof of the first recursion theorem the closure ordinal of φ, which of course is exactly the length of $\prec$, is $\leq |\tau|$ for any norm τ on φ^∞ such that

$$\varphi^\infty(\bar{x}) \Longleftrightarrow \varphi(\bar{x}, \{\bar{y} : \bar{y} <_\tau^* \bar{x}\}).$$

Moreover by the same proof φ^∞ admits such a τ which is also a Γ-norm so that the length of $\prec$ is $\leq |\tau| \leq \varkappa^\Gamma$. ⊣

The next result is easier to prove but quite useful in applications.

1.6.2. The Covering Theorem

If Γ is a Spector class on $\mathfrak{A}$, σ a regular Γ-norm on $P \in \Gamma$ and $R \subseteq P$ is in $\breve{\Gamma}$ then for some $\xi < \varkappa^\Gamma$

$$\bar{x} \in R \Rightarrow \sigma(\bar{x}) \leq \xi.$$

Proof. By the separation theorem for $\breve{\Gamma}$ find $S \in \Delta$ such that

$$R \subseteq S \subseteq P.$$

Let $S^* = \{\bar{x} \in P : \exists \bar{y} \in S(\sigma(\bar{x}) \leq \sigma(\bar{y}))\}$. Then $S^* \in \Delta$ and σ restricted to S^* is a Γ-norm, so by 1.6.1

$$\xi = \sup\{\sigma(\bar{x}) : \bar{x} \in S^*\} < \varkappa^\Gamma.$$ ⊣

We present now two applications of the preceding theorems. Let $\Gamma \prec \Gamma' \Longleftrightarrow \Gamma \subseteq \Delta'$.

1.6.3. The Spector Criterion

Let Γ, Γ' be two Spector classes on $\mathfrak{A}$. Then

$$\Gamma \subseteq \Gamma' \Rightarrow (\Gamma \prec \Gamma' \Longleftrightarrow \varkappa^\Gamma < \varkappa^{\Gamma'}).$$

Proof. Assume $\Gamma \subseteq \Gamma'$. If $\Gamma < \Gamma'$ then $\Gamma \subseteq \Delta'$ so clearly $\varkappa^{\Gamma} < \varkappa^{\Gamma'}$. Conversely assume $\varkappa^{\Gamma} < \varkappa^{\Gamma'}$. Let $P \in \Gamma$ and let σ be a regular Γ-norm on P. Then σ is also a regular Γ'-norm on P and $|\sigma| \leq \varkappa^{\Gamma} < \varkappa^{\Gamma'}$, so $P \in \Delta'$. Thus $\Gamma < \Gamma'$. ⊣

1.6.4. Theorem. Let Γ be a Spector class on $\langle \lambda, < \rangle$, where λ is a cardinal. Then $\mathrm{cof}(\lambda) = \mathrm{cof}(\varkappa^{\Gamma})$.

Proof. Let $P \in \Gamma - \Delta$, $P \subseteq \lambda$. Let σ be a regular Γ-norm on P. Then $|\sigma| = \varkappa^{\Gamma}$. We will define a map $f : \lambda \to \varkappa^{\Gamma}$ such that

(i) $\xi \leq \xi' < \lambda \Rightarrow f(\xi) \leq f(\xi')$

(ii) $\lim_{\xi<\lambda} f(\xi) = \varkappa^{\Gamma}$.

This will complete the proof.

For $\xi < \lambda$, let

$$f(\xi) = \sup\{\sigma(\eta) : \eta \in P \;\&\; \eta < \xi\}.$$

Clearly f is nondecreasing, $f(\xi) \leq \varkappa^{\Gamma}$ and $\lim_{\xi<\lambda} f(\xi) = \varkappa^{\Gamma}$. It is thus enough to show that $f(\xi) < \varkappa^{\Gamma}$. Put

$$P_{\xi} = \{\eta : \eta \in P \;\&\; \eta < \xi\};$$

P_{ξ} is in Γ and σ restricted to P_{ξ} in a Γ-norm on P_{ξ}. Let σ' be the regular Γ-norm associated with $\sigma \restriction P_{\xi}$. Then since $P_{\xi} \subseteq \xi < \lambda$ and λ is a cardinal we have $|\sigma'| < \lambda < \varkappa^{\Gamma}$, so by the Boundedness Theorem $P_{\xi} \in \Delta$. Then by the Covering Theorem $f(\xi) < \varkappa^{\Gamma}$. ⊣

2. THE MAIN EXAMPLES

2.1. Positive induction in a quantifier Q (Aczel [3]). A quantifier on a set A is a subset $\emptyset \subsetneq Q \subsetneq \mathrm{power}(A)$ which is monotone i.e. $X \in Q \;\&\; X \subseteq Y \Rightarrow Y \in Q$. We usually write interchangeably

$$X \in Q \Longleftrightarrow Q(X) \Longleftrightarrow Qx\, X(x).$$

The dual of the quantifier Q is the quantifier

$$\breve{Q}x\, X(x) \Longleftrightarrow \neg Qx \neg X(x).$$

Examples

(i) $Q = \exists$, $\breve{Q} = \forall$

(ii) Let $\langle\, \rangle : \bigcup_n A^n \xrightarrow{1-1} A$ be a tuple coding function on A. The Souslin quantifier (relative to $\langle\, \rangle$) is given by

$$\mathfrak{S}uP(u) \Longleftrightarrow \forall x_0 \forall x_1 \forall x_2 \ldots \exists n P(\langle x_0 \ldots x_n \rangle);$$

Its dual is usually denoted by $\mathfrak{A}$ and is given by

$$\mathfrak{A}u\, P(u) \Longleftrightarrow \exists x_0 \exists x_1 \exists x_2 \ldots \forall n P(\langle x_0 \ldots x_n \rangle).$$

(iii) Let $\langle\ \rangle$ be as before. The open game quantifier $\mathcal{G}_0$ is given by

$$\mathcal{G}_0 uP(u) \Leftrightarrow \exists x_0 \forall x_1 \exists x_2 \forall x_3 \ldots \exists n P(\langle x_0 \ldots x_n \rangle)$$

$\Leftrightarrow$ Player I has a winning strategy in the game where players I, II alternatively choose members $x_0, x_1, \ldots$ of A (with I picking x_0, II picking $x_1, \ldots$) and I wins iff $\exists n P(\langle x_0 \ldots x_n \rangle)$.

I	x_0		x_2		$\ldots$
II		x_1		x_3	

Since this is an open game, by the Gale-Stewart theorem, either player I or II has a winning strategy so the dual quantifier of $\mathcal{G}_0$ is the closed game quantifier $\mathcal{G}_c$ given by

$$\mathcal{G}_c uP(u) \Leftrightarrow \breve{\mathcal{G}}_0 uP(u) \Leftrightarrow \forall x_0 \exists x_1 \ldots \forall n P(\langle x_0 \ldots x_n \rangle).$$

(iv) The previous example can be generalized as follows (Aczel [3]): Let Q be an artibrary quantifier on A. The next quantifier Q^+ of Q is defined by

$$Q^+ uR(u) \Leftrightarrow Qx_0 \breve{Q}x_1 \exists x_2 \forall x_3 Qx_4 \breve{Q}x_5 \exists x_6 \forall x_7 \ldots \exists n R(\langle x_0 \ldots x_n \rangle).$$

The way a formula

(*) $$Q_0 x_0 Q_1 x_1 \ldots P(x_0, x_1, \ldots)$$

where $Q_0, Q_1, \ldots$ are quantifiers on A is interpreted is as follows (the motivation coming from the remark that $Qx\, R(x) \Leftrightarrow \exists X \in Q \forall x \in X\, R(x)$): Players I, II play the following game

I	II
X_0	
	x_0
X_1	
	x_1
$\vdots$	

I chooses $X_0 \in Q_0$, II chooses $x_0 \in X_0$, I chooses $X_1 \in Q_1$, II chooses $x_1 \in X_1, \ldots$ and I wins iff $P(x_0, x_1, \ldots)$.

Then, (*) $\Leftrightarrow$ I has a winning strategy in this game.

It is clear now that $\exists^+ = \forall^+ = \mathcal{G}_0$, so that this generalizes example (iii).

Given now a structure $\mathfrak{A}$ and a quantifier Q on A we consider the language $\mathcal{L}(Q)$ obtained from the first order language $\mathcal{L}$ of $\mathfrak{A}$ by adding the quantifiers $Q, \breve{Q}$ i.e. asserting that if $\varphi(x)$ is a formula of $\mathcal{L}(Q)$ so are $Qx\varphi(x)$ and $\breve{Q}x\varphi(x)$. We can then define what it means for a formula $\varphi(\bar{x}, S)$ in $\mathcal{L}(Q)$ to be positive in S in exactly the same way we do it for $\mathcal{L}$. Put

$$\mathrm{IND}(\mathfrak{A}, Q)$$

for the class of all inductively definable relations from operators $\varphi(\bar{x}, S)$ which are positive in S and definable in $\mathcal{L}(Q)$ (see Moschovakis [1], Ch. 9 and Aczel [3]). Then we have the following analog of 1.3.1 and 1.3.2, calling a Spector class Γ closed under Q if for all $R(\bar{x}, a)$ in Γ, $QaR(\bar{x}, a)$ is also in Γ.

2.1.1. <u>Theorem</u> (Aczel). Let $\mathfrak{A}$ be an almost acceptable structure and Q a quantifier on $\mathfrak{A}$. Then $\mathrm{IND}(\mathfrak{A},Q)$ is the smallest Spector class on $\mathfrak{A}$ closed under Q and $\breve{Q}$.

The proof is completely analogous to that of 1.3.1 and 1.3.2.

The following result is also relevant here.

2.1.2. <u>Theorem</u> (Aczel). Let Γ be a Spector class on $\mathfrak{A}$. Let Q be a quantifier on A. Then Γ is closed under $Q,\breve{Q}$ iff Γ is closed under Q^+.

<u>Proof</u>. It is assumed here that Q^+ is defined relative to a coding of tuples in Δ.

If Γ is closed under Q^+ it is trivially closed under Q and $\breve{Q}$. For the converse let $R(u,\bar{y})$ be in Γ. We have to prove that

$$Q^+uR(u,\bar{y}) \Leftrightarrow Qx_0\breve{Q}x_1\exists x_2\forall x_3 \dots Qx_{4m}\breve{Q}x_{4m+1}\exists x_{4m+2}\forall x_{4m+3} \dots \exists nR(\langle x_0,x_1,x_2,x_3,\dots,\dots,x_{4n},x_{4n+1},x_{4n+2},x_{4n+3}\rangle,\bar{y})$$

is in Γ. Let

$$\varphi(w,\bar{y},S) \Leftrightarrow \mathrm{Seq}(w) \wedge \ell h(w) = 4n \text{ for some } n \wedge \wedge \{R(w,\bar{y}) \vee Qa\breve{Q}b\exists c\forall dS(w^\frown\langle a,b,c,d\rangle,\bar{y})\},$$

where if $w = \langle w_1 \dots w_n\rangle$, $v = \langle v_1 \dots v_m\rangle$, $w^\frown v = \langle w_1 \dots w_n\ v_1 \dots v_m\rangle$. Then we prove that if $\mathrm{Seq}(w) \wedge \ell h(w) = 4n$, for some n, then

$$(*) \qquad Q^+uR(w^\frown u,\bar{y}) \Leftrightarrow \varphi^\infty(w,\bar{y})$$

so that $Q^+uR(u,\bar{y}) \Leftrightarrow \varphi^\infty(\langle\emptyset\rangle,\bar{y})$. Since φ is a monotone operator under which Γ is closed, $\varphi^\infty \in \Gamma$ by the first recursion theorem and we are done.

To prove $(*)$ we first show by an easy induction on ξ that $\varphi^\xi(w,\bar{y}) \Rightarrow Q^+uR(w^\frown u,\bar{y})$. For the other direction assume $\neg\varphi^\infty(w,\bar{y})$ but $Q^+uR(w^\frown u,\bar{y})$. Then $\neg R(w,\bar{y})$ and $\neg Qa\breve{Q}b\exists c\forall d\ \varphi^\infty(w^\frown\langle a,b,c,d\rangle,\bar{y})$, so we have $\breve{Q}aQb\forall c\exists d\neg\varphi^\infty(w^\frown\langle a,b,c,d\rangle,\bar{y})$. This allows II to answer the winning strategy of I in $Q^+uR(w^\frown u,\bar{y})$ as follows: I plays $X_0 \in Q$. Since if $X \in Q$, $Y \in \breve{Q}$ we have $X \cap Y \neq 0$, II plays $a \in X_0$ such that $Qb\forall c\exists d\neg\varphi^\infty(w^\frown\langle a,b,c,d\rangle,\bar{y})$. Then I plays $X_1 \in \breve{Q}$. II now chooses $b \in X_1$ such that $\forall c\exists d\neg\varphi^\infty(w^\frown\langle a,b,c,d\rangle,\bar{y})$. Then I plays $X_2 \neq \emptyset$ and II answers by picking $c \in X_2$ such that $\exists d\neg\varphi^\infty(w^\frown\langle a,b,c,d\rangle,\bar{y})$. Finally I plays $X_3 = A$ and II answers by picking $d \in X_3$ such that $\neg\varphi^\infty(w^\frown\langle a,b,c,d\rangle,\bar{y})$. We are now back in the original situation with w replaced by $w^\frown\langle a,b,c,d\rangle$. By repeating we can produce a series of moves for II which beat I's strategy, a contradiction. ⊣

From the previous result and a normal form theorem for positive $\mathcal{L}(Q)$-definable formulas (see Moschovakis [1], p. 169) we also have (calling a structure $\mathfrak{A}$ <u>acceptable</u> if it admits an elementary coding scheme).

2.1.3. Theorem (Aczel [3]). Let $\mathfrak{A}$ be an acceptable structure and Q a quantifier on A. Then

$$\mathrm{IND}(\mathfrak{A},Q) = \{Q^{+}u\varphi(u,\bar{x}) : \varphi \text{ is first-order definable (with parameters) on } \mathfrak{A}\}.$$

We conclude this section by stating the following representation theorem which will be proved later on (see 5.2).

Theorem (Harrington). Let $\mathfrak{A}$ be almost acceptable. If Γ is a Spector class on $\mathfrak{A}$ there is a quantifier Q on A such that

$$\Gamma = \mathrm{IND}(\mathfrak{A},Q).$$

2.2. Nonmonotone Induction

General references are Aczel [4], Richter-Aczel [5] and Moschovakis [6].

Let $\mathfrak{A}$ be a structure. Let $\mathfrak{J}$ be a collection of second order relations $\psi(\bar{x},\bar{S})$ on A. A relation R on A is called $\mathfrak{J}$-inductive if there is $\varphi(\bar{x},\bar{y},S)$ in $\mathfrak{J}$ such that for some fixed $\bar{a}$

$$R(\bar{y}) \Longleftrightarrow \varphi^{\infty}(\bar{a},\bar{y}).$$

A relation R is $\mathfrak{J}$-hyperdefinable if both R and $\neg R$ are $\mathfrak{J}$-inductive. We denote by $\mathfrak{J}$-IND and $\mathfrak{J}$-HYP these two classes of relations, resp.

A collection of second order relations $\mathfrak{J}$ is called typical, nonmonotone if

(i) All the second order relations definable by existential or universal first order formulas in the language of $\mathfrak{A}$ are in $\mathfrak{J}$.

(ii) $\mathfrak{J}$ is closed under $\vee$, $\wedge$.

(iii) $\mathfrak{J}$ is closed under trivial substitutions i.e. substitutions definable by quantifier-free formulas (with parameters) in $\mathfrak{A}$.

(iv) There is a copy of $\langle \omega, \leq \rangle$ and a pairing function $\langle\ \rangle : A^2 \xrightarrow{1-1} A$ in $\mathfrak{J}$-HYP.

(v) The $\mathfrak{J}$-IND relations are A-parametrized.

Remark. Usually in practice the following stronger condition is satisfied.

(v^*) $\mathfrak{J}$ itself is A-parametrized i.e. for each signature $\nu = (n,k_1 \ldots k_m)$ there is a relation U^{ν} of signature $(n+1,k_1 \ldots k_m)$ in $\mathfrak{J}$ such that a relation $\varphi(\bar{x},S)$ of signature ν belongs to $\mathfrak{J}$ iff for some $a \in A$, $\varphi(\bar{x},\bar{S}) \Longleftrightarrow U^{\nu}(a,\bar{x},\bar{S})$. Here the signature of a second order relation $\varphi(x_1 \ldots x_n, S_1 \ldots S_m)$ is $(n,k_1 \ldots k_m)$ where S_i is a k_i-ary relation on A.

Examples of Typical, Nonmonotone $\mathfrak{J}$.

(i) $\mathfrak{J} = \Sigma^m_k$ or Π^m_k second order relations on $\mathfrak{A}$, with $m, k \geq 1$.

(ii) $\mathfrak{J} = \Sigma^0_k(C)$ or $\Pi^0_k(C)$ second order relations on A, where C is a hyper-elementary coding scheme on $\mathfrak{A}$. Here $\Sigma^0_k(C)$ is the collection of all second order relations of the form

$$\varphi(\bar{x},\bar{S}) \Longleftrightarrow \exists \bar{z}_1 \forall \bar{z}_2 \ldots Q\bar{z}_k \Psi(\bar{x},\bar{S},\bar{z}_1 \ldots \bar{z}_k),$$

where Ψ is a formula in the language of the expended structure $\langle \mathcal{A},\mathcal{C}\rangle = \langle \mathcal{A},N,\leq,\mathrm{Seq},\ell h,(x)_i\rangle$ which contains only bounded quantifiers of the form $\exists i \leq j$, $\forall i \leq j$. Similarly for $\Pi^0_k(\mathcal{C})$. It should be noted that $\Sigma^0_k(\mathcal{C})$-IND, $\Pi^0_k(\mathcal{C})$-IND does not depend on $\mathcal{C}$.

We now have

2.2.1 <u>Theorem</u> (Aczel [4], Richter-Aczel [5], Moschovakis [6]). Let $\mathfrak{J}$ be a typical, nonmonotone class of second order relations on $\mathcal{A}$. Then $\mathfrak{J}$-IND is a Spector class on $\mathcal{A}$.

<u>Proof</u>. That $\mathfrak{J}$-IND has all the closure properties in the definition of a Spector class is obvious using the substitution theorem for nonmonotone induction (see e.g. Moschovakis [6]). All the other properties of a Spector class are trivially satisfied by the definition of a typical, nonmonotone class. Finally, the prewellordering property follows from the stage comparison theorem for nonmonotone induction: If $\varphi(\bar{x},S)$, $\psi(\bar{y},T) \in \mathfrak{J}$ and we let $|\bar{x}|_\varphi$ = least ξ such that $\bar{x} \in \varphi^\xi$ and similarly for ψ, then the following two relations are $\mathfrak{J}$-inductive:

$$\bar{x} \leq^*_{\varphi,\psi} \bar{y} \Longleftrightarrow \bar{x} \in \varphi^\infty \;\&\; [\bar{y} \notin \psi^\infty \vee |\bar{x}|_\varphi \leq |\bar{y}|_\psi],$$

$$\bar{x} <^*_{\varphi,\psi} \bar{y} \Longleftrightarrow \bar{x} \in \varphi^\infty \;\&\; [\bar{y} \notin \psi^\infty \vee |\bar{x}|_\varphi < |\bar{y}|_\psi].$$

For example, for $\leq^*_{\varphi,\psi}$ consider the simultaneous induction

$$\varphi_1(\bar{x},S_1,S_2,S_3) \Longleftrightarrow \varphi(\bar{x},S_1)$$

$$\varphi_2(\bar{y},S_1,S_2,S_3) \Longleftrightarrow \psi(\bar{y},S_2)$$

$$\varphi_3(\bar{x},\bar{y},S_1,S_2,S_3) \Longleftrightarrow \varphi(\bar{x},S_1) \;\&\; \neg S_2(\bar{y}).$$

Then show by induction on ξ that, letting $|\bar{y}|_\psi = \infty$, if $\bar{y} \notin \psi^\infty$,

$$\varphi_3^\xi(\bar{x},\bar{y}) \Longleftrightarrow |\bar{x}|_\varphi \leq \xi \;\&\; |\bar{x}|_\varphi \leq |\bar{y}|_\psi$$

so that

$$\varphi_3^\infty(\bar{x},\bar{y}) \Longleftrightarrow \bar{x} \leq^*_{\varphi,\psi} \bar{y}.$$

$\dashv$

In analogy with the minimality characterizations of the examples given in 2.1 we now have the following characterization of $\mathfrak{J}$-IND.

2.2.2 <u>Theorem</u> (Moschovakis [6]). Let $\mathfrak{J}$ be typical, nonmonotone on $\mathcal{A}$. Then $\mathfrak{J}$-IND is the smallest Spector class Γ on $\mathcal{A}$ such that $\mathfrak{J}$ is Δ on Δ and Γ is $\mathfrak{J}$-compact.

Before we proceed to the proof of this result we have to explain the notions "Δ on Δ" and "$\mathfrak{J}$-compact".

2.2.3. Defining " Γ on Δ" and "Δ on Δ" (Moschovakis [1]). Let Γ be a Spector class on $\mathfrak{a}$. An A-parametrization of Δ consists of a sequence $\{I^n, H^n, \check{H}^n\}_{n \geq 1}$ of relations in Γ such that for each n,

(i) $I^n \in \Gamma$, $H^n \in \Gamma$, $\check{H}^n \in \check{\Gamma}$.

(ii) If $a \in I^n$, then $H^n_a = \check{H}^n_a$.

(iii) If $R \subseteq A^n$ is in Δ, there is $a \in I^n$ such that $R = H^n_a$ (we refer to such an a as a code of R).

Fact. There is an A-parameterization of Δ.

Proof. Let W^n be universal for the n-ary relations in Γ. Define

$$I^n(a) \Leftrightarrow W^n_{(a)_0} \cup W^n_{(a)_1} = A.$$

Put also

$$W^n_0(a,\bar{x}) \Leftrightarrow W^n((a)_0,\bar{x}),\ W^n_1(a,\bar{x}) \Leftrightarrow W^n((a)_1,\bar{w}).$$

Reduce W^n_0, W^n_1 to U^n_1, U^n_2 in Γ. Then put

$$H^n(a,\bar{x}) \Leftrightarrow U^n_1(a,\bar{x})$$

$$\check{H}^n(a,\bar{x}) \Leftrightarrow \neg U^n_2(a,\bar{x}).$$

⊣

Consider now a second order relation $\varphi(\bar{x}, S_1 \dots S_m)$, where S_i is k_i-ary. We say that φ is Γ on Δ if the relation

$$\varphi^{\#}(\bar{x}, a_1 \dots a_m) \Leftrightarrow a_1 \in I^{k_1} \& \dots \& a_m \in I^{k_m} \& \varphi(\bar{x}, H^{k_1}_{a_1} \dots H^{k_m}_{a_m})$$

is in Γ i.e. φ is Γ "in the codes" for Δ relations. We say that φ is Δ on Δ if both φ, $\neg\varphi$ are Γ on Δ.

It is easy to check that this definition is independent of the particular A-parametrization used to define them. It is also easy to give a direct definition that does not involve parametrization. For example a relation $\varphi(\bar{x}, S)$ with S n-ary is Γ on Δ if Γ is closed (see 1.4) under φ^* where

$$\varphi^*(\bar{x}, T) \Leftrightarrow T_0 \cap T_1 \neq \emptyset \vee [T_0 \cup T_1 = A^n \& \varphi(\bar{x}, T_0)].$$

Here T has one more argument than S and $T_i = \{x : (i,x) \in T\}$, $i = 0,1$.

2.2.4 Defining "$\mathfrak{J}$-compact" (Moschovakis [6]). Let Γ be a Spector class on $\mathfrak{a}$ and $\varphi(S)$ a second order relation on A. We say that Γ is φ-compact if for every $R^0 \subseteq R$, $R^0 \in \Delta$ if $\varphi(R)$ holds, there is $R^* \in \Delta$, $R^0 \subseteq R^* \subseteq R$ such that $\varphi(R^*)$ holds. If $\mathfrak{J}$ is a collection of second order relations on A then we say that Γ is $\mathfrak{J}$-compact if it is φ-compact for each $\varphi \in \mathfrak{J}$.

The following equivalent formulation of $\mathfrak{J}$-compactness is very useful: If Γ is a Spector class, then Γ is $\mathfrak{J}$-compact iff for each $\varphi(S) \in \mathfrak{J}$, each $R \in \Gamma - \Delta$ and each regular Γ-norm σ on R, if $\varphi(R)$ holds then for some $\xi < |\sigma|$,

$\varphi(\{\bar{x} \in R : \sigma(\bar{x}) < \xi\})$ holds. That the norm-version of compactness implies the original one is almost obvious. For the converse, which actually needs that $\mathfrak{J}$ has some simple closure properties, see Moschovakis [6].

The following fact illustrates the present notions and establishes a connection with those of 2.1.

Fact. Let Γ be a Spector class on $\mathfrak{a}$. If Q is a quantifier on A then Γ is closed under Q iff Γ is Q-compact and Q is Γ on Δ.

Proof. $\Rightarrow$. We have $Q^{\#}(a) \Longleftrightarrow a \in I \,\&\, Q(H_a) \Longleftrightarrow a \in I \,\&\, QbH(a,b)$, so $Q^{\#} \in \Gamma$. $\therefore$ Q is Γ on Δ. To see that Γ is Q-compact let $R \in \Gamma - \Delta$, let σ be a regular Γ-norm on R and assume $Q(R)$ holds but for no $\xi < |\sigma|$, $Q(\{x \in R : \sigma(x) < \xi\})$ holds. Then

$$x \notin R \Longleftrightarrow Qy(y <^*_\sigma x)$$

so $R \in \Delta$, a contradiction.

$\Leftarrow$. We have for $R \in \Gamma$, $QbR(a,b) \Longleftrightarrow \exists X \in \Delta[\forall b(b \in X \Rightarrow R(a,b)) \,\&\, Q(X)] \Longleftrightarrow$ $\Longleftrightarrow (\exists c)(c \in I \,\&\, \forall b(\breve{H}(c,b) \Rightarrow R(a,b)) \,\&\, Q(H_c))$, so $QbR(a,b)$ is in Γ. $\dashv$

As an application of this fact consider the following alternative proof of the first recursion theorem:

Let $\varphi(\bar{x},S)$ be monotone and assume the Spector class Γ is closed under φ. Let $W \in \Gamma - \Delta$ and σ a regular Γ-norm on it. By the second recursion theorem find $R(a,\bar{x})$ in Γ such that

$$R(a,\bar{x}) \Longleftrightarrow (a \in W \,\&\, \sigma(a) = 0 \,\&\, \varphi(\bar{x},\emptyset)) \vee$$
$$(a \in W \,\&\, \sigma(a) > 0 \,\&\, \varphi(\bar{x},\{\bar{y} : \exists a'(a' <^*_\sigma a \wedge R(a',\bar{y}))\})).$$

Then it is easy to check by induction on $\sigma(a)$ that

$$R(a,\bar{x}) \Longleftrightarrow a \in W \,\&\, \bar{x} \in \varphi^{\sigma(a)}.$$

So

$$\bar{x} \in \varphi^{<|\sigma|} \Longleftrightarrow \exists a(a \in W \wedge R(a,\bar{x})),$$

thus $\varphi^{<|\sigma|} \in \Gamma$. It is enough therefore to show that $\varphi(\bar{x},\varphi^{<|\sigma|}) \Rightarrow \bar{x} \in \varphi^{<|\sigma|}$. Assume $\varphi(\bar{x},\varphi^{<|\sigma|})$ i.e. $\varphi(\bar{x},\{\bar{y} : \exists a(a \in W \wedge R(a,\bar{y}))\})$. By the previous fact there is $\xi < |\sigma|$ such that $\varphi(\bar{x},\{\bar{y} : \exists a(a \in W \wedge \sigma(a) < \xi \wedge R(a,\bar{y}))\})$ $\therefore$ $\bar{x} \in \varphi^{\xi} \subseteq \varphi^{<|\sigma|}$ and we are done.

It is now time to give the proof of Theorem 2.2.2.

Proof of 2.2.2. It is not hard to see that $\mathfrak{J}$-IND has the required properties (see Moschovakis [6]). The converse follows immediately from the following.

2.2.5 Nonmonotone First Recursion Theorem. Let Γ be a Spector class on $\mathfrak{a}$. If $\varphi(\bar{x},S)$ in Δ on Δ and Γ is φ-compact (for each fixed $\bar{x}$) then $\varphi^{\infty} \in \Gamma$.

Proof. We will use there the fact that Γ is φ-compact in the norm-version given in 2.2.4.

Let $W \in \Gamma - \Delta$ and let σ be a Γ-norm on W. We will use the second recursion theorem to find $R(a,\bar{x})$, $S(a,\bar{x})$ in Γ such that

$$R(a,\bar{x}) \Longleftrightarrow a \in W \wedge \bar{x} \in \varphi^{\sigma(a)}$$

$$S(a,\bar{x}) \Longleftrightarrow a \in W \wedge \bar{x} \notin \varphi^{\sigma(a)}.$$

Granting this we immediately have that $\varphi^{<|\sigma|}$ is in Γ and $|\bar{x}|_\varphi$ is a Γ-norm on $\varphi^{<|\sigma|}$. If $\varphi^{<|\sigma|} \in \Delta$ then since

$$\forall \bar{x} \in \varphi^{<|\sigma|} \exists a(a \in W \wedge R(a,\bar{x}))$$

we can find $\xi < |\sigma|$ such that

$$\forall \bar{x} \in \varphi^{<|\sigma|} \exists a(a \in W \wedge \sigma(a) < \xi \wedge R(a,\bar{x}))$$

so $\varphi^{<|\sigma|} \subseteq \varphi^{<\xi}$ $\therefore$ $\varphi^{<|\sigma|} = \varphi^{\infty}$ and we are done. Otherwise $\varphi^{<|\sigma|} \in \Gamma - \Delta$, so if $\varphi(\bar{x},\varphi^{<|\sigma|})$, we have by φ-compactness $\varphi(\bar{x},\varphi^{<\xi})$ for some $\xi < |\sigma|$ $\therefore$ $\bar{x} \in \varphi^{<|\sigma|}$ and we are done.

We define now R,S by using the second recursion theorem as follows

$$R(a,\bar{x}) \Longleftrightarrow a \in W \wedge \{[\sigma(a) = 0 \ \& \ \varphi(\bar{x},\emptyset)] \vee$$
$$\vee [\sigma(a) > 0 \ \& \ [\exists a'(a' \overset{*}{<}_\sigma a \wedge R(a',\bar{x})) \vee$$
$$\vee \varphi(\bar{x},\{\bar{y} : \exists a'(a' \overset{*}{<}_\sigma a \wedge R(a',\bar{x}))\}]]\}$$

$$S(a,\bar{x}) \Longleftrightarrow a \in W \wedge ([\sigma(a) = 0 \ \& \ \neg\varphi(\bar{x},\emptyset)] \vee$$
$$\vee [\sigma(a) > 0 \ \& \ \forall a'(a' \overset{*}{<}_\sigma a \Rightarrow S(a',\bar{x})) \ \&$$
$$\& \ \neg\varphi(\bar{x},\{\bar{y} : \exists a'(a' \overset{*}{<}_\sigma a \wedge R(a',\bar{y}))\})]\}$$

⊣

2.2.6 Theorem (Aanderaa [8]). Let $\mathfrak{J}$ be typical, nonmonotone, closed under $\exists$. If $\mathfrak{J}$ has the prewellordering property, then

$$\mathfrak{J}\text{-IND} \prec \breve{\mathfrak{J}}\text{-IND}.$$

Proof. Here $\breve{\mathfrak{J}}$ is of course the class of all negations of relations in $\mathfrak{J}$. To say that $\mathfrak{J}$ has the prewellordering property means that if $\varphi(\bar{x},\bar{S})$ is in $\mathfrak{J}$ there is $\sigma : \varphi \to$ Ordinals such that the two relations below are in $\mathfrak{J}$:

$$(\bar{x},\bar{S}) \overset{*}{\leq}_\sigma (\bar{y},\bar{T}) \Longleftrightarrow \varphi(\bar{x},\bar{S}) \ \& \ [\neg\varphi(\bar{y},\bar{T}) \vee \sigma(\bar{x},\bar{S}) \leq \sigma(\bar{y},\bar{T})]$$

$$(\bar{x},\bar{S}) \overset{*}{<}_\sigma (\bar{y},\bar{T}) \Longleftrightarrow \varphi(\bar{x},\bar{S}) \ \& \ [\neg\varphi(\bar{y},\bar{T}) \vee \sigma(\bar{x},\bar{S}) < \sigma(\bar{y},\bar{T})]$$

Using this we see immediately that if $\varphi(\bar{x},S)$ is in $\mathfrak{J}$ then we can find $\varphi^*(\bar{x},S)$ in $\breve{\mathfrak{J}}$ such that

$$\exists \bar{x}(\varphi(\bar{x},S) \ \& \ \bar{x} \notin S) \Rightarrow \exists \bar{x}(\varphi^*(\bar{x},S) \ \& \ \bar{x} \notin S)$$

and $\varphi^*(\bar{x},S) \Rightarrow \varphi(\bar{x},S)$ & $\bar{x} \notin S$. Indeed let σ be an $\mathfrak{J}$-norm on

$$\psi(\bar{x},S) \Longleftrightarrow \varphi(\bar{x},S) \ \& \ \bar{x} \notin S$$

and let $\varphi^*(\bar{x},S) \Longleftrightarrow \sigma(\bar{x},S)$ is minimal

$$\Longleftrightarrow \neg \exists \bar{y}((\bar{y},S) <^*_\sigma (\bar{x},S)).$$

Consider now the following simultaneous induction in $\breve{\mathfrak{J}}$

$$\varphi_1(\bar{x},S_1,S_2,S_3,S_4) \Longleftrightarrow \exists \bar{t}(\bar{t} \in S_3 \ \& \ (\bar{t},\bar{x}) \notin S_2)$$
$$\varphi_2(\bar{t},\bar{x},S_1,S_2,S_3,S_4) \Longleftrightarrow \varphi^*(\bar{t},S_1) \ \& \neg(\varphi(\bar{x},S_1) \vee \bar{x} \in S_1)$$
$$\varphi_3(\bar{t},S_1,S_2,S_3,S_4) \Longleftrightarrow \varphi^*(\bar{t},S_1)$$
$$\varphi_4(u,S_1,S_2,S_3,S_4) \Longleftrightarrow \forall \bar{x}(\varphi(\bar{x},S_1) \Rightarrow \bar{x} \in S_1) \wedge u = 0.$$

Then assuming without loss of generality that $\varkappa = |\varphi|$ = closure ordinal of φ, is limit we can easily show by induction that for each limit ordinal $\lambda < \varkappa$

$$\varphi_1^\lambda = \varphi^{<\lambda}$$
$$\varphi_2^\lambda = \varphi_2^{<\lambda} \cup ((\varphi_3^\lambda - \varphi_3^{<\lambda}) \times \neg(\varphi^\lambda))$$
$$\varphi_3^\lambda \subseteq \varphi^\lambda, \ \emptyset \neq \varphi_3^\lambda - \varphi_3^{<\lambda} \subseteq \varphi^\lambda - \varphi^{<\lambda}$$
$$\varphi_4^\lambda = \emptyset$$

and

$$\varphi_1^{<\varkappa} = \varphi^\infty$$
$$\varphi_4^\varkappa = \{0\}$$

so that φ^∞ is $\breve{\mathfrak{J}}$-hyperdefinable. ⊣

<u>Corollary</u>. For a countable almost acceptable structure $\mathfrak{A}$,

$$\Pi_1^1 - \text{IND} \prec \Sigma_1^1 - \text{IND}$$
$$\Sigma_2^1 - \text{IND} \prec \Pi_2^1 - \text{IND}.$$

We conclude by mentioning the following result whose proof will not be given here. We denote by $\text{IND}^2(\mathfrak{A},Q)$ the class of second order relations which are positive $\mathfrak{L}(Q)$-inductive over $\mathfrak{A}$.

<u>Theorem</u> (Harrington-Moschovakis). Let $\mathfrak{A}$ be almost acceptable, Q a quantifier on A. Let $\mathfrak{J} = \text{IND}^2(\mathfrak{A},Q)$. Then

$$\breve{\mathfrak{J}}\text{-IND} = \text{IND}(\mathfrak{A},Q^+).$$

2.3. <u>Recursion in type 2 objects</u>

Let A be an infinite set. A <u>type</u> 2 <u>object</u> over A is a subset

$$\underset{\sim}{F} \subseteq p(A)$$

of the power set of A. For the case $A = \omega$ Kleene [9] has defined the notion of a partial function

$$f : \omega^n \to \omega$$

being <u>recursive</u> in $\underset{\sim}{F}_1 \dots \underset{\sim}{F}_n$, where each $\underset{\sim}{F}_i$ is a type 2 object over ω. Call a relation on ω <u>semirecursive</u> in $\underset{\sim}{F}_1 \dots \underset{\sim}{F}_n$ if it is the domain of a partial function recursive in $\underset{\sim}{F}_1 \dots \underset{\sim}{F}_n$. Call the class of semirecursive in $\underset{\sim}{F}_1 \dots \underset{\sim}{F}_n$ relations the 1-<u>envelope</u> of $\underset{\sim}{F}_1 \dots \underset{\sim}{F}_n$, in symbols

$${}_1\mathrm{en}(\underset{\sim}{F}_1 \dots \underset{\sim}{F}_n) = \{R \subseteq \omega^n : R \text{ is semirecursive in } \underset{\sim}{F}_1 \dots \underset{\sim}{F}_n\}.$$

From the work of Gandy [10] it follows that if ${}^2\underset{\sim}{E}$ is the type 2 object embodying existential quantification over ω, i.e.

$${}^2\underset{\sim}{E}(X) \Longleftrightarrow \exists n\ (n \in X),$$

then for any $\underset{\sim}{F}_1 \dots \underset{\sim}{F}_n$,

$${}_1\mathrm{en}(\underset{\sim}{F}_1 \dots \underset{\sim}{F}_n, {}^2\underset{\sim}{E}) \text{ is a Spector class on } \langle \omega, +, \cdot \rangle.$$

Moschovakis [2] shows that ${}_1\mathrm{en}(\underset{\sim}{F}_1 \dots \underset{\sim}{F}_n, {}^2\underset{\sim}{E})$ is the smallest Spector class on $\langle \omega, +, \cdot \rangle$ for which $\underset{\sim}{F}_1 \dots \underset{\sim}{F}_n$ are Δ on Δ. For example we obtain that

$${}_1\mathrm{en}({}^2\underset{\sim}{E}) = \Pi^1_1,$$

which was first proved by Kleene [10].

More generally we have

<u>Fact</u>. Let $\mathfrak{A}$ be an almost acceptable structure and let $\underset{\sim}{F}_1 \dots \underset{\sim}{F}_n$ be a finite sequence of type 2 objects over A. Then there is a smallest Spector class Γ on $\mathfrak{A}$ such that each $\underset{\sim}{F}_i$ $(1 \le i \le n)$ is Δ on Δ.

<u>Proof</u>. Take $n = 1$ for simplicity. Define the following quantifier Q on A

$$Q(X) \Longleftrightarrow (X_0 \cap X_1 \neq \emptyset) \vee [X_0 \cup X_1 = A \ \&\ \underset{\sim}{F}(X_0)],$$

where

$$Y_i = \{a \in A : \langle i, a \rangle \in X\}, \quad i = 0, 1.$$

Then $\mathrm{IND}(\mathfrak{A}, Q)$ is the required Spector class. $\dashv$

It turns out that the Spector class asserted to exist above has a recursion theoretic characterization even on an arbitrary (almost acceptable) $\mathfrak{A}$. Indeed, let ${}^2\underset{\sim}{E}^{\#}_A$ be the functional embodying existential quantification over A and $f_1 \dots f_\ell, f_=$ the characteristic functions of $R_1, \dots, R_\ell$, $=$, where $\mathfrak{A} = \langle A, R_1 \dots R_\ell \rangle$. Then the smallest Spector class Γ on $\mathfrak{A}$ in which $\underset{\sim}{F}_1 \dots \underset{\sim}{F}_n$ are Δ on Δ is the same as the class of relations semirecursive (with parameters) in $\underset{\sim}{F}_1 \dots \underset{\sim}{F}_n, {}^2\underset{\sim}{E}^{\#}_A, f_1 \dots f_\ell, f_=$. For the relevant definitions and some generalizations of the present type of example arising in recursion relative to "normal type 2 functionals" on A, see Kechris-Moschovakis [11].

2.4 Σ-like Spector Classes

Let $\mathfrak{A}$ be a structure and B a set. A relation on A,B is any $R \subseteq A^n \times B^m$. Let Π be a collection of relations on A,B. Put

$$\exists^B \Pi = \{\exists \bar{b}\ P(\bar{a},\bar{b}) : P \in \Pi\}.$$

Call Π nice if

i) $\exists^B \Pi$ is closed under $\wedge$, $\vee$, $\exists^A, \forall^A$, contains all the first order definable in $\mathfrak{A}$ (with parameters) relations and a coding scheme on A.

ii) $\exists^B \Pi$ is A-parametrized.

iii) If $P(\bar{a},\bar{b})$ is in Π there is a norm

$$\sigma : P \to \text{Ordinals}$$

such that the two relations below are in $\exists^B \Pi$:

$$P_1(\bar{a},\bar{b},\bar{a}^*) \Leftrightarrow P(\bar{a},\bar{b})\ \&\ \forall \bar{a}' \forall \bar{b}'\ (\sigma(\bar{a}',\bar{b}') \leq \sigma(\bar{a},\bar{b}) \Rightarrow \bar{a}' \neq \bar{a}^*)$$

$$P_2(\bar{a},\bar{b},\bar{a}^*) \Leftrightarrow P(\bar{a},\bar{b})\ \&\ \forall \bar{a}' \forall \bar{b}'\ (\sigma(\bar{a}',\bar{b}') < \sigma(\bar{a},\bar{b}) \Rightarrow \bar{a}' \neq \bar{a}^*).$$

Fact. Let $\mathfrak{A}$ be a structure, B be a set and Π a nice collection of relations on A, B. Then $\exists^B \Pi$ is a Spector class on $\mathfrak{A}$.

Proof. It is enough to check that $\exists^B \Pi$ is normed. Let

$$R(\bar{a}) \Leftrightarrow \exists \bar{b}\ P(\bar{a},\bar{b}),$$

where $P \in \Pi$. Let σ be a norm on P with property (iii) as above. Put for $\bar{a} \in R$,

$$\tau(\bar{a}) = \min\{\sigma(\bar{a},\bar{b}) : P(\bar{a},\bar{b})\}.$$

Then

$$\bar{a} \leq_\tau^* \bar{a}^* \Leftrightarrow \exists \bar{b}[P(\bar{a},\bar{b})\ \&\ \forall \bar{a}' \forall \bar{b}'.\ (\sigma(\bar{a}',\bar{b}') < \sigma(\bar{a},\bar{b}) \Rightarrow \bar{a}' \neq \bar{a}^*]$$

$$\bar{a} <^* \bar{a}^* \Leftrightarrow \exists \bar{b}[P(\bar{a},\bar{b})\ \&\ \forall \bar{a}' \forall \bar{b}'\ (\sigma(\bar{a}',\bar{b}') \leq \sigma(\bar{a},\bar{b}) \Rightarrow \bar{a}' \neq \bar{a}^*],$$

so $\leq_\tau^*$, $<_\tau^*$ are in $\exists^B \Pi$ and τ is a $\exists^B \Pi$-norm. ⊣

Applications

i) (Moschovakis [1]). Let $\mathfrak{A}$ be a countable almost acceptable structure and let B be the set of all relations on A. Let Π be the collection of all Π_1^1 second order relations on $\mathfrak{A}$. Then Π is nice (see Moschovakis [1]) and so $\exists^B \Pi$, which by definition is exactly $\Sigma_2^1(\mathfrak{A})$, is a Spector class on $\mathfrak{A}$.

ii) (Addison [12]). Assume V = L. Let $\mathfrak{A} = \langle \omega,+,\cdot \rangle$, B = all relations on ω and Π = all second order Π_n^1 relations on ω, for $n \geq 2$. Then Π is nice; to see that condition (iii) is satisfied, let $\varphi(\bar{a},\bar{S})$ be Π_n^1 and put

$$\sigma(\bar{a},\bar{S}) = \text{the ordinal of } \bar{S} \text{ in the canonical wellordering of } L.$$

So $\exists^B \Pi$, i.e. the Σ_{n+1}^1 relations on ω, form a Spector class, for all $n \geq 1$. Similarly we can see that the Σ_n^1 relations on (say) $\langle \aleph_1,< \rangle$ form a Spector class for $n \geq 1$.

iii) Let $\mathfrak{A}$ be an almost acceptable structure and let M be admissible above $\mathfrak{A}$ (M may contain other relations beyond $\in$, see Barwise [13]). Assume moreover M is resolvable, i.e. there is a $\Delta_1(M)$ map

$$\tau : O(M) \to M$$

such that

$$\bigcup_\xi \tau(\xi) = M$$

Let Λ consist of all relations on A,M which are $\Delta_1(M)$ with parameters in A. Then Λ is nice. [To verify iii) let $P(\bar{a},\bar{x})$ be in Λ and put

$$\sigma(\bar{a},\bar{x}) = \text{least } \xi \text{ such that } \bar{x} \in \tau(\xi)].$$

Then $\exists^B\Lambda$, which is of course just the collection of all relations on A which are $\Sigma_1(M)$ with parameters in A, is a Spector class on $\mathfrak{A}$. For more about this example see Moschovakis [1], Barwise [13].

2.5 The Game Quantifier

Let A be an infinite set and let A^ω be the set of all infinite sequences from A. To each $X \subseteq A^\omega$ we associate the game G_X played as follows:

I	II	
		Players I, II play alternatively members of A with
a_0		I playing a_0, II playing a_1, etc. We say that I
	a_1	
a_2		wins if $\alpha = (a_0,a_1,\dots) \in X$. Otherwise II wins.
	a_3	
$\vdots$		

Put

$$\mathcal{G}\alpha X(\alpha) \Leftrightarrow \text{I has a winning strategy in } G_X.$$
$$\Leftrightarrow \exists a_0 \forall a_1 \exists a_2 \forall a_3 \dots (a_0,a_1,\dots) \in X .$$

Clearly $\mathcal{G}$ is a quantifier on A^ω, called the (unrestricted) game quantifier.

We say that $X \subseteq A^\omega$ is determined if either I or II has a winning strategy in G_X. If $\mathcal{S}$ is a class of subsets of A^ω we put

$$\mathrm{Det}(\mathcal{S}) \Leftrightarrow \forall X \in \mathcal{S} \ (X \text{ is determined}).$$

Let now $\mathcal{R}$ be a collection of relations on A, A^ω, i.e. relations of the form $P(\bar{x},\bar{\alpha})$, $\bar{x} \in A^n$, $\alpha \in (A^\omega)^m$. We put

$$\mathcal{G}\mathcal{R} = \{\mathcal{G}\alpha\, P(\bar{x},\alpha) : P \in \mathcal{R}\}.$$

Definition. Let $\mathfrak{A}$ be a structure. A collection $\mathcal{R}$ of relations on A, A^ω is called adequate if

i) $\mathcal{R}$ is closed under simple substitutions i.e. if $P(y_1 \dots y_k, \beta_1 \dots \beta_\ell)$ is in $\mathcal{R}$ and $f_1 \dots f_k$, $g_1 \dots g_\ell$ are simple functions then

$$P'(\bar{x},\bar{\alpha}) \Leftrightarrow P(f_1(\bar{x},\bar{\alpha}) \dots f_k(\bar{x},\bar{\alpha}), g_1(\bar{x},\bar{\alpha}) \dots g_\ell(\bar{x},\bar{\alpha}))$$

is also in $\mathcal{R}$. Here a simple function is a function of the form

$$f(x_1 \dots x_n, \alpha_1 \dots \alpha_m) = x_i, \qquad 1 \leq i \leq n$$

or

$$f(x_1 \dots x_n, \alpha_1 \dots \alpha_m) = \alpha_j(q), \qquad 1 \leq j \leq n,\ q \in \omega$$

or

$$f(x_1 \dots x_n, \alpha_1 \dots \alpha_m) = a\frown\alpha_j = (a, \alpha_j(0), \alpha_j(1) \dots), \qquad 1 \leq j \leq n,\ a \in A$$

or

$$f(x_1 \dots x_n, \alpha_1 \dots \alpha_m) = \alpha_j \circ t = (\alpha_j(t(0)), \alpha_j(t(1)), \dots),$$

$1 \leq j \leq n$, t a primitive recursive function on ω.

ii) $\mathfrak{R}$ is closed under $\wedge$, $\vee$ and contains all the relations definable by quantifier-free formulas with parameters from A.

iii) $\mathcal{G}\mathfrak{R}$ is A-parametrized and contains a coding scheme on A.

2.5.1 <u>Theorem</u> (Kechris-Moschovakis). Let $\mathfrak{A}$ be a structure and $\mathfrak{R}$ an adequate class of relations on A, A^ω. If $\mathfrak{R}$ is normed and $\mathrm{Det}(\mathfrak{R})$ holds, then $\mathcal{G}\mathfrak{R}$ is a Spector class.

<u>Proof</u>. To see that $\mathcal{G}\mathfrak{R}$ is closed under $\wedge$ note that

$$\mathcal{G}\alpha P(\bar{x},\alpha) \wedge \mathcal{G}\beta R(\bar{x},\beta) \Longleftrightarrow$$
$$\exists c_0 \forall c_1 \exists c_2 \forall c_3 \dots [(c_1 = 0 \Rightarrow P(\bar{x},(c_2,c_3,\dots))) \wedge$$
$$\wedge\ (c_1 = 1 \Rightarrow R(\bar{x},(c_2,c_3,\dots)))],$$

where $0 \neq 1$ are two distinct members of A. For closure under $\vee$ use the fact that

$$\mathcal{G}\alpha P(\bar{x},\alpha) \vee \mathcal{G}\beta R(\bar{x},\beta) \Leftrightarrow$$
$$\exists c_0 \forall c_1 \exists c_2 \forall c_3 \dots [(c_0 = 0 \wedge P(\bar{x},(c_2,c_3,\dots))) \vee$$
$$\vee\ (c_0 = 1 \wedge R(\bar{x},(c_2,c_3,\dots)))].$$

Similarly

$$\exists a(\mathcal{G}\alpha P(\bar{x},a,\alpha)) \Leftrightarrow \exists c_0 \forall c_1 \exists c_2 \forall c_3 \dots P(\bar{x},c_0,(c_2,c_3,\dots))$$

and

$$\forall a(\mathcal{G}\alpha P(\bar{x},a,\alpha)) \Leftrightarrow \exists c_0 \forall c_1 \exists c_2 \forall c_3 \dots P(\bar{x},c_1,(c_2,c_3,\dots)),$$

so that $\mathcal{G}\mathfrak{R}$ is closed under existential and universal quantification over A.

It only remains to prove that $\mathcal{G}\mathfrak{R}$ is normed. For that we use the same idea as in the proof of the Third Periodicity Theorem of Moschovakis [14]. Let $P(\bar{x},\alpha)$ be in $\mathfrak{R}$ and let

$$\sigma : P \to \text{Ordinals}$$

be an $\mathfrak{R}$-norm on P. Let $R(\bar{x}) \Longleftrightarrow \mathcal{G}\alpha P(\bar{x},\alpha)$. For $\bar{x},\bar{y} \in R$ put

$$\bar{x} \lesssim \bar{y} \Leftrightarrow \underline{|\exists a_0 \forall a_1 \forall b_0 \exists b_1|}\ \underline{|\exists a_2 \forall a_3 \forall b_2 \exists b_3|}\ \dots$$

$$(\bar{x},(a_0,a_1,\dots)) \leq^*_\sigma (\bar{y},(b_0,b_1,\dots)).$$

Then it can be proved that $\lesssim$ is a prewellordering on R. Let τ be its associated norm. It turns out that τ is a $\mathcal{G}\mathcal{R}$-norm, since

$$\bar{x} \leq^*_\tau \bar{y} \Leftrightarrow \underline{|\exists a_0 \forall a_1 \forall b_0 \exists b_1|}\ \underline{|\exists a_2 \forall a_3 \forall b_2 \exists b_3|} \cdots$$
$$(\bar{x},(a_0,a_1,\ldots)) \leq^*_\sigma (\bar{y},(b_0,b_1,\ldots)),$$

$$\bar{x} <^*_\tau \bar{y} \Leftrightarrow \underline{|\forall b_0 \exists b_1 \exists a_0 \forall a_1|}\ \underline{|\forall b_2 \exists b_3 \exists a_2 \forall a_3|} \cdots$$
$$(\bar{x},(a_0,a_1,\ldots)) <^*_\sigma (\bar{y},(b_0,b_1,\ldots)).$$

The proofs of these facts use essentially $\mathrm{Det}(\mathcal{R})$ and can be found in Moschovakis [14] or Kechris [15]. ⊣

Remark. One can see more generally that if

$$R(\bar{x}) \Leftrightarrow Q_0 a_0 Q_1 a_1 \ldots P(\bar{x},(a_0,a_1,\ldots)),$$

where $Q_0, Q_1, \ldots$ are arbitrary quantifiers (see 2.1), then (under appropriate determinacy hypotheses) any norm σ on P can be transferred to a norm τ on R so that

$$\bar{x} \leq^*_\tau \bar{y} \Leftrightarrow Q_0 a_0 \breve{Q}_0 b_0 Q_1 a_1 \breve{Q}_1 b_1 \ldots (\bar{x},(a_0,a_1,\ldots)) \leq^*_\sigma (\bar{y},(b_0,b_1,\ldots)),$$
$$\bar{x} <^*_\tau \bar{y} \Leftrightarrow \breve{Q}_0 b_0 Q_0 a_0 \breve{Q}_1 b_1 Q_1 a_1 \ldots (\bar{x},(a_0,a_1,\ldots)) <^*_\sigma (\bar{y},(b_0,b_1,\ldots)).$$

This allows extension of most of the results in this section to arbitrary quantifiers.

Examples

i) Let $\mathcal{A}$ be an arbitrary almost acceptable structure and let $\mathcal{R}$ be the collection of all relations of the form

$$P(\bar{x},\alpha_1 \ldots \alpha_n) \Leftrightarrow \exists k \varphi(\bar{x},\bar{\alpha}_1(k) \ldots \bar{\alpha}_n(k)),$$

where φ is a first order formula with parameters from A and

$$\bar{\alpha}(k) = \langle \alpha(0) \ldots \alpha(k-1) \rangle.$$

Then P is adequate and normed. [If $P(\bar{x},\alpha)$ is as above, take $\sigma(\bar{x},\alpha)$ = least k such that $\varphi(\bar{x},\bar{\alpha}(k))$]. Moreover, $\mathrm{Det}(\mathcal{R})$ holds since all the games in $\mathcal{R}$ are open. So $\mathcal{G}\mathcal{R}$ is a Spector class. Indeed by a basic fact on inductive relations (see Moschovakis [1] or take $Q = \exists$ in 2.1.3.) we have

$$\mathcal{G}\mathcal{R} = \mathrm{IND}(\mathcal{A}).$$

ii) Take now $\mathcal{A} = \langle \omega,+,\cdot \rangle$. As usual a relation $P(\bar{x},\bar{\alpha})$ is Σ^0_n if

$$P(\bar{x},\alpha_1 \ldots \alpha_m) \Leftrightarrow \exists k_1 \forall k_2 \ldots Q k_n\ R(\bar{x},k_1 \ldots k_n,\bar{\alpha}_1(k_n) \ldots \bar{\alpha}_m(k_n)),$$

for some recursive R. Clearly Σ^0_n is adequate and normed. By Martin's Borel Determinacy Theorem we also have $\mathrm{Det}(\Sigma^0_n)$, so $\mathcal{G}\Sigma^0_n$ is a Spector class for each $n \geq 1$. Of course

$$\mathcal{G}\,\Sigma^0_1 = \Pi^1_1 \,.$$

The next result identifies $\mathcal{G}\Sigma^0_2$.

2.5.2 <u>Theorem</u> (Solovay). $\mathcal{G}\Sigma^0_2 = \Sigma^1_1$-IND.

<u>Proof</u>. We actually show that $\mathcal{G}\Sigma^0_2 = \Sigma^{1,pos}_1$-IND, where by $\Sigma^{1,pos}_1$ we denote the class of all positive Σ^1_1 operators on ω. By a result of Grilliot $\Sigma^{1,pos}_1$-IND $= \Sigma^1_1$-IND, so $\mathcal{G}\Sigma^0_2 = \Sigma^1_1$-IND.

Let $\bar{x} \in R \Leftrightarrow \mathcal{G}\alpha \exists n \forall m\, P'(\bar{x},n,\bar{\alpha}(m))$, with P' recursive. Let

$$P(\bar{x},n,u) \Leftrightarrow \forall m' < \ell h(u) P'(\bar{x},n,\bar{u}(m')),$$

where if $u = \langle a_0 \ldots a_k\rangle$, $\ell h(u) = k+1$ and for $m \leq k$, $\bar{u}(m) = \langle a_0 \ldots a_m\rangle$. Then P is recursive, $\bar{x} \in R \Leftrightarrow \mathcal{G}\alpha \exists n \forall m\, P(\bar{x},n,\bar{\alpha}(m))$ and

$$\exists n \forall m\, P(\bar{x},n,\bar{\alpha}(m)) \Leftrightarrow \exists n\, [\text{for infinitely many } m,\ P(\bar{x},n,\bar{\alpha}(m))].$$

Put now

$$(\bar{x},s) \in R^* \Leftrightarrow s \text{ codes a finite sequence of even length } \& \ \mathcal{G}\alpha\, \exists n \forall m\, P(\bar{x},n,s^\frown\bar{\alpha}(2m)),$$

where if $t = \langle x_0 \ldots x_n\rangle$, $v = \langle y_0 \ldots y_m\rangle$ then $t^\frown v = \langle x_0 \ldots x_n\, y_0 \ldots y_m\rangle$. It is enough to show that R^* is $\Sigma^{1,pos}_1$-IND. Since $\bar{x}$ is carried through as a parameter below, we stop indicating it explicitly.

Define the operator

$$\varphi(s,X) \Leftrightarrow s \text{ codes a finite sequence of even length } \& \ \mathcal{G}\alpha \forall m\, [\exists n \leq \ell h(s)\, P(n,s^\frown\bar{\alpha}(2m)) \vee s^\frown\bar{\alpha}(2m) \in X] \,.$$

We can immediately see that φ is positive Σ^1_1 bu replacing "$\mathcal{G}\alpha \ldots$" by "I has a winning strategy in ..." above. We claim now that

$$\varphi^\infty(s) \Leftrightarrow s \in R^*,$$

which completes the proof that $\mathcal{G}\Sigma^0_2 \subseteq \Sigma^{1,pos}_1$-IND.

$\Rightarrow$. By induction on ξ we show that $\varphi^\xi \subseteq R^*$. Assume $\varphi^{<\varphi} \subseteq R^*$. Let $s \in \varphi^\xi$. Then I has a winning strategy τ in the game

$$T(\alpha) \Leftrightarrow \forall m\, [\exists n \leq \ell h(s)\, P(n,s^\frown\bar{\alpha}(2m)) \vee s^\frown\bar{\alpha}(2m) \in \varphi^{<\xi}] \,.$$

The following is then a winning strategy for I in the game

$$U(\alpha) \Leftrightarrow \exists n \forall m\, P(n,s^\frown\bar{\alpha}(2m)):$$

I plays according to τ as long as $\exists n \leq \ell h(s)\, P(n,s^\frown\bar{\alpha}(2m))$. Otherwise, for some m, $s^\frown\bar{\alpha}(2m) \in \varphi^{<\xi}$, so $s^\frown\bar{\alpha}(2m) \in R^*$ and then I continues by playing a winning strategy given to him by this fact.

$\Leftarrow$. Assume $s \in R^*$ but $s \notin \varphi^\infty$. Let σ be a winning strategy for I in the game $U(\alpha)$ as above. Since $s \notin \varphi^\infty$, II has a winning strategy in the game

$$V(\alpha) \Leftrightarrow \forall m\, [\exists n \leq \ell h(s) P(n, s^\frown\bar{\alpha}(2m)) \vee s^\frown\bar{\alpha}(2m) \in \varphi^\infty].$$

Let I play according to σ while II plays according to this strategy. After finitely many moves we obtain an m_1 such that

$$\forall n \leq \ell h(s) \neg P(n, s^\frown\bar{\alpha}(2m_1)) \wedge s^\frown\bar{\alpha}(2m_1) \notin \varphi^\infty.$$

Repeat now with $s^\frown\bar{\alpha}(2m_1)$ instead of s (while I still plays σ). At the end of the run of the game an α is produced such that for some $m_1 < m_2 < \ldots$

$$\forall n \leq \ell h(s^\frown\bar{\alpha}(m_k)) \neg P(n, s^\frown\bar{\alpha}(2m_{k+1})),$$

so that

$$\neg \exists n \forall m\, P(n, \bar{\alpha}(m)).$$

Thus I loses, following σ, a contradiction.

In order to prove that $\Sigma_1^{1,\mathrm{pos}}\text{-IND} \subseteq \mathcal{G}\Sigma_2^0$, let $\varphi(\bar{x}, S)$ be a monotone Σ_1^1 operator and notice that

$$\varphi(\bar{x}, S) \Leftrightarrow \exists T\, (T \subseteq S \wedge \varphi(\bar{x}, T)),$$

so that in order to prove that $\varphi^\infty \in \mathcal{G}\Sigma_2^0$ it will be enough (by the first recursion theorem for Spector classes) to show that $\mathcal{G}\Sigma_2^0$ is closed under the closed game quantifier

$$\mathcal{G}_C u R(u) \Leftrightarrow \forall a_0 \exists a_1 \forall a_2 \ldots \forall n\, R(\langle a_0 \ldots a_n \rangle),$$

(see 2.1); i.e. we have to show that if $P \in \Sigma_2^0$ then

$$\forall a_0 \exists a_1 \ldots \forall n\, (\exists b_0 \forall b_1 \ldots P(\langle a_0 \ldots a_n \rangle, (b_0, b_1, \ldots)))$$

is in $\mathcal{G}\Sigma_2^0$. Taking negations and noticing that when X is determined we have

$$\neg \exists a_0 \forall a_1 \ldots X((a_0, a_1, \ldots)) \Leftrightarrow \forall a_0 \exists a_1 \ldots \neg X((a_0, a_1, \ldots)),$$

we see that it is enough to show that $\mathcal{G}\Pi_2^0$ is closed under the open game quantifier. $\mathcal{G}_0$. This follows immediately from the following equivalence:

$$\exists a_0 \forall a_1 \exists a_2 \ldots \exists n\, (\exists b_0 \forall b_1 \exists b_2 \ldots P(\langle a_0 \ldots a_{2n-1} \rangle, (b_0, b_1, \ldots)))$$

$$\Leftrightarrow \underline{\exists k_0 \exists c_0} \forall c_1 \underline{\exists k_1 \exists c_2} \forall c_3 \underline{\exists k_2 \forall c_4} \ldots$$

$$[\forall j (k_j \leq 1) \wedge \forall j, j' (j \leq j' \wedge k_{j'} = 0 \Rightarrow k_j = 0)$$

$$\wedge \exists j\, (k_j = 1) \wedge \forall j [k_j = 0 \wedge k_{j+1} = 1 \Rightarrow$$

$$\Rightarrow P(\langle c_0 \ldots c_{2j-1} \rangle, (c_{2j}, c_{2j+1}, \ldots))].$$

$\perp$

Remark. The study of the higher $\mathcal{G}\Sigma_n^0$, $n \geq 3$ is the subject of much current research. Results of Martin-Solovay show that $\mathcal{G}\Sigma_3^0$ is already very big.

The process of applying the game quantifier to the Σ_n^0's can of course be iterated. Put for $m,n \geq 1$

$$\Gamma_{m,n} = \mathcal{G}^m\Sigma_n^0 = \underbrace{\mathcal{G}\mathcal{G}\cdots\mathcal{G}}_{m}\Sigma_n^0.$$

Thus a typical member of $\Gamma_{m,n}$ looks like

$$\mathcal{G}\alpha_1\mathcal{G}\alpha_2\cdots\mathcal{G}\alpha_m P(\bar{x},\alpha_1\cdots\alpha_m),$$

with $P \in \Sigma_n^0$. Using a relativized version of Theorem 2.5.1, it is easy to see that each $\Gamma_{m,n}$ is a Spector class, assuming Projective Determinacy, i.e. that every projective subset of ω^ω is determined.

We identify now the classes $\Gamma_{m,1} = \Gamma_m$. We already saw that

$$\Gamma_1 = \Pi_1^1.$$

From (a relativized version of) this fact and the definition it follows (with no determinacy hypothesis) that

$$\Gamma_2 = \mathcal{G}\,\Pi_1^1 = \Sigma_2^1;$$

then using Projective Determinacy (actually only $\mathrm{Det}(\underset{\sim}{\Delta}_2^1)$ is needed here) we have that

$$\Gamma_3 = \Pi_3^1$$

etc., so that with projective determinacy

$$\Gamma_{2n+1} = \Pi_{2n+1}^1$$

$$\Gamma_{2n+2} = \Sigma_{2n+2}^1$$

Thus we have

2.5.3 <u>Theorem</u> (Martin [16], Moschovakis [17]). Assuming Projective Determinacy, $\Pi_{2n+1}^1, \Sigma_{2n+2}^1$ are Spector classes for all $n \geq 0$.

The Spector classes $\Gamma_{m,n}$ for $n \geq 2$ lie strictly between the analytical classes as follows:

$$\Gamma_m = \Gamma_{m,1} < \Gamma_{m,2} < \Gamma_{m,3} < \Gamma_{m,4} < \cdots < \Gamma_{m+1} = \Gamma_{m+1,1}.$$

This can be easily seen by noticing that if $\mathcal{R}$ is adequate and $\mathcal{R} \subseteq \mathcal{R}'$, $\breve{\mathcal{R}} \subseteq \mathcal{R}'$ then, assuming $\mathrm{Det}(\mathcal{R}')$, we have

$$\mathcal{G}\mathcal{R} < \mathcal{G}\mathcal{R}'.$$

This is because, if $\breve{\mathcal{G}}$ is the quantifier on A^ω dual to $\mathcal{G}$, i.e.

$$\breve{\mathcal{G}}\alpha R(\alpha) \Leftrightarrow \neg\,\mathcal{G}\,\alpha\,\neg\,R(\alpha)$$

then (using $\mathrm{Det}(\mathcal{R}')$) we have

$$\mathfrak{G}\mathfrak{R}' = \breve{\mathfrak{G}}\mathfrak{R}',$$

so that

$$\breve{\mathfrak{G}\mathfrak{R}} = \breve{\mathfrak{G}}\breve{\mathfrak{R}} \subseteq \breve{\mathfrak{G}}\mathfrak{R}' = \mathfrak{G}\mathfrak{R}'.$$

2.6 Spector classes obtained by reduction.

Let $\mathfrak{A} = \langle A, R_1 \ldots R_\ell \rangle$ be a structure and let $A \subseteq B$. Let Θ be a class of relations on B such that

(i)' Θ is closed under $\wedge, \vee, \exists^A, \forall^A$.

(ii)' Θ contains all the relations on A which are first order definable with parameters in $\mathfrak{A}$.

(iii)' Θ contains a coding scheme on A.

(iv)' Θ is A-parametrized.

(v)' Θ is normed.

Put

$$\Gamma = \Theta \upharpoonright A = \{R \subseteq A^n : \text{For some } P \subseteq B^n,\ P \in \Theta,\ R = P \cap A^n\}.$$

Then Γ is a Spector class on $\mathfrak{A}$, called the reduct of Θ (to A).

Example iii) of 2.4 provides Spector classes obtained by reduction (with $\Theta = \Sigma_1(M)$ of course). Other important examples come from recursion in higher types. Here B is the set of all higher types over A

$$B = \bigcup_n A^{(n)},$$

where

$$A^{(0)} = A$$
$$A^{(j+1)} = {}^{\omega}A^{(j)}$$

(we are assuming here that $\omega \subseteq A$) and Θ is the class of all relations on B semirecursive in some appropriate higher type object or functional. For more on this see e.g. Kechris-Moschovakis [11].

2.7 Spector classes obtained by projection.

Let $\mathfrak{A} = \langle A, R_1 \ldots R_\ell \rangle$ be a structure and B a set. Let Θ be a collection of relations on B. We say that Θ is invariantly normed or has the invariant prewellordering property if for every equivalence relation $\approx$ on B which is in $\Theta \cap \breve{\Theta}$ and every $\approx$-invariant relation P in Θ (i.e. a $P \in \Theta$ such that $P(x_1 \ldots x_n) \,\&\, x_1 \approx y_1 \,\&\, x_2 \approx y_2 \,\&\, \ldots \,\&\, x_n \approx y_n \Rightarrow P(y_1 \ldots y_n)$) P admits a Θ-norm σ such that

$$x_1, \ldots, x_n \in P \,\&\, x_1 \approx y_1 \,\&\, \ldots \,\&\, x_n \approx y_n \Rightarrow \sigma(x_1 \ldots x_n) = \sigma(y_1 \ldots y_n).$$

Let now

$$\pi : C \xrightarrow{\text{onto}} A$$

be a mapping from a subset $C \subseteq A$ which is in $\Theta \cap \breve{\Theta}$ onto A such that the equivalence relation

$$x \approx_{\pi} x' \Leftrightarrow (x,x' \in C \ \& \ \pi(x) = \pi(x')) \vee (x,x' \notin C)$$

is also in $\Theta \cap \breve{\Theta}$. Let

$$\pi[\Theta] = \{R \subseteq A^n : \pi^{-1}[R] = \{(x_1 \dots x_n) \in B^n : (\pi(x_1) \dots \pi(x_n))\} \in \Theta\},$$

be the <u>projection</u> of Θ onto A via π. Assume that $\pi[\Theta]$ satisfies (i)-(iv) of the definition of a Spector class in 1.1. Then $\pi[\Theta]$ is a Spector class. Indeed, for each $R \in \pi[\Theta]$ and each $(y_1 \dots y_n) \in R$ put

$$\tau(y_1 \dots y_n) = \sigma(x_1 \dots x_n)$$

for any $x_1 \dots x_n$ such that $\pi(x_i) = y_i$, where σ is an $\approx_{\pi}$-invariant norm on $\pi^{-1}[R]$. Then it is trivial to check that τ is a $\pi[\Theta]$-norm on R.

By the method of projection it has been shown (in Kechris [18]) that under AD (the full Axiom of Determinacy) $\Pi^1_{2n+1}(\langle\aleph_1,<\rangle)$, $\Sigma^1_{2n+2}(\langle\aleph_1,<\rangle)$ are Spector classes for $n \geq 0$. Among other things this proof utilizes the fact that, assuming Projective Determinacy, for all $n \geq 0$ the class of Π^1_{2n+1} second order relations on ω has the invariant prewellordering property. This was proved by Burgess and Miller [19] for $n = 0$ (in ZF + DC only) and by Solovay in general. For further results along this line see also [20]. Harrington has also shown that it is consistent with ZFC that $\Pi^1_1(\langle\aleph_1,<\rangle)$ is a Spector class.

3. REPRESENTATION THEOREMS

3.1 Inaccessible and Mahlo Spector classes

Let $\mathfrak{A} = \langle A, R_1 \dots R_\ell\rangle$ be an infinite structure. If Γ, Γ' are two Spector classes on $\mathfrak{A}$ recall that we put

$$\Gamma' \prec \Gamma \Leftrightarrow \Gamma' \subseteq \Delta.$$

<u>Definition</u>. A Spector class Γ is called <u>Mahlo</u> iff for every $\underset{\sim}{F} \subseteq p(A)$:

$$\underset{\sim}{F} \text{ is } \Delta \text{ on } \Delta \Rightarrow \text{there is } \Gamma' \prec \Gamma \text{ such that } \underset{\sim}{F} \text{ is } \Delta' \text{ on } \Delta'.$$

The terminology is motivated by the following considerations:

Call an admissible set $\mathfrak{M} = \langle\mathfrak{A},M;\in;R\rangle$ above $\mathfrak{A}$ <u>Mahlo</u> if for all $X \subseteq A \cup M$ and $x \in A \cup M$, if $X \in \Delta_1(\mathfrak{M})$ there is $\mathfrak{N} = \langle\mathfrak{A},N;\in,R \cap (A \cup N)\rangle \in \mathfrak{M}$ admissible above $\mathfrak{A}$ such that $x \in A \cup N$ and $\langle\mathfrak{N},X \cap (A \cup N)\rangle$ is admissible. (For $\mathfrak{M}$ of the form $\langle L_\alpha,\in\rangle$ and projectible, this coincides with the definition of α being recursively Mahlo.) We now have

<u>Fact</u>. If Γ is a Spector class on $\mathfrak{A}$, then Γ is Mahlo $\Longleftrightarrow$ the companion of Γ is Mahlo.

We abbreviate below for $\underset{\sim}{F} \subseteq p(A)$:

${}_1\text{en}(\underset{\sim}{F})$ = 1-<u>envelope</u> of $\underset{\sim}{F}$ = smallest Spector class on which $\underset{\sim}{F}$ is Δ on Δ.

<u>Theorem</u> (Harrington-Kechris, [21], Simpson). A Spector class Γ on $\mathcal{A}$ is the 1-envelope of a type two object iff Γ is not Mahlo.

<u>Proof.</u> $\Rightarrow$. Obvious.

$\Leftarrow$. The proof is easy from the following

<u>Lemma.</u> If Γ is a Spector class, there is a (nonmonotone) operator $\varphi(x,S)$ which is Δ on Δ such that $\varphi^{\infty} = \varphi^{<\kappa^{\Gamma}}$ (i.e. the closure ordinal of φ is $\leq \kappa^{\Gamma}$) and φ^{∞} is Γ-complete (i.e. for all $R \in \Gamma$ there is $f \in \Delta$ such that $R(\bar{x}) \Longleftrightarrow \varphi^{\infty}(f(\bar{x})))$.

Indeed, let $\underset{\sim}{F} \subseteq p(A)$ demonstrate that Γ is not Mahlo, so that $\underset{\sim}{F}$ is Δ on Δ and $\Gamma' = {}_1\text{en}(\underset{\sim}{F}) \not\subset \Gamma$. Then by Spector's Criterion $\kappa^{\Gamma'} = \kappa^{\Gamma}$ (note that $\Gamma' \subseteq \Gamma$). Identifying now the φ that comes from the lemma with the type 2 object $H = \{\langle x,S\rangle : \varphi(x,S)\}$, where $\langle x,S\rangle = \{\langle x,a\rangle : a \in S\}$, put $G = \underset{\sim}{H} \vee \underset{\sim}{F} = \{\langle 0,X\rangle : X \in \underset{\sim}{H}\} \cup \{\langle 1,X\rangle : X \in \underset{\sim}{F}\}$. Then $\underset{\sim}{G}$ is Δ on Δ, so $\Gamma \supseteq {}_1\text{en}(\underset{\sim}{G}) = \Gamma''$ and $\kappa^{\Gamma''} = \kappa^{\Gamma}$, since $\Gamma'' \supseteq \Gamma'$. But φ is Δ'' on Δ'' so by the Proof of 2.2.5, $\varphi^{\infty} = \varphi^{<\kappa^{\Gamma''}} \in \Gamma''$ $\therefore \Gamma = \Gamma'' = {}_1\text{en}(\underset{\sim}{G})$ and we are done.

<u>Proof of Lemma.</u> Let $W \subseteq A \times A$ be universal for Γ. If $X \subseteq A$ is in Δ find $c = \langle c_0, c_1, c_2\rangle$ such that $(c_1,c_2) \in W$ and

$$a \in X \Longleftrightarrow (c_0,a) \in W$$
$$\Longleftrightarrow \sigma(c_0,a) \leq \sigma(c_1,c_2),$$

where σ is a Γ-norm on W. Let $I = \{\langle c_0,c_1,c_2\rangle : (c_1,c_2) \in W\}$ and put for $c \in I$,

$$H_c = \{a : \sigma((c)_0,a) \leq \sigma((c)_1,(c)_2)\}.$$

Then if

$$H(c,a) \Leftrightarrow ((c)_0,a) \leq^*_{\sigma} ((c)_1,(c)_2),$$
$$\check{H}(c,a) \Leftrightarrow \neg\, [((c)_1,(c)_2) <^*_{\sigma} ((c)_0,a)]$$

$\{I,H,\check{H}\}$ is an A-parametrization of the Δ subsets of A. Let $g : A \to A$ be in Δ such that for each $c \in I$,

$$j(H_c) = H_{g(c)},$$

where for any $X \subseteq A$, $j(X) \subseteq A$ is given by

$$u \in j(X) \Leftrightarrow \forall a,b,c(\langle a,u,b,c\rangle \notin X).$$

Define now

$$\varphi(x,S) \Leftrightarrow S \in \Delta \;\&\; x = \langle y,z\rangle, \text{ for some } y,z \;\&$$
$$\forall d \in I[H_d = j(S) \Rightarrow$$
$$y \leq^*_{\sigma} z \leq^*_{\sigma} ((d)_1,(d)_2)].$$

Now φ is Δ on Δ, since for $c \in I$,

$$\varphi(x,H_c) \Longleftrightarrow x = \langle y,z\rangle, \text{ for some } y,z \ \&$$
$$\forall d\,[((d)_1,(d)_2) \leq^*_\sigma ((g(c))_1,(g(c))_2) \ \&\ H_d = H_{g(c)} \Rightarrow$$
$$\Rightarrow y \leq^*_\sigma z \leq^*_\sigma ((d)_1,(d)_2)].$$

Thus $\varphi^{<\kappa^\Gamma}$ is in Γ. If we can show that $\varphi^{<\kappa^\Gamma}$ is Γ-complete then $\varphi^{<\kappa^\Gamma} \notin \Delta$ so

$$x \in \varphi^{\kappa^\Gamma} \Longleftrightarrow x \in \varphi^{<\kappa^\Gamma} \vee \varphi(x,\varphi^{<\kappa^\Gamma})$$
$$\Longleftrightarrow x \in \varphi^{<\kappa^\Gamma},$$

so $\varphi^{\kappa^\Gamma} = \varphi^{<\kappa^\Gamma} = \varphi^\infty$ and we are done.

For each $\xi < \kappa^\Gamma$ let

$$W^{\leq\xi} = \{\langle y,z\rangle : y,z \in W \ \&\ \sigma(y) \leq \sigma(z) \leq \xi\},$$

and for $\xi \leq \kappa^\Gamma$,

$$W^{<\xi} = \{\langle y,z\rangle : y,z \in W \ \&\ \sigma(y) \leq \sigma(z) < \xi\}.$$

Clearly $W^{<\kappa^\Gamma}$ is Γ-complete. We show that $\varphi^{<\kappa^\Gamma} = W^{<\kappa^\Gamma}$. Obviously $\varphi^{<\kappa^\Gamma} \subseteq W^{<\kappa^\Gamma}$. For the converse we show that

$$\forall \xi < \kappa^\Gamma \exists \xi' < \kappa^\Gamma\ (\xi \leq \xi' \ \&\ \varphi^\xi = W^{\leq\xi'}).$$

The proof is by induction on ξ. Assume it holds for all $\eta < \xi$. Let $\zeta = \sup\{\eta' + 1 : \eta < \xi\} \geq \xi$. Then

$$\varphi^{<\xi} = \bigcup_{\eta<\xi} \varphi^\eta = \bigcup_{\eta<\xi} W^{\leq\eta'} = W^{<\zeta}.$$

Then $\zeta < \kappa^\Gamma$, since $\varphi^{<\xi} \in \Delta$. Let c in I be such that $H_c = j(W^{<\zeta})$. Then $\sigma(((c)_1,(c)_2)) \geq \zeta$ since otherwise

$$u \in j(W^{<\zeta}) \Longleftrightarrow \sigma((c)_0,u) \leq \sigma((c)_1,(c)_2) < \zeta$$
$$\Longleftrightarrow \langle (c)_0,u,(c)_1,(c)_2\rangle \in W^{<\zeta},$$

a contradiction. So if $\xi' = \min\{\sigma((c)_1,(c)_2) : c \in I \ \&\ H_c = j(W^{<\zeta})\}$ then $\xi' \geq \zeta \geq \xi$ and

$$\varphi^\xi = \varphi^{<\xi} \cup \{x : \varphi(x,\varphi^{<\xi})\}$$
$$= W^{<\zeta} \cup \{x : \varphi(x,W^{<\zeta})\}$$
$$= W^{<\zeta} \cup W^{\leq\xi'}$$
$$= W^{\leq\xi'}$$

and we are done. ⊣

Corollary. Let Γ be a Spector class on $\mathfrak{A}$. If Γ is $\forall_2$-compact (see Moschovakis [2]) then Γ is Mahlo, so Γ is not the 1-envelope of a type 2 object. In particular $\mathfrak{J}$-IND with $\mathfrak{J} \supseteq \Pi^0_2(C)$, where C is a hyperelementary coding scheme on $\mathfrak{A}$, is never the 1-envelope of a type 2 object.

Proof. By Moschovakis [2] the companion of Γ is Π_3-reflecting. But every Π_3-reflecting admissible set is Mahlo. (Use the fact that admissibility is a Π_3 notion.) ⊣

Definition. A Spector class Γ on $\mathfrak{A}$ is inaccessible iff $WF = \{S \subseteq A : \neg\exists a_0, a_1, \ldots \forall i \langle a_{i+1}, a_i\rangle \in S\}$ is Δ on Δ.

Definition. An admissible set $\mathfrak{M}$ above $\mathfrak{A}$ is inaccessible iff for all $x \in \mathfrak{M}$ there is $\mathfrak{N} \in \mathfrak{M}$, $\mathfrak{N}$ admissible such that $x \in \mathfrak{N}$.

Facts. Let Γ be a Spector class on $\mathfrak{A}$. Then:

i) Γ is inaccessible $\Leftrightarrow$ the companion of Γ is inaccessible.

ii) If Γ is Mahlo, then Γ is inaccessible.

Proof of ii). Note that if $S \in \Delta$ and $X \in IND(\langle \mathfrak{A}, S\rangle)$, then $X \in \Delta$. Indeed let $X = \varphi^\infty$ with $\varphi(x,T)$ first order positive having S as a parameter. Then φ is Δ on Δ, so φ is Δ' on Δ' for some $\Gamma' < \Gamma$, therefore $X = \varphi^\infty \in \Gamma' \subseteq \Delta$.

Find now $\varphi(x,T,S)$ positive in T such that (for some constant $a \in A$)

$$WF(S) \Leftrightarrow a \in \varphi^\infty(S).$$

Then for $S \in \Delta$,

$$\neg WF(S) \Leftrightarrow \exists X \in \Delta\ (X = \varphi^\infty(S) \ \&\ a \notin X)$$

so $\neg WF(S)$ is Γ on Δ. But $WF(S)$ is also Γ on Δ and we are done. ⊣

In particular, if Γ is not inaccessible then $\Gamma = {}_1en(\underset{\sim}{F})$, for some $\underset{\sim}{F} \subseteq p(A)$.

3.2 The Harrington Representation Theorem.

Theorem (Harrington [22]). Let $\mathfrak{A}$ be almost acceptable. If Γ is a Spector class on $\mathfrak{A}$ then for some quantifier Q on A

$$\Gamma = IND(\mathfrak{A}, Q).$$

Proof. If Γ is not inaccessible then $\Gamma = {}_1en(\underset{\sim}{F}) = IND(\mathfrak{A}, Q)$, where Q is given in the proof of the fact in 2.3. So we can assume that Γ is inaccessible.

We shall construct a monotone oparator $\varphi(c,X)$ on A such that φ^∞ is Γ-complete and Γ is closed under $\varphi, \check{\varphi}$ (here $\check{\varphi}(c,X) \Leftrightarrow \neg\varphi(c, A-X)$). Having such a φ we can put

$$Q(X) \Leftrightarrow \exists c(\exists d(\langle c,d\rangle \in X) \wedge \varphi(c, X_c)),$$

where $X_c = \{d : \langle c,d\rangle \in X\}$. Then Q is a quantifier and Γ is closed under Q, $\check{Q}$ so $IND(\mathfrak{A}, Q) \subseteq \Gamma$. But also

$$\varphi(a,X) \Leftrightarrow Q(\langle a, X\rangle)$$
$$\Leftrightarrow Q(\{\langle a,d\rangle : d \in X\}),$$

so $IND(\mathfrak{A}, Q)$ is closed under φ and thus $\varphi^\infty \in IND(\mathfrak{A}, Q)$ $\therefore$ $\Gamma = IND(\mathfrak{A}, Q)$.

<u>Construction of</u> $\varphi(c,X)$. Let $P \in \Gamma$ be Γ-complete, σ a Γ-norm on P. Put $R = \{\langle 0,a,b\rangle : b \in P \ \& \ a \overset{*}{\leq}_\sigma b\} \cup \{\langle 1,a,b\rangle : b \in P \ \& \ b \overset{*}{<}_\sigma a\}$. For $c = \langle i,a,b\rangle \in R$, put $|c| = \sigma(b)$. For any Z let $Z_i = \{x : \langle i,x\rangle \in Z\}$, so that $R_0 = \{\langle a,b\rangle : b \in P \ \& \ a \overset{*}{\leq}_\sigma b\}$ and similarly for R_1. Put now

$$\varphi(c,X) \Leftrightarrow [X_0 \cap X_1 \neq \emptyset] \vee [c \in R \ \& \ \forall c' \in R(|c'| < |c| \Rightarrow c' \in X)].$$

Clearly, if $R_\xi = \{c \in R : |c| \leq \xi\}$, then $\varphi^\xi = R_\xi$ and $\varphi^\infty = R$, so φ^∞ is Γ-complete.

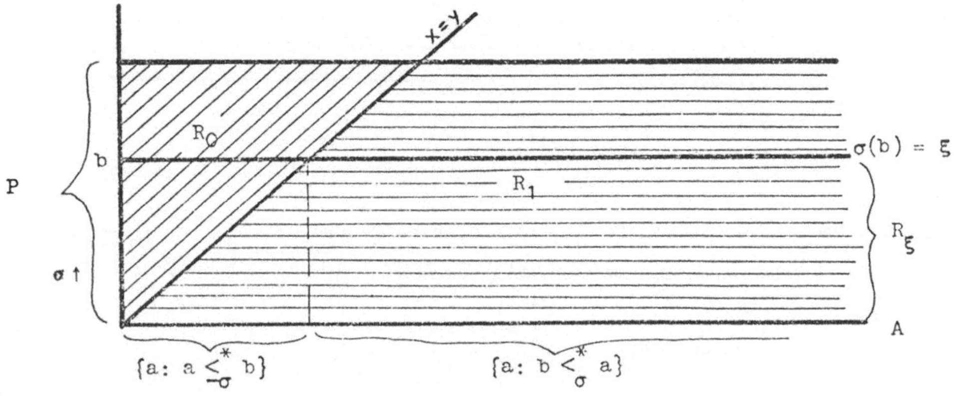

We show now that Γ is closed under both φ and $\breve{\varphi}$ which completes the proof. This is clear for φ. Consider now $\breve{\varphi}$. We have

$$\breve{\varphi}(c,Y) \Leftrightarrow \neg\, \varphi(c,A-Y)$$

$$\Leftrightarrow [Y_0 \cup Y_1 = A] \wedge [c \in R \vee \exists c' \in R(|c'| < |c| \ \& \ c' \in Y)].$$

Assume $Y \in \Gamma$ and $Y_0 \cup Y_1 = A$. If we can pass in a uniform Δ way from a Γ-code of Y to a member $d \in P$, with say $\sigma(d) = \xi$, such that

$$(c \in R \ \& \ \forall c' \in R\, [|c'| < |c| \Rightarrow c' \in Y]) \Rightarrow |c| \leq \xi$$

we are done, because then

$$c \notin R \vee \exists c' \in R[|c'| < |c| \ \& \ c' \in Y] \Longleftrightarrow$$

$$c \notin R_\xi \vee \exists c' \in R[|c'| < |c| \ \& \ c' \in Y].$$

So fix $c = \langle i,a,b\rangle \in R$ such that $\forall c' \in R[|c'| < |c| \Rightarrow c' \notin Y]$. Let $T_0 = \{\langle a',b'\rangle : a' \overset{*}{\leq}_\sigma b' \overset{*}{<}_\sigma b\}$, $T_1 = \{\langle a',b'\rangle : b' \overset{*}{<}_\sigma a', b' \overset{*}{<}_\sigma b\}$. Then $x \in T_0 \Rightarrow x \notin Y_0$ $\therefore$ $T_0 \subseteq A - Y_0 = (Y_1 \cup Y_0) - Y_0 = Y_1 - Y_0$. Similarly $T_1 \subseteq Y_0 - Y_1$. But for $b' \overset{*}{<}_\sigma b$, $(T_0)_{b'}$ $(= \{a' : \langle a',b'\rangle \in T_0\})$, $(T_1)_{b'}$ from a partition of A $\therefore$ $(T_0)_{b'} = (Y_1 - Y_0)_{b'}$, $(T_1)_{b'} = (Y_0 - Y_1)_{b'}$ $\therefore$ for $b' \overset{*}{<}_\sigma b$, $\langle a',b'\rangle \in T_0 \Longleftrightarrow \langle a',b'\rangle \in Y_1 \Longleftrightarrow \langle a',b'\rangle \notin Y_0$ and we have the picture below:

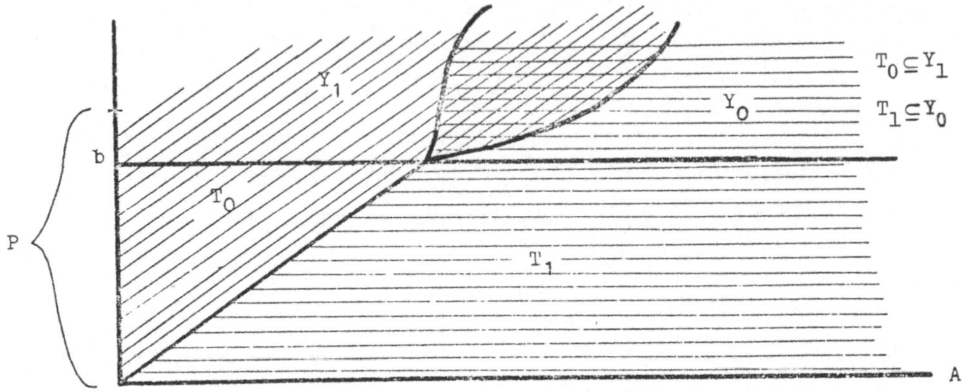

Using the reduction theorem for Spector classes reduce Y_1, Y_0 to $X_0 \subseteq Y_1$, $X_1 \subseteq Y_0$ in Γ. Since $Y_1 \cup Y_0 = A$ clearly $X_0 \cup X_1 = A$; $X_0 \cap X_1 = \emptyset$ so X_0, X_1 are in Δ. Then for $b' \overset{*}{<}_\sigma b$,

$$a' \overset{*}{\leq}_\sigma b' \Leftrightarrow \langle a',b'\rangle \in X_0,$$

$$b' \overset{*}{<}_\sigma a' \Leftrightarrow \langle a',b'\rangle \notin X_0 \Rightarrow \langle a',b'\rangle \in X_1.$$

Put for $\bar{b} \overset{*}{<}_\sigma b$

$$U^{\leq\sigma(\bar{b})} = \{\langle a',b'\rangle : a' \overset{*}{<}_\sigma b' \overset{*}{\leq}_\sigma \bar{b}\}.$$

Then

$$U^{\leq\sigma(\bar{b})} = \{\langle a',b'\rangle : \langle b',a'\rangle \notin X_0 \ \& \ \langle b',\bar{b}\rangle \in X_0\}$$
$$=^{\text{def}} W(X_0,\bar{b}).$$

Thus

$$\sigma(b) \leq \sup\{\text{length } (W(x_0,\bar{b})) : W(X_0,\bar{b}) \text{ is wellfounded}\}$$
$$\leq \text{length } (\prec_{X_0}),$$

where

$$(e,f) \prec_{X_0} (g,h) \Leftrightarrow e = g \ \& \ W(X_0,e) \text{ is wellfounded } \& \ \langle f,h\rangle \in W(X_0,e).$$

Now $\prec_{X_0}$ is in Δ and wellfounded and a Δ-code of it can be found in a Δ way from a Γ-code of Y. Thus one can find from a Γ-code of Y some $d \in P$ such that $\text{length}(\prec_{X_0}) < \sigma(d)$ and we are done. $\dashv$

REFERENCES

[1] Y. N. Moschovakis, Elementary Induction on Abstract Structures, North Holland, 1974.

[2] Y. N. Moschovakis, Structural characterizations of classes of relations, J. E. Fenstad and P. G. Hinman (Eds.), Generalized Recursion Theory, North Holland, 1974, 53-79.

[3] P. Aczel, Games, Quantifiers and Inductive Definitions, Proc. of the 3rd Scandinavian Logic Symposium, S. Kanger (Ed.), North Holland, Amsterdam, 1975, 1-14.

[4] P. Aczel, Stage Comparison theorems and game playing with inductive definitions, preprint, 1972.

[5] W. Richter and P. Aczel, Inductive definitions and reflecting properties of admissible ordinals, in Generalized Recursion Theory, J. E. Fenstad and P. G. Hinman (Eds.), North Holland, 1974, 301-381.

[6] Y. N. Moschovakis, On nonmonotone inductive definability, Fund. Math. 82 (1974), 39-83.

[7] A. S. Kechris, The theory of countable analytical sets, TAMS, 202 (1975), 259-297.

[8] S. Aanderaa, Inductive definitions and their closure ordinals, Generalized Recursion Theory (Oslo, 1972), J. E. Fenstad and P. G. Hinman (Eds.), North Holland, 1974.

[9] S. C. Kleene, Recursive functionals and quantifiers of finite type I, TAMS 91 (1959), 1-52.

[10] R. O. Gandy, General recursive functionals of finite type and hierarchies of functionals, Ann. Fac. Sci. Univ. Clermont-Ferrand, 35 (1967), 5-24.

[11] A. S. Kechris and Y. N. Moschovakis, Recursion in higher types, Handbook of Math. Logic, J. Barwise (Ed.), North Holland, 1976, p. 681-737.

[12] J. Addison, Some consequences of the axiom of constructibility, Fund. Math. 46 (1959a), 123-135.

[13] J. Barwise, Admissible sets and structures, Springer-Verlag, 1975.

[14] Y. N. Moschovakis, Descriptive Set Theory, North Holland, to appear.

[15] A. S. Kechris, Lecture notes on Descriptive Set Theory, M.I.T., 1973.

[16] D. A. Martin, The axiom of determinateness and reduction principles in the analytical hierarchy, BAMS, 74 (1968), 687-689.

[17] J. W. Addison and Y. N. Moschovakis, Some consequences of the axiom of definable determinateness, Proc. Nat. Acad. Sci., USA, 59 (1968), 708-712.

[18] A. S. Kechris, Countable ordinals and the analytical hierarchy, II, to appear.

[19] J. Burgess and D. Miller, Remarks on invariant descriptive set theory, Fund. Math. 90 (1975), 53-75.

[20] A. S. Kechris and D. A. Martin, On the theory of Π^1_3 sets of reals, Bull. Amer. Math. Soc., 84 (1978), 149-151.

[21] A. S. Kechris and L. A. Harrington, On characterizing Spector classes, J. Symbolic Logic 40 (1975), 19-24.

[22] L. A. Harrington, Monotone quantifiers and inductive definability, April 1975.

APPENDIX

The following list of problems was distributed during a very informal gathering of logicians at U.C.L.A. in January 1978. We are reproducing it here because of its obvious relevance to the contents of this volume.

The Victoria Delfino Problems

A cash prize of $100 is offered by the logicians in the Los Angeles area for the solution of each of the following five problems. This competition is financed by the Victoria Delfino Fund for the Advancement of Logic which was established by a generous contribution from Miss Victoria Delfino.

Employees of U.C.L.A. and Caltech and their immediate families (other than students) are inelligible for these prizes; competition is open to everyone else. All decisions by the judges are final. Multiple entries are allowed.

1. Projective ordinals.

For each positive integer n, let $\underset{\sim}{\delta}^1_n$ be the least nonzero ordinal not the length of a $\underset{\sim}{\Delta}^1_n$ prewellordering of the reals.

Assume AD + DC. It is known that $\underset{\sim}{\delta}^1_1 = \omega_1$, $\underset{\sim}{\delta}^1_2 = \omega_2$, $\underset{\sim}{\delta}^1_3 = \omega_{\omega+1}$, $\underset{\sim}{\delta}^1_4 = \omega_{\omega+2}$, $\underset{\sim}{\delta}^1_{2n+2} = (\underset{\sim}{\delta}^1_{2n+1})^+$, and $\underset{\sim}{\delta}^1_{2n+1}$ is always the successor (cardinal) of a cardinal of cofinality ω.

Problem: Compute $\underset{\sim}{\delta}^1_5$.

Kunen has some partial results on this problem, results which suggest the answer ω_{ω^3+1}.

The problem is related to that of whether $\underset{\sim}{\delta}^1_3 \to (\underset{\sim}{\delta}^1_3)^{\underset{\sim}{\delta}^1_3}$. Kunen has shown that $\underset{\sim}{\delta}^1_3 \to (\underset{\sim}{\delta}^1_3)^\alpha$ for each $\alpha < \underset{\sim}{\delta}^1_3$. Results of Kleinberg imply that $\underset{\sim}{\delta}^1_3$ has exactly three normal measures. It is likely that the regular cardinals between $\underset{\sim}{\delta}^1_3$ and $\underset{\sim}{\delta}^1_5$ are exactly the ultrapowers of $\underset{\sim}{\delta}^1_3$ with respect to these normal measures. This would be important in getting an upper bound on $\underset{\sim}{\delta}^1_5$ from Choice plus Det(L[R]), the hypothesis that every set of reals in L[R] is determined.

(Needless to say, the decision of the judges as to what constitutes a "computation" of $\underset{\sim}{\delta}^1_5$ will be final.)

2. The extent of definable scales.

A semiscale on a set $P \subseteq R^k$ $(R = {}^{\omega}\omega)$ is a sequence $\bar{\varphi} = \{\varphi_n\}$ of norms on P, where each $\varphi_n : P \longrightarrow \lambda$ maps P into some ordinal λ and the following convergence condition holds: If $x_0, x_1, \ldots \in P$ and for each n the sequence $\varphi_n(x_0)$, $\varphi_n(x_1)$, $\varphi_n(x_2), \ldots$ is ultimately constant, then $x \in P$. We call $\bar{\varphi}$ a scale, if under the same hypotheses we can infer that

$$\varphi_n(x) \leq \varphi_n(x_i) \quad \text{for all large } i.$$

A semiscale $\bar{\varphi}$ is in a class of relations Γ if both relations

$$U(n,x,y) \iff x \in P \ \&\ [y \notin P \vee \varphi_n(x) \leq \varphi_n(y)]$$

$$V(n,x,y) \iff x \in P \ \&\ [y \notin P \vee \varphi_n(x) < \varphi_n(y)]$$

are in Γ.

It is easy to check that a set P admits a semiscale $\bar{\varphi}$ into λ if and only if P is λ-Suslin i.e. P is the projection of some tree T on $\omega^k \times \lambda$; moreover, T is definable exactly when $\bar{\varphi}$ is definable. Sets which admit definable scales are well-behaved in many ways, e.g. we can use a scale on $P \subseteq R \times R$ to uniformize P.

Granting projective determinacy, we can prove that every projective set admits a projective scale (Moschovakis); on the other hand it is easy to check that $\{(x,y) : x \text{ is not ordinal definable from } y\}$ does not admit a scale which is OD in a real, granting only that for each y there is some x which is not OD in y. Thus not every "definable" set admits a "definable" scale.

The strongest result we can get with current methods is that inductive sets admit inductive scales, granting inductive determinacy; here P is inductive if P is Σ_1 over the smallest admissible set M which contains the reals, $R \in M$.

Problem. Assume $ZF + DC + AD + V = L[R]$; prove or disprove that every coinductive set of reals is λ-Suslin for some λ.

3. The invariance of $L[T^3]$.

Let n be an odd integer. Let P be a complete Π^1_n set of reals and assuming PD let $\bar{\varphi} = \{\varphi_m\}_{m\in\omega}$ be a Π^1_n-scale on P. (It is understood here that each φ_m maps P onto an initial segment of the ordinals.) The tree $T^n = T^n(\bar{\varphi})$ associated with this scale is defined by

$$T^n = \{(\alpha(0), \varphi_0(\alpha), \ldots, \alpha(k), \varphi_k(\alpha)) : \alpha \in P\}.$$

Let Det(L[R]) be the hypothesis that every set of reals in L[R] is determined.

Problem. Assume ZF + DC + Det(L[R]). Prove or disprove that $L[T^3] = L[T^3(\overline{\varphi})]$ is independent of the choice of the complete Π^1_3 set P and the particular Π^1_3-scale $\overline{\varphi}$ on P.

Background. It is known that $L[T^1] = L$ (Moschovakis). Also under the above hypothesis it is known that for all odd n and all $T^n = T^n(\overline{\varphi})$, $L[T^n] \cap R = \mathbf{C}_{n+1}$, where C_{n+1} is the largest countable Σ^1_{n+1} set of reals (Harrington-Kechris), so that $R \cap L[T^n]$ does not depend on the choice of T^n.

In many ways, the model $L[T^n]$ is an excellent analog of L for the (n+1)-st level of the analytical hierarchy.

4. The strength of $\mathrm{Sep}(\Sigma^1_3)$ in the presence of (#).

Let

$$(\#) \iff (\forall x \subseteq \omega)\ [x^{\#} \text{ exists}]$$

and let

$$\mathrm{Sep}(\Sigma^1_3) \iff \text{for every } x \subseteq \omega, \text{ every two disjoint } \Sigma^1_3(x) \text{ sets of reals can be separated by a } \Delta^1_3(x) \text{ set.}$$

Problem. Prove or disprove that:

$$\mathrm{ZFC} + \mathrm{Sep}(\Sigma^1_3) + (\#) \Rightarrow \text{Determinacy } (\underset{\sim}{\Delta}^1_2).$$

Background. Harrington has shown that ZFC + $\mathrm{Sep}(\Sigma^1_3)$ is consistent relative to ZFC. Using, however, Jensen's Absoluteness Theorem for the core model K (which states that if (#) holds and Σ^1_3 formulas are not absolute for K, then $0^{\dagger}$ exists) one can see that

$$\mathrm{ZF} + \mathrm{DC} + \mathrm{Sep}(\Sigma^1_3) + (\#) \Rightarrow \forall x \subseteq \omega\ (x^{\dagger} \text{ exists}).$$

5. A classification of functions on the Turing degrees.

D is the set of Turing degrees. A property P of degrees holds almost everywhere (a.e.) iff $\exists c \forall d \geq c P(d)$. For $f, g : D \to D$, let $f \leq_m g$ iff $f(d) \leq g(d)$ a.e. A function $f : D \to D$ is representable iff $\exists F : {}^{\omega}\omega \to {}^{\omega}\omega \forall x (dg(F(x)) = f(dg(x)))$.

Working in ZF + AD + DC, settle the following conjectures of D. Martin:

(a) If $f : D \to D$ is representable and $d \not\leq f(d)$ a.e., then $\exists c (f(d) = c)$ a.e.

(b) $\leq_m$ is a prewellorder of $\{f : f$ is representable $\wedge\ d \leq f(d)$ a.e.$\}$. Further, if f has rank α in $\leq_m$, then f' has rank $\alpha + 1$, where $f'(d) = f(d)'$ = the Turing jump of f(d).

Remarks. With regard to (a), it is known that if $f(d) \leq d$ and $\forall c (c \leq f(d)$ a.e.), then $f(d) = d$ a.e. It is known that conjecture (b) is true when restricted to uniformly representable f so that $d \leq f(d)$ a.e. (f is uniformly representable if $\exists F : {}^{\omega}\omega \to {}^{\omega}\omega\ (\forall x(dg(F(x)) = f(dg(x))) \wedge \exists t : \omega \to \omega \forall x \forall y\ (x \equiv_T y$ via $e \Rightarrow F(x) \equiv_T F(y)$ via $t(e)))$. It is conjectured that every representable $f : D \to D$ is uniformly representable.

A proof of conjecture (b) would yield a strong negative answer to a question of G. Sacks: is there a degree invariant solution to Post's problem?

Vol. 580: C. Castaing and M. Valadier, Convex Analysis and Measurable Multifunctions. VIII, 278 pages. 1977.

Vol. 581: Séminaire de Probabilités XI, Université de Strasbourg. Proceedings 1975/1976. Edité par C. Dellacherie, P. A. Meyer et M. Weil. VI, 574 pages. 1977.

Vol. 582: J. M. G. Fell, Induced Representations and Banach *-Algebraic Bundles. IV, 349 pages. 1977.

Vol. 583: W. Hirsch, C. C. Pugh and M. Shub, Invariant Manifolds. IV, 149 pages. 1977.

Vol. 584: C. Brezinski, Accélération de la Convergence en Analyse Numérique. IV, 313 pages. 1977.

Vol. 585: T. A. Springer, Invariant Theory. VI, 112 pages. 1977.

Vol. 586: Séminaire d'Algèbre Paul Dubreil, Paris 1975–1976 (29ème Année). Edited by M. P. Malliavin. VI, 188 pages. 1977.

Vol. 587: Non-Commutative Harmonic Analysis. Proceedings 1976. Edited by J. Carmona and M. Vergne. IV, 240 pages. 1977.

Vol. 588: P. Molino, Théorie des G-Structures: Le Problème d'Equivalence. VI, 163 pages. 1977.

Vol. 589: Cohomologie l-adique et Fonctions L. Séminaire de Géométrie Algébrique du Bois-Marie 1965–66, SGA 5. Edité par L. Illusie. XII, 484 pages. 1977.

Vol. 590: H. Matsumoto, Analyse Harmonique dans les Systèmes de Tits Bornologiques de Type Affine. IV, 219 pages. 1977.

Vol. 591: G. A. Anderson, Surgery with Coefficients. VIII, 157 pages. 1977.

Vol. 592: D. Voigt, Induzierte Darstellungen in der Theorie der endlichen, algebraischen Gruppen. V, 413 Seiten. 1977.

Vol. 593: K. Barbey and H. König, Abstract Analytic Function Theory and Hardy Algebras. VIII, 260 pages. 1977.

Vol. 594: Singular Perturbations and Boundary Layer Theory, Lyon 1976. Edited by C. M. Brauner, B. Gay, and J. Mathieu. VIII, 539 pages. 1977.

Vol. 595: W. Hazod, Stetige Faltungshalbgruppen von Wahrscheinlichkeitsmaßen und erzeugende Distributionen. XIII, 157 Seiten. 1977.

Vol. 596: K. Deimling, Ordinary Differential Equations in Banach Spaces. VI, 137 pages. 1977.

Vol. 597: Geometry and Topology, Rio de Janeiro, July 1976. Proceedings. Edited by J. Palis and M. do Carmo. VI, 866 pages. 1977.

Vol. 598: J. Hoffmann-Jørgensen, T. M. Liggett et J. Neveu, Ecole d'Eté de Probabilités de Saint-Flour VI – 1976. Edité par P.-L. Hennequin. XII, 447 pages. 1977.

Vol. 599: Complex Analysis, Kentucky 1976. Proceedings. Edited by J. D. Buckholtz and T. J. Suffridge. X, 159 pages. 1977.

Vol. 600: W. Stoll, Value Distribution on Parabolic Spaces. VIII, 216 pages. 1977.

Vol. 601: Modular Functions of one Variable V, Bonn 1976. Proceedings. Edited by J.-P. Serre and D. B. Zagier. VI, 294 pages. 1977.

Vol. 602: J. P. Brezin, Harmonic Analysis on Compact Solvmanifolds. VIII, 179 pages. 1977.

Vol. 603: B. Moishezon, Complex Surfaces and Connected Sums of Complex Projective Planes. IV, 234 pages. 1977.

Vol. 604: Banach Spaces of Analytic Functions, Kent, Ohio 1976. Proceedings. Edited by J. Baker, C. Cleaver and Joseph Diestel. VI, 141 pages. 1977.

Vol. 605: Sario et al., Classification Theory of Riemannian Manifolds. XX, 498 pages. 1977.

Vol. 606: Mathematical Aspects of Finite Element Methods. Proceedings 1975. Edited by I. Galligani and E. Magenes. VI, 362 pages. 1977.

Vol. 607: M. Métivier, Reelle und Vektorwertige Quasimartingale und die Theorie der Stochastischen Integration. X, 310 Seiten. 1977.

Vol. 608: Bigard et al., Groupes et Anneaux Réticulés. XIV, 334 pages. 1977.

Vol. 609: General Topology and Its Relations to Modern Analysis and Algebra IV. Proceedings 1976. Edited by J. Novák. XVIII, 225 pages. 1977.

Vol. 610: G. Jensen, Higher Order Contact of Submanifolds of Homogeneous Spaces. XII, 154 pages. 1977.

Vol. 611: M. Makkai and G. E. Reyes, First Order Categorical Logic. VIII, 301 pages. 1977.

Vol. 612: E. M. Kleinberg, Infinitary Combinatorics and the Axiom of Determinateness. VIII, 150 pages. 1977.

Vol. 613: E. Behrends et al., L^p-Structure in Real Banach Spaces. X, 108 pages. 1977.

Vol. 614: H. Yanagihara, Theory of Hopf Algebras Attached to Group Schemes. VIII, 308 pages. 1977.

Vol. 615: Turbulence Seminar, Proceedings 1976/77. Edited by P. Bernard and T. Ratiu. VI, 155 pages. 1977.

Vol. 616: Abelian Group Theory, 2nd New Mexico State University Conference, 1976. Proceedings. Edited by D. Arnold, R. Hunter and E. Walker. X, 423 pages. 1977.

Vol. 617: K. J. Devlin, The Axiom of Constructibility: A Guide for the Mathematician. VIII, 96 pages. 1977.

Vol. 618: I. I. Hirschman, Jr. and D. E. Hughes, Extreme Eigen Values of Toeplitz Operators. VI, 145 pages. 1977.

Vol. 619: Set Theory and Hierarchy Theory V, Bierutowice 1976. Edited by A. Lachlan, M. Srebrny, and A. Zarach. VIII, 358 pages. 1977.

Vol. 620: H. Popp, Moduli Theory and Classification Theory of Algebraic Varieties. VIII, 189 pages. 1977.

Vol. 621: Kauffman et al., The Deficiency Index Problem. VI, 112 pages. 1977.

Vol. 622: Combinatorial Mathematics V, Melbourne 1976. Proceedings. Edited by C. Little. VIII, 213 pages. 1977.

Vol. 623: I. Erdelyi and R. Lange, Spectral Decompositions on Banach Spaces. VIII, 122 pages. 1977.

Vol. 624: Y. Guivarc'h et al., Marches Aléatoires sur les Groupes de Lie. VIII, 292 pages. 1977.

Vol. 625: J. P. Alexander et al., Odd Order Group Actions and Witt Classification of Innerproducts. IV, 202 pages. 1977.

Vol. 626: Number Theory Day, New York 1976. Proceedings. Edited by M. B. Nathanson. VI, 241 pages. 1977.

Vol. 627: Modular Functions of One Variable VI, Bonn 1976. Proceedings. Edited by J.-P. Serre and D. B. Zagier. VI, 339 pages. 1977.

Vol. 628: H. J. Baues, Obstruction Theory on the Homotopy Classification of Maps. XII, 387 pages. 1977.

Vol. 629: W. A. Coppel, Dichotomies in Stability Theory. VI, 98 pages. 1978.

Vol. 630: Numerical Analysis, Proceedings, Biennial Conference, Dundee 1977. Edited by G. A. Watson. XII, 199 pages. 1978.

Vol. 631: Numerical Treatment of Differential Equations. Proceedings 1976. Edited by R. Bulirsch, R. D. Grigorieff, and J. Schröder. X, 219 pages. 1978.

Vol. 632: J.-F. Boutot, Schéma de Picard Local. X, 165 pages. 1978.

Vol. 633: N. R. Coleff and M. E. Herrera, Les Courants Résiduels Associés à une Forme Méromorphe. X, 211 pages. 1978.

Vol. 634: H. Kurke et al., Die Approximationseigenschaft lokaler Ringe. IV, 204 Seiten. 1978.

Vol. 635: T. Y. Lam, Serre's Conjecture. XVI, 227 pages. 1978.

Vol. 636: Journées de Statistique des Processus Stochastiques, Grenoble 1977, Proceedings. Edité par Didier Dacunha-Castelle et Bernard Van Cutsem. VII, 202 pages. 1978.

Vol. 637: W. B. Jurkat, Meromorphe Differentialgleichungen. VII, 194 Seiten. 1978.

Vol. 638: P. Shanahan, The Atiyah-Singer Index Theorem, An Introduction. V, 224 pages. 1978.

Vol. 639: N. Adasch et al., Topological Vector Spaces. V, 125 pages. 1978.

Vol. 640: J. L. Dupont, Curvature and Characteristic Classes. X, 175 pages. 1978.

Vol. 641: Séminaire d'Algèbre Paul Dubreil, Proceedings Paris 1976–1977. Edité par M. P. Malliavin. IV, 367 pages. 1978.

Vol. 642: Theory and Applications of Graphs, Proceedings, Michigan 1976. Edited by Y. Alavi and D. R. Lick. XIV, 635 pages. 1978.

Vol. 643: M. Davis, Multiaxial Actions on Manifolds. VI, 141 pages. 1978.

Vol. 644: Vector Space Measures and Applications I, Proceedings 1977. Edited by R. M. Aron and S. Dineen. VIII, 451 pages. 1978.

Vol. 645: Vector Space Measures and Applications II, Proceedings 1977. Edited by R. M. Aron and S. Dineen. VIII, 218 pages. 1978.

Vol. 646: O. Tammi, Extremum Problems for Bounded Univalent Functions. VIII, 313 pages. 1978.

Vol. 647: L. J. Ratliff, Jr., Chain Conjectures in Ring Theory. VIII, 133 pages. 1978.

Vol. 648: Nonlinear Partial Differential Equations and Applications, Proceedings, Indiana 1976–1977. Edited by J. M. Chadam. VI, 206 pages. 1978.

Vol. 649: Séminaire de Probabilités XII, Proceedings, Strasbourg, 1976–1977. Edité par C. Dellacherie, P. A. Meyer et M. Weil. VIII, 805 pages. 1978.

Vol. 650: C*-Algebras and Applications to Physics. Proceedings 1977. Edited by H. Araki and R. V. Kadison. V, 192 pages. 1978.

Vol. 651: P. W. Michor, Functors and Categories of Banach Spaces. VI, 99 pages. 1978.

Vol. 652: Differential Topology, Foliations and Gelfand-Fuks-Cohomology, Proceedings 1976. Edited by P. A. Schweitzer. XIV, 252 pages. 1978.

Vol. 653: Locally Interacting Systems and Their Application in Biology. Proceedings, 1976. Edited by R. L. Dobrushin, V. I. Kryukov and A. L. Toom. XI, 202 pages. 1978.

Vol. 654: J. P. Buhler, Icosahedral Golois Representations. III, 143 pages. 1978.

Vol. 655: R. Baeza, Quadratic Forms Over Semilocal Rings. VI, 199 pages. 1978.

Vol. 656: Probability Theory on Vector Spaces. Proceedings, 1977. Edited by A. Weron. VIII, 274 pages. 1978.

Vol. 657: Geometric Applications of Homotopy Theory I, Proceedings 1977. Edited by M. G. Barratt and M. E. Mahowald. VIII, 459 pages. 1978.

Vol. 658: Geometric Applications of Homotopy Theory II, Proceedings 1977. Edited by M. G. Barratt and M. E. Mahowald. VIII, 487 pages. 1978.

Vol. 659: Bruckner, Differentiation of Real Functions. X, 247 pages. 1978.

Vol. 660: Equations aux Dérivée Partielles. Proceedings, 1977. Edité par Pham The Lai. VI, 216 pages. 1978.

Vol. 661: P. T. Johnstone, R. Paré, R. D. Rosebrugh, D. Schumacher, R. J. Wood, and G. C. Wraith, Indexed Categories and Their Applications. VII, 260 pages. 1978.

Vol. 662: Akin, The Metric Theory of Banach Manifolds. XIX, 306 pages. 1978.

Vol. 663: J. F. Berglund, H. D. Junghenn, P. Milnes, Compact Right Topological Semigroups and Generalizations of Almost Periodicity. X, 243 pages. 1978.

Vol. 664: Algebraic and Geometric Topology, Proceedings, 1977. Edited by K. C. Millett. XI, 240 pages. 1978.

Vol. 665: Journées d'Analyse Non Linéaire. Proceedings, 1977. Edité par P. Bénilan et J. Robert. VIII, 256 pages. 1978.

Vol. 666: B. Beauzamy, Espaces d'Interpolation Réels: Topologie et Géometrie. X, 104 pages. 1978.

Vol. 667: J. Gilewicz, Approximants de Padé. XIV, 511 pages. 1978.

Vol. 668: The Structure of Attractors in Dynamical Systems. Proceedings, 1977. Edited by J. C. Martin, N. G. Markley and W. Perrizo. VI, 264 pages. 1978.

Vol. 669: Higher Set Theory. Proceedings, 1977. Edited by G. H. Müller and D. S. Scott. XII, 476 pages. 1978.

Vol. 670: Fonctions de Plusieurs Variables Complexes III, Proceedings, 1977. Edité par F. Norguet. XII, 394 pages. 1978.

Vol. 671: R. T. Smythe and J. C. Wierman, First-Passage Perculation on the Square Lattice. VIII, 196 pages. 1978.

Vol. 672: R. L. Taylor, Stochastic Convergence of Weighted Sums of Random Elements in Linear Spaces. VII, 216 pages. 1978.

Vol. 673: Algebraic Topology, Proceedings 1977. Edited by P. Hoffman, R. Piccinini and D. Sjerve. VI, 278 pages. 1978.

Vol. 674: Z. Fiedorowicz and S. Priddy, Homology of Classical Groups Over Finite Fields and Their Associated Infinite Loop Spaces. VI, 434 pages. 1978.

Vol. 675: J. Galambos and S. Kotz, Characterizations of Probability Distributions. VIII, 169 pages. 1978.

Vol. 676: Differential Geometrical Methods in Mathematical Physics II, Proceedings, 1977. Edited by K. Bleuler, H. R. Petry and A. Reetz. VI, 626 pages. 1978.

Vol. 677: Séminaire Bourbaki, vol. 1976/77, Exposés 489–506. IV, 264 pages. 1978.

Vol. 678: D. Dacunha-Castelle, H. Heyer et B. Roynette. Ecole d'Eté de Probabilités de Saint-Flour. VII-1977. Edité par P. L. Hennequin. IX, 379 pages. 1978.

Vol. 679: Numerical Treatment of Differential Equations in Applications, Proceedings, 1977. Edited by R. Ansorge and W. Törnig. IX, 163 pages. 1978.

Vol. 680: Mathematical Control Theory, Proceedings, 1977. Edited by W. A. Coppel. IX, 257 pages. 1978.

Vol. 681: Séminaire de Théorie du Potentiel Paris, No. 3, Directeurs: M. Brelot, G. Choquet et J. Deny. Rédacteurs: F. Hirsch et G. Mokobodzki. VII, 294 pages. 1978.

Vol. 682: G. D. James, The Representation Theory of the Symmetric Groups. V, 156 pages. 1978.

Vol. 683: Variétés Analytiques Compactes, Proceedings, 1977. Edité par Y. Hervier et A. Hirschowitz. V, 248 pages. 1978.

Vol. 684: E. E. Rosinger, Distributions and Nonlinear Partial Differential Equations. XI, 146 pages. 1978.

Vol. 685: Knot Theory, Proceedings, 1977. Edited by J. C. Hausmann. VII, 311 pages. 1978.

Vol. 686: Combinatorial Mathematics, Proceedings, 1977. Edited by D. A. Holton and J. Seberry. IX, 353 pages. 1978.

Vol. 687: Algebraic Geometry, Proceedings, 1977. Edited by L. D. Olson. V, 244 pages. 1978.

Vol. 688: J. Dydak and J. Segal, Shape Theory. VI, 150 pages. 1978.

Vol. 689: Cabal Seminar 76–77, Proceedings, 1977. Edited by A. S. Kechris and Y. N. Moschovakis. V, 282 pages. 1978.